Quantum Information and Complexity

Proceedings of the Meijo Winter School 2003

Quantum Information and Complexity

Meijo University, Nagoya, Japan
6–10 January 2003

editors

T Hida & K Saitô
Meijo University, Japan

Si Si
Aichi Prefectural University, Japan

World Scientific

NEW JERSEY • LONDON • SINGAPORE • BEIJING • SHANGHAI • HONG KONG • TAIPEI • CHENNAI

Published by

World Scientific Publishing Co. Pte. Ltd.

5 Toh Tuck Link, Singapore 596224

USA office: 27 Warren Street, Suite 401–402, Hackensack, NJ 07601

UK office: 57 Shelton Street, Covent Garden, London WC2H 9HE

British Library Cataloguing-in-Publication Data
A catalogue record for this book is available from the British Library.

QUANTUM INFORMATION AND COMPLEXITY
Meijo University, Japan 6–10 January 2003

ISBN 981-256-047-5

Printed in Singapore by World Scientific Printers (S) Pte Ltd

Preface

The Winter School and the International Conference on "Quantum Information and Complexity" was held from 6 to 10 January 2003, at Meijo University, Nagoya.

On this occasion the research agreement between Meijo University and the Solvay Institutes has been concluded, to our great pleasure, and actual research collaboration has launched. The conference, therefore, was combined with the kick-off meeting of the agreement.

The meeting was organized as follows. Morning sessions were devoted to the Winter school on Quantum Information and Complexity where the three lecturers Professors L. Accardi, I. Antoniou and M. Röckner gave series of lectures. All the participants are grateful to the lecturers for their beautiful and stimulating talks. The notes of these lectures appear in Part I of this volume. It is hoped that the notes in the present form will be useful to mathematicians and physicists alike.

Research papers were presented in the afternoon sessions. All the subjects that were discussed there are related to the current topics in the fields that are shown in the title of the conference, and all the papers presented are, of course, in a high standard. Most of them appear in Part II of this volume.

Let us say a few words on the choice of the title "Quantum Information and Complexity". These subjects have close connection, or could be said to be future directions of white noise theory, which has now extensively achieved by many authors since 1975. The Quantum Information theory or quantum probability theory is also a new field of Mathematics which is also developing. The collaboration with white noise theory is a recent trend, so that we can find, as a result, much common interest among the two, so that it is the most exciting area to be studied at present and in future.

As for the complex systems, they require basic theory of white noise analysis if the systems involve fluctuation or randomness, and the random systems in question suggest some new directions to us.

We are extremely happy for that both the Winter School and the Conference are quite successful, indeed more than expected. The papers that show the results are now widely in front of the readers, not only the participants of the meeting but also much wider class of scientists who are interested in the subjects dealt with in this publication. We are sure that this volume would give a great deal of impact to the future development of the theories of quantum information and complexity.

Once again, the organizers of the meeting, in particular, the editors wish to express their deep thanks to the lecturers, those who presented their important research papers and all the participants of the School and the Conference for their great contribution to the success of this meeting.

It is also a great pleasure to acknowledge the ideal cooperation with World Scientific Publishing Co. Pte. Ltd. for producing this useful volume.

May 2004 T. Hida, K. Saitô and Si Si

Contents

Part I

Part II

Quantum information, quantum communication and innovation

LUIGI ACCARDI

Lectures given at the Meijo winter school: Quantum information and complexity Meijo University

1. Introduction

Quantum information (QI) is a subfield of quantum engineering whose goal is to try and exploit the specific features of quantum mechanics to the advantage of information technology. It has several aspects, related to traditional computer science, information and communication theory, different fields of physics and of mathematics.

The processing of information, independently of the physical process of transmission, is based on three essential moments:

(1) coding and decoding information
(2) elaboration of information (algorithms)
(3) transmision of information (channels, capacity)

For example:

– A computer is a system planned for information processing, so its activity is related to items (1) and (2)

– Cryptography has to do with items (2) and (3)

– Teleportation has to do with item (3)

According to a call for Proposals launched, by the EC on 19 March 1999:

"... Quantum information processing and computing (QIPC) is perhaps one of the most important changes that have appeared in the basic science underlying the computing and communications field since the digital computer age dawned some 60 years ago. Potentially it is of enormous importance because it could revolutionise the way we handle, transfer and even define information. Moreover, the science involved is so close to fundamentals that it is almost certain that completely new ideas and applications, as yet unthought of, will emerge. ..."

The general objective of QIPC is to develop novel systems and techniques for information processing, transmission and security by exploiting the properties of quantum mechanical systems. The hope is that this will allow a substantial increase in power, speed and security of QIPC.

In order to realize this long term goal, QIPC has to:

i) make increasingly solid its scientific foundations. In particular by answering the following questions:

– What problems would quantum computers be able to solve better than classical?

– Can one **prove** that there exist problems whci quantum computers are able to solve better than classical?

– Are these problems of sufficient interest to justify development?

ii) to develop technologies to make practical QIPC systems. In particular by answering the following questions:

– can we make quantum computers?

– can we make quantum computers that scale up?

– can we make quantum computers that are cost effective?

Let us discuss these objectives and challenges one by one.

Can q–computers be effectively built?

The huge development of digital information processing is based on the abundance and the facility to produce systems with the following characteristics:

– they can find themselves only in two sharply differentiated states, say $0, 1$

– these states are stable

– these states can be controlled, i.e. put in one of the two basic states $0, 1$.

– large chains of these objects can be produced

– these chains are static, i.e. each of their components is in a well defined fixed position

– each component of these chains can be controlled independently of the other ones

A device to control sequentially a chain of binary systems is called a **logic gate**.

The convinction, widespread among physicists, that a quantum computer can be built is based on recent developments in the technology of **nanostructures**, i.e. the techniques to handle individual quantum systems such as:

– linear and nonlinear ion traps

– cavity QED,

– quantum dots and semiconductor quantum dots,

– single-electron transistor
– nanolasers
– nuclei NMR
– SQUID
– ...

These technologies should allow to individuate the natural candidates for the construction of elementary quantum logic gates. At the moment several candidates are available, but each of these has its pro and its contra:

1) atoms and molecules have well defined states, but are difficult to manipulate
2) spins (NMR, ESR) states are stable. However the emitted signals are weak so that many identical copies of them are needed.
3) Photons have stable binary states, but can be localized only by destroying them
4) Solid states systems
4.1) single electrons in q–dots (or He surface)
4.2) single spin: spintronics q–dot
4.3) excitons in q–dots
4.4) superconductors with JJ
4.4.1) charges of Cooper pairs (Nakamura was able to use single CP)
4.4.2) currents: single fluxons or macroscopic phase (fluxon: unit of flux)
4.4.3) superposition of two currents in opposite directions using the SQUID mechanism

In which problems can q–computers outperform c–computers?

The hope that quantum mechanics can be used to build a computer able of performances which no classical computer could achieve has been supported by theoretical studies according to which a QC would allow:

– to develop quantum algorithms, orders of magnitude more powerful than any known classical algorithm

– to realize quantum teleportation or, more generally, quantum communication (i.e. to replace the classical communication channels by quantum channels)

– to implement fully secure cryptographic systems

Quantum versus classical computer: the reversibility paradigm

A quantum computer is an **analogical machine**: if we identify

[quantum] information processing to a [quantum] dynamical system, then we are naturally lead to the idea to build a (quantum) dynamical system whose evolution corresponds to the elaboration of information.

In such a system the output is obtained by making one or more measurements which determine one of the following things:

i) the value of a quantum observable

ii) a quantum state.

According to the presently accepted paradigm only reversible, i.e. unitary, evolutions are allowed. However the theoretical foundations of this limitation are not well established and in the following we shall discuss some recent proposals to exploit and control irreversible quantum evolutions. In fact thare are concrete examples indicating that one can obtain advantages by using irreversible evolutions.

The main corollaries of the reversibility ansatz are the following:

i) the intermediate steps of a q–computation algorithm are unobservable (because in QM observation breaks reversibility of the evolution)

ii) in the reversible scheme of quantum computation there cannot exist convergent algorithms, in the sense that the system reaches a final state and stays there. In fact if T is a reversible evolution then

$$T^{n+1}x = T^n x \quad , \quad \text{if and only if } \; Tx = x$$

which means that an algorithm can arrive to a stable end point if and only if it was there from the beginning, but in this case there is no algorithm at all.

New characteristics of quantum computers

Which are the new features which should allow q–computers to outperform c–computers?

The accepted answer to this question are that the essential new ingredients for quantum computation devices are:

– quantum parallelism

– entanglement

– quantum interference

In the early years of development of quantum computer also quantum non locality was invoked but, after the experiment [AcImRe01a] these claims have been dismissed.

From the above it is clear that this enterprise has an intrinsically interdisciplinary character since it involves in an essential way physicists, mathematicians, computer scientists, engineers, ...

For this reason the governments of several industrialized countries, starting from the late 90's, are pouring an abundance of funds in the field of QIPC inducing bright scientists from several disciplines to converge in this field.

The idea of using the laws of quantum physics for information processing purposes has a old history. A pioneering role in quantum information was played by the quantum probability community, in Japan Ohya, developing earlier ideas of Umegaki and Hirota, mainly in connection with quantum communication; in Russia Stratonovich, Belavkin and Holevo; in Hungary, Petz; .... These early studies were mainly theoretical and only in the late 90's the idea to actually "build" a quantum computer begun to emerge as well as the idea that such a computer would have an intrinsic paralleism built into by the very laws of quantum mechanics.

In what follows I will discuss some selected topics in quantum information processing without aiming neither at systematicity nor at completeness. After a quick introduction to quantum mechanics in finite dimensional spaces, Lecture (I) deals with some basic ideas of cryptography and on how to simulate quantum entanglement with classical computer; Lecture (II) is devoted to an exposition of the Ohya–Masuda algorithm following the paper [AcSa00a] where the combinatorial nature of this algorithm, i.e. totally constructive and not introducing any additional "oracle" informations, was first emphasized; a precise definition of "algorithm with oracle" is attempted in Lecture III and illustrated with the simplest such algorithm; finally in Lecture IV the general approach to the teleportation problem developed in [AcOh98b] is shortly reviewed.

2. Quantum mechanics (QM) in nuce (in a nut)

In this section we recall the basic elementary rules of quantum mechanics.

The experimentally measurable quantities in (QM) are conditional probabilities of the form:

$$\text{Prob (an observable } A \text{ takes values in an interval } I \mid \text{ preparation)} \quad (1)$$

All quantities of physical interest (mean values, correlations ...) are calculated in terms of these probabilities.

Observables, such as position, momentum, spin, polarization, ... are represented by Hermitean operators in $\mathbb{C}^n$ (in QI $n < \infty$; in usual (QM) $\mathbb{C}^n$ is replaced by a general Hilbert space). Any Hermitean operator in a finite dimensional Hilbert space has the form (spectral theorem):

$$A = A^* = \sum_{j=1}^{n} \lambda_j |\psi_j\rangle\langle\psi_j|$$

The adjoint is meant with respect to the usual scalar product in $\mathbb{C}^n$:

$$\langle u, v\rangle = \sum_{1}^{n} \overline{u}_j v_j$$

Observables represented by projection operators

$$P^* = P^2 = P$$

correspond to events in the sense of classical probability, e.g.

$$[A \in I] \to E^A(I) \quad \text{spectral projection of } A \text{ in the interval } I$$

Our information on the preparation (state) of a system is coded into a density matrix (a positive definite operator ρ with trace 1).

In most quantum systems one postulates that the correspondence: states $\to$ density matrices is one-to-one. In terms of states and observables

the conditional probabilities (1) are expressed through the formula

$$Tr(\rho E^A(I)) \tag{2}$$

Time evolution: the time evolution of a state of an isolated system is described by the Heisenberg evolution

$$\rho \mapsto u_t(\rho)$$

where u_t is an affine evolution the set of states into itself

$$u_t(\text{convex combination}) = \text{convex combination } u_t(\cdot)$$

$$u_t u_s = u_{t+s}$$

Combining the probabilistic and the dynamical prescriptions one obtains the fundamental formula of QM:

$$\text{Prob } (A_t \in I \mid \rho) = Tr\left(\rho u_t(E^A(I))\right) \tag{3}$$

In many cases (always in finite dimensions) the evolution has the form

$$u_t(\rho) = U_t \rho U_t^*$$

where U_t satisfies the unitarity condition

$$U_t^* U_t = U_t U_t^* = 1 \tag{4}$$

and the conditions defining a 1–parameter unitary group:

$$U_t U_s = U_{t+s} \qquad ; \qquad U_0 = 1$$

In this case the evolution is called "inner" and U_t admit an equivlent infinitesimal characterization, which is called the Schrödinger equation:

$$\partial_t U_t = -iHU_t \qquad ; \qquad U_0 = 1$$

and the infinitesimal generator H is a self–adjoint operator. When H is bounded below (always in finite dimensions) it is called an Hamiltonian.

Pure states and superpositions

Preparation means: information available on a system at a given time.

Preparations are called **states**.

A **maximal** information is one that cannot be increased without loss of previous information (maximal compatible with Heisenberg's principle). In the mathematical model **maximal** information corresponds to **pure states** and **non maximal** information to **mixtures**.

Any state can be represented in the form (mixture)

$$\rho = \sum_{j=1}^{n} p_j |\psi_j\rangle\langle\psi_j|$$

where we use the standard notations for an orthonormal basis of $\mathbb{C}^n$

$$\langle\psi_j, \psi_k\rangle = \delta_{j,k}$$

and for the corresponding rank one projection

$$|\psi_j\rangle\langle\psi_j| : \xi \in \mathbb{C}^n \to \langle\psi_j, \xi\rangle\psi_j \in \mathbb{C}^n$$

In particular

$$\langle\psi_j, \psi_j\rangle = \| \psi_j \|^2 = 1$$

ρ is a pure state (maximal information) if and only if the sum is reduced to a single term:

$$\rho = |\psi\rangle\langle\psi|$$

In this case formula (3) reduces to

$$\text{Prob } (A_t \in I|\psi) = \langle\psi, u_t(A)\psi\rangle \tag{5}$$

and one says that the pure state $\rho = |\psi\rangle\langle\psi|$ is represented by the state vector $\psi = |\psi\rangle$. The converse statement:

pure states correspond to values of maximal observables

equivalently:

every unit vector $\psi = |\psi\rangle$ defines a pure state by formula (5).

is widely postulated but not always physically true: coherent states are eigenvectors of the annihilation operator but not of any physically meaningful observable.

A corollary of this postulate is the superposition principle: any linear combination of state vectors defines a state vector while in a mixture we only know that the pure state $|\psi_j\rangle\langle\psi_j|$ is realized with probability p_j.

Any evolution is uniquely detrmined by its restriction to the pure states $|\psi_j\rangle\langle\psi_j|$. When such an evolution is inner:

$$u_t(|\psi_j\rangle\langle\psi_j|) = U_t|\psi\rangle\rangle\psi|U_t^* =: |\psi_t\rangle\rangle\psi_t|$$

and ψ_t satisfies the Schrödinger equation:

$$\partial_t\psi_t = -iH\psi_t \qquad ; \qquad \psi_0 = \psi$$

Finally there are the measurement postulates: there are many such postulates and they have been discussed elsewhere. The most frequently used in the quantum computer literature is the Luders–Zumino postulate

$$\rho \to \frac{P\rho P}{Tr(\rho P)}$$

describing the change undergone by the density matrix ρ when the event (projection) P measured on the system.

Finally composite systems are described by tensor products and product states are those of the form

$$\psi \otimes \phi$$

(ψ, ϕ density matrices). Any (normal) linear functional ψ on a tensor product of two matrix algebras (more generally, von Neumann algebras cf. [Sak71] Definition 1.22.10) is a linear combination of product states (or limit thereof). However it is not true that, if ψ is a state, then the coefficients of the linear combination can be taken to be positive. States which are not convex combinations of tensor product states are called entangled states. An example is the single state on $M_2 \otimes M_2$. This notion of entanglement is intrinsic to the state and does not depend on any other choice.

There is another, frequently used, notion of entanglement which is synonim of linear superposition and should be better called "entanglement with respect to a given observable". This is the following: let $A = \sum_j a_j|\psi_j\rangle\langle\psi_j|$ be an observable. A state ρ is said to be entangled with respect to A if $\rho = |\psi\rangle\langle\psi|$ with ψ of the form

$$\psi = \sum_j \lambda_j\psi_j$$

If all the λ_j have the same modulus (which must then be $1/\sqrt{N}$ where N is the dimension of the space, one says that ψ is maximally entangled (with respect to A). Notice that a maximally entangled state with respect to a basis may be a product state with respect to another basis and therefore not entangled at all in the sense of the intrinsic definition.

A typical example is obtained by considering the two bases

$$\{|0\rangle, |1\rangle\} \qquad ; \qquad \{|-\rangle, |+\rangle\}$$

such that

$$|0\rangle = \frac{1}{\sqrt{2}}(|-\rangle - |+\rangle) \qquad ; \qquad |1\rangle = \frac{1}{\sqrt{2}}(|-\rangle + |+\rangle)$$

and tensor products of copies of them. For example:

$$|0, \ldots, 0\rangle = |0\rangle \otimes \cdots \otimes |0\rangle = 2^{-n/2} \sum_{\varepsilon_1, \ldots, \varepsilon_n \in \{\pm\}} |\varepsilon_1, \ldots, \varepsilon_n\rangle$$

with

$$|\varepsilon_1, \ldots, \varepsilon_n\rangle = |\varepsilon_1\rangle \otimes \cdots \otimes |\varepsilon_n\rangle$$

Qubits

The elementary unit of quantum information is the q–**bit** A single q–bit can be envisaged as a two-state system such as a spin-half particle or a two-level atom.

Identifying a system with its state space we can say that a q–bit is a copy of the space $\mathbf{C}^2$.

We will write two orthogonal states of a single q–bit as $\{|0\rangle, |1\rangle\}$.

A quantum system is said to have n q–bits if its state space is the tensor product of n q–bits

$$\mathbf{C}^2 \otimes \cdots \otimes \mathbf{C}^2 \qquad n\text{–times}$$

The basis

$$|\varepsilon_1, \ldots, \varepsilon_n\rangle = |\varepsilon_1\rangle \otimes \cdots \otimes \varepsilon_n\rangle$$

will be called th computational basis.

This means that the state space of this system has 2^n dimensions and therefore any of its bases is made of 2^n **mutually orthogonal** quantum states

$$|\varepsilon_1, \ldots, \varepsilon_n\rangle$$

where the symbols ε_j can take the values 0 or 1 in all possible ways (recall that n classical bits can represent up to 2^n different things).

Quantum gates

A device to control a chain of binary (quantum) systems is called a **logic (quantum) gate**. We want the control to be **reversible**, i.e.: if we are able to bring the system from a to b we want also to be able to bring it from b to a.

The mathematical model of a (quantum) gate is a unitary operator. The simplest case is when the chain consists of a single element, described by a single q–bit. In this case the structure of quantum gates is described by the following:

Theorem The most general unitary operator, acting on a single q–bit, has the form

$$U = e^{i\alpha} R(\theta, \phi)$$

where $\alpha, \theta, \phi \in [0, 2\pi]$ and $R(\theta, \phi)$ is a rotation

$$R(\theta, \phi) = \begin{pmatrix} \cos(\theta/2) & -ie^{-i\phi}\sin(\theta/2) \\ -ie^{i\phi}\sin(\theta/2) & \cos(\theta/2) \end{pmatrix}$$

So we see that there are infinitely single-q–bit quantum gates, in contrast to classical information theory, where only two (reversible) logic gates are possible for a single bit: the identity and the logical **not** operation.

Examples of elementary quantum gates the identity are:

$$1 := |0\rangle\langle 0| + |1\rangle\langle 1|$$

and the Pauli operator:

$$U_{NOT} := X = \sigma_x := |0\rangle\langle 1| + |1\rangle\langle 0| \quad = \text{not}$$

Notice that

$$U_{NOT}|0\rangle = |1\rangle \qquad ; \qquad U_{NOT}|1\rangle = |0\rangle$$

The (single q–bit) **Hadamard** transformation

$$H := \frac{1}{\sqrt{2}} \left[\left(|0\rangle + |1\rangle \right) \langle 0| + \left(|0\rangle - |1\rangle \right) \langle 1| \right]$$

defines the unitary transformation

$$|0\rangle \to \frac{1}{\sqrt{2}} \left(|0\rangle + |1\rangle \right) \quad ; \qquad |1\rangle \to \frac{1}{\sqrt{2}} \left(|0\rangle - |1\rangle \right)$$

which maps the $\{|0\rangle, |1\rangle\}$–basis into its canonical complementary basis. All these unitary transformations act on a single q–bit.

If U is any single-q–bit gate the two-q–bit gate

$$|0\rangle\langle 0| \otimes 1 + |1\rangle\langle 1| \otimes U$$

where 1 is the single-q–bit identity operation, is called a "controlled–U" gate, since the second q–bit undergoes a U–gate if and only if the first q–bit is in the state $|1\rangle$ and it undergoes no operation if the first q–bit is in the state $|0\rangle$.

If $U = U_{NOT}$, then we have the controlled-**not** gate whose effect on a state $|a\rangle|b\rangle$ can be written

$$|a\rangle|b\rangle \to |a\rangle|a \oplus b\rangle$$

where $\oplus$ signifies the exclusive or (**xor**) operation (i.e. the addition modulo 2):

$$y \text{ XOR } z = y'z + yz' = z\backslash y \cup y\backslash z$$

For this reason, this gate is also called the **xor** gate.

Other logical operations require further q–bits in order to guarantee unitarity.

For example, the **and** operation is achieved by use of the 3-q–bit "controlled-controlled-**not**" gate, in which the third q–bit experiences **not** if and only if both the others are in the state $|1\rangle$

$$(1 - |1\rangle\langle 1| \otimes |1\rangle\langle 1|) \otimes 1|1\rangle\langle 1| \otimes |1\rangle\langle 1| \otimes U_{NOT}$$

This gate is named a Toffoli gate, after Toffoli (1980) who showed that the classical version is universal for classical reversible computation.

Its effect on a state $|a, \langle b, 0\rangle$ is

$$a \to a, b \to b, 0 \to a \cdot b$$

In other words if the third q–bit is prepared in $|0\rangle$ then this gate computes the **and** of the first two q–bits and replaces 0 by the result. The use of three q–bits is necessary in order to permit the whole operation to be unitary.

More explicitly the Toffoli gate is the unitary operator:

$$|\varepsilon_1, \varepsilon_2, \varepsilon_3\rangle \mapsto |\varepsilon_1, \varepsilon_2, \varepsilon_3 \dot{+} \varepsilon_1\varepsilon_2\rangle = |\varepsilon_1, \varepsilon_2, \varepsilon_3 \text{XOR} \varepsilon_1\varepsilon_2\rangle \qquad (6)$$

where $\varepsilon_j \in \{0, 1\}$, $|\varepsilon_1, \varepsilon_2, \varepsilon_3\rangle = |\varepsilon_1\rangle \otimes |\varepsilon_2\rangle \otimes |\varepsilon_3\rangle$ and $\dot{+}$ means addition mod 2.

2-q–bit gates are sufficient for quantum computation in the sense that it can be shown that any $n \times n$ unitary matrix can be formed by composing

(i) rotations acting on a single q–bit, i.e.

(ii) 2-q–bit controlled-not (or **xor**)

Therefore, this pair of operations is universal for quantum computation.

3. Some basic ideas of classical cryptography

Fact (1): any message can be coded in the binary alphabet $\{0,1\}$ Consequently in the following we fix the binary alphabet to be $\{0,1\}$. Thus **message** will mean: sequence of 0's and 1's (**binary sequence**).

We will use the notation

$$\varepsilon := (\varepsilon_0, \varepsilon_1, \ldots, \varepsilon_n, \ldots) \in \{0,1\}^n$$

Fact (2): There is a one–to–one correspondence between **binary sequence** and **natural integers**. This is given by

$$\varepsilon := (\varepsilon_0, \varepsilon_1, \ldots, \varepsilon_n, \ldots) \mapsto \varepsilon_0 2^0 + \varepsilon_1 2^1 + \cdots + \varepsilon_n 2^n + \ldots$$

Thus the following are equivalent:

(i) to send a long message

(ii) to send a long sequence of 0's and 1's

(iii) to send a large integer

That is why classical cryptography is strictly related to the number theoretical analysis of large integers.

The theree basic steps of cryptography are:

1–st stage encripting

2–d stage: communication channel

3–d stage: decripting

Remark Replacing the terms "encripting", "decripting" by "encoding", "decoding" one obtains the general scheme of information transmission

1–st stage: encoding

An encoding procedure is an invertible map which transforms a given binary sequence ε into another one ε':

$$C_\varepsilon \ : \ \varepsilon := (\varepsilon_1, \varepsilon_2, \ldots, \varepsilon_n, \ldots) \mapsto C_\varepsilon(\varepsilon) := (C_\varepsilon \varepsilon_1, C_\varepsilon \varepsilon_2, \ldots, C_\varepsilon \varepsilon_n, \ldots) =:$$

$$= \varepsilon' =: (\varepsilon'_1, \varepsilon'_2, \ldots, \varepsilon'_n, \ldots)$$

We recognize that there are competing constraints:

(i) speed: both the encoding and the decoding procedures should be "quick" (say polinomyal in the length of the input sequence). In other words: both

$$C_\varepsilon(\varepsilon) \quad \text{and} \quad C_\varepsilon^{-1}(\varepsilon')$$

should be efficiently computable.

(ii) standardization: one would like to use the same encoding procedure for several binary sequences. In other words:

$$C_\varepsilon = C \text{ (independent of) } \varepsilon$$

(iii) security (most important): given ε' it should be impossible (at least extremely difficult) to reconstruct $\varepsilon = C_\varepsilon^{-1}(\varepsilon')$ (without additional information).

In other words you want that C_ε is **easy to compute**, but its inverse C_ε^{-1} is **difficult to compute.**

Functions F such that $F(x)$ is **easy to compute**, but the inverse of F is **difficult to compute** play a crucial role in cryptography so they have a special neme.

Definition. Let $\mathbf{N}$ be the set of integers. A function $f : \mathbf{N} \to Y$ is called *one way (or a hash) function* if

1) $F(x)$ can be computed in polynomial time (in the length of x)
2) $F^{-1}(x)$ cannot be computed in polynomial time.

It is not known if such functions exist. There are examples in which (1) can be proved, but (2) has not (yet) been proved.

Example (mod p)–logarithm.

Let p be a prime, then any $(0, 1 \neq)\alpha \in GF(p) := \mathbf{Z}_p := \mathbf{Z}/p\mathbf{Z}$ is such that: $\forall n \in \mathbf{Z}_p\ \exists!\, m$: $\alpha^m \equiv n \pmod p$. However it is not known a polynomial time algorithm to solve the equation $y = \alpha^x \pmod p$, given y.

Connection with the $P - NP$ problem

Recall that:

$\mathcal{P} :=$ {problems solvable in polynomial time}

$\mathcal{NP} :=$ {problems such that you can check in polynomial time if a candidate solution is a solution or not}

Theorem. If one way functions exist then $P \neq NP$.

Proof. Let F be a one way function and let $y \in \mathbf{N}$ let P denote the problem: *compute* $F^{-1}(y)$. Then by assumption P cannot be solved in polynomial time. Thus $P \notin \mathcal{P}$.

Now let x be a candidate solution. Then by assumption $F(x)$ can be computed in polynomial time. Therefore you can check in polynomial time if a candidate solution is a solution or not. Thus $P \in \mathcal{NP}$.

Therefore we conclude that: $\mathcal{NP} \neq \mathcal{P}$

Fact The converse of the above theorem is not known.

4. The role of secret keys

The three competing constraints: (i) speed, (ii) standardization, (iii) security, can be fulfilled with the help of a **secret key**

The **basic example** of secret key is a commonly shared secret sequence

$$\varepsilon'' = (\varepsilon''_1, \ldots, \varepsilon''_n, \ldots)$$

The algorithm to send a message using such a sequence is the following.

encoding: Let the (input) sequence to be encoded be $\varepsilon = (\varepsilon_1, \ldots, \varepsilon_n, \ldots)$.

Add ($mod2$) the input sequence and the secret key

$$\varepsilon'' = (\varepsilon''_1, \ldots, \varepsilon''_n, \ldots)$$

This gives the encripted sequence

$$\varepsilon' := \varepsilon \dot{+} \varepsilon''$$

where $\dot{+}$ means componentwise addition modulo 2.

decoding: since, for any binary sequence ε'' $2\varepsilon'' = 0$, to recover the original sequence we have only to add again the same ε'' to the encripted sequence

$$\varepsilon' \dot{+} \varepsilon'' = (\varepsilon \dot{+} \varepsilon'') \dot{+} \varepsilon'' = \varepsilon \dot{+} 2\varepsilon'' = \varepsilon$$

security: if a spy knows ε', she only knows the sum of two sequences therefore she knows nothing because, given the sum of two sequences, the individual summands cannot be reconstructed.

summing up:

Suppose that A (Alice) and B (Bob) are spatially separated and want to communicate with perfect security. Then, if they can share a secret binary sequence ε'' of length N, then they can exchange with perfect security any message of length N.

Problems with this approach:

(i) secret keys can be violated

(ii) one has to renew them frequently

(iii) to achieve perfect security one must change
the random sequence ε'' every new message

This leads to the problem: how to produce, at command and with perfect security, the same binary sequence ε'' in 2 spatially separated places and at the same time ?

Any algorithm used to produce such a key is called a **key distribution algorithm**.

The problem to produce such an algorithm and to effectively realize it is called the **key distribution problem**.

A brilliant idea to solve this problem was proposed in 1976 by Diffie and Hellman and gave rise to what is nowadays called **public key (or asymmetric) cryptography**.

5. Basic scheme of public key cryptography

(i) A and B start from some secret information (shared or not)

(ii) then they share publicly some open information

(iii) finally, combining (i) and (ii) they produce a joint secret key

The idea is nice, but how to concretely realize it?

The best known answer presently known to this problem and the most widely applied scheme of public key cryptography is due to R.L. Rivest, A. Shamir, L.M. Adleman (MIT), and from them takes the name: **RSA algorithm** .

Its main ingredients are:

(i) a public part made of:

– a large prime p

– a natural integer α such that, given any n, there exists a unique integer m with the property

$$\alpha^m = n \qquad ; \qquad (modp)$$

(ii) a secret part made of:

– a secret number chosen by A

$$x_A \in \mathbf{Z}_p := \{0, 1, 2, \ldots, p-1\}$$

– a secret number chosen by B

$$x_B \in \mathbf{Z}_p := \{0, 1, 2, \ldots, p-1\}$$

Remark It is important to emphasize that A needs not to know x_B and B needs not to know x_A.

The algorithm consists in the following.

Using her secret key $x_A \in \mathbf{Z}_p$ A produces

$$\alpha^{x_A} =: y_A$$

and makes it public.

Using his secret key $x_B \in \mathbf{Z}_p$ B produces

$$\alpha^{x_B} = y_B$$

and makes it public.

A knows y_B and x_A so she forms

$$y_B^{x_A} = (\alpha^{x_B})^{x_A} = \alpha^{x_A x_B}$$

B knows y_A and x_B so he forms

$$y_A^{x_B} = (\alpha^{x_A})^{x_B} = \alpha^{x_A x_B}$$

At the end of the procedure they both have **the secret key**:

$\alpha^{x_A x_B}$ = large natural number = long binary string.

The main features of the RSA approach to public key cryptography are:

(i) an high degree of mathematical sophistication

(i1) great freedom in the secret information: A does not need to know the secret key of B and conversely

(i2) no error correcting procedure needed, zero probability of error

(i3) rather easy to implement, but may need rather large computer facilities

(i4) but breakable ... in principle...

In fact it seems that in USA there are legal bounds on the length of the code, e.g. the prime number p.

The main features of the q–cryptography approach to public key cryptography are:

(ii1) an high degree of physical sophistication

(ii2) less freedom in the secret information: A and B share the same secret information.

(ii3) Secrecy is guaranteed by a basic principle of nature: Heisenberg indeterminacy principle

(ii3) complex error correcting procedure, non zero probability of error

(ii4) complex and expensive to be realized

(ii5) but unbreakable .... in principle ...

The main difference between the mathematical and the physical solutions is that:

– the mathematical solution is based on computations

– the physical solution is based on physical measurement

6. Explanation of the quantum key distribution algorithm by means of classical physics

Prepare a box with many pairs of balls there are 4 types of balls. More precisely, each ball is

– either green ($S_c = +1$) or brown ($S_c = -1$)

– either weights 10 g ($S_w = +1$) or 20 g ($S_w = -1$)

For $x = c$ (color) or $x = w$ (weight) and for each pair, denote $S_x^{(1)}$ (resp. $S_x^{(2)}$) the random variable S_x for particle 1 (resp. part 2). Moreover the preparation is such that

$$S_x^{(1)} = -S_x^{(2)} \; ; \qquad x = c, w$$

$$P(S_x^{(1)} = +) = Pr(S_x^{(2)} = +) = \frac{1}{2} \; ; \qquad x = c, w$$

In our case the source $S =$ is a Box with many pairs.

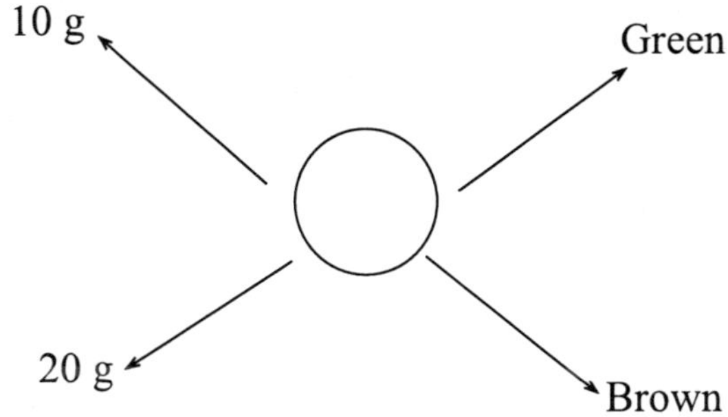

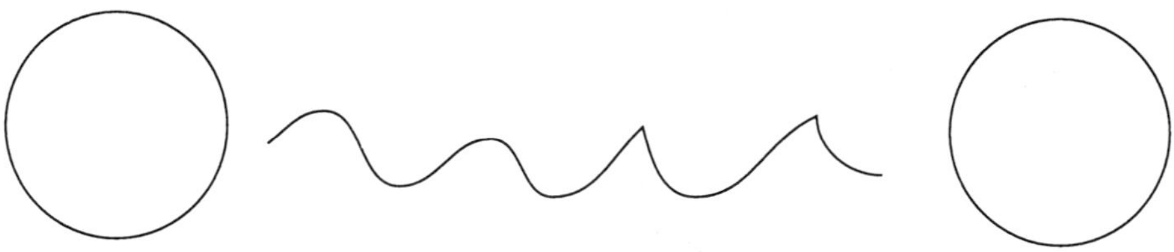

Alternatives for each pair:

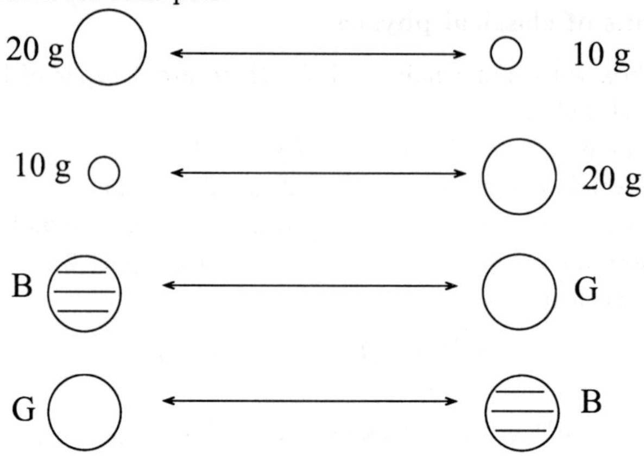

notice that in any ase one has **perfect anticorrelation**.
– pairs are separated and sent (secretly) one to A and one to B
– N pairs are sent
– both A and B choose, randomly with probability 1/2, to measure either weight or color

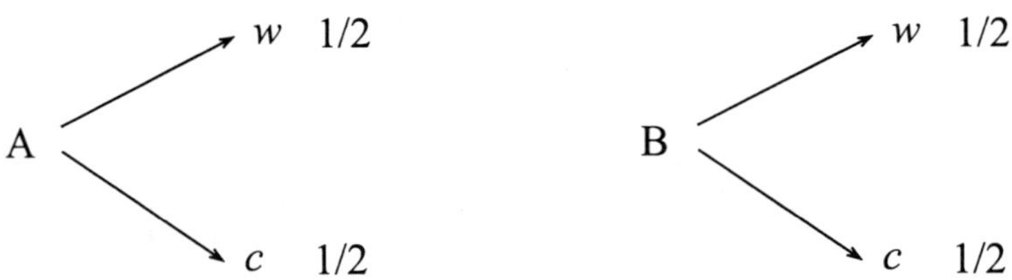

both A and B produce two secret sequences

$$A \equiv ((x_1, \varepsilon_1), \ldots, (x_N, \varepsilon_N))$$

$$B \equiv ((x'_1, \varepsilon'_1), \ldots, (x'_N, \varepsilon'_N)$$

where
– the variables x_j, x'_j stand for: w (weight) or c (color)
– $\varepsilon_j, \varepsilon'_j = +1, -1$ stand for: results of the measurement
both A and B make public half of their sequences, i.e. the x_j, x'_j part:

$$A \equiv (x_1, \ldots, x_N) \qquad ; \qquad B \equiv (x'_1, \ldots, x'_N)$$

Thus everybody can know the common sequence

$$A \cup B \equiv \{x_j = x'_j\} = (z_1, \ldots, z_N)$$

where

$$z_j = \begin{cases} x_j = x'_j & \text{if } \; x'_j = x_j \text{ same} \\ 0 & \text{if } \quad x'_j \neq x_j \text{ different} \end{cases}$$

and, erasing all the 0's, everybody can know when they made the same measurement

$$A \cap B = (x_{j_1}, \ldots, x_{j_k})$$

But only A (resp. B) knows the sequence of the corresponding results only A knows $(\varepsilon_{j_1}, \ldots, \varepsilon_{j_k})$ only B knows $(\varepsilon'_{j_1}, \ldots, \varepsilon'_{j_k})$.

Moreover, because of the conservation law: $S_x^{(1)} = -S_x^{(2)}$ B knows that $\varepsilon'_{j_\alpha} = -\varepsilon_{j_\alpha}$. Thus B changes the sign of his sequence and both A and B have the same sequence

$$(A \cap B)_{SK} \equiv (\varepsilon_{j_1}, \ldots, \varepsilon_{j_k})$$

this is the secret key

Estimate of the average length of the key

Denote ξ_j the j–th choice of A and ξ'_j the j–th choice of B. Then ξ_j, ξ'_j are independent identically distributed ± 1–valued , i.e. Bernoulli, random variables.

Therefore $\xi_j \xi'_j$ is also iid ± 1 (Bernoulli). Therefore

$$0 = E\left(\sum \xi_j \xi'_j\right) = NP(\xi_j = \xi'_j) - NP(\xi'_j = -\xi'_j) = E(N_=) - E(N_{\neq})$$

Since $N_= + N_{\neq} = N =$ (total number of measurements) it follows that

$$E(N_=) = N/2$$

Thus we conclude that:

$k :=$ average length of the key $= E(N_=) = \frac{N}{2}$.

Estimate of secrecy

Anybody may know the sequence

$$(x_{j_1}, \ldots, x_{j_k}) \qquad ; \qquad x_j = w, c$$

But for each j_α the corresponding ε_j has probability $1/2$ so a person different from A and B has 0 information.

Drawbacks of the classical physics algorithm

(i) impractical to realize

(ii) a spy can "read" the color and weight of each transmitted ball without altering them.

7. Quantum cryptography

There are two main experimental contexts to realize quantum cryptography:

(i) q–c based on weak coherent states [Wies83], [BeBr84], [BeBeBrSaSm92]

(ii) q–c based on entangled photons [Eker91], [EkRaTaPa92].

Each has advantages and disadvantages. I will discuss the second one.

The scheme is exactly as the one described for balls, the only changes being:

– balls $\to$ photons

– $S_c^{(1)}$ (color), $S_w^{(2)}$ (weight) $\to$ $S_a^{(1)}, S_b^{(2)}$ (polarization in direction $a, b, ...$) these too are $\{\pm 1\}$–valued observables

– conservation law (of color, weight, ...) $\to$ entanglement (conservation law of polarization, momentum, energy, spin, ...)

The standard approach is to consider only polarization entanglement we will follow it.

Scheme of the experiment:

laser $\to$ non linear crystal $\to$ parametric down conversion $\to$ pair of photons in singlet state

Definition. Two observables of a physical system are called compatible if they can be measured simltaneously with arbitrary precision.

Definition. A state of a physical system is a rule to assign a (joint) probability distribution to any set of compatible observables.

Definition. A state $\langle \ \cdot \rangle$ of a composite system $(1, 2)$ is called a singlet state if there exist two families of $[-1, +1]$–valued observables

$$\{S_a^{(1)} \ : \ a \text{ is a unit vector in } \mathbb{R}^3\}$$

$$\{S_b^{(2)} \ : \ a \text{ is a unit vector in } \mathbb{R}^3\}$$

such that for any unit vectors $a, b \in \mathbb{R}^3$:

(i) the pair $S_a^{(1)}, S_b^{(2)}$ is compatible

(ii) the correlation of the pair $S_a^{(1)}, S_b^{(2)}$ in the given state $\langle \ \cdot \rangle$ is

$$\langle S_a^{(1)} S_b^{(2)} \rangle = -cos(\hat{ab})$$

Remark If $a = b$ then

$$\langle S_a^{(1)} S_b^{(2)} \rangle = -1$$

i.e. perfect anticorrelation holds. Therefore

$$ {}^{(1)}_{a} = -S^{(2)}_{b} \qquad \text{(a.e.)} $$

8. A local, deterministic, reversible, classical dynamical system realizing the EPR correlation

Definition Let be given: – a measurable (state) space

$$ (\Omega_1 \times \Omega_2, \mathcal{F}_1 \times \mathcal{F}_2) $$

– for each $a, b \in [0, 2\pi]$ two ±1–valued maps (observables)

$$ S^{(1)}_a : \omega_1 \in \Omega_1 \to S^{(1)}_a(\omega_1) \in \{+1, -1\} $$

$$ S^{(2)}_b : \omega_2 \in \Omega_2 \to S^{(2)}_b(\omega_2) \in \{+1, -1\} $$

– for each $a, b \in [0, 2\pi]$ two maps (local dynamics)

$$ T_{1,a} : \Omega_1 \to \Omega_1 \quad ; \quad T_{2,b} : \Omega_2 \to \Omega_2 $$

The deterministic, reversible, classical, local, adaptive [**to be defined in the following section**] dynamical system

$$ \{(\Omega_1 \times \Omega_2, \mathcal{F}_1 \times \mathcal{F}_2, P := P_S \otimes P_{1,a} \otimes P_{1,b}) \ , \ T := T_{1,a} \otimes T_{2,b}\} $$

is said to **realize the EPR correlations** if

$$ \langle S^{(1)}_a S^{(2)}_b \rangle := \int\int_{\Omega_1 \times \Omega_2} S^{(1)}_a(\omega_1) S^{(2)}_b(\omega_2) dP \circ T^{-1}(\omega_1, \omega_2) = $$

$$ = \int\int_{\Omega_1 \times \Omega_2} S^{(1)}_a(T(\omega_1)) S^{(2)}_b(T(\omega_2)) dP(\omega_1, \omega_2) = $$

$$ = -\cos(b - a) $$

Notice that the condition $P = P_{a,b}$ (old contextuality) is not sufficient because of locality and causality. The dynamical condition $T = T_{a,b}$ (called “chameleon effect”) is really essential.

We will consider a classical deterministic local dynamical system composed by two spatially separated sub–systems $(1, M_1)$ and $(2, M_2)$ interpreted as follows:

– 1 and 2 are particles

– M_1 and M_2 are measurement apparata

The state space of $(1, M_1)$ is

$$\Omega_1 := [0, 2\pi] \times \mathbb{R}$$

also the state space of $(2, M_2)$ is

$$\Omega_1 := [0, 2\pi] \times \mathbb{R}$$

– therefore the **state space** of the whole system $(1, M_1, 2, M_2)$ is

$$\Omega_1 \times \Omega_2 := [0, 2\pi]^2 \times \mathbb{R}^2$$

The local adaptive dynamics $T_{1,a}, T_{2,b}$ are described as follows: $1, 2$ are labels for particles, a, b are labels for apparata

$$T_{1,a}, T_{2,b} \; : \; [0, 2\pi] \times \mathbb{R} \to [0, 2\pi] \times \mathbb{R} \tag{7}$$

$$T_{1,a}(\sigma_1, \lambda_1) := (s_{1,a}(\sigma_1, \lambda_1), m_{1,a}(\sigma_1, \lambda_1))$$

$$T_{2,b}(\sigma_2, \lambda_2) := (s_{2,b}(\sigma_2, \lambda_2), m_{2,b}(\sigma_2, \lambda_2))$$

$$\sigma_1, \sigma_1 \in [0, 2\pi] \qquad ; \qquad \lambda_1, \lambda_2 \in \mathbb{R}$$

$$T_{1,a}(\sigma_1, \lambda_1) := (\sigma_1 \; , \; \frac{\lambda_1 4}{\sqrt{2\pi}} |\cos(\sigma_1 - a)|)$$

$$T_{2,b}(\sigma_2, \lambda_2) := (\sigma_2 \; , \; \frac{\lambda_2}{\sqrt{2\pi}})$$

The local measures

For $(\sigma_1, \sigma_2 \in [0, 2\pi] \; , \; \lambda_1, \lambda_2 \in \mathbb{R})$ we define the local measures

$$p_S(d\sigma_1, d\sigma_2) = \frac{1}{2\pi} \delta(\sigma_1 - \sigma_2) d\sigma_1 d\sigma_2$$

which only depends on σ_1, σ_2 (hidden states of the particles)

$$p_{1,a}(\sigma_1, d\lambda_1) = \delta(m_{1,a}(\sigma_1, \lambda_1) - m_a) d\lambda_1$$

the choice of the number a is arbitrary, local, independent of the number b.

$$p_{2,b}(\sigma_2, d\lambda_2) = \delta(m_{2,b}(\sigma_2, \lambda_2) - m_b) d\lambda_2$$

the choice of b is arbitrary, local, independent of a.

The choices of a and b can be produced by a (pseudo–)random generator, exactly as in the Aspect experiment m_a, m_b are arbitrary real numbers.

The global statistics is obtained by putting together the local pieces

$$p_S(\sigma_1, \sigma_2) p_{1,a}(\sigma_1, \lambda_1) p_{2,b}(\sigma_2, \lambda_2) d\sigma_1 \sigma_2 d\lambda_1 d\lambda_2 \tag{8}$$

this gives a measure on

$$\Omega_1 \times \Omega_2 := [0, 2\pi]^2 \times \mathbb{R}^2$$

Remark In the following section it is described how the general theory of adaptive dynamical systems naturally leads to measures of this type.

The local observables

For a, b as above, define the ± 1–valued maps (**observables**)

$$S_a^{(1)}, S_b^{(2)} \ : \ [0, 2\pi] \times \mathbb{R} \to \{\pm 1\} \tag{9}$$

by ($\sigma \in [0, 2\pi], \mu \in \mathbb{R}$)

$$S_a^{(1)}(\sigma, \mu) = S_a^{(1)}(\sigma) = \operatorname{sgn}(\cos(\sigma - a)) \tag{10}$$

$$S_b^{(2)} := -S_b^{(1)}$$

Theorem. In the above notations, the measure (8) is a probability measure on $[0, 2\pi]^2 \times \mathbb{R}^2$. Moreover

$$\int S_a^{(1)} \left(S_{1,a}(\sigma_1, \lambda_1), m_{1,a}(\sigma_1, \lambda_1)\right) \tag{11}$$

$$S_b^{(2)} \left(s_{2,b}(\sigma_2, \lambda_2), m_{2,b}(\sigma_2, \lambda_2)\right)$$

$$\cdot p_S(\sigma_1, \sigma_2) p_{1,a}(\sigma_1, \lambda_1) p_{2,b}(\sigma_2, \lambda_2) d\sigma_1 d\sigma_2 d\lambda_1 d\lambda_2 =$$

$$= -\cos(a - b)$$

In other words: the local deterministic adaptive dynamical system defined above realizes the EPR correlations.

Proof. The positivity is obvious. The normalization condition

$$\int p_S(\sigma_1, \sigma_2) p_{1,a}(\sigma_1, \lambda_1) p_{2,b}(\sigma_2, \lambda_2) d\sigma_1 \sigma_2 d\lambda_1 d\lambda_2 = 1 \tag{12}$$

follows from (11) by choosing $a = b$.

To prove (11) notice that, with the choices (10), (8), the correlations (11) become

$$\int\int\int\int S_a^{(1)}(s_{1,a}(\sigma_1, \lambda_1), m_{1,a}(\sigma_1, \lambda_1)) \tag{13}$$

$$S_b^{(2)}(s_{2,b}(\sigma_2, \lambda_2), m_{2,b}(\sigma_2, \lambda_2))$$

$$\delta(m_{1,a}(\sigma_1, \lambda_1) - m_a)\delta(m_{2,b}(\sigma_2, \lambda_2) - m_b) p_S(\sigma_1, \sigma_2)$$

$$d\lambda_1 d\lambda_2 d\sigma_1 d\sigma_2$$

Changing variables

$$m_{1,a}(\sigma_1, \lambda_1) = \mu_1 \qquad ; \qquad m_{2,b}(\sigma_2, \lambda_2) = \mu_2$$

$$m'_{1,a}(\sigma_1, \lambda_1) d\lambda_1 = d\mu_1 \qquad ; \qquad m'_{2,b}(\sigma_2, \lambda_2) d\lambda_2 = d\mu_2$$

and noting that given a, $b \in [0, 2\pi]$ for almost all $a, b, \sigma_1, \sigma_2 \in [0, 2\pi]$ the functions

$$m_{1,a}(\sigma_1, \cdot),\ m_{2,b}(\sigma_2, \cdot) : \mathbb{R} \to \mathbb{R} \tag{14}$$

are a.e. invertible, one finds

$$d\lambda_1 = \frac{1}{m'_{1,a}(\sigma_1, m_{1,a}^{-1}(\sigma_1, \mu_1))} d\mu_1 =: T'_{1,a}(\sigma_1, \mu_1) d\mu_1 \tag{15}$$

$$d\lambda_2 = \frac{1}{m'_{2,b}(\sigma_2, m_{2,b}^{-1}(\sigma_2, \mu_2))} d\mu_2 =: T'_{2,b}(\sigma_2, \mu_2) d\mu_2 \tag{16}$$

After the change of variables, (13) becomes

$$\int\int\int\int S_a^{(1)}(s_{1,a}(\sigma_1, m_{1,a}^{-1}(\sigma_1, \mu_1)), \mu_1)$$

$$S_b^{(2)}(s_{2,b}(\sigma_2, m_{2,b}^{-1}(\sigma_2, \mu_2)), \mu_2)$$

$$T'_{1,a}(\sigma_1, \mu_1) T'_{2,b}(\sigma_2, \mu_2) \delta(\mu_1 - m_a) \delta(\mu_2 - m_b) p_S(\sigma_1, \sigma_2)$$

$$d\mu_1 d\mu_2 d\sigma_1 d\sigma_2 \tag{17}$$

Because of our choice of the dynamics and of the observables, the functions $S_a^{(1)}$, $S_b^{(2)}$ in (17) depend only on the variables σ_1, σ_2 respectively.

Therefore (17) becomes

$$\int\int S_a^{(1)}(\sigma_1) S_b^{(2)}(\sigma_2) T'_{1,a}(\sigma_1) T'_{1,b}(\sigma_2) p_S(\sigma_1, \sigma_2) d\sigma_1 d\sigma_2 =$$

$$= \int\int S_a^{(1)}(\sigma_1) S_b^{(2)}(\sigma_2) T'_{1,a}(\sigma_1) T'_{1,b}(\sigma_2) \frac{1}{2\pi} \delta(\sigma_1 - \sigma_2) d\sigma_1 d\sigma_2 =$$

$$= \int S_a^{(1)}(\sigma) S_b^{(2)}(\sigma) T'_{1,a}(\sigma) T'_{1,b}(\sigma) \frac{d\sigma}{2\pi} =$$

$$= -\int_0^{2\pi} \cos(\sigma - a) \operatorname{sgn}(\cos(\sigma - b)) d\sigma = -\cos(b - a)$$

which is the thesis.

9. Geometric Structure of Logical networks

Logical networks, in particular brains, of humans or of other animals, are intrinsically parallel. Serial algorithms, like Turing machine or von Neumann programming, do not capture this essential feature. The following consideartions suggest the existence of an intrinsic geometry of algorithms, a geometry of thought we might say, which seems more suitable to describe parallel processes. We will show elsewhere that a clever exploitation of this geometrical structure may lead in some cases to an exponential reduction of the complexity of some algorithms.

Definition 9.1. A logical network is a triple

$$\{\pi : \mathcal{H} \to \overline{N}, \Omega_0 \subseteq \overline{N} \times \overline{N}, \mathcal{U}\}$$

where:

$\overline{N}$, called the neural manifold, is the set

$$\overline{N} = N \cup \{-\infty, +\infty\}$$

(disjoint union)

$\pi : \mathcal{H} \to \overline{N}$

is a fibre bundle (neural bundle) over $\overline{N}$

$\pm\infty$ are privileged points in $\overline{N}$, called **outside**

The elements of the set M are called **bits** or **neurons**

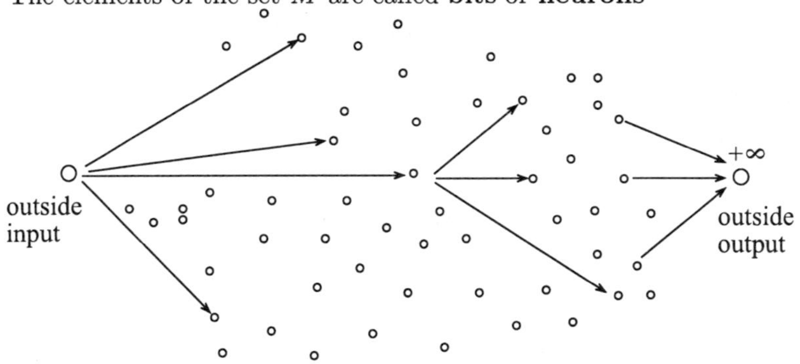

Figure 1.

The fiber

$$\mathcal{H}_0 := \pi^{-1}(m) \qquad ; \qquad m \in \overline{N}$$

is called the state space of the neurons.

We will only consider the case of a trivial bundle:

$$\mathcal{H} = \mathcal{H}_0 \times N$$

Examples:

– For a classical computer the fibre $\mathcal{H}_0$ is the set $\{0,1\}$;
– For a quantum computer it is the vector space $\mathbf{C}^2$.

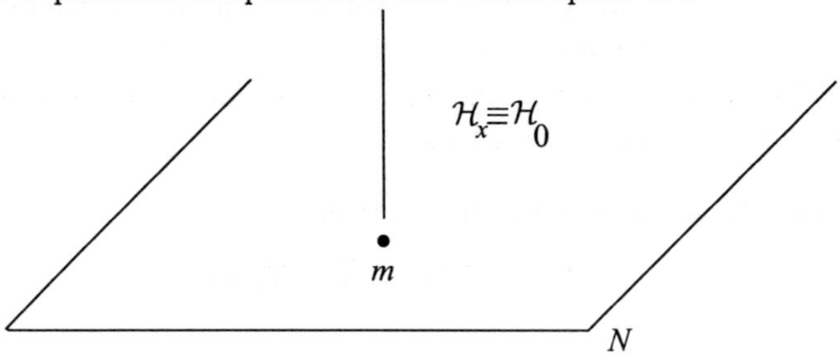

Figure 2.

(ii) Ω_0 is a subset of $\overline{N} \times \overline{N}$
(iii) For $(x,y) \in \Omega_0$, denote

$$Mor\{\mathcal{H}_x \to \mathcal{H}_y\} = Mor\{\mathcal{H}_x, \mathcal{H}_y\}$$

$\mathcal{U}$ is a family of maps

$$U : (x,y) \in \Omega_0 \to U_{(x,y)} \in Mor\{\mathcal{H}_x, \mathcal{H}_y\}$$

called connections or synapses or parallel transport

Examples:

– the family of one–to–one maps $\{0,1\}_x \to \{0,1\}_y$.
– If $\mathcal{H}_0$ is a Hilbert space say $\mathbf{C}^2$, we denote

$$Mor\{\mathcal{H}_x, \mathcal{H}_y\}$$

the family of unitary operators

$$U_{x,y} : \mathcal{H}_x \to \mathcal{H}_y$$

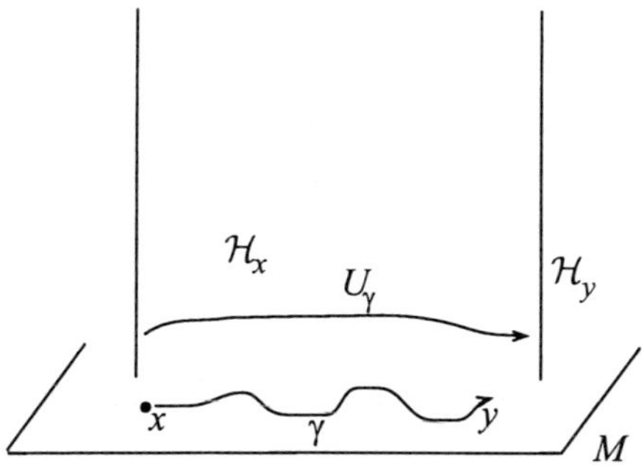

Figure 3.

10. Curves, loops and geometry of parallel algorithms

Let Ω_0 be a set of oriented curves on N without loops.

In our case $\Omega_0 \subseteq \overline{N} \times \overline{N}$.

Let Ω be the groupoid generated by composition of the curves of Ω_0.

Given a curve $\gamma \in \Omega$ we write

$\iota_\gamma =$ initial point of γ

$\varphi_\gamma =$ final point of γ

If $\gamma \in \Omega_0$ then

$$\gamma = (x, y) \in \overline{N} \times \overline{N} \tag{18}$$

The curves $\gamma \in \Omega_0$ with

$\iota_\gamma =$ initial point of $\gamma = -\infty$

are called input curves

The curves $\gamma \in \Omega_0$ with

$\varphi_\gamma =$ final point of $\gamma = +\infty$

are called output curves

The curves $\gamma \in \Omega_0$ with both

$$\iota_\gamma \neq -\infty \qquad ; \qquad \varphi_\gamma \neq +\infty$$

are called **inner strata** of the network.

We will denote Ω the groupoid generated by composition of the curves of Ω_0.

Given a curve $\gamma \in \Omega$ we write

$\iota_\gamma =$ initial point of γ

φ_γ = final point of γ
If $\gamma \in \Omega_0$ then

$$\gamma = (x, y) \in \overline{M} \times \overline{M} \tag{19}$$

Definition 10.1. An **(N–step) single input algorithm** of the logical network is a triple

$$\{\varepsilon_0 \in \mathcal{H}_0, (\mathcal{F}_1, \ldots, \mathcal{F}_N), U\} \tag{20}$$

where:

(i) $N \geq 1$ is a natural integer (the number of steps of the algorithm)
(ii) $\mathcal{F}_j$ is a family of curves in Ω_0 ($j = 1, \ldots, N$), called the strata of the algorithm, such that

(ii1) if $j \neq 1$ the curves $\gamma \in \mathcal{F}_j$ are **inner strata** of the network, i.e.

$$\gamma \in \mathcal{F}_j \text{ and } j \neq 1, N \Rightarrow \iota_\gamma \neq -\infty, \varphi_\gamma \neq +\infty \tag{21}$$

(ii2) the elements of $\mathcal{F}_1$ are input curves, i.e.

$$\gamma \in \mathcal{F}_1 \Rightarrow \iota_\gamma = -\infty \tag{22}$$

(ii3) the elements of $\mathcal{F}_N$ are output curves, i.e.

$$\gamma \in \mathcal{F}_N \Rightarrow \varphi_\gamma = +\infty \tag{23}$$

(ii4) the initial points of curves in the j–th stratum are the end points of curves in the $(j-1)$–st stratum, i.e.

$$\gamma \in \mathcal{F}_j \Rightarrow \iota_\gamma \in \mathcal{F}_{j-1}\ ; \quad j > 1 \tag{24}$$

(ii5) the single input condition is satisfied:

$$\iota_\gamma \neq \iota_{\gamma'} \Rightarrow \varphi_\gamma \neq \varphi_{\gamma'} \tag{25}$$

this means that each neuron can send many inputs but it can receive only one

(iii) ε_0 is called **initial state**.

U, called **parallel transport**, is a map

$$U : \gamma \in \bigcup_{j=1}^{N} \mathcal{F}_j \to U_\gamma \in Mor\{\mathcal{H}_{\iota_\gamma} \to \mathcal{H}_{\varphi_\gamma}) \tag{26}$$

From (23) and (19) it follows that $\mathcal{F}_1$ has the form

$$\mathcal{F}_1 = \{(\infty, x_{1,1}), \ldots, (\infty, x_{1,k_1})\} \tag{27}$$

for some $k_1 \in \mathbf{N}$ and $x_{1,j} \in N$.

The **inputs** of the network are defined by:

$$U_{(\infty, x_{1,\alpha})}\varepsilon_0 \in \mathcal{H}_{x_{1,\alpha}} \ ; \quad \alpha = 1, \ldots, k_1 \tag{28}$$

From (23) and (19) it follows that $\mathcal{F}_N$ has the form

$$\mathcal{F}_N = \{(x_{N,1}, \infty), \ldots, (x_{N,k_N}, \infty)\} \tag{29}$$

for some $k_N \in \mathbf{N}$ and $x_{N,\alpha} \in N$.

Because of (25), for any $x_{N,\alpha}$, there exists a unique path $\gamma_j^{(\alpha)}$ $(j = 1, \ldots, N-1)$ such that

$$\varphi_{\gamma_j^{(\alpha)}} = \iota_{\gamma_{j+1}^{(\alpha)}} \qquad ; \qquad j = 1, \ldots, N-1$$

The **outputs** of the network are defined by

$$U_{(x_{N,\alpha}, \infty)} U_{\gamma_{N-1}^{(\alpha)}} U_{\gamma_{N-2}^{(\alpha)}} \ldots U_{\gamma_1^{(\alpha)}} |\varepsilon_o\rangle$$

Definition 10.2. An N–step multiple input algorithm is defined as a single input algorithm but with following properties: for any $j = 1, \ldots N-1$ and for any

$$\gamma = (x_{j-1}, x_j) \in \mathcal{F}_j$$

define

$$\gamma^{-1}(j, x_j) = \{y : (y, x_j) \in \mathcal{F}_j\}$$

Then there is a map, which is not necessarily a morphism:

$$U_\gamma : \otimes_{y \in \gamma^{-1}(j, x_j)} \mathcal{H}_y \to \mathcal{H}_{x_j}$$

In the following we shall give a prescriptive condition on how to build such a map.

Local constructive conditions to build multiple input algorithms

Time is discretized $j \in \mathbf{N}$ and has an origin 0

Each neuron $x \in N$ has a clock and two activity modes:

(i) send mode

(ii) receive mode

at time 0 all neurons are in receive mode with the exception of $-\infty$

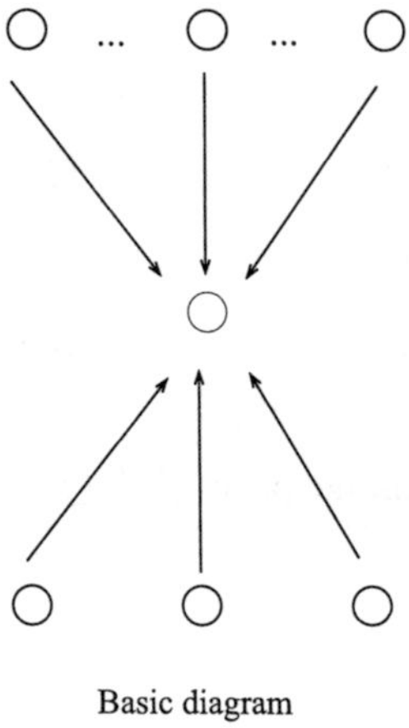

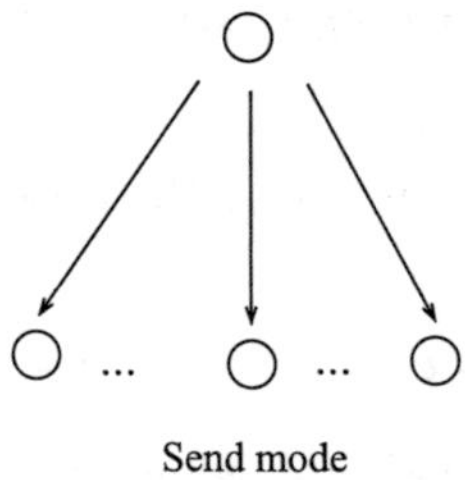

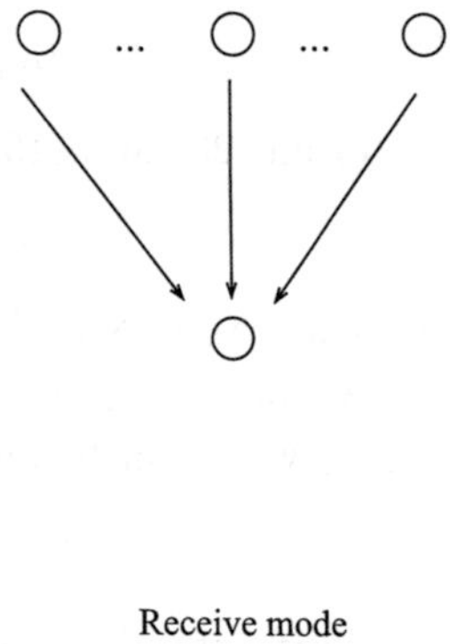

Figure 4a *Figure 4b* *Figure 4c*

$-\infty$ applies the choice function $\kappa_0(-\infty)$ which to an algorithm α associates a pair

$$\kappa_0(-\infty) \ : \ \alpha \ \rightarrow \ (\varepsilon_0 \in \mathcal{H}_{-\infty} \ , \ \mathcal{S}_1(-\infty) \subseteq N)$$

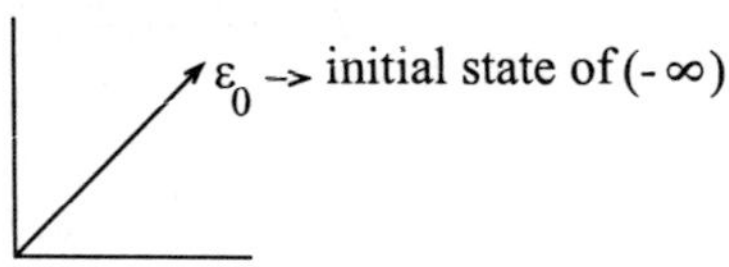

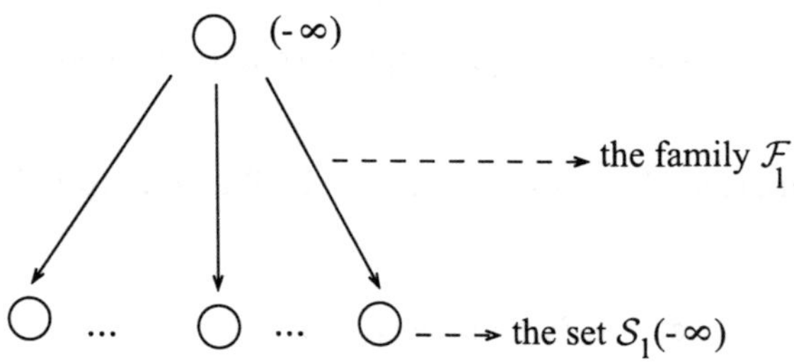

Figure 5.

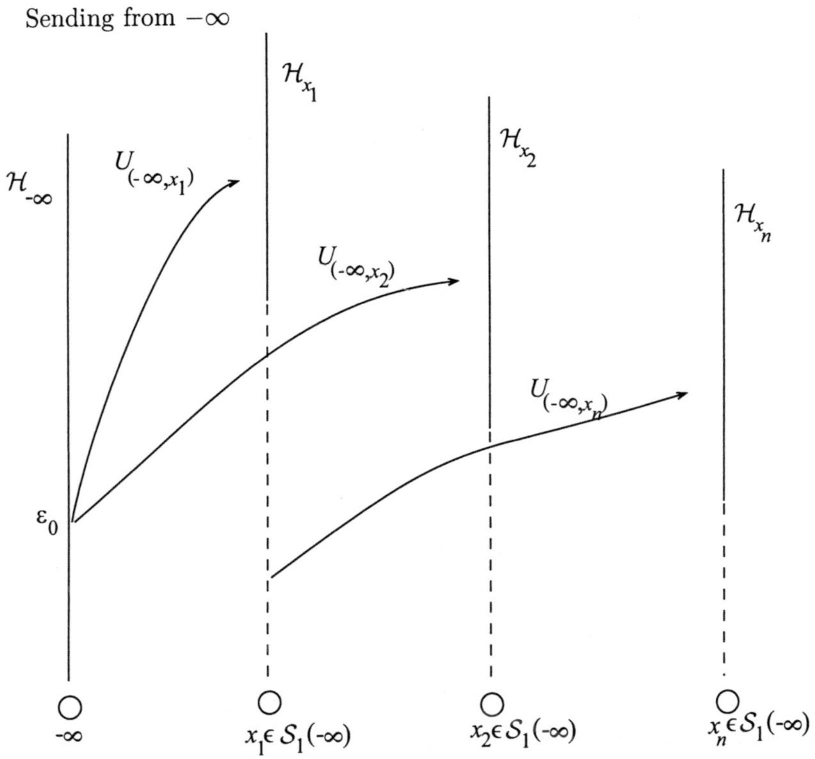

Figure 6. Sending from $-\infty$.

after that, $-\infty$ activates the send mode and sends

$$\otimes_{x\in\mathcal{S}_1(-\infty)} U_{(-\infty,x)}\varepsilon_0 = \otimes_{x\in\mathcal{S}_1(-\infty)}\varepsilon_1(x) \in \tag{30}$$

$$\in \otimes_{x\in\mathcal{S}_1(-\infty)}\mathcal{H}_x$$

this uniquely determines the family of curves

$$\mathcal{F}_1 := \{\gamma : \iota_\gamma = -\infty \ ; \ \varphi_\gamma \in \mathcal{S}_1(-\infty)\} \tag{31}$$

and, for any $x \in N$ the (possibly empty) subset of N

$$\mathcal{F}_1(x) := \{\iota_\gamma \ : \ \gamma \in \mathcal{F}_1 \ , \ \varphi_\gamma = x\} \tag{32}$$

if $\mathcal{F}_1(x)$ is empty, x remains in receive mode
if $\mathcal{F}_1(x)$ is not empty, x receives the input

$$\otimes_{\gamma:\iota_\gamma\in\mathcal{F}_1(x)}\varepsilon_1(\gamma) \in \otimes_{|\mathcal{F}_1(x)|}\mathcal{H}_x \tag{33}$$

then x applies the choice function

$$\kappa_1(x) \ : \ \{\otimes_{\gamma:\iota_\gamma \in \mathcal{F}_1(x)} \varepsilon_1(\gamma) \ , \ \mathcal{F}_1(x)\} \ \to$$

$$\to (\varepsilon_1(x) \in \mathcal{H}_x \ , \ \mathcal{S}_2(x) \subseteq N)$$

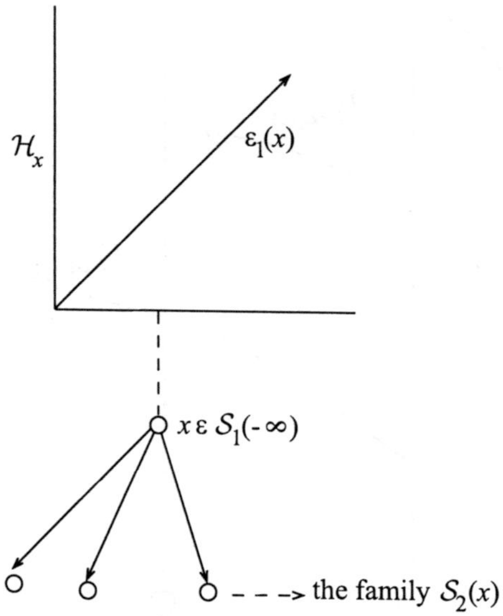

Figure 7. Choice of $\varepsilon_1(x)$ *and of* $\mathcal{S}_2(x)$.

Important:

$$\kappa(x) \equiv \kappa(x, U_{(-\infty,x)}\varepsilon_0)$$

after that x activates the send mode and sends

$$\otimes_{x_2 \in \mathcal{S}_2(x)} U_{(x,x_2)} \varepsilon_1(x) =: \otimes_{x_2 \in \mathcal{S}_2(x)} \varepsilon_2(x_2) \in \tag{34}$$

$$\in \otimes_{x_2 \in \mathcal{S}_2(x)} \mathcal{H}_{x_2}$$

this uniquely determines the family of curves

$$\mathcal{F}_2 := \{\gamma : \iota_\gamma \in \mathcal{S}_1(-\infty) \; ; \; \varphi_\gamma \in \bigcup_{x \in N} \mathcal{S}_2(x)\} \tag{35}$$

and, for any $x \in N$ the (possibly empty) subset of N

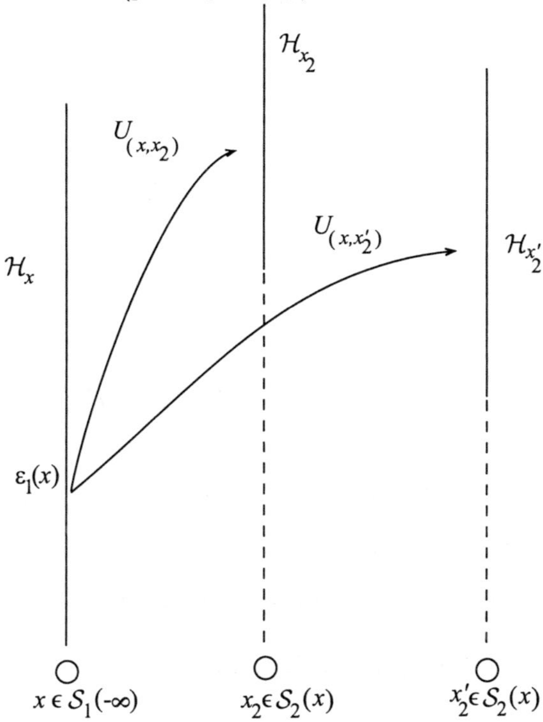

Figure 8. Sending inputs from x to $\mathcal{S}_2(x)$.

$$\mathcal{F}_2(x) := \{\iota_\gamma \;:\; \gamma \in \mathcal{F}_2 \;,\; \varphi_\gamma = x\} \tag{36}$$

having defined $\mathcal{F}_j$, $\mathcal{S}_j$

any $x \in N$ will receive from

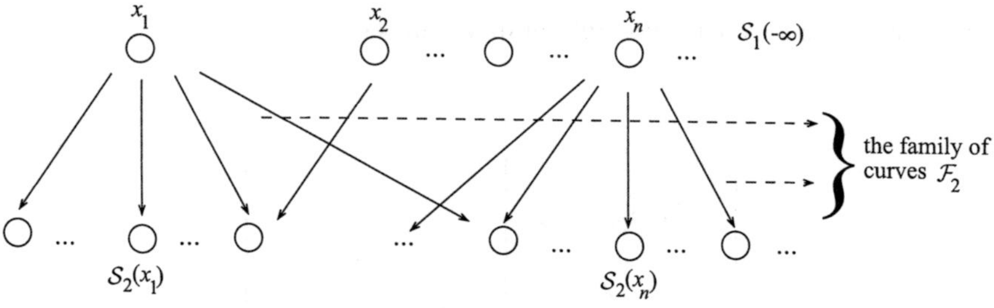

Figure 9. The second stratum: sending mode.

$$\mathcal{F}_j(x) := \{\iota_\gamma \ : \ \gamma \in \mathcal{F}_j \ , \ \varphi_\gamma = x\} \tag{37}$$

the input

$$\otimes_{\gamma:\iota_\gamma \in \mathcal{F}_j(x)} \varepsilon_j(\gamma) \in \otimes_{|\mathcal{F}_1(x)|} \mathcal{H}_x \tag{38}$$

then x applies the choice function

$$\kappa_j(x) \ : \ \{\otimes_{\gamma:\iota_\gamma \in \mathcal{F}_j(x)} \varepsilon_j(\gamma) \ , \ \mathcal{F}_j(x)\} \ \rightarrow$$

$$\rightarrow (\varepsilon_{j+1}(x) \in \mathcal{H}_x \ , \ \mathcal{S}_{j+1}(x) \subseteq N \ )$$

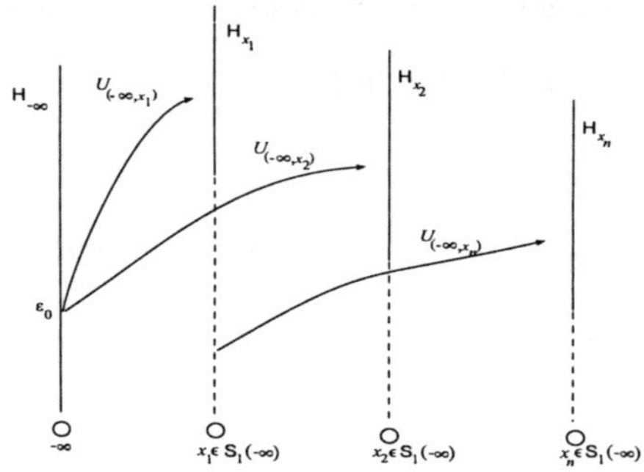

Figure 10. The second stratum: receiving mode.

$$\left.\begin{array}{llll}
x_1 & \circ\cdots\overset{\varepsilon_j(x_1)}{\longrightarrow} & \mathcal{H}x_1 \overset{U_{(x_1,x)}\varepsilon_j(x_1)}{\longrightarrow} & \mathcal{H}_x \\
 & & \otimes & \otimes \\
x_2 & \circ\cdots\overset{\varepsilon_j(x_2)}{\longrightarrow} & \mathcal{H}x_2 \overset{U_{(x_2,x)}\varepsilon_j(x_2)}{\longrightarrow} & \mathcal{H}_x \\
 & & \otimes & \\
 & & \otimes & \otimes \\
x_n & \circ\cdots\overset{\varepsilon_j(x_n)}{\longrightarrow} & \mathcal{H}x_n \overset{U_{(x_n,x)}\varepsilon_j(x_n)}{\longrightarrow} & \mathcal{H}_x \\
 & & \otimes & \otimes \\
\cdots\cdots & & \cdots\cdots\cdots &
\end{array}\right\}\ \begin{array}{c}\text{virtual fiber}\\ \text{at } x\end{array}$$

Figure 11. Sending inputs to the virtual fiber.

after that x activates the send mode and sends

$$\otimes_{x_{j+1}\in\mathcal{S}_{j+1}(x)} U_{(x,x_{j+1})}\varepsilon_{j+1}(x) =: \tag{39}$$

$$\otimes_{x_{j+1}\in\mathcal{S}_{j+1}(x)} \varepsilon_{j+1}(x_{j+1}) \in \otimes_{x_{j+1}\in\mathcal{S}_{j+1}(x)} \mathcal{H}_{x_{j+1}} \tag{40}$$

this uniquely determines the (possibly empty) subset of N

$$\mathcal{S}_{j+1} := \bigcup\{\mathcal{S}_{j+1}(x) \ : \ \textit{there exists } \gamma \in \mathcal{F}_j \ , \ \varphi_\gamma = x\} \tag{41}$$

and the family of curves

$$\mathcal{F}_{j+1} := \{\gamma : \iota_\gamma =\in \mathcal{S}_j \ ; \ \varphi_\gamma \in \mathcal{S}_{j+1}(x)\} \tag{42}$$

and, for any $x \in N$

Example: simplest quantum Markov field

$$N = \mathbf{Z}^2$$

$$U_{(x,y)} := e^{it\sigma_x^{(3)}\otimes\sigma_y^{(3)}}$$

11. Combinatorial algorithms

A general theory of logical networks can be developed, but it is more productive and gives more insight, before passing to abstractions which risk to be relatively empty, to examine in detail the real essence of the present research in quantum computer: the construction of concrete quantum algorithms.

Definition 11.1. An (single or multiple input) algorithm is said to have loops if there exist indices $j < k$ and curves $\gamma \in \mathcal{F}_j$, $\gamma' \in \mathcal{F}_k$ such that

$$\iota_\gamma = \varphi_{\gamma'}$$

An algorithm without loops is called a combinatorial algorithm.

Combinatorial algorithms are the simplest ones. Probably they will be the first to be implemented on a quantum computer.

Lecture II: The Ohya–Masuda algorithm

The Ohya–Masuda algorithm is combinatorial. It is one of the very few presently known quantum algorithm whose implementation does not need an oracle, i.e. for which the unitary operator U_f, associated to the function f to be calculated, can be explicitly constructed in terms of elementary gates.

12. Clause Spaces

Definition 12.1. A *clause* on the set

$$X := \{x_1, \ldots, x_n, x'_1, \ldots, x'_n\} \ , \tag{43}$$

$n \in \mathbf{N}$, is a subset of X.

A clause is determined by two subsets:

$$I \subseteq \{x_1, \ldots, x_n\} \tag{44}$$

$$I' \subseteq \{x'_1, \ldots, x'_n\} \tag{45}$$

i.e. a clause is uniquely fixed by an ordered pair $(I, (I')')$ of subsets of $\{1, \ldots, n\}$ where, by definition, for any subset $I' \subseteq \{x'_1, \ldots, x'_n\}$,

$$(I')' = \{x_j \in \{x_1, \ldots, x_n\} : x'_j \in I'\}$$

A clause (I, I') is said to be *minimal* if

$$I \cap (I')' = \emptyset$$

Remark Introducing the set

$$N := \{1, \ldots, n\} \backslash (I \cup (I')') \tag{46}$$

it is easy to see that minimal clauses are in 1–to–1 correspondence with the partitions

$$\cup (I')' \cup N = \{1, \ldots, n\}$$

where the possibility that any of the three sets of the partition is an empty set is not excluded. Therefore the number of minimal clauses is equal to the number of solutions in natural integrs of the equation $n_1 + n_2 + n_3 = n$, i.e.:

$$M = \binom{n+3-1}{3} = \binom{n+2}{3} = \frac{(n+2)(n+1)n}{6} \sim \frac{n^3}{6} \tag{47}$$

13. Truth Functions

The set $\{0,1\}$ is a boolean algebra with the operations

$$\varepsilon \vee \varepsilon' := \max\{\varepsilon, \varepsilon'\} \tag{48}$$

$$\varepsilon \wedge \varepsilon' := \min\{\varepsilon, \varepsilon'\} \tag{49}$$

where $\varepsilon, \varepsilon' \in \{0,1\}$.

Definition **13.1.**
A clause truth function on the set $\{x_1, \ldots, x_n, x_1', \ldots, x_n'\}$ is a boolean algebra homomorphism

$$t: \text{ Parts of } \{x_1, \ldots, x_n, x_1', \ldots, x_n'\} \to \{0,1\} \tag{50}$$

with the property (*principle of the excluded third*):

$$t(x_j) \vee t(x_j') = 1 \quad ; \quad t(x_j') = 1 - t(x_j) \quad ; \quad \forall j = 1, \ldots, n \tag{51}$$

Because of (51), such a function is uniquely determined by the values

$$\{t(x_1), \ldots, t(x_n)\}$$

Conversely given any n–ple

$$\varepsilon = (\varepsilon_1, \ldots, \varepsilon_n) \in \{0,1\}^n$$

there exists only one truth function on $\{x_1, \ldots, x_n\}$, with the property that

$$t(x_j) = \varepsilon_j \ ; \qquad \forall j = 1, \ldots, n$$

Hence such truth functions are 2^n. For this reason, in the following, we will simply say *truth function on* $\{x_1, \ldots, x_n\}$ meaning by this a truth function on the clauses of the set $\{x_1, \ldots, x_n, x_1', \ldots, x_n'\}$. In the following, given a truth function t, we will denote ε_t the string in $\{t(x_1), \ldots, t(x_n)\}$ uniquely associated to that function.

Lemma 13.1. *Let $C \subseteq X$ be a clause and I, I' the sets associated to it through the procedure explained in § (1). Let t be a truth function on $\{x_1, \ldots, x_n\}$. Then*

$$t(C) = \left[\bigvee_{i \in I} t(x_i)\right] \vee \left[\bigvee_{j \in I'} (1 - t(x_j))\right]$$

Proof. By definition

$$C = \left(\bigcup_{i \in I} \{x_i\} \right) \cup \left(\bigcup_{j \in I'} \{x'_j\} \right)$$

The thesis follows because t is a boolean algebra homomorphism.

Definition 13.2. A set of clauses $\mathcal{C}_0$ is said to be SAT (satisfiable) if there exists a truth function t, on $\{x_1, \ldots, x_n\}$ such that all the clauses in $\mathcal{C}_0$) are t–true, i.e.

$$\wedge_{C \in \mathcal{C}_0} t(C) = \prod_{C \in \mathcal{C}_0} t(C) =: t(\mathcal{C}_0) = 1 \tag{52}$$

The SAT problem consists in the following:
given a set of clauses $\mathcal{C}_0$, decide if it is SAT or not.

Remark. Notice that, by considering, in the notation (52), the function

$$\tau : t \in \mathcal{T} \; := \; \text{the set of all truth functions} \; \mapsto t(\mathcal{C}_0)$$

the SAT problem becomes equivalent to the following:
decide if the $\{0, 1\}$–valued function $\tau_{\mathcal{C}_0}$ is constant or not.

In fact, without changing the complexity of the problem, one can always suppose that the family $\mathcal{C}_0$) contains the clause $x \vee x'$, which is true for every truth function. Under this assumption, the SAT problem is equivalent to the fact that $\tau_{\mathcal{C}_0}$ is the constant function equal to 1.

Remark. It is clear that there exist sets of non satisfiable clauses. For example the set of clauses:

$$\{\{x_1\}, \{x'_1\}\}$$

It can be proved that the above example is in some sense general in the sense that, by an appropriate eleimination procedure, every non SAT problem can be reduced to the above form.

14. Quantum formulation of the SAT Problem

Lemma 14.1. *Let $\mathcal{T}$ be the set of truth functions on $\{x_1, \ldots, x_n\}$. The function*

$$t \in \mathcal{T} \mapsto |t(x_1), \ldots, t(x_n)\rangle \in \otimes^n \mathbf{C}^2$$

defines a one–to–one correspondence between $\mathcal{T}$ and the computational basis of $\otimes^n \mathbf{C}^2$.

Proof. From what was said above the map

$$t \in \mathcal{T} \mapsto (t(x_1), \ldots, t(x_n)) \in \{0,1\}^n$$

is a one–to–one correspondence. Let

$$\mathcal{C}_0 = \{C_1, \ldots, C_m\} \tag{53}$$

be a set of clauses on X. Let t be a truth function and $\varepsilon_t \in \{0,1\}^n$ the corresponding binary string. Ohya and Masuda proposed a technique that succeeds in solving the following problem: *to build.*

(i) A Hilbert space

$$\mathcal{H} = \otimes^{n+\mu} \mathbf{C}^2$$

where μ is a number to be determined as small as possible.

(ii) a unitary operator:

$$U_{\mathcal{C}_0} : \mathcal{H} \to \mathcal{H}$$

with the following property: for any truth function t

$$U_{\mathcal{C}_0} |\varepsilon_t, 0_\mu\rangle = |\varepsilon_t, x^{\varepsilon_t}_{\mu-1}, t(\mathcal{C}_0)\rangle \tag{54}$$

where 0_μ (resp. $x^{\varepsilon}_{\mu-1}$) is a string of μ zeros (resp. a string of $(\mu-1)$ binary symbols depending on ε)

$$t(\mathcal{C}_0) = \prod_{C_j \in \mathcal{C}_0} t(C_j)$$

(iii) Furthermore $U_{\mathcal{C}_0}$ is a product of *gates*, namely of unitary operators that act at most on two q–bits each time.

Definition 14.1. Any operator U that obeys the three previous conditions will be called *an Ohya–Masuda (OM) unitary operator* for the SAT problem $(\mathcal{C}_0)$.

15. Degrees of satisfiability

Let $\mathcal{C}_0$ and $U_{\mathcal{C}_0}$ be as in (53), (54) respectively and, for every $\varepsilon \in \{0,1\}^n$, let t_ε be the corresponding truth function. Define the vector:

$$\psi_0 := \frac{1}{2^{n/2}} \sum_{\varepsilon \in \{0,1\}^n} |\varepsilon, 0_\mu\rangle \tag{55}$$

and apply the unitary operator $U_{\mathcal{C}_0}$ to it:

$$U_{\mathcal{C}_0} \psi_0 = \frac{1}{2^{n/2}} \sum_{\varepsilon \in \{0,1\}^n} |\varepsilon, x^{\varepsilon}_{\mu-1}, t_\varepsilon(\mathcal{C}_0)\rangle$$

Theorem 15.1. *$\mathcal{C}_0$ is SAT if and only if, denoting*

$$P_{n+\mu,1} := 1_{n+\mu-1} \otimes |1\rangle\langle 1| \quad ; \quad P_{n+\mu,0} := 1_{n+\mu-1} \otimes |0\rangle\langle 0|$$

the projector on to the subspace of $\mathcal{H}$ spanned by the vectors

$$|\varepsilon_n, \varepsilon_{\mu-1}, 1\rangle$$

one has

$$P_{n+\mu,1}\psi_0 \neq 0 \tag{56}$$

Proof. Condition (56) means that exists $t \in \mathcal{T}$ that $t(\mathcal{C}_0) = 1$. Conversely, if such a t exists then (56) holds.

Now, in the notations (55), denote

$$\rho = |\psi_o\rangle\langle\psi_o| \tag{57}$$

After a totally incomplete measurement of the observable with spectral projections $P_{n+\mu,0}, P_{n+\mu,1}$ the state (57) becomes, after application of the extended von Numann formula for measurement,

$$\rho \to P_{n+\mu,1}\rho P_{n+\mu,1} + P_{n+\mu,0}\rho P_{n+\mu,0} =: \rho'$$

so that the probability of $P_{n+\mu,1}$ is

$$Tr\rho' P_{n+\mu,1} = \|P_{n+\mu,1}\psi\|^2 = \frac{|T(\mathcal{C}_0)|}{2^n}$$

where $|T\mathcal{C}_0)|$ is the cardinality of the set $T(\mathcal{C}_0)$, of all the truth functions t such that $t(\mathcal{C}_0) = 1$.

This shows that the Ohya–Masuda algorithm can in fact do more than simply verifying satisfiabilty: in fact it allows to introduce a kind of *absolute measure of truth* for different set of clauses, measured in terms of the cardinality of the set of truth functions which give value 1 to the given set of clauses. We can introduce a total order in the family of these sets by saying that a family $\mathcal{C}$ is *more true* than another one $\mathcal{C}'$ if

$$|T(\mathcal{C})| \geq |T(\mathcal{C}'|)$$

To give a combinatorial estimate of this cardinality is quite difficult (in fact much more difficult than the SAT problem itself), however the Ohya–Masuda algorithm shows that one can get a probabilistic estimate by using a quantum computer.

16. Locality of the Ohya–Masuda Unitary Operators

The following exposition emphasizes the local, sequential and constructive nature of the Ohya–Masuda algorithm: the unitary operator is constructed step by step by purely local operations and without any additional "oracle" informations.

Let us start to consider the case of a single clause C and let (I, I') be a pair of subsets of $\{1, \ldots, n\}$, related to C, as described in § (1). Then, given a truth function

$$t_\varepsilon \equiv (\varepsilon_1, \ldots, \varepsilon_n) \tag{58}$$

we have that

$$t_\varepsilon(C) = \left(\bigvee_{i\in I} \varepsilon_i\right) \vee \left(\bigvee_{j\in I'} (1-\varepsilon_j)\right) \tag{59}$$

Therefore, if U_C is an OM unitary operator for the SAT problem (C), we must have:

$$U_C|\varepsilon, 0_\mu\rangle = |\varepsilon, x^\varepsilon_\mu, \left(\bigvee_{i\in I} \varepsilon_i\right) \vee \left(\bigvee_{j\in I'} (1-\varepsilon_j)\right)\rangle \tag{60}$$

where the x^ε_μ are additional bits produced by the algorithm, which are introduced to keep the unitarity of the operator. To implement in a local way such an unitary operator U_C *three kinds of gates* will be needed:

(I.) Flip (or NOT) gates
(II.) localised OR gates
(III.) localised AND gates,

These are standard classical and quantum gates. While NOT is localized on single bit, AND and OR are localized on three bits

17. Flip (or NOT) gates

A flip, or NOT gate, localised in a point h, is denoted N_h and is characterised by the property

$$N_h := (\otimes_{j\neq h} 1_j) \otimes (U_{\mathrm{NOT}})_h := \left(\bigotimes_{j\neq h} 1_j\right) \otimes (|0\rangle\langle 1| + |1\rangle\langle +0|) = (\otimes_{j\neq h} 1_j) \otimes N_h \tag{61}$$

its action on a binary string $|\varepsilon\rangle$ consists in flipping the h–th bit.

$$N_h|\varepsilon\rangle = |\varepsilon_1, \ldots, \varepsilon_{h-1}, 1-\varepsilon_h, \varepsilon_{h+1}, \ldots\rangle \tag{62}$$

Since $h \neq k$, N_h and N_k act on different q–bits hence, they commute and can, therefore, be parallelly implemented. Given the clause $C \equiv (I, I')$, we define

$$N_{I'} := \otimes_{h\in I'} N_h \tag{63}$$

where N_h is as in (61). Therefore

$$N_{I'}|\varepsilon, 0_\mu\rangle = |(\varepsilon_i)_{i\notin I'}, (1-\varepsilon_j)_{j\in I'}, 0_\mu\rangle =: |\varepsilon_{I'}, 0_\mu\rangle \tag{64}$$

where $\varepsilon_{I'}$, is obtained from ε by changing in $1-\varepsilon_j$ all the ε_j with $j \in I'$.

18. Localised OR, AND Gates

Definition 18.1. In the notation of Lemma (14.1), the quantum AND gate corresponds to the choices:

$$S = \{0,1\}^{2n}$$

$$f : (\varepsilon_1, \varepsilon_2) \in \{0,1\}^{2n} \to f(\varepsilon_1, \varepsilon_2) := \varepsilon_1 \wedge \varepsilon_2$$

where the $\wedge$–operation is meant componentwise. Explicitly

$$A := U_{AND} := |\varepsilon_1, \varepsilon_2, \varepsilon_1 \wedge \varepsilon_2\rangle\langle 0, \varepsilon_1, \varepsilon_2| + |\varepsilon_1, \varepsilon_2, 1-(\varepsilon_1 \wedge \varepsilon_2)\rangle\langle 1, \varepsilon_1, \varepsilon_2|$$

The OR gate corresponds to the choice:

$$O := U_{OR} := |\varepsilon_1, \varepsilon_2, \varepsilon_1 \vee \varepsilon_2\rangle\langle 0, \varepsilon_1, \varepsilon_2| + |\varepsilon_1, \varepsilon_2, 1-(\varepsilon_1 \vee \varepsilon_2)\rangle\langle 1, \varepsilon_1, \varepsilon_2|$$

Remark The quantum AND gate is also called a Toffoli gate.
Remark When an AND (OR) is implemented on a pair of bits, an additional dust–qbit is needed to keep unitarity.

Definition 18.2. A localised *NOT* (or *AND*, or *localised OR)* gate is an operator of the form:

$$N_h := (\otimes_{j\neq h} 1_j) \otimes N \tag{65}$$

$$A_{i,j,k} = \left(\bigotimes_{h\neq i,j,k} 1_h\right) \otimes A \tag{66}$$

$$O_{i,j,k} = \left(\bigotimes_{h \neq i,j,k} 1_h \right) \otimes O \tag{67}$$

where (1) means that N acts on the h–th factor of $\otimes^N \mathbf{C}^2$ and (2), (3) that A and O act on the factor product i, j, k.

In the following it will be convenient to consider also a modified localised OR operator:

$$\overline{O}_{i,j,k} := N_i O_{i,j,k} N_i \tag{68}$$

This new operator first changes the i–th 0 into 1 (or conversely) then acts with the ordinary $O_{i,j,k}$. Explicitly:

$$\overline{O}_{i,j,k} |\varepsilon_i, \varepsilon_j, 0_k\rangle = O_{i,j,k} |1 - \varepsilon_i, \varepsilon_j, 0_k\rangle$$

19. The Case of Only One Clause

In the notations of § (1) and (1a), given a clause $C \equiv (I, I')$ and a truth function $t_\varepsilon \equiv \varepsilon \in \{0,1\}^{2n}$ Assuming that

$$I = \{1 \le j_1 < j_2 < \cdots < j_i \le n\} \tag{69}$$

$$I' = \{1 \le j'_1 < j'_2 < \cdots < j'_{i'} \le n\} \tag{70}$$

define the vector:

$$\psi^0_\varepsilon := |\varepsilon, 0_{2n}\rangle \in (\mathbf{C}^2)^{3n}$$

Then we have:

$$O_{j_1, j_2, n+1} \psi^0_\varepsilon = |\varepsilon, \varepsilon_{j_1} \vee \varepsilon_{j_2}, 0_{2n-1}\rangle =: \psi^1_\varepsilon$$

$$O_{j_3, n+1, n+2} \psi^1_\varepsilon = |\varepsilon, \varepsilon_{j_1} \vee \varepsilon_{j_2}, \varepsilon_{j_1} \vee \varepsilon_{j_2} \vee \varepsilon_{j_3}, 0_{2n-2}\rangle =: \psi^2_\varepsilon$$

generally speaking, for $1 \le k \le i \le n$ we have:

$$O_{j_k, n+(k-2), n+(k-1)} \psi^{k-1}_\varepsilon = |\varepsilon, \varepsilon_{j_1} \vee \varepsilon_{j_2}, \varepsilon_{j_1} \vee \varepsilon_{j_2} \vee \varepsilon_{j_3}, \ldots, \varepsilon_{j_1} \vee \cdots \vee \varepsilon_{j_k}, 0_{2n-k}\rangle =: \psi^k_\varepsilon$$

After i steps we arrive to:

$$\left(\overleftarrow{\prod_{k=3,\ldots,i}} O_{j_k, n+(k-2), n+(k-1)} \right) O_{j_1, j_2, n+1} \psi^0_\varepsilon =$$

$$= |\varepsilon, \varepsilon_{j_1} \vee \varepsilon_{j_2}, \ldots, \varepsilon_{j_1} \vee \cdots \vee \varepsilon_{j_i}, 0_{2n-i}\rangle =: \psi^i_\varepsilon$$

where the symbol $\overleftarrow{\prod}$ means that the product is ordered from right to left, namely

$$\ldots O_{j_k,n+(k-2),n+(k-1)} \cdots O_{j_3,n+1,n+2} O_{j_1,j_2,n+1}$$

At this point the operation is repeated with the overlined OR $\overline{O}_{j_k,n+(k-2),n+(k-1)}$ (cf. (68) and, with the usual notation $\varepsilon' := 1-\varepsilon$, we find:

$$\left(\overleftarrow{\prod_{k=3,\ldots,i'}} \overline{O}_{j'_k,n+(k-2),n+(k-1)}\right) \overline{O}_{j'_1,j'_2,n+1} \psi^i_\varepsilon =$$

$$= |\varepsilon, \varepsilon_{j_1} \vee \varepsilon_{j_2}, \ldots, \varepsilon_{j_1} \vee \cdots \vee \varepsilon_{j_i}, \varepsilon'_{j'_1} \vee \varepsilon'_{j'_2}, \ldots,$$

$$\ldots, \varepsilon'_{j'_1} \vee \cdots \vee \varepsilon'_{j'_{i'}}, (\varepsilon_{j_1} \vee \cdots \vee \varepsilon_{j_i}) \vee (\varepsilon'_{j_1}, \vee \cdots \vee \varepsilon'_{j'_{i'}}), 0_{2n-i-i'}\rangle =: \psi^{i+i'}_\varepsilon =$$

$$= |\varepsilon, \varepsilon_{j_1} \vee \varepsilon_{j_2}, \ldots, t_\varepsilon(C), 0_{2n-i-i'}\rangle$$

Then the unitary operator associated to a clause C will be

$$U_C := \left(\overleftarrow{\prod_{k=3,\ldots,i'}} \overline{O}_{j'_k,n+(k-2),n+(k-1)}\right) \overline{O}_{j'_1,j'_2,n+1} \left(\overleftarrow{\prod_{k=3,\ldots,i}} O_{j_k,n+(k-2),n+(k-1)}\right) O_{j_1,j_2,n+1} \tag{71}$$

20. More Clauses: General Case

Let

$$\mathcal{C}_0 = \{C_1, \ldots, C_m\} \tag{72}$$

be an (ordered) set of clauses and, using the notation of § (1), for every $j = 1, \ldots, n$,

$$C_j \equiv (I_j, I'_j) \tag{73}$$

Assuming that:

$$I_k = \{1 \le j_{k1} < j_{k2} < \cdots < j_{ki_k} \le n\} \tag{74}$$

$$I'_k = \{1 \le j'_{k1} < j'_{k2} < \cdots < j'_{ki'_k} \le n\} \tag{75}$$

For every clause we need at most $2n$ dust q–bits. Furthermore to realise AND between the clauses truth values, we need at most m new dust q–bits.

Eventually we need at most $2mn + m$ dust q–bits and therefore the final space will be

$$(\mathbf{C}^2)^{2mn+m+n} = \mathcal{H}_\varepsilon \otimes \mathcal{H}_D^1 \otimes [\mathcal{H}_D^2 \otimes \mathbf{C}^2] \otimes \cdots \otimes [\mathcal{H}_D^m \otimes \mathbf{C}^2] \tag{76}$$

where $\mathcal{H}_\varepsilon \sim (\mathbf{C}^2)^n$ codes the truth function ε, $\mathcal{H}_D^k \sim (\mathbf{C}^2)^{2n}$ is the dust space of the k–th clause due to the OR operations and the additional factors $\mathbf{C}^2$ are needed for the AND operations. The unitary operator for the k–the clause C_k will then be of the form

$$\hat{U}_{C_k} := (U_{C_k})_{\varepsilon, D_k} \otimes 1_{others} \quad ; \quad k = 1, 2, \ldots, m$$

where U_{C_k} is defined as in the previous section and the index (ε, D_k), in the right hand side means that the operator acts on $\mathcal{H}_\varepsilon \otimes \mathcal{H}_D^k$. The unitary operator for the AND operation will be of the form

$$A_{C_o} := A_{c_{m-1},\alpha_m,c_m} \cdots A_{c_3,\alpha_4,c_5} A_{c_2,\alpha_3,c_3} A_{\alpha_1,\alpha_2,c_2}$$

where the indices $\alpha_1, \alpha_2, \ldots \alpha_m$ refer to the q–bits encoding the truth value $t_\varepsilon(C_1), \ldots, t_\varepsilon(C_m)$ and the indices c_k to the $\mathbf{C}^2$ dust factor of the k–th clause in (76).

The unitary operator for the set (1) of clauses will then be of the form

$$U := A_{C_o} \prod_{k=1,\ldots,m} \hat{U}_{C_k}$$

Notice that no order is required in the above product because the unitary operators $\hat{U}_{C_k}$ commute.

21. Conclusions

In the original OM algorithm the number of dust q–bits is of order mn, where m is the number of clauses and n is the cardinality of the alphabet (i.e. the number of x's not including their negation).

Since the number of clauses can be of order 2^n, this means that the dimension of the network can be exponential.

However, keeping in mind that, once the truth value

$$(\varepsilon_{j_1} \vee \cdots \vee \varepsilon_{j_i}) \vee (\varepsilon'_{j'_1} \vee \cdots \vee \varepsilon'_{j_{i'}}) = t_\varepsilon(c)$$

has been calculated for the clause

$$\{j_1, \ldots, j_i, j'_1, \ldots, j'_{i'}\}$$

the clause itself need not to be memorized, and the fact that different clauses are orthogonal, one can iterate the procedure by passing from

$$|\varepsilon, t_\varepsilon(c), O_{2n-(i+i')}\rangle$$

to

$$|\varepsilon', \ldots, 0, O_{2n-(i+i')}\rangle$$

which is unitarily implemented because $|\varepsilon\rangle \perp |\varepsilon'\rangle$. This unitary involves at least $2n - 1$ flips.

In this way one decreases exponentially the dimensions of the network. The price to pay for this is that the number of flips increases exponentially even if each block of $2n - 1$ can be implemented in a parallel way. In fact this still leaves m (which can be of order 2^n) flips, i.e. one for each clause.

This concrete analysis shows that a reliable measure of complexity for quantum algorithms should keep into account not only the number of steps but also the dimension of the network and the number of parallel flips (or localized operations).

Lecture III: Algorithms with oracle

Let f be a function. We say that an algorithm involving f has an oracle if it involves an unitary operator U_f such that:

(i) U_f depends on all the values of f

(ii) no explicit estimate is given of the complexity of U_f, measured for example in terms of the minimum number of factors needed to express f as a product of 2–q–bits unitaries.

22. The Deutsch–Josza algorithm

In this section we analyse the simplest algorithms with oracle: the Deutsch–Josza algorithm. A similar analysis can be carried out for all algorithms of this type.

A function

$$f : S := \{1 \dots, N\} \to \{0, 1\}$$

with N even, is called balanced if

$$|A| := |f^{-1}(0)| = |f^{-1}(1)| = |A^c|$$

Deutsch–Josza [DeuJo92] consider the following problem:

Given f decide if it constant or balanced.

Suppose that the computation of each value of f is polynomially complex. Classically one would choose at random (i.e. with independent trials) k points $\omega_1, \dots, \omega_k$ from S and, if f is balanced, the probability that all these points are in $A := f^{-1}(0)$ is 2^{-k}. Therefore one can solve the problem in k steps with probability

$$1 - 2^{-k}$$

We code the numbers $\{1 \dots, N\}$ in binary expansion and we introduce the standard basis $|0\rangle, |1\rangle$ in $\mathbf{C}^2$, so that we can work in $\otimes^n \mathbf{C}^2 \equiv \mathbf{C}^{2n}$ with the computational basis $\{|\varepsilon_1, \dots, \varepsilon_n\rangle\}$. Let H be the Hadamard gate, defined as in Parthasarathy's lectures, and consider $H_n = \otimes^n H$

$$H_n|0, \dots, 0\rangle = 2^{-n/2} \sum_{\varepsilon_1, \dots, \varepsilon_n} |\varepsilon_1, \dots, \varepsilon_n\rangle =: \psi_0$$

Let U_f be defined by

$$U_f|\varepsilon_1, \dots, \varepsilon_n\rangle = U_f|\varepsilon\rangle := (-1)^{f(\varepsilon)}|\varepsilon\rangle$$

Then

$$U_f H_n|0\rangle =: U|0\rangle = 2^{-n/2}\sum(-1)^{f(\varepsilon)}|\varepsilon\rangle = 2^{-n/2}\left(\sum_{f^{-1}(0)}|\varepsilon\rangle - \sum_{f^{-1}(1)}|\varepsilon'\rangle\right) \tag{77}$$

Now define (recall that $f^{-1}(0) =: A$)

$$\psi_A := \sum_{\varepsilon\in A}|\varepsilon\rangle$$

so that

$$\psi_0 = 2^{-n/2}(\psi_A + \psi_{A^c})$$

By assumption f is balanced or constant. Therefore, is f is constant then:

$$U|0\rangle = \pm 2^{-n/2}(\psi_A + \psi_{A^c}) = \pm\psi_0$$

and

$$|\langle\psi_0, U0\rangle|^2 = |\langle\psi_0, \psi_0\rangle|^2 = 1$$

If f non constant then it is balanced, i.e. $|A| = |A^c|$. Therefore

$$U|0\rangle = 2^{-n/2}(\psi_A - \psi_{A^c})$$

and

$$|\langle\psi_0, U0\rangle|^2 = 2^{-n/2}|\langle\psi_A + \psi_{A^c}, U0\rangle|^2 = 2^{-n}|\langle\psi_A + \psi_{A^c}, \psi_A - \psi_{A^c}\rangle|^2 =$$

$$= 2^{-n}(\|\psi_A\|^2 - \|\psi_{A^c}\|^2) = 0$$

Thus, after 1 step, the problem is reduced to discriminate between two orthogonal states which can be solved with a single measurement.

Lecture IV: Teleportation

23. The BBC teleportation protocol: qualitative description

Teleportation was invented by Benett and Brassard and its channel theoretical formulation of the quantum gates was introduced by Ohya and Watanabe [OhWa97] and then applied to the teleportation problem by Inoue Ohya Suyari [InOhSu98]. Accardi and Ohya used this scheme to construct a general method to solve the teleportation problem in spaces of arbitrary finite dimensions. In fact, as we shall see, the finite dimensionality of the space is a necessary condition for the existence of the solution of the original B–B problem. Kimble's method allies to infinite dimensional spaces, but the solution it leads to is approximate. The method of [AcOh91] was then modified to be applied directly to the Hilbert space (rather than algebra) level by Albeverio and Fei [AlFe00a], [AlFe00b] (see also [Sabb00]).

The teleportation algorithm is the following.

STEP 0:

Alice (A) wants to teleport to Bob (B) an unknown state $\rho^{(1)}$ of a given quantum system (system 1)

STEP 1:

introduce two auxiliary systems denoted $2, 3$ and localized near A and near B respectively. Then fix a state $\sigma_k^{23}k$ in the $(2,3)$–system and prepare an ensemble of $(1,2,3)$–systems in the state

$$\rho^{(123)} := \rho^{(1)} \otimes \sigma_k^{(23)} \tag{78}$$

The elements of the $(2,3)$–system must be pairs (typically entangled pairs), i.e. if A makes a measurement on an element E_2 of a pair (E_2, E_3) in the ensemble $\sigma_k^{(23)}$, B must be able to recognize the companion element E_3.

STEP 2:

Fix an observable A with spectral decomposition

$$A := \sum_k \lambda_k F_k^{(12)}$$

Having arbitrarily fixed one value of A, say λ_k, Alice performs a first kind measurement, involving only the $(1,2)$ system, which filters the chosen value λ_k, i.e. after the measurement on the given ensemble (2), of identically prepared systems, only those with $A = \lambda_k$ are allowed to pass.

In other words Alice performs an incomplete measurement corresponding to the orthogonal projection

$$F_k^{(12)} \otimes 1_3 \tag{79}$$

Let us denote $\pi_k^*\left(\rho_k^{(123)}\right)$ the state of the $(1, 2, 3)$–system becomes after Alice's measurement.

STEP 3:

Bob is informed, by classical tools (e.g. telephone) and without disturbance, which measurement was done by Alice, i.e. which eigenvalue of A was chosen.

STEP 4:

Bob reconstructs the state $\rho^{(3)}$, induced by the state $\pi_k^*\left(\rho_k^{(123)}\right)$ on the system 1, by making local measurements on the system 3, i.e. the system localized near him.. Notice that this state may be unknown Alice, because the initial ensemble (78) has not necessarily been prepared by her.

Since Bob and Alice are spatially separated Bob can perform his experiments on the same ensemble of systems found by Alice as a result of her measurement or at least on a sub–ensemble of it (in our case the system 3). This situation corresponds to the original BBC protocol and requires the physical transmission of system 3 from A to B. Such a physical transmission is necessary in order to exploit the entanglement of the state $\sigma_k^{(23)}$ which, as we shall see, plays an important role in the protocol.

The crucial point of the construction is that, knowing the information transmitted by Alice about which measurement was done by her, Bob is able to reconstruct in a unique way the original state $\rho^{(1)}$, of system 1, from the state $\rho^{(3)}$, of system 3.

When the state $\sigma_k^{(23)}$ is independent of k, the above problem reduces to the channel theoretical formulation of [InOhSu98].

24. The teleportation protocol: channel theoretical formulation

Recall that a quantum state is described by a density operator on a Hilbert space $\mathcal{H}$ and that the set of all these density operators is denoted by $Dens(\mathcal{H})$. In other words, denoting $B(\mathcal{H})$ the set of all bounded operators on $\mathcal{H}$, on has:

$$Dens(\mathcal{H}) = \{\rho \in B(\mathcal{H}) \ ; \ \rho^* = \rho, \ \rho \geq 0, \ \text{tr}\rho = 1\}$$

With these notations the channel theoretical formulation of the teleportation problem goes as follows.

Introduce two auxiliary Hilbert spaces $\mathcal{H}_2$ and $\mathcal{H}_3$. Initially $\mathcal{H}_3$ is localized near A but there is a dynamics, not explicit in the protocol, which after some time brings system 3, hence its state space $\mathcal{H}_3$ to be localized near B.

Fix a set of states

$$\sigma_k^{23} \in Dens\,(\mathcal{H}_2 \otimes \mathcal{H}_3) \tag{80}$$

in the (2, 3)–space. Then prepare, in the initial stage of the experiment, an ensemble of systems in the $(1,2,3)$–space in the state

$$\rho_k^{(123)} := \rho^{(1)} \otimes \sigma_k^{(23)} \tag{81}$$

The above procedure can be realized by a channel (dual of a completely positive map)

$$\Lambda_k^* : Dens\,(\mathcal{H}_1) \longrightarrow Dens\,(\mathcal{H}_1)$$

composed of the following four channels:

i) a set of trivial (i.e. product) liftings in the sense of [AcOh91]

$$\gamma_k^* : Dens\,(\mathcal{H}_1) \longrightarrow Dens\,(\mathcal{H}_1 \otimes \mathcal{H}_2 \otimes \mathcal{H}) \tag{82}$$

$$\gamma_k^* \left(\rho^{(1)}\right) = \rho^{(1)} \otimes \sigma_k^{(23)} \qquad ; \qquad \forall \rho^{(1)} \in Dens\,(\mathcal{H}_1) \tag{83}$$

expressing the independent coupling of the initial state $\rho^{(1)}$ with the state $\sigma_k^{(23)}$.

(ii) according to the Zumino–Luders projection postulate, after Alice's measurement, the state of the $(1,2,3)$–system becomes

$$\rho_k^{(123)} := \frac{\left(F_k^{(12)} \otimes 1_3\right) \rho_k^{(123)} \left(F_k^{(12)} \otimes 1_3\right)}{\mathrm{tr}_{123} \left(F_k^{(12)} \otimes 1_3\right) (\rho^{(1)} \otimes \sigma_k^{(23)}) \left(F_k^{(12)} \otimes 1_3\right)} \tag{84}$$

tr_{123} hence the second step is described by a measurement type channel

$$\pi_k^* : Dens\,(\mathcal{H}_1 \otimes \mathcal{H}_2 \otimes \mathcal{H}_3) \longrightarrow Dens\,(\mathcal{H}_1 \otimes \mathcal{H}_2 \otimes \mathcal{H}_3) \tag{85}$$

of the form

$$\pi_k^* \left(\rho_k^{(123)}\right) := \frac{\left(F_k^{(12)} \otimes 1_3\right) \rho_k^{(123)} \left(F_k^{(12)} \otimes 1_3\right)}{\mathrm{tr}_{123} \left(F_k^{(12)} \otimes 1_3\right) \rho_k^{(123)} \left(F_k^{(12)} \otimes 1_3\right)}$$

where $\rho_k^{(123)} \in \mathcal{B}\,(\mathcal{H}_1 \otimes \mathcal{H}_2 \otimes \mathcal{H}_3)$, corresponding to an incomplete first kind measurement describing the state change determined by Alice's filtering of the λ_k eigenvalue of the observable A. It should be emphasised that, by construction, the observable A is not maximal, hence the projection postulate we are applying here is not the original von Neumann projection postulate but the Zumino–Luders one.

(iii) the third step is defined by the channel $a^* : Dens\,(\mathcal{H}_1 \otimes \mathcal{H}_2 \otimes \mathcal{H}_3) \longrightarrow Dens\,(\mathcal{H}_3)$ defined by

$$\rho_k^{(3)} = a^*\left(\rho_k^{(123)}\right) = \mathrm{tr}_{12}\rho_k^{(123)} \quad , \quad \forall \rho_k^{(123)} \in Dens\,(\mathcal{H}_1 \otimes \mathcal{H}_2 \otimes \mathcal{H}_3)$$

Here tr_{12} is the partial trace on the Hilbert space $\mathcal{H}_1 \otimes \mathcal{H}_2$

$$\langle \Phi_1, \mathrm{tr}_{12}\, Q\Phi_2 \rangle \equiv \sum_n \langle \Psi_n \otimes \Phi_1,\, Q\Psi_n \otimes \Phi_2 \rangle,\ Q \in B(\mathcal{H}_1 \otimes \mathcal{H}_2 \otimes \mathcal{H}_3)$$

for any CONS $\{\Psi_n\}$ of $\mathcal{H}_1 \otimes \mathcal{H}_2$ and any $\Phi_1, \Phi_2 \in \mathcal{H}_3$. This channel a^* corresponds to Bob's partial measurement over the system 3 and describes the reduction, from the state $\rho^{(123)}$ obtained after Alice's measurement, to the state $\rho^{(3)}$, obtained by Bob.

Thus the whole teleportation process is written by the channel

$$\Lambda_k^* : Dens\,(\mathcal{H}_1) \longrightarrow Dens\,(\mathcal{H}_1)$$

$$\Lambda_k^* = \Lambda_{k,A\to B}^* = a^* \circ \pi_k^* \circ \gamma^*$$

The above subscript "$[k]$" means that the channels Λ_k^* and $\Lambda_{k,A\to B}^*$ depend on the choice of Alice's measurement $F_k^{(12)}$. More precisely, $\forall \rho^{(1)} \in Dens\,(\mathcal{H}_1)$

$$\Lambda_k^* \rho^{(1)} \equiv \mathrm{tr}_{12}\left[\frac{\left(F_k^{(12)} \otimes 1_3\right)\left(\rho^{(1)} \otimes \sigma_k^{(23)}\right)\left(F_k^{(12)} \otimes 1_3\right)}{\mathrm{tr}_{123}\left(F_k^{(12)} \otimes 1_3\right)\left(\rho^{(1)} \otimes \sigma_k^{(23)}\right)\left(F_k^{(12)} \otimes 1_3\right)}\right] \tag{86}$$

Note that the channel Λ_k^* is generally non linear.

With these notations we can formulate the general mathematical problem of teleportation as follows.

Given the initial Hilbert space $\mathcal{H}_1$, find:

(1) two auxiliary Hilbert spaces $\mathcal{H}_2$, $\mathcal{H}_3$
(2) a family of states $\sigma_k^{(23)} \in \mathcal{H}_2 \otimes \mathcal{H}_3$
(3) a family of mutually orthogonal projections $\left\{F_k^{(12)}\right\}$ acting on $\mathcal{H}_1 \otimes \mathcal{H}_2$

(4) for each k a unitary operator U_k such that the associated unitary channel

$$u_k^* : Dens\,(\mathcal{H}_3) \longrightarrow Dens\,(\mathcal{H}_1)$$

$$u_k^*\left(\rho^{(3)}\right) = U_k \rho^{(3)} U_k^* \quad \forall \rho^{(3)} \in Dens\,(\mathcal{H}_3)$$

satisfies the identity

$$\Lambda_k^* \rho^{(1)} = U_k^* \rho^{(1)} U_k = u_k^*\left(\rho^{(1)}\right) \tag{87}$$

for any k and for any state $\rho^{(1)} \in Dens(\mathcal{H}_1)$ or at least for $\rho^{(1)}$ in a preassigned subset of $Dens(\mathcal{H}_1)$.

If condition (3), (4)above are replaced by the weaker ones:

(3') a single projection $F^{(12)}$ acting on $\mathcal{H}_1 \otimes \mathcal{H}_2$
(4') a single unitary operator U such that the identity

$$\Lambda_k^* \rho^{(1)} = U^* \rho^{(1)} U \tag{88}$$

holds for any state $\rho^{(1)} \in Dens(\mathcal{H}_1)$ then we speak of the *weak teleportation problem.* The connection between the weak and the general teleportation problem is the following. Given a family $\{\sigma_k^{(23)}, F_k^{(12)}, U_k\}$ of solutions of the weak teleportation problem such that the projections $F_k^{(12)}$ are mutually orthogonal, then this family provides a solution of the general teleportation problem.

In the following section we shall solve the weak teleportation problem and then we shall use this result to solve the general problem.

25. Solution of the weak teleportation problem

In the notations of the previous section, we shall assume that

$$N = \dim \mathcal{H}_1 < +\infty \tag{89}$$

Under this assumption we shall look for a solution of the weak teleportation problem in which

$$N = \dim \mathcal{H}_1 = \dim \mathcal{H}_2 = \dim \mathcal{H}_3 \tag{90}$$

$$\sigma^{(23)} = |\psi\rangle\langle\psi| \tag{91}$$

$$F := |\xi\rangle\langle\xi| \tag{92}$$

where $\psi \in \mathcal{H}_2 \otimes \mathcal{H}_3$ and $\xi \in \mathcal{H}_1 \otimes \mathcal{H}_2$ are unit vectors In the following we identify F with

$$F = |\xi\rangle\langle\xi| \otimes 1_3 \in \text{ Proj } (\mathcal{H}_1 \otimes \mathcal{H}_2 \otimes \mathcal{H}_3) \tag{93}$$

and we look for a unitary transformation $U : \mathcal{H}_3 \to \mathcal{H}_1$ such that for any density matrix $\rho \in$ Dens $(\mathcal{H}_1)$ one has

$$U \cdot \frac{\overline{Tr}_{1,2}(F(\rho \otimes |\psi\rangle\langle\psi|)F)}{Tr(F(\rho \otimes |\psi\rangle\langle\psi|)F)} U^* = \rho \tag{94}$$

Under these assumptions, let us fix three arbitrary orthonormal bases:

$$(\varepsilon_\alpha) \qquad \text{of} \qquad \mathcal{H}_3 \tag{95}$$

$$(\varepsilon'_h) \qquad \text{of} \qquad \mathcal{H}_2 \tag{96}$$

$$(\varepsilon''_n) \qquad \text{of} \qquad \mathcal{H}_1 \tag{97}$$

Lemma (1) In the notations and assumptions of this section fix an arbitrary $N \times N$ complex unitary matrix $(\lambda_{\gamma\alpha})$ and define

$$\psi := \sum \lambda_{h\alpha} |\varepsilon'_h\rangle \otimes |\varepsilon_\alpha\rangle \in \mathcal{H}_2 \otimes \mathcal{H}_3 \tag{98}$$

$$\xi = \frac{1}{N^{1/2}} \sum_\mu \varepsilon''_\mu \otimes \varepsilon'_\mu \tag{99}$$

Then if $U : \mathcal{H}_3 \to \mathcal{H}_1$ is the unique unitary operator such that

$$\overline{\lambda}_{h\alpha} \varepsilon''_h = U \varepsilon_\alpha \tag{100}$$

(existing because of our assumption on the $\lambda_{k\beta}$'s), the triple ψ, ξ, U satisfies

$$Tr_{(1,2)}\Big(F(\rho \otimes |\psi\rangle\langle\psi|)F\Big) = \frac{1}{N} U^* \rho U \tag{101}$$

for any choice of the density operator $\rho \in Dens(\mathcal{H}_1)$ and with F given by (4), (5). In particular it satisfies (6).

Proof. Notice that in conditions (8) and (9) we use in a crucial way the finite dimensionality of $\mathcal{H}_2$ and $\mathcal{H}_3$. In particular

$$\|\psi\|^2 = \sum_{h,\alpha} |\lambda_{h\alpha}|^2 = \sum_h 1 = \dim \mathcal{H}_2 = \dim \mathcal{H}_3 \tag{102}$$

We do not normalize ψ because in all the formulae the corresponding rank one projection will enter both in the numerator and in the denominator. We use this procedure because it suggests a natural infinite dimensional generalization of the present protocol which will be discussed elsewhere.

For F as in (5) one has, using from now on the convenction of summation over repeated indices

$$F = \frac{1}{N}|\varepsilon''_{\mu}\rangle\langle\varepsilon''_{\mu'}| \otimes |\varepsilon'_{\mu}\rangle\langle\varepsilon'_{\mu'}| \otimes 1_3$$

therefore

$$F(\rho \otimes |\psi\rangle\langle\psi|)F = \lambda_{h\alpha}\overline{\lambda}_{k\beta}F(\rho \otimes |\varepsilon'_h\rangle\langle\varepsilon'_k|)F \otimes |\varepsilon_\alpha\rangle\langle\varepsilon_\beta| =$$

$$= \frac{1}{N^2}|\varepsilon''_{\mu}\rangle\langle\varepsilon''_{\mu'}, \rho\varepsilon''_{\nu}\rangle\langle\varepsilon''_{\nu'}| \otimes |\varepsilon'_{\mu}\rangle\langle\varepsilon'_{\mu'}, \varepsilon'_h\rangle\langle\varepsilon'_k, \varepsilon'_\nu\rangle\langle\varepsilon'_{\nu'}| \otimes |\varepsilon_\alpha\rangle\langle\varepsilon_\beta| =$$

$$= \frac{1}{N^2}\lambda_{h\alpha}\overline{\lambda}_{k\beta}\delta_{\mu',h}\delta_{k,\nu}\langle\varepsilon''_{\mu'}, \rho\varepsilon''_{\nu}\rangle|\varepsilon''_{\mu}\rangle\langle\varepsilon''_{\nu'}| \otimes |\varepsilon'_{\mu}\rangle\langle\varepsilon'_{\nu'}| \otimes |\varepsilon_\alpha\rangle\langle\varepsilon_\beta| =$$

$$= \frac{1}{N^2}\lambda_{h\alpha}\overline{\lambda}_{k\beta}\langle\varepsilon''_h, \rho\varepsilon''_k\rangle|\varepsilon''_{\mu}\rangle\langle\varepsilon''_{\nu'}| \otimes |\varepsilon'_{\mu}\rangle\langle\varepsilon'_{\nu'}| \otimes |\varepsilon_\alpha\rangle\langle\varepsilon_\beta|$$

Taking trace over $\mathcal{H}_1 \otimes \mathcal{H}_2$ this gives

$$= \frac{1}{N}\lambda_{h\alpha}\overline{\lambda}_{k\beta}\langle\varepsilon''_h, \rho\varepsilon''_k\rangle|\varepsilon_\alpha\rangle\langle\varepsilon_\beta| = \frac{1}{N}\langle\overline{\lambda}_{h\alpha}\varepsilon''_h, \rho\overline{\lambda}_{k\beta}\varepsilon''_k\rangle|\varepsilon_\alpha\rangle\langle\varepsilon_\beta|$$

Therefore, with the choice of U given by (10), (2) becomes

$$Tr_{(1,2)}\Big(F(\rho \otimes |\psi\rangle\langle\psi|)F\Big) = \frac{1}{N}\langle\varepsilon_\alpha, U^*\rho U\varepsilon_\beta\rangle|\varepsilon_\alpha\rangle\langle\varepsilon_\beta| = \frac{1}{N}U^*\rho U$$

which is (101).

26. A necessary condition

In the notations and assumptions of the previous section, let us suppose that the normalized state vector ξ has the form

$$\xi = \sum t_\mu \varepsilon''_\mu \otimes \varepsilon'_\mu \tag{103}$$

and let us look for conditions under which the map (3.6) becomes linear.
Lemma (1). Given ξ as in (1) and F as in (2) the trace

$$Tr_{123}(F(\rho \otimes |\psi\rangle\langle\psi|F)$$

is independent of ρ if and only if the coefficients t_k, of ξ, have the following form

$$t_k = e^{i\theta_k}/\sqrt{N} \tag{104}$$

In particular, if condition (104) above is satisfied then the map (3.6) linearizes.

Proof. One has:

$$F(\rho\otimes|\psi\rangle\langle\psi|)F = \lambda_{h\alpha}\overline{\lambda}_{k\beta}F(\rho\otimes|\varepsilon'_h\rangle\langle\varepsilon'_k|)F\otimes|\varepsilon_\alpha\rangle\langle\varepsilon_\beta|$$

$$= \lambda_{h\alpha}\overline{\lambda}_{k\beta}t_\mu\bar{t}_{\mu'}t_\nu\bar{t}_{\nu'}|\varepsilon''_\mu\rangle|\varepsilon'_\mu\rangle\langle\varepsilon'_{\mu'}|(\rho\otimes|\varepsilon'_h\rangle\langle\varepsilon'_k|)|\varepsilon''_\nu\rangle|\varepsilon'_\nu\rangle\langle\varepsilon''_{\nu'}|\langle\varepsilon'_{\nu'}|\otimes|\varepsilon_\alpha\rangle\langle\varepsilon_\beta|$$

$$= \lambda_{h\alpha}\overline{\lambda}_{k\beta}t_\mu\bar{t}_{\mu'}t_\nu\bar{t}_{\nu'}|\varepsilon''_\mu\rangle\langle\varepsilon''_{\mu'},\rho,\varepsilon''_\nu\rangle\langle\varepsilon''_{\nu'}|\otimes|\varepsilon'_\mu\langle\varepsilon'_{\mu'},\varepsilon'_h\rangle\langle\varepsilon'_k,\varepsilon'_\nu\rangle\langle\varepsilon'_{\nu'}|\otimes|\varepsilon_\alpha\rangle\langle\varepsilon_\beta|$$

$$= \lambda_{h\alpha}\overline{\alpha}_{k\beta}t_\mu\bar{t}_h t_k\bar{t}_{\nu},\langle\varepsilon'_h,\rho\varepsilon''_k\rangle|\varepsilon''_\mu\rangle\langle\varepsilon''_{\nu'}|\otimes|\varepsilon'_\mu\rangle\langle\varepsilon'_{\nu'}|\otimes|\varepsilon_\alpha\rangle\langle\varepsilon_\beta|$$

Tracing over (1, 2)

$$\lambda_{h\alpha}\overline{\lambda}_{k\beta}t_\mu\bar{t}_h t_k\bar{t}_\mu\langle\varepsilon''_h,\rho\varepsilon''_k\rangle|\varepsilon_\alpha\rangle\langle\varepsilon_\beta|$$

Since $\sum_\mu |t_\mu|^2 = 1$, this is equal to

$$\lambda_{h\alpha}\overline{\lambda}_{k\beta}\bar{t}_h t_k\langle\varepsilon''_h,\rho\varepsilon''_k\rangle|\varepsilon_\alpha\rangle\langle\varepsilon_\beta|$$

Taking $\mathcal{H}_3$–trace we find, because of the unitarity of $\lambda_{\alpha,\beta}$,

$$\lambda_{h\alpha}\overline{\lambda}_{k\alpha}\bar{t}_h t_k\langle\varepsilon''_h,\rho\varepsilon''_k\rangle = |t_h|^2\langle\varepsilon''_h,\rho\varepsilon''_h\rangle$$

If follows that the problem linearizes if and only if

$$\sum|t_k|^2|\varepsilon''_k\rangle\langle\varepsilon''_k| = c\cdot 1 = \text{ multiple of } 1$$

and this is equivalent to:

$$|t_k|^2 = c\ ; \quad \forall\, k = 1,\ldots,N$$

Consequently $c = N$ and (2) follows.

27. Solution of the general teleportation problem

The remark at the end of section (24) and the result of section (26) show that, in order to apply the solution of the weak teleportation problem to the solution of the general one, one has to solve the the following problem: find an o.n. basis f_n of a finite dimensional Hilbert space $\mathcal{H}$, identified to $\mathbf{C}^N$, such that one can determine the phases $\sigma_{\alpha,n}$ ($n = 1,\ldots,N$) so that the vectors

$$g_n := \frac{1}{N^{1/2}}\sum_{k=1}^{N} e^{i\sigma_{\alpha,k}} f_k$$

are still an o.n. basis of $\mathbf{C}^N$.

In other words the bases f_n and g_n should be complementary in the sense introduced in [Ac82c]. It is well known how to construct such bases. One way to achieve this goal is to start from an arbitrary orthonormal basis f_n and to take as g_n the eigenvectors of the unitary operator implementing the cyclic transformation

$$f_1 \to f_2 \to \cdots \to f_n \to f_1$$

In this case the coefficients are given by the coefficents of the N–dimesnional Fourier transform.

28. Uniqueness of the key

Theorem (1). Let $\rho = \sum p_\gamma P_\gamma$ be the spectral decomposition of ρ. Then U and V satisfy equation (101) with the above ρ, if and only if one has $VU^* = \sum_\gamma W_\gamma$ with

$$W_\gamma = W_\gamma P_\gamma = P_\gamma W_\gamma \qquad ; \qquad W_\gamma W_\gamma^* = \delta_{\gamma\gamma'} P_\gamma$$

Proof. Suppose U and V are two solutions of equation (101), then

$$U^* \rho U = V^* \rho V$$

or, equivalently

$$VU^* \rho = \rho VU^* \tag{105}$$

This means that

$$VU^* = S : \mathcal{H}_1 \to \mathcal{H}_1$$

is in the commutant of ρ. Now, in the above notations, suppose U is a unitary operator commuting with ρ then

$$UP_\gamma = UP_\gamma$$

therefore

$$U = \sum P_\gamma U P_\gamma = \sum U_\gamma : U_\gamma P_\gamma = P_\gamma U_\gamma = U_\gamma$$

Conversely if U has this form it will commute with ρ. Notice that, under this assumption:

$$U_\gamma U_{\gamma'}^* = \delta_{\gamma\gamma'} P_\gamma$$

in fact

$$1 = \sum_{\gamma\gamma'} U_\gamma^* U_{\gamma'} = \sum_{\gamma\gamma'} U_\gamma^* P_\gamma P_{\gamma'} U_\gamma = \sum_\gamma U_\gamma^* U_\gamma$$

So

$$P_\gamma = U_\gamma^* U_\gamma$$

Corollary (2). Let $\mathcal{H}$ be an Hilbert space of arbitrary dimensions. If U and V are two solutions of the weak teleportation problem corresponding to the same F and ψ then they coincide up to multiplication by a number of modulus one.

Proof. Equation (105) above implies that in this case the operator VU^* commutes with all density operators, since they are unitary, it follows that $V = e^{i\theta}U$ for some real number θ.

29. Bibliography

[AcLuVo02] Accardi L., Lu Y.G., Volovich I.: QUANTUM THEORY AND ITS STOCHASTIC LIMIT, Springer Verlag (2002)

[AcImRe01a] Accardi L., Imafuku K., Regoli M.: ON THE PHYSICAL MEANING OF THE EPR–CHAMELEON EXPERIMENT, Infinite dimensional analysis, quantum probability and related topics, 5 N. 1 (2002) 1–20 quant-ph/0112067 Volterra Preprint N. 494 December (2001) N. 494

[AcSa00a] ACCARDI L., SABBADINI R.: *On the Ohya–Masuda quantum SAT algorithm*, Volterra preprint N. 432 (2000)

[AcSa00b] ACCARDI L., SABBADINI R.: *A Generalization of Grover's Algorithm*, Proceedings International Conference: Quantum Information III, (Meijo University, Nagoya, 27–31 mARCH 2001) World Scientific (2002) to appear, qu-phys 0012143, Preprint Volterra, N. 444 (2001)

[AcOhSa01] ACCARDI L., OHYA M., SABBADINI R.: In preparation.

[AcOh98b] Accardi L., Ohya M.: Teleportation of general quantum states, in: Quantum Information, T. Hida, K. Saito (eds.) World Scientific (1999) 59–70 Invited talk to the: International Conference on quantum information and computer, Meijo University 1998

[Ac82c] ACCARDI L.: "Some trends and problems in quantum probability", In: *Quantum probability and applications to the quantum the-*

ory of irreversible processes, L. Accardi, A. Frigerio and V. Gorini (eds.) Springer LNM N. 1055 (1984) 1-19

[BeBr84] BENNETT C.H., BRASSARD G.: in *Proc. of the IEEE Int. Conf. on Computers, Systems, and Signal Processing*, Bangalore, India IEEE, New York (1984) 175-179

[BeBeBrSaSm92] BENNETT C.H., BESSETTE F., BRASSARD G., SALVAIL L., SMOLIN J.: "Experimental Quantum Cryptography", *Journal of Cryptology* **5** (1992) 3-28

[DeuJo92] DEUTSCH D., JOZSA R.: "Rapid solution of problems by quantum computation", *Proc. Roy. Soc. Lond* **A 439** (1992) 553–558

[Eker91] EKERT A.K.: "Quantum cryptography based on Bell's theorem", *Phys. Rev. Lett.* **67** (1991) 661-663

[EkRaTaPa92] EKERT A.K., RARITY J.G., TAPSTER P.R., PALMA G.M.: "Practical Quantum Cryptography based on Two-Photon Interferometry", *Physical Review Letters* **69** (1992) 1293-1295

[Gro97] GROVER LOV K. : "Quantum Mechanics helps in searching for a needle in a haystack", *Phys. Rev. Lett.* **79** 325–328

[BBH96] BOYER M., BRASSARD G., HOYER P., TAPP A.: *Tight bounds on quantum searching*, (Preprint quant-ph/9605034)

[Ki95] KITAEV A.YU.: *Quantum measurement and the abelian stabilizer problem*, quant-ph/9511026, 20-11 (1995)

[OhMa98] OHYA M., MASUDA N.: *NP problem in Quantum Algorithm*, quant–ph/9809075

[OhVo00] OHYA M., VOLOVICH I.V.: *Quantum computing, NP–complete problems and chaotic dynamics*, Volterra preprint N. 426 (2000)

[Sak71 Sakai S.: C^*–ALGEBRAS AND W^*–ALGEBRAS, Springer (1971)

[Wies83] WIESNER S.: SIGACT News, **15** 78 (1983) original manuscript written $\sim$ (1970)

ON THE QUANTUM LIOUVILLE SPACE

I. Antoniou[1,2,3] and Z. Suchanecki[2,3,4]

[1] Department of Mathematics, Aristoteles University of Thessaloniki
54124, Thessaloniki, Greece

[2] International Solvay Institutes for Physics and Chemistry
C.P. 231, Campus Plaine ULB
Bd. du Triomphe, 1050 Brussels, Belgium

[3] Theoretische Natuurkunde
Free University of Brussels

[4] Institute of Mathematics University of Opole
ul. Oleska 48, 45-052 Opole, Poland

1 Introduction

A pure state in quantum mechanics is represented by one dimensional projections in the Hilbert Space of wavefunctions, being solutions of Schrödinger's equation. When information is distributed in an ensemble of pure states, the state is represented by a statistical mixture, i.e. a convex combination of pure states. In mathematical terms mixed states are bounded positive operators with trace one. The spectral representation of trace class operators [1] allows to express any mixed state as a mixture of pure states [2]. The set of state operators is a subset of the space of bounded operators. The Hilbert space spanned by state operators is called the *Liouville space.*

There exists a one to one correspondence between the observables self adjoint operators A on the Hilbert space of wavefunctions and observables on the corresponding Liouville space established by the adjoint map $A \longmapsto [A, \ ]$, where $[A, B] = AB - BA$ is the commutator of the operators A and B. However, this mapping is not onto and there exists many self adjoint operators on the Liouville space that do not have a counterpart on the Hilbert space. This fact already implies that Liouville space has richer structure than the Hilbert space. Other reasons for which Liouville space has this richer structure will be presented below.

There exits a description of quantum mechanics on Liouville space that simply lifts the formalism of Hilbert space to Liouville space [2]. This description does not take into account the peculiarities of Liouville space, which in fact opens new possibilities. Let us mention, as an example, that there exist observables which have continuous singular spectra on Hilbert space, but the corresponding observables on Liouville space have absolutely continuous spectra. This has a very interesting consequence in scattering theory as there are situations in quantum scattering for which scattering states exist in Liouville space but not in the Hilbert space. This also means that the study of quantum mechanics in Liouville space has added value. In addition, since Liouville space provides a unified representation for mixtures and pure states, and deals directly with this representation, it is the adequate framework for the formalization of quantum statistical mechanics [3]. One of the purposes of this paper is to present a summary of the general properties of Liouville space and show differences, sometimes very profound, between the Hamiltonian and Liouville formalisms. Another important subject we will discuss here is decay in Liouville space.

According to the traditional approach to decay (Weisskopf and Wigner [4]) the probability that the atom will remain in its initial state decreases exponentially with time. The decay law of a state can be related with its energy distribution function [5]. The exponential decay means actually that the energy distribution is Lorentzian. However, as it was shown by Khalfin [6], the "natural" assumption of exponential decay is inconsistent with the Hilbert space formulation of quantum mechanics because it is not compatible with the physical requirement that the total energy is bounded from below.

A possible way to avoid these incompatibilities is to go from the Hilbert space of pure states to the Liouville space where the evolution of states is governed by the Liouville operator. Since the spectrum of the Liouvillian in the Hilbert-Schmidt space, corresponding to a Hamiltonian with a spectrum on the half line, is unbounded, the above mentioned inconsistency does not appear in the statistical formulation.

Concerning exponential decay of pure states, the two formulations are actually equivalent. Exponentially decaying states have absolutely continuous distributions of energy (their densities exist). If a Hamiltonian has only absolutely continuous spectrum, then the spectral properties are preserved in the transition from Hamiltonian to the Liouvillian description. The problem arises when we admit states with singular energy distributions, which we shall call briefly singular states. Contrary to the convensional wisdom, there is a profound difference in the physics in Hilbert and Liouville space, if singular states are present.

There is yet another subject which we would like to present in this article. It is associated with the well-known problem of irreversibility versus dynamics or, in other words, with the incompleteness of the conventional formulation of classical and quantum dynamics. We proposed recently [7, 8] an extension of the conventional formulation of dynamics beyond the Hilbert-Schmidt formulation of the Liouville space in order to include irreversible processes. Following the algebraic approach to dynamical systems we have constructed an extended algebra of observables, called the algebra with diagonal singularity. This algebra gives, among others, a rigorous meaning to diagonal singularities of operators discovered earlier by van Hove [9, 10].

The main idea of the algebraic approach is that states correspond to the normalized positive linear functionals over a suitably chosen algebra of operators representing the observables of a dynamical system. In the classical algebraic formulation of quantum theory [20], the observables form a C*-algebra of bounded operators. However the algebra of operators with diagonal singularity turns out to be an involutive Banach algebra which is not a C*-algebra. This more general structure allows us to incorporate the above mentioned states with diagonal singularity which cannot be described in terms of the conventional [24-26] or generalized [24,27,16] formulation of quantum theory.

The paper is divided into four sections. In Section 2, we introduce the definitions of the Liouville space and the Liouvillian of an Hamiltonian operator. Then we present the relation between the spectral properties of the Liouville operator and the corresponding Hamiltonian in the Hilbert-Schmidt space. In particular, we will show that if the Hamiltonian has a purely singular continuous spectrum, the corresponding Liouvillian may have absolutely continuous spectrum.

In Section 3 we characterize decaying states in the classical Hilbert space formulation of quantum mechanics. We summarize the complete spectral characterization of decay of pure states, which we have obtained recently [11, 12]. Before, it was only known that states with absolutely continuous energy distribution function decay, and that decay implies continuity of the distribution function. None of the above conditions is necessary and sufficient at the same time. We proved that the full information whether a state decays or not can be extracted from the spectrum of the Hamiltonian. Among other results, we proved that the decaying states form a closed subspace of a Hilbert space of all pure states. Then we present the main results concerning the spectra of Liouville operators and we discuss the decay in Liouville space.

Section 4 is devoted to the quantum systems with diagonal singularity, their algebras of observables, their state spaces, and their temporal evolution. First we define the operators with diagonal singularity and the corresponding algebra of obsrvables. Then we formulate the time-evolution and the spectral decompositions of Liouville operators.

2 The Spectrum of the Liouville Operator

Let $\mathcal{H}$ denote the Hilbert space of pure states of some physical system with the scalar product $\langle\cdot|\cdot\rangle$. A bounded linear operator A on $\mathcal{H}$ is called *Hilbert-Schmidt* (HS) if for some orthonormal basis $\{\psi_i\}$

$$\sum_i \langle A\psi_i|A\psi_i\rangle < \infty\,. \tag{1}$$

One can prove [1, 2] that this sum does not depend on the chosen orthonormal basis and that (1) defines the norm in the class of all HS operators. In fact HS operators form a Hilbert space with the scalar product

$$(A,B) = \sum_i \langle A\psi_i|B\psi_i\rangle\,.$$

We denote this space by $\mathcal{B}^2_{\mathcal{H}}$. The HS operators have the following properties:

i) All HS operators are bounded. Therefore, the adjoint $A^\dagger$ of any HS operator A always exist and it is also a HS operator. The product of two HS operators also belongs to $\mathcal{B}^2_{\mathcal{H}}$. However, the identity operator on $\mathcal{H}$ is not in $\mathcal{B}^2_{\mathcal{H}}$. As a consequence, the HS class is an involutive algebra without identity. The involution is given by the map $A \longmapsto A^\dagger$.

ii) If $A \in \mathcal{B}^2_{\mathcal{H}}$ and B is a bounded operator in $\mathcal{H}$, both AB and BA are in $\mathcal{B}^2_{\mathcal{H}}$. Thus the HS class is an ideal in the algebra of bounded operators in $\mathcal{H}$.

A bounded linear operator A on $\mathcal{H}$ is called *nuclear* (or *trace class*) if

$$\sum_i |\langle A\psi_i|\psi_i\rangle| < \infty \tag{2}$$

for some orthonormal basis $\{\psi_i\}$ in $\mathcal{H}$. One can prove that this sum does not depend on the choice of ψ_i. Therefore for a given nuclear operator we can correctly define its trace

$$\mathrm{tr}A = \sum_i \langle A\psi_i|\psi_i\rangle \tag{3}$$

We denote by $\mathcal{B}^1_{\mathcal{H}}$ the trace class which is a Banach space with the norm

$$\|A\|_{\mathrm{tr}} = \sum_i \|A\psi_i\| \,.$$

An operator A belongs to the trace class if and only if $\mathrm{tr}\,|A|$ is finite. The trace class has similar algebraic properties as the HS class. Thus, it is a vector space, an algebra and an ideal in the algebra of bounded operators on $\mathcal{H}$. Moreover the trace is a continuous linear functional on $\mathcal{B}^1_{\mathcal{H}}$. It is easy to prove that any nuclear operator is HS.

Using the notion of trace we can expand the list of properties of HS class

iii) If $A, B \in \mathcal{B}^2_{\mathcal{H}}$, then $\mathrm{tr}\,AB$ exists and is finite. Moreover the scalar product on $\mathcal{B}^2_{\mathcal{H}}$ can be expressed through the trace as follows

$$(A, B) = \mathrm{tr}\,(A^\dagger B)\,. \tag{4}$$

Before going further let us introduce the dyadic notation. We shall denote by kets $|\psi\rangle$ elements of $\mathcal{H}$ and by bra $\langle\varphi|$ elements of the dual $\mathcal{H}^\times$, i.e. continuous linear functionals on $\mathcal{H}$. The action of $\langle\psi|$ onto an arbitrary vector (ket) $|\varphi\rangle$ is naturally given by the scalar product $\langle\psi|\varphi\rangle$. It is well known that $\mathcal{H}^\times$ is a Hilbert space which is isomorphic to $\mathcal{H}$.

A bounded operator A is called normal if it commutes with its adjoint: $AA^\dagger = A^\dagger A$. A is positive if $\langle\psi|A\psi\rangle \geq 0$ for all $\psi \in \mathcal{H}$. Positive operators are self adjoint and hence normal. If A is normal and HS, then there exists an orthonormal basis of $\mathcal{H}$ formed by eigenvectors of A. If the eigenvectors are $\{|\psi_i\rangle\}$ and the corresponding eigenvalues are $\{\lambda_i\}$, then we have:

$$A = \sum_i \lambda_i\,|\psi_i\rangle\,\langle\psi_i| \quad \text{with} \quad \sum_i |\lambda_i|^2 < \infty \tag{5}$$

We recall that in Quantum Mechanics a state is represented by a positive trace class operator ρ with trace one. Each state has an orthonormal basis formed by its eigenvectors. Its eigenvalues are always real nonnegative numbers with sum equal to one. The set of states is a convex subset of $\mathcal{B}^2_{\mathcal{H}}$. This means that all sums of the form $\sum_i \lambda_i \rho_i$ where λ_i are nonnegative real numbers with $\sum_i \lambda_i = 1$ are also states. The minimal Hilbert space containing all the states is again $\mathcal{B}^2_{\mathcal{H}}$. This is the reason why it is convenient to use $\mathcal{B}^2_{\mathcal{H}}$ in quantum mechanics and quantum statistical mechanics.

State ρ, as a vector in Hilbert space $\mathcal{B}^2_{\mathcal{H}}$, can be written in the form (5) where $|\psi_i\rangle$ are eigenvectors of ρ, λ_i the corresponding eigenvalues and $\operatorname{tr}\rho = \sum_i \lambda_i = 1$. Each dyad $|\psi_i\rangle\langle\psi_i|$ represents the orthogonal projection onto the one dimensional subspace spanned by $|\psi_i\rangle$.

Consider now the tensor product $\mathcal{H}\otimes\mathcal{H}^\times$. It is generated by the elements of the form $|\psi\rangle \otimes \langle\varphi|$ that we usually write as $|\psi\rangle\,\langle\varphi|$. Linear combinations of these dyads form a vector space. A scalar product can be introduced for dyads

$$(|\eta\rangle\,\langle\xi|, |\psi\rangle\,\langle\varphi|) = \langle\eta|\psi\rangle\,\langle\varphi|\xi\rangle \tag{6}$$

We extend this product to linear combinations of dyads by linearity obtaining a vector space with a scalar product. If we now add all limit points we obtain a Hilbert space. This is what we understand by the Hilbert space $\mathcal{H}\otimes\mathcal{H}^\times$. It has the following important property: if $\{|\psi_i\rangle\}$ and $\{|\varphi_j\rangle\}$ are orthonormal bases for $\mathcal{H}$, then, $\{|\psi_i\rangle\,\langle\varphi_j|\}$ is an orthonormal basis for $\mathcal{H}\otimes\mathcal{H}^\times$. If both bases coincide, then we obtain the basis $\{|\psi_i\rangle\,\langle\psi_j|\}$. The norm of the dyad $|\psi\rangle\,\langle\varphi|$ is $||\psi||\,||\varphi||$, as a consequence of the definition of the scalar product.

One can prove that the Hilbert spaces $\mathcal{B}^2_{\mathcal{H}}$ and $\mathcal{H}\otimes\mathcal{H}^\times$ can be identified. Thus we shall work either with $\mathcal{B}^2_{\mathcal{H}}$ or $\mathcal{H}\otimes\mathcal{H}^\times$ whichever is more convenient. Observe also that the scalar product given by (6) can be written as a trace:

$$\operatorname{tr}(|\eta\rangle\,\langle\xi|)^\dagger\,(|\psi\rangle\,\langle\varphi|) = \operatorname{tr}(|\xi\rangle\,\langle\eta|\psi\rangle\,\langle\varphi|) = \langle\eta|\psi\rangle\,\langle\varphi|\xi\rangle \tag{7}$$

We shall call $\mathcal{B}^2_{\mathcal{H}} \approx \mathcal{H}\otimes\mathcal{H}^\times$ the *Liouville space* associated with the Hilbert space $\mathcal{H}$.

Let H be a Hamiltonian on $\mathcal{H}$ (we always assume that Hamiltonians are self adjoint operators). We define the following operator $\mathbb{L}$ on $\mathcal{H}\otimes\mathcal{H}^\times$. The domain of $\mathbb{L}$ is the vector space of HS operators ρ such that both $H\rho$ and ρH are HS operators. Note that this is true for all HS operators if and only if H is bounded. We call this domain $\mathcal{D}(\mathbb{L})$. Then we define the operator $\mathbb{L}$ as:

$$\mathbb{L}\rho = [H,\rho] = H\rho - \rho H \quad , \quad \forall\rho\in\mathcal{D}(\mathbb{L}) \tag{8}$$

Since H is self adjoint, the operator $\mathbb{L}$ is essentially self adjoint, i.e. it has a unique self adjoint extension, which will be also denoted by $\mathbb{L}$.

Let us now consider the relation between the spectral family of a Hamiltonian and the spectral family of the corresponding Liouvillian. We begin with the well know fact that for any self adjoint operator H with spectrum $\sigma(H)$, there exists a spectral family of orthogonal projections E_λ on $\mathcal{H}$, labelled by the points $\lambda \in \sigma(H)$, such that

$$\langle\psi|H\psi\rangle = \int_{\sigma(H)} \lambda\, d\langle\psi|E_\lambda\psi\rangle\,, \tag{9}$$

for any $|\psi\rangle$ in the domain of H. Note that $d\langle\psi|E_\lambda\psi\rangle$ is a well defined positive measure on $\sigma(H)$, for all $\psi \in \mathcal{H}$, but the above integral converges if and only if ψ is in the domain of H. Spectral family $\{E_\lambda\}$ can be extended to all the values of the real line in such a way that measure $d\langle\psi|E_\lambda\psi\rangle$ is zero outside $\sigma(H)$. Then, the integral limits in (9) can be changed to be from $-\infty$ to ∞. Spectral decomposition (9) is usually written as

$$H = \int_{\sigma(H)} \lambda\, dE_\lambda \qquad \text{or} \qquad H = \int_{-\infty}^{\infty} \lambda\, dE_\lambda \tag{10}$$

Consider now the Liouvillian $\mathbb{L}$ corresponding to $\mathcal{H}$. Since $\mathbb{L}$ is also self adjoint, it has its own spectral family $\{\mathbb{E}_\omega\}$ so that

$$\mathbb{L} = \int_{\sigma(\mathbb{L})} \omega\, d\mathbb{E}_\omega = \int_{-\infty}^{\infty} \omega\, d\mathbb{E}_\omega\,. \tag{11}$$

The relation between the spectral families $\{E_\lambda\}$ and $\{\mathbb{E}_\omega\}$ is the following

$$(|\psi\rangle\,\langle\varphi|\,,\, \mathbb{E}_\omega\,|\psi\rangle\,\langle\varphi|) = \int_{-\infty}^{\infty} \langle\psi|E_{\omega+\lambda}\psi\rangle\, d\langle\varphi|E_\lambda\varphi\rangle \tag{12}$$

which can be written in the following form:

$$\mathbb{E}_\omega\,|\psi\rangle\,\langle\varphi| = \int_{-\infty}^{\infty} |E_{\omega+\lambda}\psi\rangle\, d\langle E_\lambda\varphi| \tag{13}$$

We want to discuss now how to construct the spectrum of the Liouvillian $\mathbb{L}$ of a given self adjoint Hamiltonian H. This problem has been treated in some publications in the past [2, 13] although not always with the desirable correctness.

The point spectrum of $\mathbb{L}$ is easily obtained if we know the eigenvalues of H. In fact, it is easy to check that

$$\sigma_p(\mathbb{L}) = \{\lambda_i - \lambda_j :\ \lambda_i\,, \lambda_j \in \sigma_p(H)\} \tag{14}$$

where $\sigma_p(H)$ denotes the set of eigenvalues of the operator H. If $H|e_i\rangle = \lambda_i |e_i\rangle$ for all i, we have the following spectral equation in $\mathcal{H} \otimes \mathcal{H}^\times$:

$$\mathbb{L} |e_i\rangle \langle e_j| = (\lambda_i - \lambda_j) |e_i\rangle \langle e_j| \tag{15}$$

More generally, it is well known that the spectrum $\sigma(H)$ of H determines the spectrum $\sigma(\mathbb{L})$ of $\mathbb{L}$ [2, 13]:

$$\sigma(\mathbb{L}) = \sigma(H) - \sigma(H) \tag{16}$$

This shows partially the importance of the Liouville space in physics, because we never observe energy values but, instead, as spectroscopic measurements show, differences between energy values. Due to this observation, the formulation of Quantum Mechanics in Liouville space seems more natural than the standard one in Hilbert space.

Denote by $\mathcal{H}_\mathrm{p}$ the closed linear hull of all eigenvectors of H and by $\mathcal{H}_\mathrm{c} = \mathcal{H} \ominus \mathcal{H}_\mathrm{p}$ the *continuous subspace* of $\mathcal{H}$, i.e. the orthogonal complement of $\mathcal{H}_p$ in $\mathcal{H}$. Furthermore, let $\mathcal{H}_\mathrm{sc}$ be the *singular continuous subspace* of $\mathcal{H}_\mathrm{c}$. Recall that the singular continuous subspace $\mathcal{H}_\mathrm{sc}$ of $\mathcal{H}_\mathrm{c}$ consists of all $\psi \in \mathcal{H}_\mathrm{c}$ for which there exists a Borel set B_0 of Lebesgue measure zero such that $\int_{B_0} dE_\lambda \psi = \psi$. By $\mathcal{H}_\mathrm{ac} = \mathcal{H}_\mathrm{c} \ominus \mathcal{H}_\mathrm{sc}$ we shall denote the *absolutely continuous subspace* of $\mathcal{H}_\mathrm{c}$. Recall also that $\mathcal{H}_\mathrm{p}$, $\mathcal{H}_\mathrm{c}$, $\mathcal{H}_\mathrm{sc}$ and $\mathcal{H}_\mathrm{ac}$ are closed linear subspaces of $\mathcal{H}$ which reduce[1] the operator H. The spectra of the corresponding reductions will be called respectively *point, continuous, singular continuous and absolutely continuous spectrum* of H, and will be denoted by $\sigma(H)$ with appropriate subscripts [27].

Let $\mu = \mu_\psi$ denote, for a given $\psi \in \mathcal{H}$, the spectral measure on $\sigma(H)$ determined by the nondecreasing function $\lambda \longmapsto \langle\psi|E_\lambda\psi\rangle$. Let $\psi = \psi_\mathrm{p} + \psi_\mathrm{sc} + \psi_\mathrm{ac}$ be the decomposition of ψ in the direct sum $\mathcal{H}_\mathrm{p} \oplus \mathcal{H}_\mathrm{sc} \oplus \mathcal{H}_\mathrm{ac}$. Putting $\mu_\mathrm{p} = \mu_{\psi_\mathrm{p}}$, $\mu_\mathrm{sc} = \mu_{\psi_\mathrm{sc}}$ and $\mu_\mathrm{ac} = \mu_{\psi_\mathrm{ac}}$ we obtain the decomposition of μ

$$\mu = \mu_\mathrm{p} + \mu_\mathrm{sc} + \mu_\mathrm{ac} \tag{17}$$

onto the point, singular continuous and absolutely continuous component. Conversely, having given three finite Borel measures μ_p, μ_sc and μ_ac, where μ_p is concentrated on a countable set of points, and the other two measures are respectively singular and absolutely continuous, one can always construct a Hilbert space $\mathcal{H}$ and a selfadjoint operator H such that these measures

[1] H reduces $\mathcal{H}_\alpha$ if $H\varphi \in \mathcal{H}_\alpha$ for any vector φ in the intersection of $\mathcal{H}_\alpha$ with the domain of H.

are spectral measures associated with some $\psi \in \mathcal{H}$. Moreover, the point, singular and absolutely continuous spectrum of H coincides with $\mu_{\rm p}$, $\mu_{\rm sc}$ and $\mu_{\rm ac}$ respectively. This can be proved by taking as $\mathcal{H}$ the direct sum $L^2(\mathbf{R}, \mu_{\rm sc}) \oplus L^2(\mathbf{R}, \mu_{\rm sc}) \oplus L^2(\mathbf{R}, \mu_{\rm sc})$ and as H the operator of multiplication by x (see [1]).

It was mentioned already that $\sigma(\mathbb{L}) = \sigma(H) - \sigma(H)$. It is therefore natural to expect [13] that a similar relation holds separately for each part of the spectrum of the Hamiltonian and the Liouvillian, namely for the point spectrum, the absolutely continuous spectrum and the singular continuous spectrum:

$$\sigma_{\rm p}(\mathbb{L}) = \{\lambda - \lambda' : \ \lambda, \lambda' \in \sigma_{\rm p}(H)\} \tag{18}$$

$$\sigma_{\rm ac}(\mathbb{L}) = \{\lambda - \lambda' : \ \lambda, \lambda' \in \sigma_{\rm ac}(H)\} \tag{19}$$

$$\sigma_{\rm sc}(\mathbb{L}) = \{\lambda - \lambda' : \ \lambda, \lambda' \in \sigma_{\rm sc}(H)\} \tag{20}$$

However, as we have shown recently [14, 15] formulas (19) and (20) are not true. More precisely, it is not true that if the set $\sigma_{\rm sc}(H)$ has Lebesgue measure zero then $\{\lambda - \lambda' : \ \lambda, \lambda' \in \sigma_{\rm sc}(H)\}$ has also Lebesgue measure zero. This can be shown on a simple example of an operator with the Cantor spectrum.

To see it, let us consider a Hamiltonian operator H for which $\sigma(H)$ is a Cantor set. The existence of such H is guaranteed by the presented above construction of the Hamiltonian with a given spectral measure. Let us note, however, that there are more "physical" examples of Hamiltonians with Cantor like spectra. For example, there is an absolutely summable sequence $\{a_n\}$ such that the spectrum of the Hamiltonian

$$H = -\frac{d^2}{dx^2} + \sum_{n=0}^{\infty} a_n \cos(x2^{-n})$$

is the Cantor set (see [13, 16] an references therein). Suppose that $\sigma(H) = C$, where C is the Cantor set on the interval $[0,1]$. According to the general property of spectra of functions of the Hamiltonian (see [1]) we have (18).

However, according to a theorem of Steinhaus [17] we have

$$\{\lambda - \lambda' : \ \lambda, \lambda' \in C\} = [-1, 1]\,, \tag{21}$$

i.e. the complex difference of two Cantor sets covers the whole interval and is thus the set of nonzero Lebesgue measure. Property (21) does not, however, imply that the spectrum of the Liouvillian is absolutely continuous. To show this let us consider the well known "devil's staircase" distribution function

$F(x)$ on the Cantor set C which has constant value equal $k/2^n$ (k are such that the fraction $k/2^n$ is non reducible) on each interval which is removed in the n-th step of the construction of the Cantor set [18]. It is well known and not difficult to check that $F(x)$ is a non decreasing continuous function such that $F(x) = 0$ for $x \leq 0$ and $F(x) = 1$ for $x \geq 1$. Therefore $F(x)$ is a probability distribution function.

We shall show that $F(x)$ is also a spectral distribution function, i.e. that there exists a selfadjoint operator $H = \int_{\sigma(H)} \lambda \, dE_\lambda$ on some Hilbert space $\mathcal{H}$ and an element $e \in \mathcal{H}$ such that

$$F(x) = \langle e|E_x e\rangle \,, \quad \text{for } \; x \in \mathbf{R} \,. \tag{22}$$

Indeed let us take as the Hamiltonian H the multiplication operator

$$Hf(\lambda) = \lambda f(\lambda) \tag{23}$$

on the Hilbert space $\mathcal{H} = L^2([0,1], dF)$. The spectral projectors E_λ of H are:

$$E_\lambda f(\lambda') = \mathbf{1}_{[0,\lambda)} f(\lambda') \,, \tag{24}$$

where

$$\mathbf{1}_{[0,\lambda)}(\lambda') = \begin{cases} 1, & 0 \leq \lambda' < \lambda \\ 0, & \text{otherwise.} \end{cases}$$

The spectral measure dF corresponds to the cyclic vector $e = 1$:

$$F(\lambda) = \langle e|E_\lambda e\rangle \,.$$

Consider the Liouvillian $\mathbb{L} = H \otimes I - I \otimes H$ corresponding to the just defined Hamiltonian H acting on $\mathcal{H} \otimes \mathcal{H}^\times$. It follows from (12) that the spectral distribution function

$$\mathbb{F}(\omega) = (|e\rangle\langle e|, \mathbb{E}_\omega |e\rangle\langle e|) \,.$$

satisfies

$$\mathbb{F}(\omega) = \int_{-\infty}^{\infty} \langle e|E_{\omega+\lambda} e\rangle \, d\langle e|E_\lambda e\rangle = \int_{-\infty}^{\infty} F(\omega+\lambda) \, dF(\lambda) \,. \tag{25}$$

It can be shown [15] that the function $\mathbb{F}(\omega)$ is continuous but not absolutely continuous. The continuity is a straightforward consequence of general properties of convolutions. The second property can be seen by considering the Fourier-Stieltjes transform

$$\widehat{F}(t) = \int_0^1 e^{it\lambda} dF(\lambda) \,. \tag{26}$$

Using the fact that $F(\lambda)$ satisfies functional equation of the De Rham type

$$F(\lambda)=\begin{cases} \frac{1}{2}F(3\lambda), & 0\le\lambda<\frac{1}{3} \\ \frac{1}{2}, & \frac{1}{3}\le\lambda<\frac{2}{3} \\ \frac{1}{2}F(3\lambda-2)+\frac{1}{2}. & \frac{2}{3}\le\lambda\le 1 \end{cases} \tag{27}$$

we find the explicit expression for $\widehat{F}(t)$:

$$\widehat{F}(t)=\prod_{k=1}^{\infty}\left[1+e^{\frac{2it}{3}}\right]. \tag{28}$$

Choosing $t_n=\pi 3^n$ we see that

$$\widehat{F}(t_n)=\prod_{k=n+1}^{\infty}\frac{1}{2}\left[1+e^{\frac{2\pi i}{3^{k-n}}}\right]=\prod_{k=1}^{\infty}\frac{1}{2}\left[1+e^{\frac{2\pi i}{3^k}}\right]. \tag{29}$$

does not depend on n. Finally we prove that $\widehat{F}(t_n)$ is non zero (see [15] for details). This means that the Fourier transform $\widehat{F}(t)$ does not converge to zero as $t\to\infty$. The same is true for the Fourier transform $\widehat{\mathbb{F}}(t)$ of $\mathbb{F}(\omega)$ because $|\widehat{\mathbb{F}}(t)|=|\widehat{F}(t)|^2$. Therefore, in view of the Riemann-Lebesgue lemma $F(\lambda)$ as well as $\mathbb{F}(\omega)$ cannot be purely absolutely continuous. In fact one can show moreover that $\mathbb{F}(\omega)$ is purely singular continuous [15]. Therefore the singular continuous spectrum has Lebesgue measure zero.

In this way we show the singular continuous spectrum $\sigma_{\rm sc}(\mathbb{L})$ of the Liouville operator, being a set with Lebesgue measure zero, is properly included in the set $\sigma_{\rm sc}(H)-\sigma_{\rm sc}(H)=[-1,1]$. Therefore formula (20) for the singular continuous spectrum of $\mathbb{L}$ should be corrected to

$$\sigma_{\rm sc}(\mathbb{L})\subset\{\sigma_{\rm sc}(H)-\sigma_{\rm sc}(H)\}. \tag{30}$$

The above example shows that Hamiltonians with singular spectra can have Liouvillians with much bigger spectra, which can even contain absolutely continuous spectra. We shall continue the discussion on this subject in the next section.

3 Decay in Liouville space

Let us now consider a quantum dynamical system in which time evolution is given by a time independent self adjoint Hamiltonian H. The time evolution of a pure state $\psi\in\mathcal{H}$ is governed by the unitary group

$$U_t=e^{-itH},\quad t\in\mathbb{R}, \tag{31}$$

on $\mathcal{H}$, where H is the Hamiltonian. Denoting by σ the spectrum of H and by $\{E_\lambda\}$ its spectral family we can write

$$U_t = \int_\sigma e^{-i\lambda t}\, dE_\lambda \,.$$

A pure state $\psi \in \mathcal{H}$ is called a *decaying state* if its *survival amplitude* decays, namely:

$$\langle \psi | U_t \psi \rangle = \int_\sigma e^{-i\lambda t} d\langle \psi | E_\lambda \psi \rangle \longrightarrow 0\,, \quad \text{as } t \to \infty\,, \tag{32}$$

The *survival probability*, i.e. the probability that at time t the state ψ has not yet decayed is:

$$p(t) = |\langle \psi | U_t \psi \rangle|^2\,. \tag{33}$$

The survival amplitude and probability are correctly defined for each element $\psi \in \mathcal{H}$. Therefore in the space $\mathcal{H}$ we can distinguish the class $\mathcal{H}^{\mathrm{D}}$ of all decaying elements. The following result has been proved in Ref. [15]:

Proposition 1 *The space $\mathcal{H}^{\mathrm{D}}$ is a closed linear subspace of $\mathcal{H}$.*

Because of the above proposition we can introduce another space, the space $\mathcal{H}^{\mathrm{ND}}$ of *non-decaying states*, as the orthogonal complement of $\mathcal{H}^{\mathrm{D}}$. Thus we have the direct sum decomposition

$$\mathcal{H} = \mathcal{H}^{\mathrm{D}} \oplus \mathcal{H}^{\mathrm{ND}}\,. \tag{34}$$

In other words each pure state can be uniquely decomposed on its decaying and non decaying part. Moreover, we have the following [15]

Proposition 2 *The spaces $\mathcal{H}^{\mathrm{D}}$ and $\mathcal{H}^{\mathrm{ND}}$ reduce the Hamiltonian operator H and, therefore, the unitary evolution U_t* (31) *generated by H.*

An important consequence of the latter proposition is that we can decompose the spectrum of the Hamiltonian on two disjoint parts σ^{D} and σ^{ND},

$$\sigma = \sigma^{\mathrm{D}} \cup \sigma^{\mathrm{ND}}\,, \tag{35}$$

corresponding to the reductions H^{D} and H^{ND} of the total Hamiltonian.

Recall (see Section 2) that traditionally we distinguish three disjoint parts of the spectrum σ of a selfadjoint operator H. The point spectrum σ_{p}, the singular σ_{sc} and the absolutely continuous spectrum σ_{ac}

$$\sigma = \sigma_{\mathrm{p}} \cup \sigma_{\mathrm{sc}} \cup \sigma_{\mathrm{ac}}\,. \tag{36}$$

The novelty in the decomposition (35) of the spectrum of the Hamiltonian concerns in fact its singular continuous component since, as it follows from the Riemann-Lebesgue lemma, each absolutely continuous component decays, i.e. $\sigma_{\rm ac} \subset \sigma^{\rm D}$. The essence of (35) is that through a refinement of the singular continuous part in the spectral decomposition of the Hamiltonian we characterized completely the decaying pure states as elements of the Hilbert space

$$\mathcal{H}^{\rm D} = \mathcal{H}_{\rm ac} \oplus \mathcal{H}_{\rm sc}^{\rm D} . \tag{37}$$

Each element $h \in \mathcal{H}$ determines the spectral measure μ_ψ

$$\mu_\psi(\Delta) = \int_\Delta d\langle\psi|E_\lambda\psi\rangle \tag{38}$$

defined for each Borel subset Δ of σ. Let $\psi = \psi_{\rm p} + \psi_{\rm sc} + \psi_{\rm ac}$ be the decomposition of ψ corresponding to the direct sum $\mathcal{H}_{\rm p} \oplus \mathcal{H}_{\rm sc} \oplus \mathcal{H}_{\rm ac}$. Putting $\mu_{\rm p} = \mu_{\psi_{\rm p}}$, $\mu_{\rm sc} = \mu_{\psi_{\rm sc}}$ and $\mu_{\rm ac} = \mu_{\psi_{\rm ac}}$ we obtain the Jordan decomposition of μ

$$\mu = \mu_{\rm p} + \mu_{\rm sc} + \mu_{\rm ac}$$

onto the point, singular continuous and absolutely continuous components.

The space $\mathcal{H}_{\rm sc}^{\rm D}$ consists of all vectors $\psi \in \mathcal{H}_{\rm sc}$ such that the corresponding measure $\mu = \mu_\psi$ is singular with respect to the Lebesgue measure and its Fourier transform decays to zero. $\mathcal{H}_{\rm sc}^{\rm D}$ is also a closed linear subspace of $\mathcal{H}_{\rm sc}$. This allows to define, as the orthogonal complement of $\mathcal{H}_{\rm sc}^{\rm D}$ in $\mathcal{H}_{\rm sc}$, the space $\mathcal{H}_{\rm sc}^{\rm ND}$ consisting of all non decaying singular states. Both spaces $\mathcal{H}_{\rm sc}^{\rm D}$ and $\mathcal{H}_{\rm sc}^{\rm ND}$ are also H invariant and reduce the Hamiltonian. Therefore we have the direct sum decomposition $\mathcal{H}_{\rm sc} = \mathcal{H}_{\rm sc}^{\rm D} \oplus \mathcal{H}_{\rm sc}^{\rm ND}$. This leads to the following direct sum decomposition of the whole space $\mathcal{H}$:

$$\mathcal{H} = \mathcal{H}_{\rm p} \oplus \mathcal{H}_{\rm sc}^{\rm D} \oplus \mathcal{H}_{\rm sc}^{\rm ND} \oplus \mathcal{H}_{\rm ac} . \tag{39}$$

The non decaying states are therefore in the Hilbert space

$$\mathcal{H}^{\rm ND} = \mathcal{H}_{\rm p} \oplus \mathcal{H}_{\rm sc}^{\rm ND} . \tag{40}$$

Consider once more Liouvillian $\mathbb{L} = H \otimes I - I \otimes H$ corresponding to Hamiltonian H defined on the Hilbert-Schmidt space $\mathcal{H} \otimes \mathcal{H}^\times$. Let $\{E_\lambda\}$ be the spectral family of $\mathcal{H}$, $H = \int_{-\infty}^{\infty} \lambda\, dE_\lambda$, and let $\{\mathbb{E}_\lambda\}$ be the the spectral family of the Liouvillian. Recall also the equation

$$(|\psi\rangle\langle\psi|, \mathbb{E}_\omega|\psi\rangle\langle\psi|) = \int_{-\infty}^{\infty} \langle\psi|E_{\omega+\lambda}\psi\rangle\, d\langle\psi|E_\lambda\psi\rangle , \tag{41}$$

where $\psi \in \mathcal{H}$, which follows from (12).

Similarly as in Section 2 we have the following relation between the spectral measure $\mu_{\boldsymbol{L}}$ corresponding to state $|\psi\rangle\langle\psi|$ and Liouville operator $\boldsymbol{L}$ and the spectral measure μ corresponding to state ψ and Hamiltonian operator H:

Proposition 3 *The spectral measure $\mu_{\boldsymbol{L}}$ is the convolution of the spectral measure μ with its reflection $\bar{\mu}$ with respect to 0, i.e.*

$$\mu_{\boldsymbol{L}} = \mu * \bar{\mu}\,, \tag{42}$$

where $\bar{\mu}(\Delta) = \mu(-\Delta) = \mu(\{-x:\ x \in \Delta\})$, for each Borel set Δ.

In particular, if the measure μ is symmetric with respect to 0 then the spectral measure of the Liouvillian is the convolution square of the spectral measure of the Hamiltonian $\mu_{\boldsymbol{L}} = \mu * \mu$.

Proposition 3 implies the following relation between the Fourier transforms:

$$\hat{\mu}_{\boldsymbol{L}}(t) = |\hat{\mu}(t)|^2\,. \tag{43}$$

Therefore going from the Hamiltonian to the Liouvillian formulation the property of decay of pure states is preserved. However, as we have seen in Section 2, there is no simple relation between the singular continuous spectrum of the Hamiltonian and the spectrum of the Liouvillian. Still more surprising fact is that Hamiltonians with purely singular spectrum may give rise to Liouvillians with absolutely continuous spectrum.

In order to show the latter fact it is enough to apply a famous theorem by Wiener and Wintner [19] (see also [20] (10.12)), which says that there exists a non-negative continuous normalized measure μ on the interval $[-\pi, \pi]$ such that

$$\int_{-\pi}^{\pi} e^{-inx} d\mu(x) = O(n^{-\frac{1}{2}+\varepsilon})\,, \tag{44}$$

for each $\varepsilon > 0$. Then let us consider a Hamiltonian with the spectral measure μ, e.g. $H\,f(\lambda) = \lambda\,f(\lambda)$ (see (23)) on the Hilbert space $L^2([-\pi,\pi],\mu)$. Its spectral resolution $H = \int_{\sigma(H)} \lambda\,dE_\lambda$ can be obtain explicitly. The spectral projectors E_λ are, like (24), the operators of multiplication by the indicators $\mathbf{1}_{[-\pi,\lambda]}$. Also $\psi_1 \equiv 1$ is the corresponding cyclic vector and $\mu([-\pi,\lambda]) = \langle\psi_1|E_\lambda\psi_1\rangle$.

The Liouvillian corresponding to H has the spectral projectors $\boldsymbol{E}_\lambda$ and spectral measure $(|\psi_1\rangle\langle\psi_1|, \boldsymbol{E}_\lambda|\psi_1\rangle\langle\psi_1|)$ which is, according to Proposition 3, the convolution $\mu * \mu$. It is easy to see that the convolution $\mu * \mu$ is absolutely continuous, because its Fourier-Stieltjes coefficients, which are equal to (44) are square summable. Therefore the measure $\mu * \mu$ has square summable density (Radon-Nikodym derivative). In fact, it can be proved [21] that for the density of $\mu * \mu$ all moments are finite:

$$\int_{-\pi}^{\pi} \left| \frac{d\mu * \mu}{dx} \right|^p dx < \infty\,, \quad \text{for all} \ \ p \geq 1\,. \tag{45}$$

Although the Liouville space does not change the decay properties of pure states it also contains new decaying states. Let us denote the Liouville space by $\mathcal{L}$ and by $\mathcal{L}^{\rm D}$ and $\mathcal{L}^{\rm ND}$ its decaying and non decaying parts respectively. Thus $\mathcal{L} = \mathcal{L}^{\rm D} \oplus \mathcal{L}^{\rm ND}$. We shall show that

$$\mathcal{L}^{\rm D} = (\mathcal{H}^{\rm D} \otimes \mathcal{H}^{\rm D^\times}) \oplus (\mathcal{H}^{\rm ND} \otimes \mathcal{H}^{\rm D^\times}) \oplus (\mathcal{H}^{\rm D} \otimes \mathcal{H}^{\rm ND^\times})$$

and

$$\mathcal{L}^{\rm ND} = (\mathcal{H}^{\rm ND} \otimes \mathcal{H}^{\rm ND^\times})\,.$$

Indeed, the evolution of mixtures in Liouville space is [14]

$$e^{-i\boldsymbol{L}t} = e^{-iHt} \otimes e^{+iHt} \tag{46}$$

Let ψ, φ be vectors in the Hilbert space $\mathcal{H}$. Then, (46) implies that (see [14] for details):

$$(e^{-i\boldsymbol{L}t}\,\psi \otimes \varphi, \psi \otimes \varphi) = (e^{-iHt} \otimes e^{+iHt}(\,\psi \otimes \varphi), \psi \otimes \varphi) = \langle \psi | e^{iHt}\,\psi \rangle\, \langle \varphi | e^{-iHt}\,\varphi \rangle \tag{47}$$

If one of the vectors ψ or φ is decaying then the expression (47) tends to 0, as $t \to \infty$. To end the proof it is enough to note that the Hilbert tensor product is the closure of linear combinations of simple tensors and apply Proposition 1. We see therefore that the Liouville space contains decaying states in $\mathcal{H}^{\rm D} \otimes \mathcal{H}^{\rm ND^\times}$ or $\mathcal{H}^{\rm ND} \otimes \mathcal{H}^{\rm D^\times}$, which are not convex combinations of decaying pure states.

The state ψ_1 constructed in the above example is non decaying. There may exist, however, decaying states ψ, $\hat{\mu}_\psi \to 0$, as $t \to \infty$, in the singular continuous subspace of $\mathcal{H}$ as well. Such states can also be constructed in a similar way, through the distribution functions on a Cantor-like set but with a different than 1/3 ratio of dissection. In fact, if the ratio of dissection is any rational number $k/n < 1/2$, where k and n are integers different from 1, then the corresponding cyclic state is decaying.

The result of Wiener and Wintner contributes moreover to the physically important question how fast the decay in the singular and absolutely continuous spectra can be. One can find in literature [22] wrong statements that in the singular continuous case the decay is very slow compared with the absolutely continuous case. This may be so sometimes. However, we can also demonstrate the opposite by constructing one state for the continuous subspace and another for the singular continuous subspace that decay with almost the same rate.

In fact, consider the spectral measure μ_1 that satisfies (44) with $\varepsilon > 0$. Then μ_1 is singular continuous and the rate of convergence of its Fourier coefficients is of the order $n^{-\frac{1}{2}+\varepsilon}$. On the other hand, consider the sequence $a_n = n^{-\frac{1}{2}-\varepsilon}, n = 1, 2, \ldots$. Because $\{a_n\}$ is square summable then, by the Riesz-Fischer theorem [20], there is a function $f \in L^2_{[-\pi,\pi]}$such the the Fourier coefficients of f coincide with $\{a_n\}$. Then define the measure $\mu_2(dx) = f(x)\,dx$. Of course μ_2 is absolutely continuous. However, the Fourier-Stieltjes coefficients of μ_1 and μ_2 have rates of convergence of the same order as the difference between the powers of n can be arbitrary small. Therefore, it is enough to apply the above method and associate with μ_1 and μ_2 two quantum states with almost identical decay rates.

Another interesting question is how big the support of the singular measure μ must be in order that $\mu * \mu$ should be absolutely continuous. It turns out that the support of μ can be arbitrary small. Namely, we proved [14] that it is possible to find purely singular spectral measures concentrated on a set of the Hausdorff dimension zero such that the associated Liouvillian has non trivial absolutely continuous subspace.

4 Quantum Systems With Diagonal Singularity

4.1 Algebra of observables

We consider quantum systems associated with a Hamiltonian operator H on a Hilbert space $\mathcal{H}$ which can be decomposed into a diagonal part H_0 and an off-diagonal part, or a perturbation, V

$$H = H_0 + \lambda V\,. \tag{48}$$

A rigorous meaning of this decomposition will be given below. The motivation and historical remarks can be found in Ref. [7, 8]. Let us only remark that, for $\mathcal{H}$ finite dimensional, decomposition (48) means that operator H_0 is diagonal in some orthonormal basis.

If $\mathcal{H}$ is infinite dimensional then it is also assumed that operator H_0 is diagonal. However, if the spectrum of H_0 is absolutely continuous, the term diagonal refers to the spectral resolution $H = \int \alpha \, dE_\alpha$, which can be written, using Dirac's notation, as

$$\int_{\sigma(H)} \omega_\alpha |\alpha\rangle\langle\alpha| \, d\alpha \tag{49}$$

in terms of generalized eigenvectors.

Physically speaking, (49) means that H_0 is diagonal with respect to the non-interacting entities $|\alpha\rangle$. The perturbation λV (λ is a dimensionless parameter) couples these entities and, in the case of integrable systems, it gives rise to new independent conserved entities associated with the total Hamiltonian H. This can be symbolically written as:

$$H = \sum_a z_a \, |a\rangle\langle a| \ . \tag{50}$$

Rigorously speaking V may represent, e.g. a kernel operator with the kernel $V(\alpha, \alpha')$ being a function on $\sigma(H) \times \sigma(H)$.

In the representation provided by continuous bases like $\{\ |\alpha\rangle\}$, van Hove [9, 10] encountered operators represented by the kernels

$$A_{\alpha\alpha'} \equiv \langle\alpha| \ A \ |\alpha'\rangle = A^d_\alpha \delta(\alpha - \alpha') + A^c_{\alpha\alpha'} \ . \tag{51}$$

The operator A is said to have diagonal singularity with respect to the basis $\{\ |\alpha\rangle\}$ if the diagonal part A^d_α does not vanish. The diagonal function A^d_α has no singularities while the off-diagonal part $A^c_{\alpha\alpha'}$ may have certain weaker singularities as in the case of continuous degeneracies appearing in scattering or in coupled fields. In such cases one may decompose further the off-diagonal part $A^c_{\alpha\alpha'}$. Here we shall assume the simplest case where $A^c_{\alpha\alpha'}$ are kernels of compact operators. The reason is that the states on the algebra of compact operators correspond to trace class operators, because the dual of the algebra $\mathcal{B}^\infty_\mathcal{H}$ of compact operators is isomorphic to the trace class operators $\mathcal{B}^1_\mathcal{H}$. Tracial states are the traditional states of quantum statistical physics [23, 24].

A reasonable choice for the algebra of operators with diagonal singularity is the sum

$$\mathcal{A} \equiv \mathcal{A}^d + \mathcal{B}^\infty_\mathcal{H} \, . \tag{52}$$

The diagonal part is the maximal Abelian von Neumann algebra generated by the chosen complete system of commuting observables [25, 26]. The algebra $\mathcal{A}^d$ is a Banach algebra with respect to the norm

$$\|A^d\| \equiv \sup_{\psi \neq 0} \frac{\|A\psi\|_H}{\|\psi\|_H}\,, \quad \text{any} \quad A^d \in \mathcal{A}^d\,. \tag{53}$$

If some observables correspond to unbounded self-adjoint operators, as in the case of potential scattering, the algebra $\mathcal{A}^d$ is generated by their spectral projections which are bounded operators.

The operator H_0 as well as any other unbounded operator are affiliated with the algebra $\mathcal{A}^d$, if their spectral projections are included in $\mathcal{A}^d$. In the case of unbounded operators, Antoine et al. [28] considered more general structures like V^*-algebras instead of maximal Abelian von Neumann algebras.

The non-diagonal part $\mathcal{B}^\infty_\mathcal{H}$ is the C*-algebra [29] of compact operators on the Hilbert Space $\mathcal{H}$. It was proven by Antoniou and Suchanecki [7] that the algebras $\mathcal{A}^d$ and $\mathcal{B}^\infty_\mathcal{H}$ have no common elements. Consequently, any operator A in the algebra is uniquely represented as the sum of the diagonal and off-diagonal parts

$$A = A^d + A^c\,. \tag{54}$$

A natural norm on $\mathcal{A}$ which captures the physical properties of these operators can be defined by

$$\|A\|_\mathcal{A} = \|A^d\| + \|A^c\|\,, \tag{55}$$

where $\|\cdot\|$ is the operator norm defined in (40).

The properties of the algebra $\mathcal{A}$ are summarized (see [7, 8] for details) by the following

Theorem 1

1. *$\mathcal{A}$ is an involutive Banach algebra of bounded operators which includes the identity operator.*

2. *$\mathcal{A}$ is not a C^*-algebra.*

The generalized basis $|\alpha\rangle$ for $\mathcal{H}$ leads to a natural basis for the algebra $\mathcal{A}$ constructed as follows. The diagonal part $\mathcal{A}^d$ can be expanded in terms of operators $|\alpha) \equiv |\alpha)^d \equiv |\alpha\rangle\langle\alpha|$

$$A^d = \int_\sigma d\alpha A^d_\alpha \;\; |\alpha)\,, \tag{56}$$

where A^d_α is an essentially bounded measurable function [30, 26, 24] on the spectrum σ.

Physicists usually expect that the off-diagonal compact part can be expanded in terms of the operators $|\alpha\alpha') \equiv |\alpha\rangle\langle\alpha'|$. If this is the case, for example if A^c is Hilbert-Schmidt, we have

$$A = \int_\sigma d\alpha A_\alpha \, |\alpha) + \int_{\sigma\times\sigma} d\alpha d\alpha' A^c_{\alpha\alpha'} \, |\alpha\alpha') \ , \tag{57}$$

where $A^c_{\alpha\alpha'}$ is the kernel function function on the product space $\sigma \times \sigma$.

The basis $\{ \ |\alpha), \ |\alpha\alpha')\}$ may be used for the expansion of unbounded operators which are not in the algebra $\mathcal{A}$. In such cases, we have more general structures like the partial involutive algebra [31, 32]. For example, in the case of kernel operators, which of course include the the Hilbert-Schmidt operators, one can show that the decomposition (44) is unique [7]. Moreover, a different choice for the complete system of commuting observables gives rise to a different generalized basis $|\alpha\rangle$ for the Hilbert space $\mathcal{H}$ and therefore to a different algebra $\mathcal{A}$. However, both bases are related by a unitary operator which induces a Banach algebra isomorphism.

4.2 States With Diagonal Singularity

The states ρ of the system correspond to the normalized positive linear functionals $(\rho|$ over the algebra $\mathcal{A}$, i.e., they satisfy the properties

1. Linearity:
$$(\rho|z_1A_1 + z_2A_2) = z_1(\rho|A_1) + z_2(\rho|A_2) \ . \tag{58}$$

2. Positivity:
$$(\rho|A^\dagger A) \geq 0 \ . \tag{59}$$

3. Normalization:
$$(\rho|I) = 1 \ . \tag{60}$$

As $\mathcal{A}$ is an involutive Banach algebra, the states ρ are bounded with norm [29]

$$\|\rho\| = (\rho|I) = 1 \, . \tag{61}$$

We shall adopt here the following bra/ket notation in connection with the states/observables duality.

$$\rho(A) \equiv (\rho|A) \equiv \langle A\rangle_\rho \ . \tag{62}$$

The bracket $(\rho|A)$ represents the expectation value of the observable A in the state ρ. The bras $(\rho|$ are to be understood as linear functionals over the algebra $\mathcal{A}$ of observables. The reader should not confuse the bras and kets $(\cdot|$ and $|\cdot)$ expressing the state/observables duality with the Dirac bracket notation $\langle\cdot|$ and $|\cdot\rangle$ for the linear and antilinear functionals representing the wave functions [33]. Our bras $(\cdot|$ and kets $|\cdot)$ notation is just Dirac's notation in the Liouville space.

The states ρ form a convex subset of the Banach dual $\mathcal{A}^\times$ of the algebra $\mathcal{A}$,

$$\mathcal{A}^\times = \mathcal{A}^{d\times} \oplus (\mathcal{B}_H^\infty)^\times = \mathcal{A}^{d\times} \oplus \mathcal{B}_H^1 \,. \tag{63}$$

Any state ρ has therefore a diagonal and an off-diagonal part, ρ^d and ρ^c correspondingly

$$\rho = \rho^d + \rho^c \,, \tag{64}$$

with

$$\begin{aligned} (\rho^d|A) &= (\rho^d|A^d) \,, \\ (\rho^c|A) &= (\rho^c|A^c) \,, \end{aligned} \tag{65}$$

for any operator $A \in \mathcal{A}$.

As the dual space of $\mathcal{B}_H^\infty$ is the space of trace class $\mathcal{B}_H^1$, the off diagonal states ρ correspond to the tracial functionals

$$(\rho^c|A) = \mathrm{tr}\hat{\rho}\hat{A} \,, \tag{66}$$

where $\hat{\rho}$ is the density operator representing the state ρ.

The dual space $(\mathcal{L}_\sigma^\infty)^\times$ is isomorphic to the space of complex additive measures of bounded total variation and absolutely continuous with respect to a suitable [34] extension $(\widetilde{\mathcal{B}}, \widetilde{\mu})$ of the Lebesgue measure structure $(\mathcal{B}, \mu)$ associated with the spectrum σ. The diagonal states ρ^d are therefore represented in terms of these normalized positive measures or their Radon-Nikodym derivatives ρ_α^d.

The diagonal elements ρ_α^d represent the probability densities associated with the states $|\alpha\rangle$ while the off-diagonal elements $\rho_{\alpha\alpha'}^c$ describe correlations or coherent superpositions of the states $|\alpha\rangle$.

For observables A admitting a kernel representation, the expectation value in the state ρ can be written

$$(\rho|A) = \langle A\rangle_\rho = \int_\sigma d\alpha\; \rho_\alpha^d A_\alpha^d + \int_{\sigma\times\sigma} d\alpha d\alpha'\; \rho_{\alpha\alpha'}^{c*} A_{\alpha\alpha'}^c \,. \tag{67}$$

In view of the duality relation (54) we define the action of the states ρ on the basis $|\alpha)$, $|\alpha\alpha')$ as

$$\begin{aligned} (\rho|\alpha) &\equiv \rho_\alpha^{0*} = \rho_\alpha^0 \,, \\ (\rho|\alpha\alpha') &\equiv \rho_{\alpha\alpha'}^{c*} \,. \end{aligned} \tag{68}$$

These equations suggest the definition of a dual basis $(\alpha|, (\alpha\alpha'|$ for the space of states

$$\begin{aligned} (\alpha|A) &\equiv (\alpha|A^d) \equiv A_\alpha^d \\ (\alpha\alpha'|A) &\equiv (\alpha\alpha'|A^c) \equiv A_{\alpha\alpha'}^c \end{aligned} \tag{69}$$

The duality relation (54) suggests the expansion of any state ρ

$$(\rho| = \int_\sigma d\alpha \rho_\alpha^d (\alpha| + \int_{\sigma\times\sigma} d\alpha d\alpha' \rho_{\alpha\alpha'}^{c*} (\alpha\alpha'| \tag{70}$$

The normalized positive linear functionals over the algebra $\mathcal{A}$ include not only the conventional quantum states but also the states with diagonal singularity which cannot be discussed in terms of the Hilbert space formulation. In fact, the reason for the algebraic formulation of quantum theory [35, 37, 36, 38] is the possibility to discuss states of physical interest which cannot be formulated in terms of the conventional Hilbert space language. For example, the dual $\mathcal{B}_\mathcal{H}^\times$ of the von Neumann algebra $\mathcal{B}_\mathcal{H}$ of bounded operators on the Hilbert space $\mathcal{H}$ includes not only the density operators which exist in the trace class space $\mathcal{B}_\mathcal{H}^1$ but also states which vanish with respect to the compact operators $\mathcal{B}_\mathcal{H}^\infty$. In fact we have the isomorphism (see [39] p. 51)

$$\mathcal{B}_\mathcal{H}^\times \cong \mathcal{B}_\mathcal{H}^1 \oplus (\mathcal{B}_\mathcal{H}^\infty)^\perp \tag{71}$$

The norm of the state ρ extends in a natural way the trace concept because

$$\|\rho\| = (\rho|I) = \int_\sigma d\alpha\, \rho_\alpha^d \,. \tag{72}$$

The microcanonical and canonical equilibrium states for the evolution generated by H_0 have a natural description within our formalism

$$(\rho_{\text{micro}}| = \frac{1}{\mathcal{N}_{\text{micro}}} \int d\alpha \delta(\omega_\alpha - E)(\alpha| \,, \tag{73}$$

$$(\rho_{\text{can}}| = \frac{1}{\mathcal{N}_{\text{can}}} \int d\alpha e^{-\beta\omega_\alpha} (\alpha| \,, \tag{74}$$

where $\mathcal{N}_{\rm micro}$, $\mathcal{N}_{\rm can}$ are the normalization constants determined by the condition

$$(\rho_{\rm micro}|I) = (\rho_{\rm can}|I) = 1\,. \tag{75}$$

We see that $\rho_{\rm micro}$ and $\rho_{\rm can}$ cannot be described in terms of the conventional formulation of quantum physics because they are purely diagonal states.

Respecting the tradition of statistical physics we work with states although, whenever convenient, some calculations may be performed with observables using the duality of states and observables. The action of a linear operator U on the state ρ satisfies the duality relation

$$(U\rho|A) = (\rho|VA) \tag{76}$$

for all observables A. The operator V acting on the observables A is the predual of U, i.e. $V = U^+$ or equivalently U is the dual of V, i.e. $U = V^\times$. This is just the duality between the Schrödinger and the Heisenberg pictures.

4.3 Time Evolution

The Liouville-von Neumann evolution equation for states is given by

$$(\partial_t\rho|A) = (\rho|i[H,A]) = (-iL^\times\rho|A)\,. \tag{77}$$

If the state ρ is tracial then the evolution is implemented by the evolution on the Hilbert space and $L^\times$ is the Liouville-von Neumann operator of the conventional formulation of quantum statistical physics [23, 36, 38]. However, for simplicity we shall use the symbol L instead of $L^\times$, keeping in mind that it acts on states.

The Liouville operator may lead out of the algebra $\mathcal{A}$. Indeed for Hamiltonians of the form

$$H = H_0 + V\,, \tag{78}$$

with

$$H_0 = \int_\sigma d\alpha\omega_\alpha\ |\alpha\rangle\langle\alpha|\ , \tag{79}$$

$$V = \int_{\sigma\times\sigma} d\alpha d\alpha' V_{\alpha\alpha'}\ |\alpha\rangle\langle\alpha'|\ , \tag{80}$$

the Liouville operator acts on the algebra as

$$\begin{aligned} LA = [H,A] &= [H_0, A^c] + [V, A^d] + [V, A^c] \\ &= \int_{\sigma^2} d\alpha d\alpha'(\omega_\alpha - \omega_{\alpha'})|\alpha\alpha')(\alpha\alpha'| \end{aligned}$$

$$+ \int_{\sigma^3} d\alpha d\alpha' d\alpha'' \Big[V_{\alpha''\alpha} |\alpha''\alpha')(\alpha\alpha'| - V_{\alpha'\alpha''} |\alpha\alpha'')(\alpha\alpha'| \Big]$$
$$+ \int_{\sigma^2} d\alpha d\alpha' \Big[V_{\alpha'\alpha} |\alpha'\alpha)(\alpha| - V_{\alpha\alpha'} |\alpha\alpha')(\alpha| \Big] . \qquad (81)$$

It is now clear that if $V_{\alpha'\alpha}$ contains singularities as in case of the van Hove condition discussed above, then L leads out of the algebra and we have to extend the algebra accordingly.

The solution of the eigenvalue problem may lead to a spectral decomposition of the Liouville operator

$$L = \sum_{\nu} z_{\nu} |F_{\nu})(f_{\nu}| , \qquad (82)$$

with f_{ν}, F_{ν} a biorthonormal dual pair.

The evolution of states can be decomposed in terms of the eigencomponents

$$(e^{-iL^{\times}t}\rho| = \sum_{\nu} e^{iz_{\nu}t} (\rho|F_{\nu})(f_{\nu}| . \qquad (83)$$

Our extended formulation allows z_{ν} to be complex. We expect that the imaginary part represents decay rates or diffusion coefficients. In the conventional Hilbert-Schmidt formulation, of course, z_{ν} is real and $f_{\nu} \equiv F_{\nu}$ because L is a self-adjoint operator on the Hilbert-Schmidt space.

Acknowledgements

We thank Profs I. Prigogine, A. Bohm, S. Shkarin, M. Gadella and L. Accardi for several discussions, Prof T. Hida for the invitation and kind support and Mejo University for their hospitality during the Winter School "Quantum Information and Complexity".

References

[1] M. Reed and B. Simon *Methods of Modern Mathematical Physics, Vol I: Functional Analysis*, Academic Press, New York 1980.

[2] E. E. Prugovecki *Quantum Mechanics in Hilbert Space*, Academic Press, New York 1981.

[3] I. Prigogine *Non-Equilibrium Statistical Mechanics*, Wiley, New York 1962.

[4] V.F. Weisskopf and E.P. Wiegner, Berechnung der natürlichen Linienbreite auf Grund der Diracschen Lichttheorie, *Z. Phys.* **63** 54-73 (1930).

[5] N.S. Krylov and V.A. Fock, On two basic interpretations of the energy-time uncertainty relation, Zh. Eksp. Teor. Fiz. bf 17 93-107 (1947).

[6] L.A. Khalfin, On the theory of the decay of a quasi-stationary state *Sov. Phys. Dokl.* **2** 340-344 (1957).

[7] I. Antoniou and Z. Suchanecki, Quantum systems with diagonal singularity, *Adv. Chem. Phys.* **99** , ed. I. Prigogine and S. Rice, Wiley 1997, pp. 299-332.

[8] I. Antoniou, Z. Suchanecki, R. Laura and S. Tasaki, Intrinsic irreversibility of quantum systems with diagonal singularity", *Physica A* **241** 737-772 (1997).

[9] L. van Hove, Energy corrections and persistent perturbation effects in continuous spectra, *Physica* **21** 901-923 (1955).

[10] L. van Hove, Energy corrections and persistent perturbation effects in continuous spectra. II. The perturbed stationary states, *Physica* **22** 343-354 (1956).

[11] I. Antoniou and Z. Suchanecki, Spectral characterization of decay in quantum mechanics, *Trends in Quantum Mechanics*, eds. H.-D. Doebner, S.T. Ali, M. Keyl and R.F. Werner, World Scientific, Singapore 2000, pp. 158-166.

[12] I. Antoniou and S. Shkarin, Decay spectrum and decay subspace of normal operators *Proc. Royal Soc. Edinburgh* **131A** 1245-1255 (2001).

[13] H. Spohn, The spectrum of the Liouville-von Neumann operator, *J. Math Phys.* **17** 57-60 (1976).

[14] I. Antoniou, M. Gadella and Z. Suchanecki, General properties of the Liouville operator, *Int. J. Theor. Phys.* **37** 1641-1654 (1998).

[15] I. Antoniou, S. Shkarin and Z. Suchanecki, The spectrum of the Liouville-von Neumann operator in the Hilbert-Schmidt space, *J. Math. Phys.* **40** 4106-4118 (1999).

[16] J. Bellisard *Schroedinger Operators with Almost Periodic Potential: an Overview*, Lecture Notes in Physics **153**, 356-363 (1982).

[17] H. Steinhaus *A new property of the Cantor set*b

[18] P. Billingsley *Probability and measure*, Wiley, New York 1985.

[19] N. Wiener and A. Wintner, Fourier-Stieltjes transforms and singular infinite convolutions, *Amer. J. Math.* **60** 513-522 (1938).

[20] A. Zygmund *Trigonometric Series* vol. I,II, Cambridge, University Press 1968.

[21] E. Hewitt and H.S. Zuckerman, Singular measures with absolutely continuous convolution squares, *Proc. Camb. Phil.Soc.* **62** 399-420 (1966).

[22] B. Milek and P. Seba, Singular continuous quasienergy spectrum in the kicked rotator with separable perturbation: Possibility of the onset of quantum chaos, *Phys. Rev. A* **42** 3213-3220 (1990).

[23] J. von Neumann *Mathematical Foundations of Quantum Mechanics*, Princeton Univ. Press, Princeton, 1955.

[24] E. Prugovecki *Quantum Mechanics in Hilbert Space*, Academic Press, New York 1981.

[25] J.M. Jauch, Systems of observables in quantum mechanics, *Helv. Phys. Acta* **33** 711-726 (1960).

[26] B. de Dormale and H.-F. Gautrin, Spectral representation and decomposition of self-adjoint operators, *J. Math. Phys.* **16** 2328-2332 (1975).

[27] H. Weidmann *Linear Operators in Hilbert space*, Springer Verlag 1980.

[28] J.-P. Antoine, G. Epifanio and C. Trapani, Complete sets of unbounded observables, *Helv. Phys. Acta* **56** 1175-1186 (1983).

[29] J. Dixmier *C^*-algebras*, North-Holland, Amsterdam 1977.

[30] J.M. Jauch and B. Misra, The spectral representation, *Helv. Phys. Acta* **38** 30-52 (1965).

[31] J.-P. Antoine and Karwowski, Partial *-algebras of closed linear operators in Hilbert space, *Publ. R.I.M.S. Kyoto Univ.* **21** 205-236 (1985).

[32] G. Epfanio and C. Trapani, Partial *-algebras and Volterra convolution of distribution kernels, *J. Math. Phys.* **32** 1096-1101 (1991).

[33] A. Bohm *Quantum Mechanics, Foundations and Applications*, Springer, Berlin 1986.

[34] N. Dunford and J. Schwartz *Linear Operators I : General Theory*, Wiley, New York 1988.

[35] I. Segal, Postulates for general quantum mechanics, *Annals Math* **48** 930-948 (1947).

[36] O. Bratteli and D. Robinson *Operator algebras and quantum statistical mechanics vol. I, II*, Springer, New York 1979, 1981.

[37] G. Emch, *Algebraic Methods in Statistical Mechanics and Quantum Field Theory*, Wiley-Interscience 1972.

[38] R. Haag *Local quantum physics*, Springer, Berlin 1992.

[39] J. Dixmier *Von Neumann Algebras*, North-Holland, Amsterdam 1981.

L^1–THEORY FOR THE KOLMOGOROV OPERATORS OF STOCHASTIC GENERALIZED BURGERS EQUATIONS

MICHAEL RÖCKNER

Michael Röckner
Fakultät für Mathematik
Universität Bielefeld
Postfach 10 01 31
33501 Bielefeld, Germany
E-mail: roeckner@mathematik.uni-bielefeld.de

ZEEV SOBOL

Zeev Sobol
Fakultät für Mathematik
Universität Bielefeld
Postfach 10 01 31
33501 Bielefeld, Germany
E-mail: sobol@mathematik.uni-bielefeld.de

This paper contains supplementary results to the recent paper [21] by the two authors. It focuses on the L^1–theory of a class of Kolmogorov operators L in infinitely many variables which e.g. are associated to stochastic generalized Burgers equations. Their L^1–theory is developed with respect to a whole class of reference measures identified in this paper which contains in particular infinitesimally invariant measures for L. Essential maximal dissipativity for L with initial domain given by C^2–smooth bounded cylinder functions is proved to hold on $L^1(\nu)$ for all measures ν in this class. The obtained respective C_0–semigroup on $L^1(\nu)$ is proved to come from the semigroup of kernels constructed in [21]. Finally a measure is constructed in this class which is of full topological support, i.e. charges every non–empty open set of the underlying infinite dimensional space, which here is $L^2(0,1)$.

1. Introduction

Consider the following stochastic partial differential equation on $X := L^2(0,1) = L^2((0,1), dr)$ (where dr denotes Lebesgue measure)

$$\begin{aligned} dx_t &= (\Delta x_t + F(x_t))\, dt + \sqrt{A}\, dw_t \\ x_0 &= x \in X. \end{aligned} \tag{1.1}$$

Here $A : X \to X$ is a nonnegative definite symmetric operator of trace class, $(w_t)_{t\geqslant 0}$ a cylindrical Brownian motion on X, Δ denotes the Dirichlet Laplacian (i.e. with Dirichlet boundary conditions) on $(0,1)$, and $F : H_0^1 \to X$ is a measurable vector field satisfying certain conditions specified below. Here $H_0^1 := H_0^1(0,1)$ denotes the Sobolev space of order 1 in $L^2(0,1)$ with Dirichlet boundary conditions. As a special case SPDE (1.1) contains so–called stochastic generalized Burgers equations (cf. [21], see also [20], where $(0,1)$ is replaced by an open set in $\mathbb{R}^2$).

A heuristic (i.e. not worrying about existence of solutions) application of Itô's formula to (1.1) implies that the corresponding generator or Kolmogorov operator L on smooth cylinder functions $u : X \to \mathbb{R}$, i.e.

$$u \in \mathcal{D} := \mathcal{F}C_b^2 := \{u = g \circ P_N \mid N \in \mathbb{N}, g \in C_b^2(E_N)\},$$

is of the following form:

$$\begin{aligned} Lu(x) &:= \frac{1}{2}\operatorname{Tr}\big(AD^2u(x)\big) + \big(\Delta x + F(x), Du(x)\big) \\ &= \frac{1}{2}\sum_{i,j=1}^{\infty} A_{ij}\partial_{ij}^2 u(x) + \sum_{k=1}^{\infty}\big(\Delta x + F(x), \eta_k\big)\partial_k u(x), \quad x \in H_0^1. \end{aligned} \tag{1.2}$$

Here $\eta_k(r) := \sqrt{2}\sin(\pi k r)$, $k \in \mathbb{N}$, is the eigenbasis of Δ in $L^2(0,1)$, equipped with the usual inner product $(\ ,\ )$, $E_N := \operatorname{span}\{\eta_k | 1 \leqslant k \leqslant N\}$, P_N is the corresponding orthogonal projection, and $A_{ij} := (\eta_i, A\eta_j)$, $i,j \in \mathbb{N}$. Finally, Du, D^2u denote the first and second Fréchet derivatives, $\partial_k := \partial_{\eta_k}$, $\partial_{ij}^2 := \partial_{\eta_i}\partial_{\eta_j}$ with $\partial_y :=$ directional derivative in direction $y \in X$ and $(\Delta x, \eta_k) := (x, \Delta\eta_k)$ for $x \in X$.

Hence the Kolmogorov equations corresponding to SPDE (1.1) are given by

$$\begin{aligned} \frac{dv}{dt}(t,x) &= \bar{L}v(t,x), \quad x \in X, \\ v(0,\cdot) &= f, \end{aligned} \tag{1.3}$$

where the function $f : X \to \mathbb{R}$ is a given initial condition for this parabolic PDE with variables in the infinite dimensional space X. We emphasize that (1.3) is only reasonable for some extension $\bar{L}$ of L (whose construction is an essential part of the entire problem) since even for $f \in \mathcal{D}$, it will essentially never be true that $v(t,\cdot) \in \mathcal{D}$.

Because of the lack of techniques to solve PDE's in infinite dimensions in situations as described above, the "classical" approach to solve (1.3) was to first solve (1.1) and then show in what sense the transition probabilities of the solution solve (1.3) (cf. e.g. [12], [2], [8], [14], [15], [18], [22], [5] and the references therein). Since about 1998, however, a substantial part of recent work in this area (cf. e.g. [9] [23], [24], and one of the initiating papers, [19]) is based on the attempt to solve Kolmogorov equations in infinitely many variables (as (1.3) above) directly and, reversing strategies, use the solution to construct weak solutions, i.e., solutions in the sense of a martingale problem as formulated by Stroock and Varadhan (cf. [25]), for SPDE's as (1.1) above, even for very singular coefficients (naturally appearing in many applications).

In [21] a new method was presented to solve (1.3) for all $x \in X$ (or an explicitly described subset thereof). It is based on finite dimensional approximation, obtaining a solution which despite of the lack of (elliptic and) parabolic regularity results on infinite dimensional spaces will nevertheless have regularity properties. More precisely, setting $X_p := L^p((0,1), dr)$, we shall construct a semigroup of Markov probability kernels $p_t(x, dy)$, $x \in X_p$, $t > 0$, on X_p such that for all $u \in \mathcal{D}$ we have $t \mapsto p_t(|Lu|)(x)$ is locally Lebesgue integrable on $[0, \infty)$ and

$$p_t u(x) - u(x) = \int_0^t p_s(Lu)(x)\, ds \quad \forall\, x \in X_p. \tag{1.4}$$

Here as usual for a measurable function $f : X_p \to \mathbb{R}$ we set

$$p_t f(x) := \int f(y)\, p_t(x, dy), \quad x \in X_p,\ t > 0, \tag{1.5}$$

if this integral exists. p has to be large enough compared to the growth of F. As a second step in [21] a conservative strong Markov process with weakly continuous paths was constructed, which is unique under a mild growth condition and which solves the martingale problem given by L as in (1.2), and hence also (1.1) weakly, for every starting point $x \in X_p$. Also an invariant measure for this process was constructed in [21].

The present paper can be considered as a supplement to [21], focussing on the $L^1(\nu)$–theory of (1.3) with respect to suitably chosen reference measures ν on X, one of which is the mentioned invariant measure. We shall thus concentrate on (1.3) and refer for treating SPDE (1.1) to [21]. The advantage of an $L^1(\nu)$–theory for (1.3) is that one really gets solutions of (1.3) in its original differential formulation rather than merely its integral (or mild) formulation (1.4). The disadvantage is that statements are considerably weakened to merely ν–a.e. statements, i.e. allowing ν–zero sets as exceptional sets of points in X where the equation does not hold. Nevertheless, one gets useful information. Such $L^1(\nu)$ (or even $L^p(\nu)$)–theory has been developed in [23], [24], [19] and more recently in [9], [1], [3], [6], [7], [10], [11],

[16], in partly more special cases than ours or in other situations. But ν was always chosen to be an infinitesimally invariant measure of L, i.e. ν is a probability measure on X solving the equation $L^*\nu = 0$ (cf. Theorem 3.2(i) below). Our main aim in this paper is to specify a large class $\mathfrak{M}$ of probability measures ν on X for which the results in [21] imply that (L, D) is well–defined and closable on $L^1(X,\nu)$, and that its closure $(\overline{L}_1^\nu, D(\overline{L}_1^\nu))$ generates a C_0–semigroup on $L^1(X,\nu)$ which then by definition gives a solution for (1.3) in $L^1(X,\nu)$. Because there is no Lebesgue measure on X, it is particularly important to find substitutes. One important feature of such a substitute should be that this measure should have full topological support, i.e. it should be strictly positive on every non–empty open set, to have that continuous representatives of $L^1(X,\nu)$–classes are unique. This issue has hardly been addressed in the above mentioned literature. In this paper for the first time we construct a measure ν_0 in the above class $\mathfrak{M}$ with full topological support on X.

We emphasize that the methods to establish an L^1-theory for Kolmogorov operators of type (1.2) developed in this paper work in general and are not restricted to the case of underlying domains which are in $\mathbb{R}$ as $(0,1)$ above, but also extend to d-dimensional domains.

The organization of this paper is as follows. In Section 2 we recall the framework from [21], keeping the notation introduced there, and describe examples. In Section 3 we summarize those results from [21] used subsequently. In Section 4 we define the said class $\mathfrak{M}$ of probability measures and show well–definedness, closability of $(L, \mathcal{D})$ on $L^1(X,\nu)$ as well as that its closure generates a Markov C_0–semigroup on $L^1(X,\nu)$ for all $\nu \in \mathfrak{M}$. Section 5 is devoted to the construction of the specific reference measure ν_0 of full topological suppor mentioned above.

This paper in connection with [21] (see also [20]) covers a major part of the contents of the lecture series given by the first named author during the conference "Quantum Information and Complexity" held at Meijo University, Nagoya, in January 2003. We refer to the references, quoted in the text below, for other material touched upon in the lectures.

It is a great pleasure for the first named author to thank Professor Takeyuki Hida for organizing the above mentioned conference and for creating such a nice scientifically stimulating atmosphere among the participants. Thanks also go to all Japanese colleagues and friends who supported him, in particular Professor K. Saito, and also Meijo University and the staff involved. We hope that also in the future Professor Hida will find all necessary support for such conferences to be able to provide both his extraordinary scientific input as well as his warmhearted hospitality to the participating mathematical community.

2. Framework and main examples

Let us recall the framework and notation from [21].

For a σ-algebra $\mathcal{B}$ on an arbitrary set E we denote the space of all bounded resp. positive real-valued $\mathcal{B}$-measurable functions by $\mathcal{B}_b$, $\mathcal{B}^+$ respectively. If E is equipped with a topology, then $\mathcal{B}(E)$ denotes the corresponding Borel σ-algebra. The spaces $X = L^2(0,1)$ and H_0^1 are as in the introduction and they are equipped with their usual norms $|\cdot|_2$ and $|\cdot|_{1,2}$; so we define for $x : (0,1) \to \mathbb{R}$, measurable,

$$|x|_p := \left(\int_0^1 |x(r)|^p \, dr \right)^{1/p} \quad (\in [0,\infty]), \quad p \in [1,\infty),$$

$$|x|_\infty := \operatorname{ess\,sup}_{r\in(0,1)} |x(r)|,$$

and define $X_p := L^p((0,1), dr)$, $p \in [2,\infty]$, so $X = X_2$. If $x, y \in H_0^1$, set

$$|x|_{1,2} := |x'|_2, \quad (x,y)_{1,2} := (x', y'),$$

where $x' := \frac{d}{dr}x$ is the weak derivative of x. We shall use this notation from now on and we also write $x'' := \frac{d^2}{dr^2}x = \Delta x$.

Let H^{-1} with norm $|\cdot|_{-1,2}$ be the dual space of H_0^1. We always use the continuous and dense embeddings

$$H_0^1 \subset X \equiv X' \subset H^{-1}, \tag{2.1}$$

so ${}_{H_0^1}\langle x, y \rangle_{H^{-1}} = (x,y)$ if $x \in H_0^1$, $y \in X$. The terms "Borel-measurable" or "measure on X, H_0^1, H^{-1} resp." will below always refer to their respective Borel σ-algebras, if it is clear on which space we work. We note that since $H_0^1 \subset X \subset H^{-1}$ continuously, by Kuratowski's Theorem $H_0^1 \in \mathcal{B}(X)$, $X \in \mathcal{B}(H^{-1})$ and $\mathcal{B}(X) \cap H_0^1 = \mathcal{B}(H_0^1)$, $\mathcal{B}(H^{-1}) \cap X = \mathcal{B}(X)$. Furthermore, the Borel σ–algebras on X and H_0^1 corresponding to the respective weak topologies coincide with $\mathcal{B}(X)$, $\mathcal{B}(H_0^1)$ respectively.

For a function $V : X \to (0,\infty]$ having weakly compact level sets $\{V \leq$

$c\}$, $c \in \mathbb{R}_+$, we define:

$$WC_V := \Big\{ f : \{V < \infty\} \to \mathbb{R} \,\Big|\, f \text{ is continuous on each } \{V \leqslant R\},\ R \in \mathbb{R}, \tag{2.2}$$
$$\text{in the weak topology inherited from } X,$$
$$\text{and } \lim_{R\to\infty} \sup_{\{V\geqslant R\}} V^{-1}|f| = 0 \Big\},$$

equipped with the norm $\|f\|_V := \sup_{\{V<\infty\}} V^{-1}|f|$. Obviously, WC_V is a Banach space with this norm. We are going to consider various choices of V, distinguished by respective subindices, namely we define for $\kappa \in (0, \infty)$

$$\begin{aligned} V_\kappa(x) &:= e^{\kappa|x|_2^2}, \quad x \in X, \\ \Theta_\kappa(x) &:= V_\kappa(x)(1 + |x'|_2^2), \quad x \in H_0^1, \end{aligned} \tag{2.3}$$

and for $p > 2$

$$\begin{aligned} V_{p,\kappa}(x) &:= e^{\kappa|x|_2^2}(1 + |x|_p^p), \quad x \in X, \\ \Theta_{p,\kappa}(x) &:= V_{p,\kappa}(x)(1 + |x'|_2^2) + V_\kappa(x)\big|\big(|x|^{\frac{p}{2}}\big)'\big|_2^2, \quad x \in H_0^1. \end{aligned} \tag{2.4}$$

Clearly, $\{V_{p,\kappa} < \infty\} = X_p$ and $\{\Theta_{p,\kappa} < \infty\} = H_0^1$. Each $\Theta_{p,\kappa}$ is extended to a function on X by defining it to be equal to $+\infty$ on $X \setminus H_0^1$. Abusing notation, for $p = 2$ we also set $V_{2,\kappa} := V_\kappa$ and $\Theta_{2,\kappa} := \Theta_\kappa$. For abbreviation, for $\kappa \in (0, \infty)$, $p \in [2, \infty)$ we set

$$WC_{p,\kappa} := WC_{V_{p,\kappa}}, \quad W_1C_{p,\kappa} := W_1C_{\Theta_{p,\kappa}}, \tag{2.5}$$

and we also abbreviate the norms correspondingly,

$$\|\cdot\|_{p,\kappa} := \|\cdot\|_{V_{p,\kappa}}, \quad \text{and} \quad \|\cdot\|_{1,p,\kappa} := \|\cdot\|_{\Theta_{p,\kappa}}. \tag{2.6}$$

All these norms are, of course, well defined for any function on X with values in $[-\infty, \infty]$. And therefore we shall apply them below not just for functions in $WC_{p,\kappa}$ or $W_1C_{p,\kappa}$. For $p' \geqslant p$ and $\kappa' \geqslant \kappa$ by restriction $WC_{p,\kappa}$ is continuously and densely embedded into $WC_{p',\kappa'}$ and into $W_1C_{p,\kappa}$ (see Korollary 5.6 in [21]), as well is the latter into $W_1C_{p',\kappa'}$. V_κ and $V_{p,\kappa}$ will serve as convenient Lyapunov functions for L and Θ_κ, $\Theta_{p,\kappa}$ naturally appear as parts of functions bounding LV_κ, $LV_{p,\kappa}$, respectively (cf. condition (F2) and Example 2.1 below, as well as Lemma 4.6 in [21]). Note that the level sets of $\Theta_{p,\kappa}$ are even strongly compact in X.

For a function $V\colon X \to (1, \infty]$, we also define spaces $Lip_{l,p,\kappa}$, $p \geqslant 2$, $\kappa > 0$, consisting of functions on X which are locally Lipschitz continuous in the norm $|(-\Delta)^{-l/2}\cdot|_2$, $l\in\mathbb{Z}_+$. The respective semi-norms are defined as follows:

$$(f)_{l,p,\kappa} := \sup_{y_1,y_2\in X_p} \left(V_{p,\kappa}(y_1)\vee V_{p,\kappa}(y_2)\right)^{-1} \frac{|f(y_1)-f(y_2)|}{|(-\Delta)^{-l/2}(y_1-y_2)|_2} \quad (\in [0,\infty]). \tag{2.7}$$

For $l\in\mathbb{Z}_+$ we define

$$Lip_{l,p,\kappa} := \left\{f : X_p \to \mathbb{R} \;\middle|\; \|f\|_{Lip_{l,p,\kappa}} < \infty\right\}, \tag{2.8}$$

where $\|f\|_{Lip_{l,p,\kappa}} := \|f\|_{p,\kappa}+(f)_{l,p,\kappa}$. When X is of finite dimension, $(f)_{l,p,\kappa}$ is a weighted norm of the generalised gradient of f (cf. Lemma 3.6 in [21]). Also, $(Lip_{l,p,\kappa}, \|\cdot\|_{Lip_{l,p,\kappa}})$ is a Banach space (cf. Lemma 5.7 in [21]) and $Lip_{l,p,\kappa}\subset Lip_{l',p',\kappa'}$ if $l'\leqslant l$, $p'\geqslant p$, $\kappa'\geqslant\kappa$. In this paper we shall only deal with the case $l=0$.

Obviously, each $f\in Lip_{0,p,\kappa}$ is uniformly $|\cdot|_2$–Lipschitz continuous on every $|\cdot|_p$-bounded set and by restriction for $p'\in[p,\infty)$, $\kappa'\in[\kappa,\infty)$

$$\mathcal{B}_b(X_p)\cap Lip_{0,p,\kappa}\subset W_1C_{p',\kappa'}. \tag{2.9}$$

Besides the space $\mathcal{D} := \mathcal{F}C_b^2$ defined in the introduction, other test function spaces $\mathcal{D}_{p,\kappa}$ on X will turn out to be convenient. They are for $\kappa\in(0,\infty)$ defined as follows:

$$\mathcal{D}_\kappa := \Big\{u = g\circ P_N \;\Big|\; N\in\mathbb{N},\ g\in C^2(\mathbb{R}^N), \\ \|u\|_{2,\kappa} + \big\||Du|_2\big\|_{2,\kappa} + \big\|Tr(AD^2u)\big\|_{2,\kappa} < \infty\Big\}. \tag{2.10}$$

Obviously, $\mathcal{D}_\kappa\subset WC_{2,\kappa}$ and $\mathcal{D}_\kappa\subset\mathcal{D}_{\kappa'}$ if $\kappa'\in[\kappa,\infty)$. We extend the definition (1.2) of the Kolmogorov operator L for all $u\in\mathcal{F}C^2 := \{u = g\circ P_N \,|\, N\in\mathbb{N}, g\in C^2(\mathbb{R}^{\mathbb{N}})\}$. So, L can be considered with domain $\mathcal{D}_\kappa$.

Now let us collect our precise hypotheses on the Kolmogorov operator (1.2). First we recall that in the entire paper $\Delta = x''$ is the Dirichlet Laplacian on $(0,1)$. Consider the following condition on the map $A : X\to X$:

(A) A is a nonnegative symmetric linear operator from X to X of trace class such that $A_N := P_NAP_N$ is an invertible operator represented by a diagonal matix on E_N for all $N\in\mathbb{N}$.

Here E_N, P_N are as defined in the introduction. Furthermore, we set

$$a_0 := \sup_{x\in H_0^1\setminus\{0\}} \frac{(x,Ax)}{|x'|_2^2} = |A|_{H_0^1\to H^{-1}}, \tag{2.11}$$

where $|\cdot|_{H_0^1 \to H^{-1}}$ denotes the usual operator norm on bounded linear operators from H_0^1 into its dual H^{-1}.

Consider the following condition for a map $F : H_0^1 \to X$.

(F2) For every $k \in \mathbb{N}$ the map $F^{(k)} := (F, \eta_k) : H_0^1 \to \mathbb{R}$ is $|\cdot|_2$-continuous on $|\cdot|_{1,2}$-balls and there exists a sequence $F_N : E_N \to E_N$, $N \in \mathbb{N}$, of bounded locally Lipschitz continuous vector fields satisfying the following conditions:

(F2a) There exist $\kappa_0 \in (0, \frac{1}{4a_0}]$ and a set $Q_{\text{reg}} \subset [2, \infty)$ such that $2 \in Q_{\text{reg}}$ and for all $\kappa \in (0, \kappa_0)$, $q \in Q_{\text{reg}}$ there exist $m_{q,\kappa} > 0$ and $\lambda_{q,\kappa} \in \mathbb{R}$ such that for

$$L_N u(x) := \frac{1}{2} \operatorname{Tr}\big(A_N D^2 u\big)(x) + \big(x'' + F_N(x), Du(x)\big),$$
$$u \in W_{loc}^{2,1}(E_N), x \in E_N, N \in \mathbb{N}, \quad (2.12)$$

we have for all $N \in \mathbb{N}$

$$L_N V_{q,\kappa} := L_N\big(V_{q,\kappa}\restriction_{E_N}\big) \leqslant \lambda_{q,\kappa} V_{q,\kappa} - m_{q,\kappa}\Theta_{q,\kappa} \quad \text{on } E_N. \quad (2.13)$$

(F2b) For all $\varepsilon \in (0,1)$ there exists $C_\varepsilon \in (0,\infty)$ such that for all $N \in \mathbb{N}$ and dx-a.e. $x \in E_N$ (where dx denotes Lebesgue measure on E_N)

$$\big(DF_N(x)y, y\big) \leqslant |y'|_2^2 + \big(\varepsilon|x'|_2^2 + C_\varepsilon\big)|y|_2^2 \quad \forall\, y \in E_N.$$

(F2c) $\lim_{N\to\infty} \big|P_N F - F_N \circ P_N\big|_2(x) = 0 \ \forall\ x \in H_0^1$, and $\lim_{N\to\infty} \big|(\eta_k, F_N) - F^{(k)}\big| = 0$ uniformly on H_0^1-balls for all $k \in \mathbb{N}$.

(F2d) For κ_0 and Q_{reg} as in (F2a), there exist $\kappa \in (0, \kappa_0)$, $p \in Q_{\text{reg}}$ such that for some $C_{p,\kappa} > 0$ and some $\omega : [0,\infty) \to [0,1]$ vanishing at infinity

$$\big|F_N \circ P_N\big|_2(x) \leqslant C_{p,\kappa}\Theta_{p,\kappa}(x)\omega\big(\Theta_{p,\kappa}(x)\big) \quad \forall\, x \in H_0^1,\ N \in \mathbb{N}.$$

We note that the second part of (F2c) was not assumed to hold in [21], except for the last part of its Appendix.

Although condition (F2) looks abstract and at first sight artificial, it is exactly what is needed for our analysis and what is satisfied in many situations as the following example shows.

Example 2.1. Consider the following condititons on the map $F : H_0^1 \to X$:

(F1)

$$F(x) = \frac{d}{dr}\big(\Psi \circ x\big)(r) + \Phi\big(r, x(r)\big), \quad x \in H_0^1(0,1),\ r \in (0,1), \tag{2.14}$$

where $\Psi : \mathbb{R} \to \mathbb{R}$, $\Phi : (0,1)\times\mathbb{R} \to \mathbb{R}$ satisfy the following conditions:

(Ψ) $\Psi \in C^{1,1}(\mathbb{R})$ (i.e. Ψ is differentiable with locally Lipschitz derivative) and there exist $C \in [0,\infty)$ and a bounded, Borel-measurable function $\omega : [0,\infty) \to [0,\infty)$ vanishing at infinity such that

$$|\Psi_{xx}|(x) \le C + \sqrt{|x|}\,\omega(|x|) \quad \text{for } dx\text{-a.e. } x \in \mathbb{R}.$$

($\Phi 1$) Φ is Borel-measurable in the first and continuous in the second variable and there exist $g \in L^{q_1}(0,1)$ with $q_1 \in [2,\infty]$ and $q_2 \in [1,\infty)$ such that

$$|\Phi(r,x)| \le g(r)(1 + |x|^{q_2}) \quad \text{for all } r \in (0,1),\ x \in \mathbb{R}.$$

($\Phi 2$) There exist $h_0, h_1 \in L^1_+(0,1)$, $|h_1|_1 < 2$, such that for a.e. $r \in (0,1)$

$$\Phi(r,x)\,\operatorname{sign} x \le h_0(r) + h_1(r)|x| \quad \text{for all } x \in \mathbb{R}.$$

($\Phi 3$) There exist $\rho_0 \in (0,1]$, $g_0 \in L^1_+(0,1)$, $g_1 \in L^{p_1}_+(0,1)$ for some $p_1 \in [2,\infty]$, and a function $\omega : [0,\infty) \to [0,\infty)$ as in (Ψ) such that with $\sigma : (0,1)\times\mathbb{R} \to \mathbb{R}$, $\sigma(r,x) := \frac{|x|}{\sqrt{r(1-r)}}$ for a.e. $r \in (0,1)$

$$\Phi(r,y) - \Phi(r,x) \le \Big[g_0(r) + g_1(r)\big|\sigma(r,x)\big|^{2-\frac{1}{p_1}}\omega\big(\sigma(r,x)\big)\Big](y-x)$$

for all $x, y \in \mathbb{R}$, $0 \leqslant y - x \leqslant \rho_0$.

Then we have that (F1) implies (F2). More precisely, (F2a) holds with $\kappa_0 := \frac{2-|h_1|_1}{8a_0}$, $Q_{\mathrm{reg}} := [2,\infty)$ and (F2d) holds with $p \in [2,\infty) \cap \big(q_2 - 3 + \frac{2}{q_1}, \infty\big)$ and any $\kappa \in (0,\kappa_0)$. The proof of this fact is extremely involved. All details can be found in Section 4 in [21]. It is obviously very easy to find plenty of examples for functions Ψ, Φ as in (F1). For instance, one can take $\Psi(x) := \alpha x^2$, $\alpha \in [0,\infty)$, and $\Phi(r,x) := -\beta x^m$, $\beta \in [0,\infty)$, $m \in \mathbb{N}$, m odd. If $\alpha := \frac{1}{2}$, $\beta := 0$, SPDE (1.1) is just the classical stochastic Burgers equation, and if $\alpha := 0$, $\beta > 0$, it is just a classical stochastic reaction diffusion equation of Ginsburg–Landau type.

3. Main background results

In this section we list a number of results from [21], which we shall use below, but in appropriately shortened form. For the complete formulations and detailed proofs we refer to [21].

Theorem 3.1. *("Pointwise solutions of the Kolmogorov equations"). Suppose (A) and (F2) hold. Let κ_0, Q_{reg} be as in (F2a), $\kappa \in (0, \kappa_0)$ and $p \in Q_{reg}$ be as in (F2d). Let $\kappa^* \in (\kappa, \kappa_0)$, $\kappa_1 \in (0, \kappa^* - \kappa]$. Then there exists a semigroup $(p_t)_{t>0}$ of probability kernels on X_p, independent of κ^*, having the following properties:*

(i) ("Existence") *Let $u \in \mathcal{D}_{\kappa_1}$. Then $t \mapsto p_t(|Lu|)(x)$ is locally Lebesgue integrable on $[0, \infty)$ and*

$$p_t u(x) - u(x) = \int_0^t p_s(Lu)(x)\, ds \quad \text{for all } x \in X_p, \tag{3.1}$$

where as usual

$$p_t u(x) := \int u(y)\ p_t(x, dy)\ . \tag{3.2}$$

In particular, for all $s \in [0, \infty)$

$$\lim_{t \to 0} p_{s+t} u(x) = p_s u(x) \quad \text{for all } x \in X_p.$$

(ii) There exists $\lambda_{\kappa^*} \in (0, \infty)$ such that

$$\int_0^\infty e^{-\lambda_{\kappa^*} s} p_s(\Theta_{p,\kappa^*})(x)\, ds < \infty \quad \text{for all } x \in X_p. \tag{3.3}$$

(iii) ("Uniqueness") *Let $(q_t)_{t>0}$ be a semigroup of probability kernels on X_p satisfying (i) with $(p_t)_{t>0}$ replaced by $(q_t)_{t>0}$. If in addition,* (3.3) *holds with $(q_t)_{t>0}$ replacing $(p_t)_{t>0}$ for some $\kappa \in (0, \kappa_0)$ replacing κ^*, then $p_t(x, dy) = q_t(x, dy)$ for all $t > 0$, $x \in X_p$.*

Proof. See Proposition 6.7 in [21]. □

Theorem 3.2. *("Invariant measure"). Assume that (A) and (F2) hold. Let p, κ^* be as in Theorem 3.1.*

(i) There exists a probability measure μ on H_0^1 which is "L-infinitesmally invariant", i.e. $Lu \in L^1(H_0^1, \mu)$ and

$$\int Lu\, d\mu = 0 \quad \text{for all } u \in \mathcal{D} \tag{3.4}$$

($L^*\mu = 0$ for short). Furthermore,

$$\int \Theta_{p,\kappa^*}\, d\mu < \infty. \tag{3.5}$$

(ii) μ, extended by zero to all of X_p, is $(p_t)_{t>0}$-invariant, i.e. for all $f : X \to \mathbb{R}$, bounded, measurable, and all $t > 0$,

$$\int p_t f\, d\mu = \int f\, d\mu.$$

(with $(p_t)_{t>0}$ from Theorem 3.1).

Proof. See Appendix in [21]. The method is taken from [4] which we also refer to as a general reference for infinitesimally invariant measures. □

Now for $(p_t)_{t>0}$ as above consider the corresponding resolvent of kernels $g_\lambda(x, dy)$ defined for $\lambda \in (0, \infty)$, $x \in X_p$ as follows: for $f \in \mathcal{B}_b(X_p)$

$$\begin{aligned} g_\lambda f(x) &:= \int f(y)\, g_\lambda(x, dy) \\ &:= \int_0^\infty e^{-\lambda t}\, p_t f(x) dt\ . \end{aligned} \tag{3.6}$$

We extend the measures $g_\lambda(x, dy)$, $x \in X_p$, $\lambda \in (0, \infty)$, by zero to all of X.

Prop 3.1. Suppose (A), (F2) hold and let p, κ^*, κ_1 be as in Theorem 3.1. Then there exists $\lambda_* \in (0, \infty)$ such that for $\lambda > \lambda_*$

$$g_\lambda\big((\lambda - L)u\big) = u \text{ for all } u \in \mathcal{D}_{\kappa_1}\ .$$

Furthermore, there exists $m_{p,\kappa^*} \in (0, \infty)$ such that

$$g_\lambda \Theta_{p,\kappa^*} \le \frac{1}{m_{p,\kappa^*}}\, V_{p,\kappa^*}\ . \tag{3.7}$$

In particular, $g_\lambda(x, X \setminus H_0^1) = 0$ for all $x \in X_p$.

Proof. Theorem 6.4 and Proposition 6.7(i) in [21]. □

Prop 3.2. Suppose (A), (F2) hold and let p, κ^*, κ_1 be as in Theorem 3.1. Let $N \in \mathbb{N}$. For L_N as in (2.12) let $(\mathcal{R}_\lambda^{(N)})_{\lambda>0}$ be the corresponding Markovian pseudo–resolvent on $L^\infty(E_N)$ (which exists according to Proposition 3.1 in [21]. For $f \in \mathcal{B}_b(X)$ define

$$G_\lambda^{(N)} f := \big(\mathcal{R}_\lambda^{(N)}(f\restriction_{E_N})\big) \circ P_N\ .$$

Then there exists $\lambda_* \in (0, \infty)$ such that for all $f \in Lip_{0,2,\kappa_1} \cap \mathcal{B}_b(X_p)$ the following assertions hold

(i) $\lambda g_\lambda f = \lim_{m\to\infty} \lambda G_\lambda^{(m)} f$ weakly in $WC_{p,\kappa*}$ (hence pointwise on X_p) uniformly in $\lambda \in [\lambda_*, \infty)$.

(ii) For all $\lambda > \lambda_*$, $G_\lambda^{(N)} f \in \cap_{\varepsilon>0}\mathcal{D}_{\kappa_1+\varepsilon}$ and, provided $f \in \mathcal{D}$, $G_\lambda^{(N)} f \in \bigcap_{\varepsilon>0} \mathcal{D}_\varepsilon$.

(iii) For all $\lambda > \lambda_*$ and all $x \in H_0^1$

$$\begin{aligned}&\big|(\lambda - L)G_\lambda^{(N)} f(x) - (f \circ P_N)(x)\big| \\ &\leqslant \frac{1}{\lambda - \lambda_*}\big|P_N F - F_N \circ P_N\big|_2(x)\, V_{2,\kappa_1}(x)\, (f)_{0,2,\kappa_1}\end{aligned}$$

and

$$\lim_{N\to\infty} \sup_{\lambda\geq\lambda_*} \lambda\big|(\lambda - L)G_\lambda^{(N)} f - f\big|(x) = 0 \ .$$

Proof. See Theorem 6.4 and Corollary 4.2, in particular, (4.5), in [21]. □

Corollary 3.1. *Consider the situation of Proposition 3.2 and let* $f \in Lip_{0,2,\kappa_1} \cap \mathcal{B}_b(X_p)$. *Let* $x \in \bigcup_{n\in\mathbb{N}} P_N X$. *Then*

$$\lim_{\lambda\to\infty} \lambda g_\lambda f(x) = f(x) \ .$$

Proof. We have $x = P_N x$ for some $N \in \mathbb{N}$. Hence for all $m \geq N$, $x = P_m x$, and

$$\begin{aligned}&\big|\lambda g_\lambda f(x) - f(x)\big| \\ &\quad \leq \big|\lambda g_\lambda f(x) - \lambda G_\lambda^{(m)} f(x)\big| + \big|\lambda \mathcal{R}_\lambda^{(m)}(f{\restriction}_{E_m})(P_m x) - f(P_m x)\big|. \end{aligned} \tag{3.8}$$

By Proposition 3.2(i) the first term in the right hand side of (3.8) converges to zero as $m \to \infty$ uniformly in $\lambda \in [\lambda_*, \infty)$. But the second term converges to zero as $\lambda \to \infty$ by Proposition 3.4(i) in [21] for all $m \geq N$. Hence the assertion follows. □

4. A class of natural reference measures

In all of this section we assume (A) and (F2) to hold. Let p, κ be as in (F2d) and let $\kappa^* \in (\kappa, \infty)$. Let $\mathfrak{M}$ denote the set of all probability measures ν on X such that

$$\int \Theta_{p,\kappa^*}\, d\nu < \infty \ , \tag{4.1}$$

and ν is *"infinitesimally L–excessive"*, i.e. for some $\lambda_\nu \in (0,\infty)$

$$\int Lu\, d\nu \le \lambda_\nu \int u\, d\nu \quad \text{for } u \in \mathcal{D} \text{ such that } u \geqslant 0 \ \nu\text{-a.e.} \tag{4.2}$$

Note that by (4.1) we have that $\nu(H_0^1) = 1$ for every $\nu \in \mathfrak{M}$ and by (F2d) that $Lu \in L^1(X,\mu)$ for all $u \in \mathcal{D}$. Obviously, an L-infinitesimally invariant measure satisfying (4.1) is in $\mathfrak{M}$. We also emphasize that for $\nu \in \mathfrak{M}$ there might exist a non–empty open set $U \subset X$ such that $\nu(U) = 0$, so $u \in \mathcal{D}$ with $u = 0$ ν–a.e. might not be identically equal to zero. So, the following proposition is crucial to consider L as an operator on $L^1(X,\nu)$.

Prop 4.1. Assume that (A), (F2) hold with p,κ as in (F2d) and let $\nu \in \mathfrak{M}$. Then

$$u \in \mathcal{D}, u = 0 \ \nu\text{–a.e.} \quad \Rightarrow \quad Lu = 0 \ \nu\text{–a.e.} \tag{4.3}$$

Proof. First we note that for $u, v \in \mathcal{D}$

$$L(uv) = uLv + vLu + 2\big(A^{1/2}Du, A^{1/2}Dv\big) \ . \tag{4.4}$$

So, if $u = 0$ ν–a.e., then by (4.2) and (4.4) with $v := u$ it follows that

$$0 \leqslant 2\int \big|A^{1/2}Du\big|_2^2\, d\nu = \int Lu^2\, d\nu \leqslant \lambda_\nu \int u^2\, d\nu = 0\ ,$$

so $A^{1/2}Du = 0$ ν–a.e. Applying (4.4) again with $v \in \mathcal{D}$ arbitrary we find, since $u \cdot (\pm 1) \cdot v = 0$ ν–a.e., that

$$\int (\pm 1)\cdot v\, Lu\, d\nu = \int L(uv)\, d\nu \leqslant \lambda_\nu \int uv\, d\nu = 0,$$

so,

$$\int v\, Lu\, d\nu = 0 \quad \text{for all } v \in \mathcal{D}.$$

Since $\mathcal{D}$ is closed under multiplication and generates $\mathcal{B}(X)$, by a monotone class argument it follows that $Lu = 0$ ν–a.e. □

Let $\nu \in \mathfrak{M}$. We define $\mathcal{D}^\nu$ to be the set of all ν–equivalence classes determined by $\mathcal{D}$. $\mathcal{D}_\kappa^\nu$ for $\kappa \in (0,\infty)$ is defined correspondigly. By Proposition 4.1 $\big(L, \mathcal{D}^\nu_{(\kappa_1)}\big)$ is a well-defined operator on $L^1(X,\nu)$. Our aim in this section is to prove that $(L, \mathcal{D}^\nu)$ is essentially maximal dissipative on $L^1(X,\nu)$.

Lemma 4.1. *Assume that (A) and (F2) hold with p,κ as in (F2d). Let $\kappa_1 \in (0, \kappa^* - \kappa]$ and $\nu \in \mathfrak{M}$. Then:*

(i) $Lu \in L^1(X,\nu)$ *for all* $u \in \mathcal{D}_{\kappa_1}$.
(ii) $(L - \lambda_\nu, \mathcal{D}^\nu)$ *is dissipative, hence* $(L, \mathcal{D}^\nu)$ *is closable on* $L^1(V,\nu)$.
(iii) *For the closure* $(\overline{L}_1^\nu, \mathcal{D}(\overline{L}_1^\nu))$ *it follows that* $\mathcal{D}_{\kappa_1}^\nu \subset D(\overline{L}_1^\nu)$ *and* $\overline{L}_1^\nu u$ *is given by formula (1.2). In particular,* (4.2) *holds for all* $u \in \mathcal{D}_{\kappa_1}$.

Proof.

(i): Since $\kappa^* > \kappa$, the assertion follows immediately by (4.1).

(ii): (L, D^ν) is a diffusion operator in the sense of Appendix B in [13]. So, the assertion follows by Appendix B, Lemma 1.8 in [13].

(iii): Let $u \in \mathcal{D}_{\kappa_1}$ and $N \in \mathbb{N}$ be such that $u = u \circ P_N$. Choose $\varphi \in C^\infty(\mathbb{R})$ such that $\varphi' \le 0$, $0 \le \varphi \le 1$, $\varphi = 1$ on $[0,1]$ and $\varphi = 0$ on $(2,\infty)$. For $n \in \mathbb{N}$ let $\varphi_n(x) := \varphi(\frac{|P_N x|_2^2}{n^2})$, $x \in X$, $u_n := \varphi_n u$. Then $u_n \in \mathcal{D}$ and

$$Lu_n = \varphi_n Lu + uL\varphi_n + 2(Du, A_N D\varphi_n).$$

Note that for $i,j = 1,\dots,N$ there are $c_j, c_{ij} \in (0,\infty)$ such that

$$|\partial_j \varphi_n| \le \frac{c_j}{n} 1_{\{|P_N x|_2 < 2n\}}, \; |\partial_{ij}^2 \varphi_n| \le \frac{c_{ij}}{n^2} 1_{\{|P_N x|_2 < 2n\}}.$$

Then $0 \le \varphi_n \uparrow 1$ as $n \to \infty$, $|A_N D\varphi_n| \le \frac{\max c_j}{n}$, and $|L\varphi_n(x)| \le \frac{c}{n}(|x'|_2 + |P_N F|_2) \le \frac{2c}{n}\Theta_{p,\kappa^*}(x)$ for all $x \in H_0^1$ and some $c \in (0,\infty)$ independent of x and n by (F2c) and (F2d). So $u_n \to u$ and $Lu_n \to Lu$ pointwise on H_0^1, up to a constant uniformly bounded by Θ_{p,κ^*}. Hence the assertion follows by (4.1) and Lebesgue's dominated convergence theorem. □

Theorem 4.1. *Assume that (A) and (F2) hold with* p, κ *as in (F2d) and let* λ_* *be as in Proposition 3.1. Let* $\nu \in \mathfrak{M}$. *Then:*

(i) *Let* $\lambda > \lambda_*$. *Then* $(\lambda - L)(\mathcal{D}^\nu)$ *is dense in* $L^1(X,\nu)$. *In particular, the closure* $(\overline{L}_1^\nu, \mathcal{D}(\overline{L}_1^\nu))$ *of* $(L, \mathcal{D}^\nu)$ *generates a* C_0*–(i.e. strongly continuous) semigroup* $(e^{t\overline{L}_1^\nu})_{t>0}$ *on* $L^1(X,\nu)$ *and* $(\overline{L}_1^\nu, \mathcal{D}(\overline{L}_1^\nu))$ *is the only closed extension of* $(L, \mathcal{D}^\nu)$ *with this property.*
(ii) $(e^{t\overline{L}_1^\nu})_{t>0}$ *is Markov.*
(iii) *Let* $(p_t)_{t>0}$ *be as in Theorem 3.1. Then for all* $t > 0$, $f \in \mathcal{B}^+(X)$

$$\int p_t f \, d\nu \le e^{\lambda_\nu t} \int f \, d\nu \tag{4.5}$$

and $p_t f$ *is a* ν*–version of* $e^{t\overline{L}_1^\nu} f$ *for all* $t > 0$, $f \in \mathcal{B}_b(X)$.

Proof.

(i) Let $u \in \mathcal{D}$. By Proposition 3.2(ii) we know that $G_\lambda^{(N)} u \in \mathcal{D}_{\kappa_1}$ for all $N \in \mathbb{N}$. Furthermore, Proposition 3.2(iii), (F2d), and (4.1) imply that $(\lambda - L)G_\lambda^{(N)} f \to f$ as $N \to \infty$ in $L^1(X, \nu)$ by Lebesgue's dominated convergence theorem. Since $\mathcal{D}$ is dense in $L^1(X, \nu)$, the first part of the assertion follows by Lemma 4.1(ii). The rest is then a consequence of the classical Lumer–Phillips theorem.

(ii) The assertion follows by (i) and Lemma 1.9 in Appendix B of [13].

(iii) Let $t > 0$. To prove (4.5), since $\mathcal{D}$ is dense in $L^1(X, \nu)$, we may assume that $f \in \mathcal{D}$, $f \geq 0$. Then for large enough $\lambda > 0$ and all $N \in \mathbb{N}$ by Proposition 3.2(ii) we have $G_\lambda^{(N)} f \in \mathcal{D}_{\kappa_1}$ and hence by Proposition 3.1 it follows that

$$\begin{aligned}\int \lambda g_{\lambda+\lambda_\nu}\left((\lambda + \lambda_\nu - L)G_\lambda^{(N)} f\right) d\nu &\\ = \lambda \int G_\lambda^{(N)} f \, d\nu &\\ \leq \int (\lambda + \lambda_\nu - L) G_\lambda^{(N)} f \, d\nu, &\end{aligned}$$

where the last step follows by the last part of Lemma 4.1(iii). Letting $N \to \infty$ we conclude by Proposition 3.2(iii), (F2d), (3.7), (4.1), and Lebesgue's dominated convergence theorem that

$$\int \lambda g_{\lambda+\lambda_\nu} f \, d\nu \leq \int f \, d\nu \, .$$

Hence by (3.6) and Fubini's theorem

$$\int_0^\infty e^{-\lambda t} \int e^{-\lambda_\nu t} p_t f \, d\nu \, dt \leq \int_0^\infty e^{-\lambda t} \int f \, d\nu \, dt$$

for all $\lambda > \lambda_0$ with $\lambda_0 \in (0, \infty)$ sufficiently large. So, for $g(t) := \int f \, d\nu - \int e^{-\lambda_\nu t} p_t \, f \, d\nu$ we have that

$$h(\lambda) := \int_0^\infty e^{-\lambda t} g(t) \, dt \geq 0 \text{ for all } \lambda > \lambda_0 \, .$$

Since h is obviously completely monotone, by the Hausdorff–Bernstein–Widder theorem (see e.g. Theorem 3.1 in the Appendix of [17]) it follows that $g \geq 0$ dt-a.e., hence $g \geq 0$ every where by right–continuity (cf. Theorem 3.1(i) above). Hence (4.5) is proved. In particular, p_t extends to a bounded linear operator on $L^1(X, \nu)$ with operator norm less than $e^{\lambda_\nu t}$

for all $t > 0$, which we denote by T_t. Clearly, $(T_t)_{t>0}$ is a semigroup of operators and (by the last part of Theorem 3.1(i)):

$$\lim_{t\to\infty} T_t f = f \quad \text{in } L^1(X,\nu)$$

if $f \in \mathcal{D}$. But then a 2ε-argument proves that $(T_t)_{t>0}$ is strongly continuous on $L^1(X,\nu)$. Let $\tilde{L}$ be the corresponding generator. We have to show that

$$T_t = e^{t\bar{L}_1^\nu} \quad \text{tor all } t > 0. \tag{4.6}$$

Let us consider the corresponding resolvents

$$G_\lambda^{(1)} f := \int e^{-\lambda t} T_t f \, dt,$$

and

$$G_\lambda^{(2)} f := \int e^{-\lambda t} e^{t\bar{L}_1^\nu} f \, dt = \left(\lambda - \bar{L}_1^\nu\right)^{-1} f$$

for $\lambda > \lambda_\nu$, $f \in L^1(X,\nu)$. It is easy to check (again using the uniqueness of the Laplace transform) that $g_\lambda f$ is a ν-version of $G_\lambda^{(1)} f$ for all $\lambda > \lambda_\nu$, $f \in L^1(X,\nu)$. Hence for all $u \in \mathcal{D}^\nu$ and λ large enough we have by Proposition 3.1

$$G_\lambda^{(1)}\big((\lambda - L)u\big) = u = G_\lambda^{(2)}\big((\lambda - L)u\big).$$

So, by continuity and since $(\lambda - L)(\mathcal{D}^\nu)$ is dense in $L^1(X,\nu)$ by assertion (i), it follows that $G_\lambda^{(1)} = G_\lambda^{(2)}$ for large enough λ. Hence (4.6) follows. □

Remark 4.1. Consider the situation of Theorem 4.1. By definition for $t \geqslant 0$ and $f \in D(\bar{L}_1^\nu)$ we have

$$\frac{d}{dt} e^{t\bar{L}_1^\nu} f = \bar{L}_1^\nu e^{t\bar{L}_1^\nu} f = e^{t\bar{L}_1^\nu} \bar{L}_1^\nu f, \tag{4.7}$$

where $\frac{d}{dt}$ is taken in the norm in $L^1(X,\nu)$. Since $p_t f$ is a ν-version of $e^{t\bar{L}_1^\nu} f$ by Theorem 4.1(iii), we see that (4.7) is an infinitesimal version of (3.1), however, only in an $L^1(X,\nu)$-sense, in particular, valid only outside a ν-zero-set of points in X.

5. A suitable reference measure

Let (A), (F2) hold with p, κ as in (F2d). In this section we shall construct a measure ν_0 in $\mathfrak{M}$ such that $\nu_0(U) > 0$ for every non-empty open set $U \subset X$.

Let $\{x_n \mid n \in \mathbb{N}\} \subset \bigcup_{N \in \mathbb{N}} P_N X$ be a dense subset of X. Let $\lambda_0 \in (0, \infty)$ bigger than the λ_* in Propositions 3.1 and 3.2. Set

$$\beta_n := \lambda_0 \big(2^n + \lambda_0 g_{\lambda_0} \Theta_{p,\kappa^*}(x_n)\big)^{-1}, \quad n \in \mathbb{N},$$

and define a measure ν_0 on X by

$$\nu_0(B) := Z^{-1} \sum_{n=1}^{\infty} \beta_n g_{\lambda_0}(x_n, B), \quad B \in \mathcal{B}(X), \tag{5.1}$$

where $Z \in (0, \infty)$ is chosen so that ν_0 is a probability measure.

Theorem 5.1. *Let (A), (F2) hold with p, κ as in (F2d) and let ν_0 be defined as in* (5.1)*. Then $\nu_0 \in \mathfrak{M}$ with $\lambda_{\nu_0} = \lambda_0$ and $\nu_0(U) > 0$ for every non-empty open set $U \subset X$.*

Proof. Let $u \in \mathcal{D}$, $u \geqslant 0$. Then by Proposition 3.1

$$\begin{aligned} \int (\lambda_0 - L)u \, d\nu_0 &= Z^{-1} \sum_{n=1}^{\infty} \beta_n g_{\lambda_0}\big((\lambda_0 - L)u\big)(x_n) \\ &= Z^{-1} \sum_{n=1}^{\infty} \beta_n u(x_n) \quad \geqslant \quad 0, \end{aligned}$$

and by construction

$$\int \Theta_{p,\kappa^*} \, d\nu_0 < \infty.$$

So, both (4.1) and (4.2) are satisfied, hence $\nu_0 \in \mathfrak{M}$. To show the last part of the assertion, let $\emptyset \neq U \subset X$, U open such that $\nu_0(U) = 0$. We may assume that U is a ball in X of radius $r > 0$ and centre y_0. Define

$$f(x) := \frac{\operatorname{dist}(x, U^c)}{r}$$

with $\operatorname{dist}(x, U^c) := \inf\{|x - y|_2 \mid y \in U^c\}$. Then f is a bounded Lipschitz function, hence $f \in Lip_{0,2,\kappa_1}$, $0 \leqslant f \leqslant 1$, and $f(y_0) = 1$. Then

$$0 = \nu_0(U) \geqslant \int f \, d\nu_0 \geqslant 0,$$

so for all $n \in \mathbb{N}$, $\lambda \geqslant \lambda_0$,

$$0 = g_{\lambda_0} f(x_n) \geqslant g_\lambda f(x_n) \geqslant 0.$$

Therefore, by Corollary 3.1

$$0 = \lim_{\lambda \to \infty} \lambda g_\lambda f(x_n) = f(x_n) \quad \text{for all } n \in \mathbb{N}.$$

Since $\{\kappa_n \mid n \in \mathbb{N}\}$ is dense in X, it follows that $f = 0$ on X, in contradiction to $f(y_0) = 1$. Hence such a ball U does not exist. □

Acknowledgement

Financial support of the BiBoS-Research Centre and the DFG-Research Group "Spectral Analysis, Asymptotic Distributions and Stochastic Dynamics" is gratefully acknowledged.

References

1. S. Albeverio and B. Ferrario, *Uniqueness results for the generators of the two-dimensional Euler and Navier-Stokes flows. The case of Gaussian invariant measures*, J. Funct. Anal. **193** (2002), no. 1, 77–93.
2. V. Barbu and G. Da Prato, *A phase field system perturbed by noise*, Nonlinear Anal. **51:6** (2002), 1087–1099.
3. V. Barbu, G. Da Prato and A. Debusche, *The transition semigroup of stochastic Navier-Stokes equations in 2-D*, Preprints di Matematica – n. 11, Scuola Normale Superiore, Pisa, 2002.
4. V. Bogachev and M. Röckner, *Elliptic equations for measures on infinite dimensional spaces and applications*, Probab. Theory Relat. Fields **120**(2001), 445–496.
5. S. Cerrai, *Second order PDE's in finite and infinite dimensions: a probabilistic approach*, Lect. Notes Math. **1762**, Springer, 2001.
6. G. Da Prato, *Elliptic operators with unbounded coefficients: construction of a maximal dissipative extension*, J. Evol. Equ. **1** (2001), no. 1, 1–18.
7. G. Da Prato and A. Debusche, *Maximal dissipativity of the Dirichlet operator corresponding to the Burgers equation*, Stochastic Processes, Physics and Geometry: New Interplays, vol. 1, AMS for the CMS, Providence (Rhode Island) 2000, pp. 85–98.
8. G. Da Prato and A. Debusche, *Ergodicity for the 3D stochastic Navier-Stokes equations*, Prépublication 03-01, Institut de Recherche Mathématique de Rennes, 2003.
9. G. Da Prato and M. Röckner, *Singular dissipative stochastic equations in Hilbert spaces*, Prob. th. Rel. Fields **124** (2002), pp. 261–303.
10. G. Da Prato and L. Tubaro, *Some results about dissipativity of Kolmogorov operators*, Czechoslovak Math. J. **51 (126)** (2001), no. 4, 685–699.
11. G. Da Prato and V. Vespri, *Maximal L^p regularity for elliptic equations with unbounded coefficients*, Nonlin. Anal. **49** (2002), 747–755.
12. G. Da Prato and J. Zabczyk, *Stochastic equations in infinite dimensions*, Encyclopedia of Mathematics and its Applications, vol. 44, Cambridge University Press, Cambridge, 1992.
13. A. Eberle *Uniqueness and non-uniqueness of semigroups generated by singular diffusion operators*, Lecture Notes in Mathematics, 1718, Springer-Verlag, Berlin, 1999.

14. F. FLANDOLI AND D. GĄTAREK, *Martingale and stationary solutions for stochastic Navier-Stokes equations*, Prob. Th. Rel. Fields **102** (1995), no. 3, 367–391.
15. F. FLANDOLI AND F. GOZZI, *Kolmogorov equation associated to a stochastic Navier-Stokes equation*, J. Funct. Anal. **160** (1998), 312–336.
16. H. LONG AND I. SIMÃO, *Kolmogorov equations in Hilbert spaces with application to essential self-adjointness of symmetric diffusion operators*, Osaka J. Math. **37** (2000), no. 1, 185–202.
17. Z.-M. MA AND M. RÖCKNER, *Introduction to the theory of (non-symmetric) Dirichlet forms*, Springer-Verlag, Berlin–Heidelberg–New York–Tokyo, 1992.
18. D. NUALART AND F. VIENS, *Evolution equation of a stochastic semigroup with white-noise drift*, Ann. Prob. **28** (2000), no. 1, 36–73.
19. M. RÖCKNER, *L^p-analysis of finite and infinite dimensional diffusion operators*, Stochastic PDE's and Kolmogorov's equations in infinite dimensions (G. Da Prato, ed.), Lect. Notes Math. **1715**, Springer, Berlin, 1999, pp. 65–116.
20. M. RÖCKNER AND S. SOBOL, *A new analytic approach to the Kolmogorov equations of the stochastic 2D-Navier-Stokes equations*, Preprint 2003.
21. M. RÖCKNER AND S. SOBOL, *Kolmogorov Equations in Infinite Dimensions: Well-Posedness and Regularity of Solutions, with Applications to Stochastic Generalized Burgers Equations*, Preprint 2003.
22. I. V. SADOVNICHAYA, *The direct and the inverse Kolmogorov equation for the stochastic Schrödinger equation*, Vestnik Moskov. Univ. Ser. I Mat. Mekh. **2000** no. 6, 15–20, 86; translation in Moscow Univ. Math. Bull. **55** (2000) no. 6, 15–19.
23. W. STANNAT, *The theory of generalized Dirichlet forms and its applications in analysis and stochastics*, Memoirs of the Am. Math. Soc. **142 (678)** (1999), pp. 1–101.
24. W. STANNAT, *(Nonsymmetric) Dirichlet operators on L^1: Existence, uniqueness and associated Markov processes*, Ann. Scuola Norm. Sup. Pisa Cl. Sci., Serie IV **28** (1999), pp. 99–140.
25. D. W. STROOCK AND S. R. S. VARADHAN, *Multidimensional Diffusion Processes*, Springer, 1979.

FROM CLASSICAL TO QUANTUM QUADRATIC COST CONTROL

LUIGI ACCARDI AND ANDREAS BOUKAS

ABSTRACT. Controlling the size of the solution of a (deterministic, stochastic or quantum stochastic) differential equation, by minimizing an appropriate cost functional, is very important in classical and quantum engineering. Of particular importance is the case of linear differential equations and quadratic cost functionals, since in that case the control processes can be explicitly calculated. In this paper we review some basic aspects of the classical theory and we present our results in the quantum case, obtained over the past few years.

1. Classical Linear Control

1.1. **Deterministic Control.** In the classical deterministic case we consider a system whose evolution over a finite time interval is modelled by the solution $x = \{x_t : t \in [0,T]\} \in C([0,T], \mathbb{R}^n)\}$ of an ordinary differential equation of the form [38]

$$dx_t = (A\,x_t + u_t)\,dt \tag{1.1}$$

$$x_0 = x, \; t \in [0,T] \tag{1.2}$$

where $A \in \mathcal{B}(\mathbb{R}^n)$, the space of bounded linear operators on $\mathbb{R}^n$, and $u \in L_\infty([0,T], \mathbb{R}^n)$ or $L_2([0,T], \mathbb{R}^n)$. Although we consider here the finite-dimensional case, the concepts and the results can be extended from $\mathbb{R}^n$ to any Hilbert space $\mathcal{H}$.

We assume that we can interfere with the performance of the model by choosing the "control process" $u = \{u_t : t \in [0,T]\}$ so as to minimize a certain "performance (or cost) functional" $J(u)$. There is a wide variety of such functionals designed for specific models. However, the most computationally accessible one is the "quadratic" performance functional of the form

$$J(u) = \textstyle\int_0^T (< x_t, Q\,x_t > + < u_t, u_t >)\,dt + < x_T, \Pi\,x_T > \tag{1.3}$$

where $< \cdot, \cdot >$ denotes the usual inner product in the Euclidean space $\mathbb{R}^n$, $\Pi \in \mathcal{B}(\mathbb{R}^n)$, $Q \in \mathcal{B}(\mathbb{R}^n)$, $\Pi \geq 0$, $Q \geq 0$. If the size of $x = \{x_t : t \in [0,T]\}$ is small, the performance functional (1.3) can serve as an approximation to many other functionals which are more adapted to the specific problem considered but also more computationaly complex. Before

---

1991 *Mathematics Subject Classification.* Primary 81S25 ; Secondary 93E20.

The second author express his gratitude to Professor Luigi Accardi for his support and guidance as well as for the hospitality of the Centro Vito Volterra of the Universita di Roma TorVergata on several occasions over the past 15 years.

one looks for the optimal control process, the system to be controlled must be "observable", "controllable", and "stabilizable". The definition of these concepts is as follows [35, 36, 38]:

Observability: Since the state x_t of the system may only be accessible through an observation process $y_t = P\,x_t$, where $P \in \mathcal{B}(\mathbb{R}^n)$, we must be able to recreate x_t (or equivalently x_0) from y_t. If $P = 1$ we speak of a "completely observable" system. Otherwise the system is only "partially observable".

Controllability: Given an initial state x_0, we should be able to choose the control process $u = \{u_t : t \in [0,T]\}$ so that the system will be steered in a finite time $t_1 \in [0,T]$ to a desired state x_1.

Stabilizability: In order to consider large terminal times T, we need the system to exhibit good long-run behavior i.e to eventually settle down to some steady-state behavior. From the mathematical point of view, this amounts to the asymptotic stability of the initial state of $x = \{x_t : t \in [0,T]\}$ or, equivalently, to the existence of a "feedback" control $u_t = K\,x_t$, where $K \in \mathcal{B}(\mathbb{R}^n)$, for which the system (1.1)-(1.2) is asymptotically stable.

The performance functional (1.3) is particularly useful in the case when a system must operate at or near a particular state, chosen here to be the origin. We can think of the first term of (1.3) as a penalty for being too far away from the origin on $(0,T)$, the second as a penalty for using too much control and the third as a penalty for being too far away from the target at the final time T. The main result in the completely observable, classical case is the following:

Theorem 1. *The performance functional (1.3) associated with the system (1.1)-(1.2) is minimized by the feedback control process*

$$u_t = -\Pi_t\,x_t \tag{1.4}$$

where $\{\Pi_t : t \in [0,T]\}$ *is the solution of the Riccati differential equation*

$$\frac{d}{dt}\Pi_t + A^*\,\Pi_t + \Pi_t\,A + Q - \Pi_t^2 = 0 \tag{1.5}$$

$$\Pi_T = \Pi. \tag{1.6}$$

If we restrict to $u_t = -K\,x_t$, i.e to feedback controls with a time-independent coefficient, then equations (1.5)-(1.6) are replaced by the "algebraic" Riccati equation

$$A^*\,\Pi + \Pi\,A + Q - \Pi^2 = 0. \tag{1.7}$$

1.2. **Stochastic Control.** In this case we consider systems whose time evolution is affected by noise. We assume that the noise can be accurately described by Brownian motion. Specifically, we consider systems whose time-evolution is described by the solution $x = \{x_t : t \in [0,T]\}$ of a stochastic differential equation of the form

$$dx_t = (A\,x_t + u_t)\,dt + C\,dB_t \tag{1.8}$$
$$x_0 = x,\ t \in [0,T] \tag{1.9}$$

where A and u are as in (1.1)-(1.2) with the added assumtion that u is a stochastic process, $C \in \mathcal{B}(\mathbb{R}^n)$, and $B = \{B_t : t \geq 0\}$ is a vector (in this case n-dimensional) Brownian motion. The performance functional (1.3) takes the form

$$J(u) = E(\int_0^T (< x_t, Q\,x_t > + < u_t, u_t >)\,dt + < x_T, \Pi\,x_T >) \tag{1.10}$$

where E denotes mathematical expectation.

For completely observable systems, Theorem 1 remains true in the stochastic case. For partially observable systems, i.e when $x = \{x_t : t \in [0,T]\}$ is available only through an observation process $y = \{y_t : t \in [0,T]\}$ satisfying

$$dy_t = H\,x_t\,dt + dW_t \tag{1.11}$$

where $H \in \mathcal{B}(\mathbb{R}^n)$, and $W = \{W_t : t \in [0,T]\}$ is a vector (in this case n-dimensional) Brownian motion independent of $B = \{B_t : t \in [0,T]\}$, the main result is as follows:

Theorem 2. *The performance functional (1.10) associated with the system (1.8)-(1.9) and (1.11) is minimized by the feedback control process*

$$u_t = -\Pi_t\,\hat{x}_t \tag{1.12}$$

where $\{\Pi_t : t \in [0,T]\}$ *is the solution of the Riccati equation*

$$\tfrac{d}{dt}\Pi_t + A^*\,\Pi_t + \Pi_t\,A + Q - \Pi_t^2 = 0 \tag{1.13}$$
$$\Pi_T = \Pi \tag{1.14}$$

and $\hat{x} = \{\hat{x}_t : t \in [0,T]\}$ *is the minimum mean-square estimate of* $x = \{x_t : t \in [0,T]\}$ *given* $\{y_s : s \leq t\}$, *obtained through the Bucy-Kalman filter* [36, 38, 39].

The field of classical deterministic and stochastic control is very well developed. For proofs, details and more information we refer to [34, 35, 36, 38, 39].

2. Quantum Evolutions and Langevin Equations

In the Schrödinger picture of quantum mechanics, the state of a quantum system is described (in Dirac's notation) by a time dependent ket vector $|\psi_t\rangle$ evolving in accordance to Schrödinger's equation

$$i\,\hbar\,\tfrac{d}{dt}\,|\psi_t\rangle = H_t\,|\psi_t\rangle \tag{2.1}$$

where H_t is the Hamiltonian of the system. Quantum mechanical observables are represented by time independent self adjoint operators X_0 acting on the state space. In the Heisenberg picture, the state of the system is represented by a time independent state vector $|\psi_0\rangle$ and it is the observables X_t that vary with time. The connection between the two pictures is provided by the unitary "time evolution operators" U_t that satisfy

$$X_t = j_t(X_0) = U_t^* \, X_0 \, U_t \tag{2.2}$$

and

$$|\psi_t\rangle = U_t \, |\psi_0\rangle. \tag{2.3}$$

In the conservative (i.e time independent Hamiltonian) case

$$U_t = e^{-\frac{i}{\hbar} t H}. \tag{2.4}$$

The differential equation satisfied by the U_t's is called an "evolution" equation, while the one satisfied by the X_t's is called a Langevin equation. In general, the Hamiltonian operator is a sum of "creation", "annihilation" and "number" operators and the above mentioned equations contain noise terms defined in terms of such operators. In what follows we will consider systems and noises represented as operators acting on a Hilbert space $\mathcal{H}$ and Boson Fock space $\Gamma(L^2(\mathbb{R}_+, \mathcal{K}))$ respectively. The Boson Fock space $\Gamma = \Gamma(L^2(\mathbb{R}_+, \mathcal{K}))$ can be defined as the Hilbert space completion of the linear span of the "exponential vectors" $\psi(f)$ under the inner product

$$< \psi(f), \psi(g) >= e^{<f,g>} \tag{2.5}$$

where $f, g \in L^2(\mathbb{R}_+, \mathcal{K})$. The set $\mathcal{K}$ will be $\mathbb{C}$ or $l_2(\mathbb{N})$ depending on what kind of noise we consider. Noise will be defined as time dependent operators on the Fock space $\Gamma(L^2(\mathbb{R}_+, \mathcal{K}))$ with differentials defined in terms of the Hida white noise functionals $a_t^\dagger$ and a_t.

For $f \in L^2(\mathbb{R}_+, \mathcal{K})$ and an adjointable linear operator F on $L^2(\mathbb{R}_+, \mathcal{K})$, the "annihilation", "creation" and "conservation" operators $A(f)$, $A^\dagger(f)$ and $\Lambda(F)$ respectively, are defined on the exponential vectors of Γ by

$$A(f)\psi(g) =< f, g > \psi(g) \tag{2.6}$$

$$A^\dagger(f)\psi(g) = \tfrac{\partial}{\partial \epsilon}|_{\epsilon=0}\psi(g + \epsilon f) \tag{2.7}$$

$$\Lambda(F)\psi(g) = \tfrac{\partial}{\partial \epsilon}|_{\epsilon=0}\psi(e^{\epsilon F} g) \tag{2.8}$$

where F must be such that the exponential $e^{\epsilon F}$ is defined.

The Itô multiplication table associated with $A(\cdot)$, $A^\dagger(\cdot)$ and $\Lambda(\cdot)$ is

$\cdot$	$dA_t^\dagger(f_1)$	$d\Lambda_t(F_1)$	$dA_t(f_1)$	dt
$dA_t^\dagger(f_2)$	0	0	0	0
$d\Lambda_t(F_2)$	$dA_t^\dagger(F_2 f_1)$	$d\Lambda_t(F_2 F_1)$	0	0
$dA_t(f_2)$	$< f_2, f_1 > dt$	$dA_t(F_1^* f_2)$	0	0
dt	0	0	0	0.

2.1. **First Order White Noise.** The first order (Hudson-Parthasarathy) quantum stochastic differentials dA_t, $dA_t^\dagger$, and $d\Lambda_t$ are defined by

$$dA_t = A(\chi_{[t,t+dt]}) \tag{2.9}$$

$$dA_t^\dagger = A^\dagger(\chi_{[t,t+dt]}) \tag{2.10}$$

$$d\Lambda_t = \Lambda(\chi_{[t,t+dt]}). \tag{2.11}$$

In terms of white noise, the basic noise differentials are

$$dA_t = a_t\, dt,\ dA_t^\dagger = a_t^\dagger\, dt,\ d\Lambda_t = a_t^\dagger\, a_t\, dt. \tag{2.12}$$

Quantum evolutions in the tensor product $\mathcal{H} \otimes \Gamma(L^2(\mathbb{R}_+, \mathcal{K}))$ have the form

$$dU_t = -((iH + \tfrac{1}{2}\, L^* L)\, dt + L^*\, W\, dA_t - L\, dA_t^\dagger + (1 - W)\, d\Lambda_t)\, U_t \tag{2.13}$$

$$t \in [0, T],\ U_0 = 1 \tag{2.14}$$

where H, L, W are in $\mathcal{B}(\mathcal{H})$, with U unitary and H self-adjoint. The corresponding Langevin equation is

$$dj_t(X) = j_t(i[H, X] - \tfrac{1}{2}(L^* L X + X L^* L - 2 L^* X L))\, dt \tag{2.15}$$
$$+ j_t([L^*, X]\, W)\, dA_t + j_t(W^*\, [X, L])\, dA_t^\dagger + j_t(W^*\, X\, W - X)\, d\Lambda_t$$

$$j_0(X) = X,\ t \in [0, T]. \tag{2.16}$$

2.2. **Square of White Noise.** The square of white noise (SWN) commutation relations are a functional extension of the $sl(2;\mathbb{R})$ commutation relations

$$[B^-, B^+] = M,\ [M, B^+] = 2B^+,\ [M, B^-] = -2B^- \tag{2.17}$$

where

$$(B^-)^* = B^+,\ M^* = M. \tag{2.18}$$

Following "renormalization", the SWN noise differentials are initially defined by

$$dB_t^- = a_t^2\, dt,\ dB_t^+ = {a_t^\dagger}^2\, dt,\ dM_t = a_t^\dagger\, a_t\, dt. \tag{2.19}$$

A representation of the $sl(2;\mathbb{R})$ Lie algebra on $l_2(\mathbb{N})$ is defined by

$$\rho^+(B^{+^n} M^k B^{-^l})\, e_m = \theta_{n,k,l,m}\, e_{n+m-l}. \tag{2.20}$$

where e_m, $m = 0, 1, 2, \cdots$ is any orthonormal basis of $l_2(\mathbb{N})$,

$$\theta_{n,k,l,m} := H(n+m-l)\sqrt{\frac{m-l+n+1}{m+1}}\, 2^k (m-l+1)_n (m+1)^{(l)} (m-l+1)^k, \tag{2.21}$$

$H(x)$ is the Heaviside function ($H(x) = 0$ for $x < 0$; $H(x) = 1$ for $x \geq 0$),

$$0^0 = 1, \ \ (B^+)^n = (B^-)^n = N^n = 0, \quad \text{for } n < 0,$$

and "factorial powers" are defined by

$$\begin{aligned} x^{(n)} &= x(x-1)\cdots(x-n+1) \\ (x)_n &= x(x+1)\cdots(x+n-1) \\ (x)_0 &= x^{(0)} = 1. \end{aligned}$$

Using this representation we obtain

$$dM_t = d\Lambda_t(\rho^+(M)) + dt \tag{2.22}$$

$$dB_t^+ = d\Lambda_t(\rho^+(B^+)) + dA_t^\dagger(e_0) \tag{2.23}$$

$$dB_t^- = d\Lambda_t(\rho^+(B^-)) + dA_t(e_0). \tag{2.24}$$

To obtain a closed Itô multiplication table we use as basic SWN differentials

$$d\Lambda_{n,k,l}(t) = d\Lambda_t(\rho^+(B^{+^n} M^k B^{-^l})) \tag{2.25}$$

$$dA_m(t) = dA_t(e_m) \tag{2.26}$$

$$dA_m^\dagger(t) = dA_t^\dagger(e_m). \tag{2.27}$$

where $n, k, l, m \in \{0, 1, ...\}$, with Itô multiplication table

$$d\Lambda_{\alpha,\beta,\gamma}(t)\, d\Lambda_{a,b,c}(t) = \sum c_{\beta,\gamma,a,b}^{\lambda,\rho,\sigma,\omega,\epsilon}\, d\Lambda_{a+\alpha-\gamma+\lambda,\omega+\sigma+\epsilon,\lambda+c}(t) \tag{2.28}$$

$$d\Lambda_{\alpha,\beta,\gamma}(t)\, dA_n^\dagger(t) = \theta_{\alpha,\beta,\gamma,n}\, dA_{\alpha+n-\gamma}^\dagger(t) \tag{2.29}$$

$$dA_m(t)\, d\Lambda_{a,b,c}(t) = \theta_{c,b,a,m}\, dA_{c+m-a}(t) \tag{2.30}$$

$$dA_m(t)\, dA_n^\dagger(t) = \delta_{m,n}\, dt \tag{2.31}$$

where

$$c^{\lambda,\rho,\sigma,\omega,\epsilon}_{\beta,\gamma,a,b} = \binom{\gamma}{\lambda}\binom{\gamma-\lambda}{\rho}\binom{\beta}{\omega}\binom{b}{\epsilon}2^{\beta+b-\omega-\epsilon}S_{\gamma-\lambda-\rho,\sigma}a^{(\gamma-\lambda)}(a+\lambda-1)^{(\rho)}(a-\gamma+\lambda)^{\beta-\omega}\lambda^{b-\epsilon}, \tag{2.32}$$

$S_{\gamma-\lambda-\rho,\sigma}$ are the Stirling numbers of the first kind and $\sum$ in (2.28) denotes the finite sum

$$\sum_{\lambda=0}^{\gamma}\sum_{\rho=0}^{\gamma-\lambda}\sum_{\sigma=0}^{\gamma-\lambda-\rho}\sum_{\omega=0}^{\beta}\sum_{\epsilon=0}^{b}.$$

All other products of differentials are equal to zero.
Quantum evolutions are of the form

$$dU_t = ((-\tfrac{1}{2}(D_-^*|D_-^*) + iH)\,dt + d\mathcal{A}_t(D_-) + d\mathcal{A}_t^\dagger(-r(W)D_-^*) + d\mathcal{L}_t(W-I))\,U_t \tag{2.33}$$

$$U_0 = 1 \tag{2.34}$$

while Langevin equations are of the form

$$\begin{aligned} dj_t(X) = {} & j_t(i\,[X,H] - \tfrac{1}{2}\{(D_-^*|D_-^*)\,X\} + (r(W)D_-^*|X\,r(W)D_-^*))\,dt \\ & + j_t(d\mathcal{A}_t^\dagger(D_-^*\,X - r(W^*\,X\circ W)D_-^*)) \\ & + j_t(d\mathcal{A}_t(X\,D_-l(W^*\circ X\,W)D_-)) \\ & + j_t(d\mathcal{L}_t(W^*\,X\circ W - X)) \end{aligned} \tag{2.35}$$

$$j_0(X) = X,\ t\in[0,T]. \tag{2.36}$$

where H is a bounded self-adjoint system operator, W is a $\circ$-product (see (2.44) for the definition of the $\circ$-product) unitary operator and $D_- = \sum_m D_{-,m}\otimes e_m$, where the D_m's are bounded system operators.

In equations (2.33)-(2.34) and (2.35)-(2.36) we have used

(i) evolution coefficients:

$$D_+ = \textstyle\sum_n D_{+,n}\otimes e_n \tag{2.37}$$

$$D_- = \textstyle\sum_m D_{-,m}\otimes e_m \tag{2.38}$$

$$D_1 = \textstyle\sum_{\alpha,\beta,\gamma} D_{1,\alpha,\beta,\gamma}\otimes \rho^+(B^{+\alpha}M^\beta B^{-\gamma}) \tag{2.39}$$

$$E_1 = \textstyle\sum_{a,b,c} E_{1,a,b,c}\otimes \rho^+(B^{+a}M^b B^{-c}) \tag{2.40}$$

where the left hand sides of the tensor products corespond to bounded system operators

(ii) module operators $\mathcal{A}$, $\mathcal{A}^\dagger$ and $\mathcal{L}$ generically defined by:

$$\mathcal{A}(a \otimes \xi) = a \otimes A(\xi) \tag{2.41}$$

$$\mathcal{A}^{\dagger}(a \otimes \xi) = a \otimes A^{\dagger}(\xi) \tag{2.42}$$

$$\mathcal{L}(a \otimes T) = a \otimes \Lambda(T) \tag{2.43}$$

and

(iii) basic operations:

$$D_1 \circ E_1 = \sum\nolimits_{\alpha,\beta,\gamma,a,b,c} \sum c^{\lambda,\rho,\sigma,\omega,\epsilon}_{\beta,\gamma,a,b} D_{1,\alpha,\beta,\gamma} E_{1,a,b,c} \otimes \rho^{+}(B^{+a+\alpha-\gamma+\lambda} M^{\omega+\sigma+\epsilon} B^{-\lambda+c}) \tag{2.44}$$

$$(D_{-}^{*} | D_{+}) = \sum\nolimits_{n} D_{-,n}\, D_{+,n} \otimes 1 \tag{2.45}$$

$$r(D_1) D_{+} = \sum\nolimits_{n,\alpha,\beta,\gamma} D_{1,\alpha,\beta,\gamma} \theta_{\alpha,\beta,\gamma,n-\alpha+\gamma} D_{+,n-\alpha+\gamma} \otimes e_n \tag{2.46}$$

$$l(E_1) D_{-} = \sum\nolimits_{n,\alpha,\beta,\gamma} D_{,n+\alpha-\gamma} \theta_{\gamma,\beta,\alpha,n+\alpha-\gamma} E_{1,\alpha,\beta,\gamma} \otimes e_n \tag{2.47}$$

where $\sum$ is as in (2.28). The SWN Ito table can be concisely written as

$$d\mathcal{A}_t(D_{-})\, d\mathcal{A}_t^{\dagger}(D_{+}) = (D_{-}^{*} | D_{+})\, dt \tag{2.48}$$

$$d\mathcal{L}_t(D_1)\, d\mathcal{L}_t(E_1) = d\mathcal{L}_t(D_1 \circ E_1) \tag{2.49}$$

$$d\mathcal{L}_t(D_1)\, d\mathcal{A}_t^{\dagger}(D_{+}) = d\mathcal{A}_t^{\dagger}(r(D_1) D_{+}) \tag{2.50}$$

$$d\mathcal{A}_t(D_{-})\, d\mathcal{L}_t(E_1) = d\mathcal{A}_t(l(E_1) D_{-}). \tag{2.51}$$

For details we refer to [2]-[10], [12], [15]-[24],[26]-[28], [33], [37], [39].

3. Quantum Control

3.1. **First order white noise.** We consider Langevin equations of the form (2.15)-(2.16) associated with evolution equations of the form (2.13)-(2.14). Generalizing from the clasical case, we consider quadratic cost functionals of the form

$$J_{\xi,T}(L, W) = \int_0^T [\, \| j_t(X)\, \xi \|^2 + \tfrac{1}{4} \| j_t(L^* L)\, \xi \|^2 \,]\, dt + \tfrac{1}{2} \| j_T(L)\, \xi \|^2 \tag{3.1}$$

where T is an arbitrary terminal time, $\xi = u \otimes \psi(f) \in \mathcal{H} \otimes \Gamma(L^2(\mathbb{R}_+, \mathcal{K}))$ is arbitrary, and the system operators L, W are viewed as controls, chosen to minimize $J_{\xi,T}(L, W)$. This cost functional is obtained from the cost functional

$$Q_{\xi,T}(u) = \int_0^T [< U_t\, \xi, X^2\, U_t\, \xi > + < u_t\, \xi, u_t\, \xi >]\, dt - < u_T\, \xi, U_T\, \xi > \tag{3.2}$$

associated with (2.13)-(2.14) as in the classical case, by restricting to feedback controls $u_t = -\frac{1}{2} L^* L\, U_t$. The main result is as follows:

Theorem 3. *Let X be a system space observable such that the pair $(i\,H,\ X)$ is stabilizable (i.e $\exists\ K \in \mathcal{B}(\mathcal{H})$ such that $i\,H + KX$ is the generator of an asymptotically stable semigroup $\mathcal{F}_t$ i.e $\exists\ M > 0$ and $\omega < 0$ such that $||\mathcal{F}_t|| \leq M\, e^{\omega t}$). The quadratic performance functional $J_{\xi,T}(L, W)$ is minimized by*

$$L = \sqrt{2}\,\Pi^{1/2}\,W_1 \ \ (\textit{polar decomposition of } L) \tag{3.3}$$

and

$$W = W_2 \tag{3.4}$$

where Π is the positive self-adjoint solution of the algebraic Riccati equation

$$i\,[H, \Pi] + \Pi^2 + X^2 = 0 \tag{3.5}$$

and W_1, W_2 are bounded unitary system operators commuting with Π. Moreover

$$\min_{L,W}\ J_{\xi,T}(L, W) =< \xi, \Pi\,\xi > \tag{3.6}$$

independent of T.

Some early work on the quadratic cost control of quantum processes was done by V.P. Belavkin [24] who considered evolutions with scalar coefficients, and classical noise without jumps i.e W was 1, and the coefficents of dA_t and $dA_t^\dagger$ were equal, thus amounting to stochastic differential equations driven by classical Brownian motion $A_t + A_t^\dagger$.

3.2. **Square of white noise.** In this case, corresponding to (2.33)-(2.34) and (2.35)-(2.36), we consider the cost functional

$$J_{\xi,T}(D_-, W) = \int_0^T [\, \|j_t(X)\,\xi\|^2 + \tfrac{1}{4}\|j_t((D_-^*|D_-^*))\,\xi\|^2\,]\,dt + \tfrac{1}{2} < \xi, j_T((D_-^*|D_-^*))\,\xi > \tag{3.7}$$

where D_-, W are the controls, to be chosen so as to minimize $J_{\xi,T}(D_-, W)$. The main result has as follows:

Theorem 4. *Let X be a bounded self-adjoint system operator such that the pair $(i\,H,\ X)$ is stabilizable. The performance functional $J_{\xi,T}(D_-, W)$ is minimized by choosing*

$$D_- = \sum_n D_{-,n} \otimes e_n \tag{3.8}$$

and

$$W = \sum_{\alpha,\beta,\gamma} W_{\alpha,\beta,\gamma} \otimes \rho^+(B^{+\alpha} M^\beta B^{-\gamma}) \tag{3.9}$$

such that

$$\frac{1}{2}\,(D_-^*|D_-^*) = (\tfrac{1}{2}\sum_n D_{-,n}\,D_{-,n}^*) \otimes 1 = \Pi, \tag{3.10}$$

and

$$[D_{-,n}, D_{-,m}] = [D_{-,n}, D_{-,m}^*] = 0 \tag{3.11}$$

$$[D_{-,n}, W_{\alpha,\beta,\gamma}] = [D_{-,n}, W_{\alpha,\beta,\gamma}^*] = 0 \tag{3.12}$$

for all $n, m, \alpha, \beta, \gamma$, where Π is the positive self-adjoint solution of the algebraic Riccati equation

$$i\,[H, \Pi] + \Pi^2 + X^2 = 0. \tag{3.13}$$

Moreover

$$\min_{D_-,W}\; J_{\xi,T}(D_-, W) =< \xi, \Pi\,\xi > \tag{3.14}$$

independent of T.

For proofs and details on the material presented in this section we refer to [1, 3, 11, 13, 14, 25, 29, 30, 31, 32].

4. Appendix: *Mathematica* algorithms

The calculation of the product of two stochastic differentials, in the case of the square of white noise, can be made easy with the use of symbolic programming. Following is a *Mathematica* algorithm for computing (2.28)-(2.31). The algorithm also computes the product of the Hudson-Parthasarathy differentials (2.9)-(2.11) as a special case (rf. [7]).

Algorithm 1. *This algorithm computes the, in general, noncommutative products of the generalized SWN stochastic differentials $d\Lambda_{n,k,l}(t)$, $dA_m(t)$ and $dA_m^\dagger(t)$, where $n, k, l, m = 0, 1, ...$, and "time" dt. Each sentence corresponds to a new input. Inputs are separated by space.*

$p[x\_, y\_] =$ **If** $[x == y == 0, 1, x\hat{\ }y]$

$u[x\_, n\_] =$ **Product**$[x - i + 1, \{i, 1, n\}]$

$v[x\_, n\_] =$ **Product**$[x + i - 1, \{i, 1, n\}]$

$\theta[n\_, k\_, l\_, m\_] =$ **If** $[n + m - 1 <$
$0, 0, Sqrt[(m - l + n + 1)/(m + 1)]2\hat{\ }k\, v[m - l + 1, n]\, u[m + 1, l]\, p[m - l + 1, k]]$

$c[\beta\_, \gamma\_, a\_, b\_, \lambda\_, \rho\_, \sigma\_, \omega\_, \epsilon\_] =$
Binomial$[\gamma, \lambda]$ **Binomial**$[\gamma - \lambda, \rho]$ **Binomial**$[\beta, \omega]$ **Binomial**$[b, \epsilon]\; 2\hat{\ }(\beta + b - \omega -$
$\epsilon)$ **StirlingS1**$[\gamma - \lambda - \rho, \sigma]\; u[a, \gamma - \lambda]\; u[a + \lambda - 1, \rho]\; p[a - \gamma + \lambda, \beta - \omega]\; p[\lambda, b - \epsilon]$

$\mathbf{NCM}[d\Lambda[\alpha\_, \beta\_, \gamma\_], d\Lambda[a\_, b\_, s\_]] = \mathbf{Sum}[c[\beta, \gamma, a, b, \lambda, \rho, \sigma, \omega, \epsilon]\, d\Lambda[a + \alpha - \gamma + \lambda, \omega + \sigma + \epsilon, \lambda + s], \{\lambda, 0, \gamma\}, \{\rho, 0, \gamma - \lambda\}, \{\sigma, 0, \gamma - \lambda - \rho\}, \{\omega, 0, \beta\}, \{\epsilon, 0, b\}]$

$\mathbf{NCM}[d\Lambda[a\_, b\_, c\_], dA^\dagger[m\_]] = \theta[a, b, c, m] dA^\dagger[a + m - c]$

$\mathbf{NCM}[dA[m\_], d\Lambda[a\_, b\_, c\_]] = \theta[c, b, a, m] dA[c + m - a]$

$\mathbf{NCM}[dA[m\_], dA^\dagger[n\_]] = \mathbf{KroneckerDelta}[m, n] dt$

$\mathbf{NCM}[dA[m\_], dA[n\_]] = 0$

$\mathbf{NCM}[dA^\dagger[m\_], dA^\dagger[n\_]] = 0$

$\mathbf{NCM}[dA^\dagger[m\_], dA[n\_]] = 0$

$\mathbf{NCM}[dA^\dagger[m\_], d\Lambda[\alpha\_, \beta\_, \gamma\_]] = 0$

$\mathbf{NCM}[d\Lambda[\alpha\_, \beta\_, \gamma\_], dA[m\_]] = 0$

$\mathbf{NCM}[d\Lambda[\alpha\_, \beta\_, \gamma\_], dt] = 0$

$\mathbf{NCM}[dt, d\Lambda[\alpha\_, \beta\_, \gamma\_]] = 0$

$\mathbf{NCM}[dA[m\_], dt] = 0$

$\mathbf{NCM}[dt, dA[m\_]] = 0$

$\mathbf{NCM}[dA^\dagger[m\_], dt] = 0$

$\mathbf{NCM}[dt, dA^\dagger[m\_]] = 0$

$\mathbf{NCM}[dt, dt] = 0$

For example, using the above algorithm to compute $d\Lambda_{4,1,2}(t)\, d\Lambda_{1,2,1}(t)$ we obtain

$\mathrm{NCM}[d\Lambda[4,1,2], d\Lambda[1,2,1]] = 8d\Lambda[4,1,2] + 16d\Lambda[4,2,2] + 10d\Lambda[4,3,2] + 2d\Lambda[4,4,2] + 32d\Lambda[5,0,3] + 32d\Lambda[5,1,3] + 10d\Lambda[5,2,3] + d\Lambda[5,3,3]$

while for $d\Lambda_{4,2,1}(t)\, dA_2^\dagger(t)$ we obtain

$\mathrm{NCM}[d\Lambda[4,2,1], dA^\dagger[2]] = 48\sqrt{5}\, dA^\dagger[5]$

References

[1] L. Accardi, *Quantum Kalman filtering*, Mathematical system theory. The influence of R.E Kalman, Springer 135-143, 1991.

[2] ———, *A mathematical theory of quantum noise*, Proceedings of the first world congress of the Bernoulli Society, Ed. Prohorov and Sazonov, vol.1 (1987)

[3] ———, *Quantum Probability: an introduction to some basic ideas and trends*, Course given at the VI Symposium de probabilidad y Procesos Estocasticas, CIMAT, Guanajuato, 24–27 May 2000, Mexico

[4] L. Accardi, A. Boukas, *Unitarity conditions for stochastic differential equations driven by nonlinear quantum noise*, Random Operators and Stochastic Equations, vol. 10, no.1, pp. 1-12 (2002).

[5] ———, *Stochastic evolutions driven by non-linear quantum noise*, Probability and Mathematical Statistics, vol.22.1 (2002).

[6] ———, *Stochastic evolutions driven by non-linear quantum noise II*, Russian Journal of Mathematical Physics, v.8, no.4, (2001).

[7] ———, *Square of white noise unitary evolutions on Boson Fock space*, International conference on stochastic analysis in honor of Paul Kree, Hammamet, Tunisie, October 22-27, 2001.

[8] ———, *Unitarity conditions for the renormalized square of white noise*, Trends in Contemporary Infinite Dimensional Analysis and Quantum Probability, Natural and Mathematical Sciences Series 3, 7-36, Italian School of East Asian Studies, Kyoto, Japan (200).

[9] ———, *The semi-martingale property of the square of white noise integrators*, Stochastic Partial Differential Equations and Applications, pp. 1-19, eds G. Da Prato and L. Tubaro, Marcel Dekker, Inc. (2002).

[10] ———, *Control of elementary quantum flows*, Proceedings of the 5th IFAC symposium on nonlinear control systems, July 4-6, 2001, St. Petersburg, Russia".

[11] ———, *The unitarity conditions for the square of white noise*, to appear in Infinite Dimensional Anal. Quantum Probab. Related Topics (2003).

[12] ———, *Quadratic control of quantum processes*, Russian Journal of Mathematical Physics (2002).

[13] ———, *Control of quantum Langevin equations*, Open Systems and Information Dynamics.9:1-15, 2002.

[14] L. Accardi,A. Boukas, H.H. Kuo, *On the unitarity of stochastic evolutions driven by the square of white noise*, Infinite Dimensional Analysis, Quantum Probability, and Related Topics, vol. 4, no. 4, pp.1-10, (2001).

[15] L. Accardi, F. Fagnola, J. Quaegebeur, *A representation free quantum stochastic calculus*, Journal of Functional Analysis, 104 (1) (1992), 149-197.

[16] L. Accardi, U. Franz, M. Skeide, *Renormalized squares of white noise and non- Gaussian noises as Levy processes on real Lie algebras*, Comm. Math. Phys. 228 (2002), no. 1, 123-150.

[17] L. Accardi, T. Hida, H.H Kuo, *The Itô table of the square of white noise*, Infinite Dimensional Analysis, Quantum Probability, and Related Topics, 4, (2001) 267-275.

[18] L. Accardi, Y.G Lu, I.V Volovich, *Quantum theory and its stochastic limit* , Springer 2002.

[19] ———, *White noise approach to classical and quantum stochastic calculi*, Lecture Notes of the Volterra International School of the same title, Trento, Italy, 1999, Volterra Center preprint 375.

[20] ———, *Non-linear extensions of classical and quantum stochastic calculus and essentially infinite dimensional analysis*, Volterra Center preprint no.268, 1996.

[21] L. Accardi, N. Obata, *Towards a non-linear extension of stochastic calculus*, Publications of the Research Institute for Mathematical Sciences, Kyoto, RIMS Kokyuroku 957, Obata N. (ed.), (1996), 1-15.

[22] L. Accardi, M. Skeide, *On the relation of the square of white noise and the finite difference algebra*, Infinite Dimensional Analysis, Quantum Probability, and Related Topics, 3, (2000), 185-189.

[23] ———, *Hilbert module realization of the square of white noise and the finite difference algebra*, Volterra Center preprint no. 384 (1999).

[24] V. Belavkin, *Non-demolition measurement and control in quantum dynamical systems*, Proceedings of the conference on "Information Complexity and Control in Quantum Physics", Udine 1985, Springer-Verlag, p331-336.

[25] A. Boukas, *Quantum stochastic analysis: a non Brownian case*, Ph.D Thesis, Southern Illinois University, 1998.

[26] ———, *An example of a quantum exponential process*, Monatsh. Math., 112(3), 209-215, 1991.

[27] ———, *Stochastic calculus on the finite difference Fock space*, Quantum Probability and related topics, vol VI, p.205-218, ed. L.Accardi, World Scientific 1991.

[28] ———,*Linear Quantum Stochastic Control*, Quantum Probability and related topics, 105-111, QP -PQ IX, World Scientific Publishing, River Edge NJ,1994.

[29] ———,*Application of Operator Stochastic Calculus to an Optimal Control problem*, Mat. Zametki 53, (1993), no5, 48-56 Russian).Translation in Math.Notes 53 (1993), No 5-6, 489 -494, MR 96a 81070.

[30] ———, *Operator valued stochastic control in Fock space with applications to noise filtering and orbit tracking*, Journal of Probability and Mathematical Statistics, Vol .16, 1, 1994.
[31] ———, *Stochastic Control of operator-valued processes in Boson Fock space*, Russian Journal of Mathematical Physics ,4 (1996) , no2, 139-150, MR 97j 81178
[32] A. Chebotarev, *Lecture notes in quantum probability*, Mexican Mathematical Society.
[33] R. F. Curtain, A. J. Pritchard, *The infinite dimensional Riccati equation*, J. Math. Anal. and Appl. (1974), **47**
[34] ———, *Functional analysis in modern applied mathematics*, Academic Press 1977.
[35] M. H. A. Davis, *Linear estimation and stochastic control*, Chapman and Hall 1977.
[36] P. J. Feinsilver, *Discrete analogues of the Heisenberg-Weyl algebra*, Monatsh. Math., 104, 89-108, 1987.
[37] W. Flemming, R. Rishel, *Deterministic and stochastic optimal control*, Springer-Verlag 1975.
[38] G. Kallianpur, *Stochastic filtering theory*, Springer-Verlag 1980.
[39] K. R. Parthasarathy, *An introduction to quantum stochastic calculus*, Birkhauser Boston Inc., 1992.

Centro Vito Volterra, Università di Roma Tor Vergata, Via Columba, 2– 00133 Roma, Italy
E-mail address: `accardi@volterra.mat.uniroma2.it`

Department of Mathematics, American College of Greece, Aghia Paraskevi, Athens 15342, Greece
E-mail address: `gxk-personnel@ath.forthnet.gr`

TRANSCENDING THE LIMITS OF TURING COMPUTABILITY *

VADIM A. ADAMYAN

Department of Theoretical Physics
Odessa National University
ul. Dvoryanskaya 2, 65026 Odessa, Ukraine
E-mail: vadamyan@paco.net.

CRISTIAN S. CALUDE

Department of Computer Science
The University of Auckland
Private Bag 92019, Auckland, New Zealand
E-mail: cristian@cs.auckland.ac.nz

BORIS S. PAVLOV

Department of Mathematics
University of Auckland
Private Bag 92019, Auckland, New Zealand
E-mail: pavlov@math.auckland.ac.nz.

Hypercomputation or super-Turing computation is a "computation" that transcends the limit imposed by Turing's model of computability. The field still faces some basic questions, technical (can we mathematically and/or physically build a hypercomputer?), cognitive (can hypercomputers realize the AI dream?), philosophical (is thinking more than computing?). The aim of this paper is to address the question: can we mathematically build a hypercomputer? We will discuss the solutions of the Infinite Merchant Problem, a decision problem equivalent to the Halting Problem, based on results obtained in Refs. 9 and 2. The accent will be on the new computational technique and results rather than formal proofs.

*VAA is grateful for partial support from the USA Civil Research and Development Foundation (CRDF grant UM1-2090); CSC and BSP have been partially supported by the Auckland University Vice-Chancellor's Development Fund 23124; BSP was also partially supported by JSPS(102704) and RFFI(01149) grants.

1. Introduction

Hypercomputation or super-Turing computation is a "computation" that transcends the limit imposed by Turing's model of computability; for a recent perspective see the special issue of the journal *Minds and Machines* (12, 4, 2002). Currently there are various proposals to break Turing' Barrier by showing that certain classes of *computing procedures* have super-Turing power (see Refs. 19,32,6,7,31,11 and 12). A specific class of computing procedures [15,9,22,2] make essential use of some physical theory, relativity theory in Ref.15, quantum theory, Refs. 9 and 22; they all reflect an attitude advocated by Landauer [23,24] (*information is inevitably physical*) and Deutsch [13,14] (*the reason why we find it possible to construct, say, electronic calculators, and indeed why we can perform mental arithmetic ... is that the laws of physics "happen" to permit the existence of physical models for the operations of arithmetic*).

The aim of the present paper is to revisit the solutions offered in Refs. 9 and 2. We will focus on the novelty of the approach and we will discuss its power and its limits. No proofs will be offered.

2. The Classical Merchant Problem

Recall that in the classical version of the Merchant Problem we have 10 stacks of coins, each stack containing 100 coins, and we know that *at most one stack contains only false coins, weighting* 1.01 g; true coins weight 1 g. The problem is to find the stack with false coins (if any) *by only one weighting.* The classical solution reduces the problem to the weighting of a special combination of coins: one coin from the first stack, two coins from the second stack, ... , ten coins from the tenth stack. If the false coins are present in the N-th stack, then the weight of the combination will be $55 + \frac{N}{100}\,g$; otherwise the weight is just 55 g. The Merchant Problem quoted above was widely spread in allies armies during the Second World War, cf. Ref.21. Probably the elegant solution described above was the very first solution of a computational problem bearing typical features of quantum computing, see an extended discussion in Ref.9.

3. The Infinite Merchant Problem

In what follows we are going to consider the following generalization of the problem, the Infinite Merchant Problem: we assume that *we have countable many stacks, given in some computable way, all of them, except at most one, containing true coins only.* True coins weight 1 and false coins weight $1 + 2^{-j}$, $j > 0$. Again we are allowed to take a coin from each stack and we want to determine whether all coins are true or there is a stack of false coins.

Next we will show that the Infinite Merchant Problem is classically undecidable by reducing it to the Halting Problem, i.e. the problem to decide whether an arbitrary Turing machine (TM) halts on an arbitrary input. Assume that a TM operates on positive integers and suppose, for the sake of contradiction, that there exists a TM HALT which can decide whether a TM T (given by its code $\#T$, a positive integer) eventually stops on input x:

$$\mathtt{HALT}(\#T, x) = \begin{cases} 1, \text{ if } T(x) \text{ stops,} \\ 0, \text{ otherwise.} \end{cases}$$

We construct a TM Q

$$Q(x) = \begin{cases} 1, & \text{if } \mathtt{HALT}(x, x) = 0, \\ \text{loops forever, otherwise,} \end{cases}$$

and deduce the contradiction:

$$\mathtt{HALT}(\#Q, \#Q) = 1 \text{ iff } \mathtt{HALT}(\#Q, \#Q) = 0.$$

We next describe the reduction. Assume that we have a classical solution of the Infinite Merchant Problem and we are given a TM T and an input x for T. We construct a computable sequence $q_1, q_2, \ldots q_i \ldots$ as follows: if the computation of $T(x)$ did not stop till the i-th step, then we put $q_i = 1$; if the computation halted at step i_0, then we put $q_{i_0} = 1 + 2^{-j}$ and $q_r = 1$, for all $r > i_0$. The sequence q_i satisfies all conditions of the Infinite Merchant Problem and $T(x)$ halts if and only if there is a false coin, i.e., $q_{i_0} = 1 + 2^{-j}$, for some j. This shows that the Infinite Merchant Problem is undecidable as the Halting Problem is undecidable.

In fact the two problems are equivalent. Indeed, assume that we could classically solve the Halting Problem. To every sequence (q_i) satisfying the conditions of the Infinite Merchant Problem we associate the TM T such that $T(i) = 1$ if $q_i = 1$, and $T(i) = 0$ otherwise. The TM T' defined by $T'(0) = \min\{i \mid T(i) = 0\}$ halts at 0 if and only if there an i_0 such that $q_{i_0} = 1 + 2^{-j}$, i.e., $T'(0)$ halts if and only if there are false coins in the system. Hence, a classical solution of the Halting Problem will produce a classical solution for the Infinite Merchant Problem.

The above discussion shows that undecidability is determined by the impossibility to decide in a finite time the answers to an infinite number of questions, "does the first stack contain a false coin?", "does the second stack contain a false coin?", etc. This might be caused either by the fact that the time of the computation grows indefinitely or by the fact that the space of computation grows indefinitely or both. The classical theories of computability and complexity (see, for example, Ref.5) do not give any indication in this respect. In the following section we will show that time can be made finite provided we use a specific probabilistic strategy.

4. A Probabilistic Solution

In this section we present, in a slightly different way, the probabilistic solution proposed in Ref.9. We will adopt the following strategy. We are given a probability $\theta = 2^{-n}$ and we assume that we work with a "device" described below[a] with sensitivity given by a real $\varepsilon = 2^{-m}$. Then, we compute classically a time $T = T_{\theta,\varepsilon}$ and run the "device" on a random input for the time T. If we get a click, then the system has false coins; if we don not get a click, then we conclude that with probability greater than $1 - \theta$ all coins are true. An essential part of the method is the requirement that the time limit T is *classically computable.*

The "device" (with sensitivity ε) will distinguish the values of the iterated quadratic form $\langle \mathbf{Q}^t(\mathbf{x}), \mathbf{x}\rangle = \sum_{i=1}^{\infty} q_i^t \,|x_i|^2$, by observing the difference between averaging over trajectories of two discrete random walks with two non-perturbed and perturbed sequences t_l, $\tilde{t}_l$ of "stops". The non-perturbed sequence corresponds to equal steps $\delta_m = 1$, $t_l = \sum_{m=0}^{l} \delta_m$, and the perturbed corresponds to the varying steps Δ_m, $0 < \Delta_m < \delta_m$, $\tilde{t}_l = \sum_{m=0}^{l} \Delta_m$. We work with the intersections of l_2 with the discrete Sobolev class l_2^1 of square-summable sequences with the square norm

$$| \mathbf{x} |_1^2 = \sum_{m=1}^{\infty} |x_m - x_{m-1}|^2, \tag{1}$$

and the discrete Sobolev class $\tilde{l}_2^1$ of weighted-summable sequences with the square norm

$$\| \mathbf{x} \|_1^2 = \sum_{m=1}^{\infty} \frac{1-\Delta_m}{\Delta_m} \, |x_m - x_{m-1}|^2. \tag{2}$$

[a]As in Ref.9 we use quotation marks when referring to our mathematical "device".

By natural extension from cylindrical sets we can define the Wiener measures $\tilde{W}$ and W on the spaces of trajectories of the perturbed and non-perturbed random walks respectively and use the absolute continuity $\tilde{W}$ with respect to W: that is for every W–measurable set Ω,

$$\tilde{W}(\Omega) = \frac{1}{\prod_{l=1}^{\infty} \sqrt{\Delta_l}} \int_{\Omega} e^{-\sum_{m=1}^{\infty} \frac{1-\Delta_m}{\Delta_m} |x_m - x_{m-1}|^2} dW.$$

Assume that the "device" revealing the exponential growth of the quadratic form of the iterations $\langle \mathbf{Q}^t(\mathbf{x}), \mathbf{x} \rangle$ clicks if

$$\langle \mathbf{Q}^t(\mathbf{x}), \mathbf{x} \rangle \geq \| \mathbf{x} \|^2 + \varepsilon \| \mathbf{x} \|_1^2 .$$

Thus the "device" sensitivity is defined in terms of the Sobolev norm.

Two cases may appear. If for some $T > 0$, $\langle \mathbf{Q}^T(\mathbf{x}), \mathbf{x} \rangle \geq \| \mathbf{x} \|^2 + \varepsilon \| \mathbf{x} \|_1^2$, then the "device" has clicked and we know for *sure* that there exist false coins in the system. However, it is possible that at some time $T > 0$ the "device" hasn't (yet?) clicked because $\langle \mathbf{Q}^t(\mathbf{x}), \mathbf{x} \rangle < \| \mathbf{x} \|^2 + \varepsilon \| \mathbf{x} \|_1^2$. This may happen because either all coins are true, i.e., $\langle \mathbf{Q}^t(\mathbf{x}), \mathbf{x} \rangle < \| \mathbf{x} \|^2 + \varepsilon \| \mathbf{x} \|_1^2$, for all $t > 0$, or because at time T the growth of $\langle \mathbf{Q}^T(\mathbf{x}), \mathbf{x} \rangle$ hasn't yet reached the threshold $\| \mathbf{x} \|^2 + \varepsilon \| \mathbf{x} \|_1^2$. In the first case the "device" will *never* click, so at each stage t the test-vector $\mathbf{x}$ produces "true" information; we can call $\mathbf{x}$ a "true" vector. In the second case, the test-vector $\mathbf{x}$ is "lying" at time T as we *do* have false coins in the system, but they were not detected at time T; we say that $\mathbf{x}$ produces "false" information at time T.

If we assume that there exist false coins in the system, say at stack j, but the "device" does not click at the moment T, then the test-vector $\mathbf{x}$ belongs to the *indistinguishable set*

$$\mathcal{F}_{\varepsilon,T} = \{ \mathbf{x} \in l_2^1 \mid ((1+\gamma)^T - 1) \mid x_j \mid^2 < \varepsilon \| \mathbf{x} \|_1^2, \text{ for some } j \}.$$

In Ref.9 it was proven that the Wiener measure of the indistinguishable set tends to zero as $T \to \infty$:

$$\tilde{W}(\mathcal{F}_{\varepsilon,T}) \leq \left(\frac{\varepsilon}{((1+\gamma)^T - 1 - \varepsilon) \cdot \prod_{m=1}^{\infty} \Delta_m} \right)^{1/2} .$$

This fact is not enough to realize the scheme described in the beginning of this section: we need a more precise result, namely we have to prove that $\tilde{W}(\mathcal{F}_{\varepsilon,T})$ *converges computably to zero.* And, indeed, this is true because:

$$\tilde{W}(\mathcal{F}_{\varepsilon,T}) \leq \eta, \text{ provided } t > \log_{1+\gamma} \left(\frac{\varepsilon}{\eta^2 \prod_{m=1}^{\infty} \Delta_m} + 1 + \varepsilon \right).$$

Denote by $P(\mathcal{N})$ the *a priori* probability of absence of false coins in the system. Then, the *a posteriori probability that the system contains only true coins, when the "device" did not click after running the experiment for the time* T, *is*

$$\mathbf{P}_{\text{non-click}}(\mathcal{N}) > 1 - \frac{1 - P(\mathcal{N})}{P(\mathcal{N})} \cdot \frac{\sqrt{\varepsilon}}{\sqrt{(1+\gamma)^T - 1 - \varepsilon \sqrt{\prod_{m=1}^{\infty} \Delta_m}}} .$$

5. A Brownian Solution Based on Resonance Amplification

In Ref.1 the idea to consider a single act of quantum computation as a scattering process was suggested.

We will first illustrate the method by describing a simple quantum scattering system realizing the quantum C_{NOT} gate, i.e., a quantum gate satisfying exactly to the same truth-table as the classical controlled-NOT gate. The C_{NOT} device has two input and output channels. Each channel can be only in two different states, say $|0\rangle, |1\rangle$. The *in* and *out* states of the control-channel are the same, $|I_{in}\rangle = |I_{out}\rangle$, but the *in* and *out* states of the current-channel may be different, $|J_{in}\rangle \neq |J_{out}\rangle$, depending upon the state $|J_{in}\rangle$ and the control-channel state. The classical controlled-NOT gate has the following truth-table:

I_{in}	J_{in}	I_{out}	J_{out}
0	0	0	0
0	1	0	1
1	0	1	1
1	1	1	0

which describes the effect of the device on the above *in* states,

$\lvert I_{in}\ J_{in}\rangle$	$\longrightarrow$	$\lvert I_{out}\ J_{out}\rangle$
$\lvert 0\ 0\rangle$	$\longrightarrow$	$\lvert 0\ 0\rangle$
$\lvert 0\ 1\rangle$	$\longrightarrow$	$\lvert 0\ 1\rangle$
$\lvert 1\ 0\rangle$	$\longrightarrow$	$\lvert 1\ 1\rangle$
$\lvert 1\ 1\rangle$	$\longrightarrow$	$\lvert 1\ 0\rangle$

The quantum C_{NOT} gate operates not only on the "classical" states $|0\rangle$ and $|1\rangle$, ($C_{\text{NOT}}|ij\rangle = |ik\rangle$, where $i, j \in \{0,1\}$ $k = i \oplus j \pmod 2$), but also on all their linear combinations,

$$\alpha_{00}|00\rangle + \alpha_{01}|01\rangle + \alpha_{10}|10\rangle + \alpha_{11}|11\rangle$$

$$\longrightarrow \alpha_{00}|00\rangle + \alpha_{01}|01\rangle + \alpha_{10}|11\rangle + \alpha_{11}|10\rangle.$$

This quantum transformation can be presented via the unitary matrix

$$U = \begin{pmatrix} 1\,0\,0\,0 \\ 0\,1\,0\,0 \\ 0\,0\,0\,1 \\ 0\,0\,1\,0 \end{pmatrix}, \tag{3}$$

with respect to the canonical basis $(e_1, e_2, e_3, e_4) = (|00\rangle, |01\rangle, |10\rangle, |11\rangle)$. More importantly, $C_{\rm NOT}$ in combination with all 1-qubit gates is *universal* [b] and *it cannot be written as a tensor product of two binary operators* $U = U_1\, U_2,\ U_l \neq I$.

We claim that the matrix U in (3) can be realized as a *scattering matrix* of a special *quantum dot.* First, here is the motivation. Consider Figure 1 in which two isolated quantum wires are placed in proximity and there is a window region in which the two wires are coupled. An electron moving in the window region oscillates between the two quantum wires and the probability of the electron exiting into a specific quantum wire depends on the length of the window. This "switching phenomenon" was discovered by del Alamo and Eugster [3] and intensely discussed in literature, see for instance Refs. 16 and 4. We can arrange the setup in such a way that, under normal conditions, the electron exits from the same wire it enters, but switches to the other wire when a classical extra potential is applied, a realization of the $C_{\rm NOT}$ relay. If the control is quantum too, then we obtain the $C_{\rm NOT}$ gate.

We continue with the mathematical model and assume that the quantum dot is inserted in an one-dimensional quantum wire $\mathbf{R} = (-\infty, \infty) = \mathbf{R}_+ \cup \mathbf{R}_-$ between the wires $\mathbf{R}_+$, $\mathbf{R}_-$ and a single electron may be redirected according to the state of the dot.

We assume also that the *inner* Hamiltonian of the quantum dot is presented by a finite or an infinite diagonal matrix $A = \operatorname{diag}\left\{\alpha_1^2, \alpha_2^2, \dots, \alpha_N^2, \dots\right\}$ with positive diagonal elements $\alpha_1^2 < \alpha_2^2 < \alpha_3^2 < \dots < \alpha_N^2 < \dots$. We assume that the quantum dot is inserted in an one-

[b]Every classically computable function can be computed by a small universal set of gates like {OR, NOT} or {NAND}. A set of quantum gates $\mathcal{S}$ is called universal if any unitary operation can be approximated with an arbitrary accuracy by a quantum circuit involving gates in $\mathcal{S}$; see more in Refs.17,18,20 and 8.

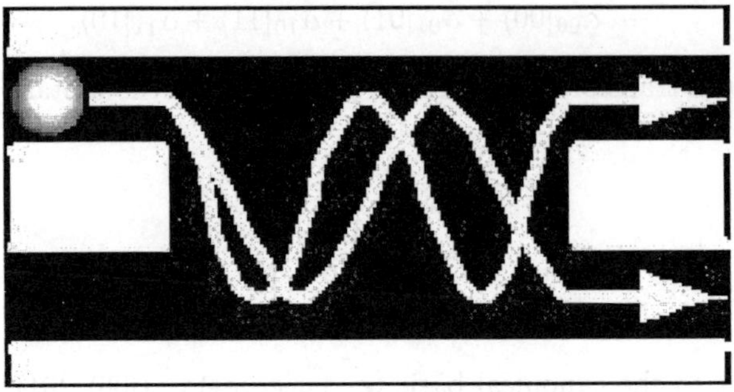

Figure 1. C_{NOT} as quantum dot

dimensional quantum wire $-\infty < x < \infty$ at the origin and a proper boundary condition is satisfied (see (9)) for connecting it with the Schrödinger operator on the wire defined in the space of square-integrable vector-functions $L_2(\mathbf{R}_+, \mathcal{E})$ with values in the infinitely-dimensional Hilbert space $\mathcal{E}$

$$l = -\frac{d^2}{dx^2} \cdot \tag{4}$$

One could assign to the above quantum system a product space $H_e \otimes H_d$ constituted respectively by the states of the electron and the states of the dot, and consider an evolution of the system generated by the total Hamiltonian $\mathcal{H}_e + \mathcal{H}_d + \mathcal{H}_{int}$ with a proper interaction term. This would lead to a quite sophisticated problem of quantum mechanics, similar to three-body problems, see for instance Ref.27. We assume now that *the state of the dot is selected independently* and thus reduce the above problem to the corresponding *one-body* problem for an electron scattered in the quantum wire depending on the state of the dot. The corresponding device should be called rather *quantum relay* rather than *quantum gate*; however, it may be transformed into a quantum gate if the state of the dot is obtained as a quantum state with finite life-time. Practically the model suggested below is acceptable if the life-time of the state of the dot is long enough during the scattering experiment.The corresponding general "zero-range" quantum Hamiltonian (solvable model) is described as a self-adjoint extension $\mathbf{A}_\beta$ of the orthogonal sum $l \oplus A$ restricted to $l_0 \oplus A_0$ in $L_2(\mathbf{R}, \mathcal{E}) \oplus E$ onto a proper domain; here $\mathcal{E}$ is the infinitely- dimensional input space and E is the inner space (with dim $(E) \geq 2$). The *spectrum* σ_β of the operator $\mathbf{A}_\beta$ is absolutely-continuous and fills the positive half-axis $\lambda \geq 0$ with multiplicity dim $(\mathcal{E})$. The role of eigen-functions of the spectral point $p^2 = \lambda > 0$ is played by the *scattered waves* $\overrightarrow{\Psi}_\nu$, $\overleftarrow{\Psi}_\nu$, labeled with vectors $\nu \in \mathcal{E}$. The components of the scattered waves $\Psi_\nu(p)$ in the *outer space* $L_2(\mathbf{R})$ are presented as linear combinations of exponentials:

$$\overrightarrow{\Psi}_{\nu,p}(x) = \begin{cases} e^{-ipx}\nu + e^{+ipx}\overleftarrow{R}(p)\nu, & x < 0, \\ e^{-ipx}\overrightarrow{T}(p)\nu, & x > 0, \end{cases}$$

$$\overleftarrow{\Psi}_{\nu,p}(x) = \begin{cases} e^{ipx}\nu + e^{-ipx}\overrightarrow{R}(p)\nu, & x > 0, \\ e^{ipx}\overleftarrow{T}(p)\nu, & x < 0. \end{cases} \tag{5}$$

The matrix

$$\mathbf{S}_\beta(p) = \begin{pmatrix} \overrightarrow{T}(p) & \overrightarrow{R}(p) \\ \overleftarrow{R}(p) & \overleftarrow{T}(p) \end{pmatrix}, \tag{6}$$

is called the *scattering matrix* of the operator $\mathbf{A}_\beta$.[c]

The evolution of the *wave function* of the quantum mechanical system with Hamiltonian $\mathbf{A}_\beta$ given by the equation

$$\frac{1}{i}\frac{\partial \Psi}{\partial t} = \mathbf{A}_\beta \Psi, \tag{7}$$

and proper initial condition

$$\Psi\Big|_{t=0} = \Psi_0,$$

can be described by the correspopnding evolution operator constructed from the above scattered waves and square-integrable bound states Ψ_s which satisfy the homogeneous equation

$$\mathbf{A}_\beta \Psi_s = \lambda_s \Psi_s,$$

with negative eigen-values λ_s. Bound states do not play an essential role in our construction, so we may assume that the initial state Ψ_0 is orthogonal to all bound states and may be expanded in an analog of Fourier integral over the scattered waves

$$\Psi_0 = \frac{1}{2\pi}\int_R \sum_\nu \Psi_{\nu,p}\langle \Psi_{\nu,p}, \Psi_0\rangle dp.$$

[c]The transmission coefficients appear on the main diagonal of the matrix to fit the physical meaning of the scattering matrix for small values of $|\beta|$, when it is reduced to the undisturbed transmission $\mathbf{S}(p) = I$.

Then the evolution described by the solution of the equation (7) and the above initial data can be presented as a (continuous) linear combination

$$\Psi(t) = \frac{1}{2\pi} \int_{\mathbf{R}} \sum_{\nu} e^{ip^2 t} \Psi_{\nu,p} \langle \Psi_{\nu,p}, \Psi_0 \rangle dp$$

of *modes* incoming from infinity on the left $(-\infty)$ and on the right $(+\infty)$, and outgoing modes *scattered* to both directions $\pm\infty$ according to the solution of the time-dependent Schrödinger equation $\frac{1}{i}\frac{\partial \mathbf{U}}{\partial t} = \mathbf{AU}$:

$$\mathbf{S}_\beta(p) : e^{ip^2 t} \begin{pmatrix} e^{-ipx} \nu_{left} \\ e^{ipx} \nu_{right} \end{pmatrix} \longrightarrow e^{ip^2 t} \begin{pmatrix} e^{-ipx} (\overrightarrow{T}(p) \nu_{left} & + \overrightarrow{R}(p) \nu_{right}) \\ e^{ipx} (\overleftarrow{R}(p) \nu_{left} & + \overleftarrow{T}(p) \nu_{right}) \end{pmatrix}. \tag{8}$$

The analytic structure of the scattering matrix depends upon the structure of the inner Hamiltonian and a sort of interaction between the inner E and outer $L_2(\mathbf{R}, \mathcal{E})$ spaces. To avoid the discussion of the general situation we use here a scattering matrix for a solvable model of the quantum dot which is based on zero-range potential with inner structure, introduced in Ref.28 and already used in Ref.4 for the description of typical features of nano-devices, see also Ref.2.

If we choose an infinitely-dimensional input space $\mathcal{E} \oplus \mathcal{E}$ with components corresponding to the amplitudes of the scattered waves at $\pm\infty$ and an one-dimensional *deficiency subspace* $\mathcal{N}_i$ spanned by the normalized vector $e = e_i \in \mathcal{E}$ (see Ref.2) and introduce the scalar function

$$\mathcal{M} = \langle \frac{I + \lambda A}{A - \lambda I} e, e \rangle,$$

then using the interaction defined by the boundary conditions (9) depending on a vector $\beta \in \mathcal{E}$ imposed on the boundary values (the jump $[u'](0)$ and the value $u(0)$ at the origin) of the component of the wave-function in the outer space and the *symplectic coordinates* $\vec{\xi}_\pm \in \mathcal{E}$ of the inner component of the wave-function (see Ref.29):

$$\begin{pmatrix} [u'](0) \\ -\vec{\xi}_- \end{pmatrix} = \begin{pmatrix} 0 & \beta \\ \beta^+ & 0 \end{pmatrix} \begin{pmatrix} u(0) \\ \vec{\xi}_+ \end{pmatrix}, \tag{9}$$

we obtain the scattering matrix in the form

$$\mathbf{S}_\beta(p) = \begin{pmatrix} \overrightarrow{T}(p) & \overrightarrow{R}(p) \\ \overleftarrow{R}(p) & \overleftarrow{T}(p) \end{pmatrix},$$

with equal transmission and reflection coefficients $\overrightarrow{T}, \overleftarrow{T}, \overrightarrow{R}, \overleftarrow{R}$:

$$\overrightarrow{T}(p) = \overleftarrow{T}(p) = \mathbf{P}_\beta^\perp + \frac{2ip}{2ip + |\beta|^2 \mathcal{M}^{-1}} \mathbf{P}_\beta,$$

$$\overrightarrow{R}(p) = \overleftarrow{R}(p) = -\frac{|\beta|^2 \mathcal{M}^{-1}}{2ip + |\beta|^2 \mathcal{M}^{-1}} \mathbf{P}_\beta.$$

Here $\mathbf{P}_\beta$ is the orthogonal projection of $\mathcal{E}$ onto the one-dimensional space spanned by the vector β, and $\mathbf{P}_\beta^\perp = I - \mathbf{P}_\beta$ is the orthogonal projection on the complimentary space. The constructed solvable model reveals the role of zeroes of the scattering matrix – the *resonances* – in implementing the switching function.

We observe first the behaviour of the scattering matrix at the resonance energy $\alpha_1^2 > 0$ in case the resonance level α_1^2 in the quantum dot is vacant as well as all levels α_l^2 above α_1^2. In this case we have

$$\mathcal{M}_1 = \frac{1 + \alpha_1^2 \lambda}{\alpha_1^2 - \lambda} |e_1|^2 + \sum_{l=2}^{N} \frac{1 + \alpha_l^2 \lambda}{\alpha_l^2 - \lambda} |e_l|^2 = \frac{1 + \alpha_1^2 \lambda}{\alpha_1^2 - \lambda} |e_1|^2 + \mathcal{M}_3,$$

where $|e_l|^2$ are the squares of the Fourier coefficients of the deficiency vector e with respect to the eigen-vectors of the operator A.

Next we consider the case when the resonance level α_1^2 is occupied. In this case

$$\mathcal{M}_2 = \frac{1 + \alpha_2^2 \lambda}{\alpha_2^2 - \lambda} |e_2|^2 + \sum_{l=3}^{N} \frac{1 + \alpha_l^2 \lambda}{\alpha_l^2 - \lambda} |e_l|^2 = \frac{1 + \alpha_2^2 \lambda}{\alpha_2^2 - \lambda} |e_2|^2 + \mathcal{M}_3.$$

In the above analysis we have ignored the electron spin, that is we have assumed that all electrons have the same constant spin on the quantum circuit $\mathbf{R}_- \cup \mathbf{R}_+$, with the quantum dot attached.[d] We may assume that the circuit lies on the surface of a semiconductor with Fermi-level α_1^2 (see Ref.26).

We assume that the state of the dot with the level α_1^2 vacant corresponds to $I_{in} = I_{out} = 0$ and the state of the dot with the level α_1^2 occupied corresponds to $I_{in} = I_{out} = 1$. We identify these states of the system as state S_1 and state S_2, respectively.

---

[d]Our hypothesis is satisfied in case the travelling electrons are polarized and the electron on the level α_1^2 is polarized. Note that Pauli's exclusion principle is still valid, but with only one electron on each orbital: the magnetic field is absent, so the polarization is not changed during the experiment.

For every vector β the transmission coefficients on the resonance electron's energy $\lambda = \alpha_1^2$ can be expressed as (see Ref.2):

$$\overrightarrow{T}(p) = \overleftarrow{T}(p) = \mathbf{P}_\beta^\perp + \frac{2ip}{2ip + |\beta|^2 \mathcal{M}_1^{-1}} \mathbf{P}_\beta = I. \ \overrightarrow{R}(p) = \overleftarrow{R}(p) = 0,$$

(at the resonance energy we have $\mathcal{M}_1^{-1} = 0$), so the scattering matrix becomes the identity when α_1^2 *is not occupied.*

In the second case, when the resonance level $p^2 = \lambda = \alpha_1^2$ is occupied, we obtain (due again to Pauli's exclusion principle) the following expression for the transmission coefficients of passing electrons with resonance energy:

$$\overrightarrow{T}(p) = \overleftarrow{T}(p) = \frac{2ip\mathcal{M}_2(\lambda)}{2ip\mathcal{M}_2(\lambda) - |\beta|^2}$$

$$= \frac{2ip\left(\frac{1+\alpha_2^2\alpha_1^2}{\alpha_2^2-\alpha_1^2}|e_2|^2 + \sum_{l=3}^N \frac{1+\alpha_l^2\alpha_1^2}{\alpha_l^2-\alpha_1^2}|e_l|^2\right)}{2ip\left(\frac{1+\alpha_2^2\alpha_1^2}{\alpha_2^2-\alpha_1^2}|e_2|^2 + \sum_{l=3}^N \frac{1+\alpha_l^2\alpha_1^2}{\alpha_l^2-\alpha_1^2 a}|e_l|^2\right) + |\beta|^2},$$

and the corresponding expressions for the reflection coefficients

$$\overrightarrow{R}(p) = \overleftarrow{R}(p) = -\frac{|\beta|^2}{2ip\mathcal{M}_2 + |\beta|^2}\mathbf{P}_\beta,$$

which can be approximated, for large enough β, as

$$\overrightarrow{T}(\alpha_1) = \overleftarrow{T}(\alpha_1) \approx \mathbf{P}_\beta^\perp, \ \overrightarrow{R}(\alpha_1) = \overleftarrow{R}(\alpha_1) \approx -\mathbf{P}_\beta.$$

Hence the scattering matrix is equal to

$$\mathbf{S}_\beta(\alpha_1^2) = \begin{pmatrix} \mathbf{P}_\beta^\perp & -\mathbf{P}_\beta \\ -\mathbf{P}_\beta & \mathbf{P}_\beta^\perp \end{pmatrix} = I - \begin{pmatrix} \mathbf{P}_\beta & \mathbf{P}_\beta \\ \mathbf{P}_\beta & \mathbf{P}_\beta \end{pmatrix}, \tag{10}$$

for relatively large enough β, if the resonance level α_1^2 *is occupied.*

We continue by showing how the probabilistic approach discussed in the previous section can be realised.We consider now an imaginable quantum scattering system with an infinitely-dimensional input-space and, in particular, with the infinite dimensional space $\mathcal{E}$ for which a solvable model was described above as an extension of the orthogonal sum $l_0 \oplus A_0$ with boundary condition (9), in which , however, the β-channel connecting the outer subspace $L_2(\mathbf{R}, \mathcal{E})$ with the inner subspace E is as before two- dimensional. We associate these extensions with two states (S_1, S_2) of the total quantum system combined of the inner and outer components, with the interaction respectively *switched on* via the boundary condition (9), $\beta \neq 0$, or *switched off*, $\beta = 0$, and we interpret the Halting Problem in a probabilistic setting as the problem of distinguishing of the states (S_1, S_2) of the quantum system via a scattering experiment with a random input.

Following the probabilistic strategy in Section 4, we compare the scattering matrices $\mathbf{S}_\beta(\alpha_1^2)$ in states S_1 and S_2. In the first state this matrix coincides with identity, hence

$$\frac{1}{2}(I + \mathbf{S}_\beta(\alpha_1^2)) = I.$$

In the second state we have

$$\frac{1}{2}(I + \mathbf{S}_\beta(\alpha_1^2)) = \begin{pmatrix} I & 0 \\ 0 & I \end{pmatrix} + \begin{pmatrix} \mathbf{P}_\beta & 0 \\ 0 & \mathbf{P}_\beta \end{pmatrix} \mathcal{P}_{sym},$$

where $\mathcal{P}_{sym} = \frac{1}{2}\begin{pmatrix} I & I \\ I & I \end{pmatrix}$ is the projection onto the symmetric subspace of the input space $\mathcal{E} \oplus \mathcal{E}$ consisting of vectors with equal componenets $\begin{pmatrix} e \\ e \end{pmatrix}$. The projections $\begin{pmatrix} \mathbf{P}_\beta & 0 \\ 0 & \mathbf{P}_\beta \end{pmatrix}$ and $\mathcal{P}_{sym}$ commute and their product gives the projection onto the space spanned by $\overrightarrow{\beta} = \begin{pmatrix} \beta \\ \beta \end{pmatrix}$. Hence,

$$\frac{1}{2}(I + \mathbf{S}_\beta(\alpha_1^2)) = \begin{pmatrix} I & 0 \\ 0 & I \end{pmatrix} - \mathcal{P}_\beta,$$

where $\mathcal{P}_\beta$ is the orthogonal projection on the subspace $\mathcal{E}$ which is collinear to the vector $\overrightarrow{\beta}$ in the input space. Therefore, in the second case, for every vector $e \in \mathcal{E} \oplus \mathcal{E}$ we have:

$$\frac{1}{2}(\langle e,\, e\rangle + \langle \mathbf{S}_\beta(\alpha_1^2)e,\, e\rangle) = |e|^2 - |\mathcal{P}_\beta e|^2.$$

The expectation is that if the probability of the event $\mathcal{P}_\beta\, e = 0$ is zero, then by choosing a random test-vector, with probability 1 the above correlation is strictly less than 1. To obtain the corresponding quantitative result we will assume that we have a testing "device" distinguishing between the two states of the system , which "clicks" if

$$|\mathcal{P}_\beta e|^2 > \varepsilon |e|^2. \tag{11}$$

Unfortunately, the above "device" is not sensitive enough to derive proper estimates for probabilities and we need another norm in the right-hand side of the last inequality. In our case the input space $\mathcal{E}$ is l_2 with the standard orthogonal basis $\mathbf{x} \longrightarrow \{x_m\}_{m=0}^{\infty}$. Following Ref.9 we consider the discrete Sobolev classes and norms introduced in Section 4, (1) and (2) in order to define the case when the "device" clicks. Next we assume that the (complex) increments $x_m - x_{m-1}$ are independent. We are going to use, together with l_2 two more spaces of test-vectors. Both are stochastic spaces of all trajectories $\mathbf{x}(t)$ of a Brownian particle on the complex plane along different discrete sequences of intermediate moments of time ("stops"): the equidistant sequence $t_l = \sum_{m=1}^{l} \delta_m$, for the first space, and the perturbed sequence $\tilde{t}_l = \sum_{m=1}^{l} \Delta_m$, for the second space. Both spaces are equipped with proper Wiener measures W, $\tilde{W}$ (see Ref.30). The measure W is defined on the algebra of all finite-dimensional cylindrical sets $C^{t_1,t_2,\ldots,t_N}_{\Delta_1,\Delta_2,\ldots,\Delta_N}$ of trajectories with fixed initial point $x_0 = 0$ and "gates" Δ_l, $l = 1,\ldots,N$ (which are open discs in the complex plane):

$$C^{t_1,t_2,\ldots,t_N}_{\Delta_1,\Delta_2,\ldots,\Delta_N} = \{\mathbf{x} \mid x_{t_l} \in \Delta_l, l = 1,2,\ldots,N\},$$

via multiple convolutions of the Green functions $G(x_{l+1}, t_{l+1} \big| x_l, t_l)$ corresponding to the sequence $\delta_{l+1} = t_{l+1} - t_l$:

$$W(C^{t_1,t_2,\ldots,t_N}_{\Delta_1,\Delta_2,\ldots,\Delta_N}) =$$

$$\frac{\int \cdots \int_{\Delta_N,\Delta_{N-1},\ldots,\Delta_1} \frac{dx_1 dx_2 \ldots dx_N}{\pi^{\frac{N}{2}} \sqrt{\delta_N \delta_{N-1} \cdots \delta_1}} e^{-\frac{|x_N - x_{N-1}|^2}{\delta_N}} \cdots e^{-\frac{|x_1 - x_0|^2}{\delta_1}}}{\int \cdots \int_{\mathbf{R}_N,\mathbf{R}_{N-1},\ldots,\mathbf{R}_1} \frac{dx_1 dx_2 \ldots dx_N}{\pi^{\frac{N}{2}} \sqrt{\delta_N \delta_{N-1} \cdots \delta_1}} e^{-\frac{|x_N - x_{N-1}|^2}{\delta_N}} \cdots e^{-\frac{|x_1 - x_0|^2}{\delta_1}}}, \tag{12}$$

where $\mathbf{R}_N = \mathbf{R}_{N-1} = \cdots = \mathbf{R}_1 = \mathbf{R}$. Using the convolution formula, the denominator of (12) can be reduced to the Green function $G(x_N, t_N \mid 0, 0)$, for any $\tau \in (s,t)$:

$$G(x,t \mid y,s) = \int_{-\infty}^{\infty} G(x,t \mid \xi,\tau) G(\xi,\tau \mid y,s) d\xi.$$

In a similar way we can define the Wiener measure for trajectories corresponding to the "perturbed" sequence $\tilde{t}_l$.

In what follows we are going to use the *absolute continuity* of the *perturbed* Wiener measure $\tilde{W}$ with respect to the non-perturbed one W: for every W–measurable set Ω,

$$\tilde{W}(\Omega) = \frac{1}{\prod_{l=1}^{\infty} \sqrt{\Delta_l}} \int_{\Omega} e^{-\sum_{m=1}^{\infty} \frac{1-\Delta_m}{\Delta_m} |x_m - x_{m-1}|^2} dW. \tag{13}$$

Further we consider the class of *quasi-loops*, that is the class of all trajectories of the "perturbed process" which begin from $(x_0, \tilde{t}_0) = (0, 0)$ and for any t, $\max_{0<s<t} |x_s|^2 < Ct$. We note that

- every $\mathbf{x} \in l_2^1$ is a quasi-loop (with $C = |\, \mathbf{x}\, |_1^2$),
- due to the reflection principle (see Ref.30, p. 221), the class of all quasi-loops has Wiener measure one, both in respect of W, $\tilde{W}$.

We assume that the device clics, if the result of averaging exceeds a certain level defined by the above norm $\| \cdot \|_1^2$:

$$|\mathcal{P}_\beta \mathbf{x}|^2 > \ \varepsilon \parallel \mathbf{x} \parallel_1^2 .$$

This device cannot identify the state of the system from the observation of the Breit-Wigner averaged correlation between the input and output of a single act of scattering when presented a randomly chosen input $\mathbf{x} \in \mathcal{E}$ if $|\mathcal{P}_\beta \mathbf{x}|^2 < \varepsilon \parallel \mathbf{x} \parallel_1^2$. This means that the test-vector $\mathbf{x}$ belongs to the *indistinguishable set*

$$\mathcal{F}_\varepsilon = \left\{ \mathbf{x} \in l_2 \cap l_2^1 \,,\, |\mathcal{P}_\beta \mathbf{x}|^2 < \ \varepsilon \ \left(\sum_{m=1}^{\infty} \frac{1 - \Delta_m}{\Delta_m} |x_m - x_{m-1}|^2 \right) \right\} \tag{14}$$

$$= \{ \mathbf{x} \in l_2 \cap l_2^1 \,,\, |\mathcal{P}_\beta \mathbf{x}|^2 < \ \varepsilon \parallel \mathbf{x} \parallel_1^2 \}.$$

Though technically we may easily consider, with Breit-Wiegner averaging, the iterated scattering processes described by the powers $\mathbf{S}^m$ of the scattering matrix, we will analyze now the independent single acts of scattering. In this case the indistinguishable set depends only upon the positive number ε, the vector $\overrightarrow{\beta} \in \mathcal{E} \oplus \mathcal{E}$ defining the interaction in the quantum system, and the sequence Δ. Without loss of generality we may assume that the vector $\overrightarrow{\beta}$ has all non-zero components $\overrightarrow{\beta}_l \neq 0$. We assume that the vector $\mathbf{b} = \{b_l\}_{l=1}^{\infty}$, $b_l = \sum_{m=l}^{\infty} \overrightarrow{\beta}_m$ belongs to l_2:

$$|b|_l^2 = \sum_{m=1}^{\infty} m^2 |\overrightarrow{\beta}_m|^2 < \infty. \tag{15}$$

Our main result reads: *If the condition (15) is satisfied, then the Wiener probability $\tilde{W}(\mathcal{F}_{\varepsilon,1})$ of the indistinguishable set $\mathcal{F}_{\varepsilon,1}$ corresponding to a single act of scattering is finite and is estimated as*

$$\tilde{W}(\mathcal{F}_\varepsilon) < \frac{\sqrt{\varepsilon}\, |\overrightarrow{\beta}|}{\prod_{l=1}^{\infty} \sqrt{\Delta_l} \sqrt{\varepsilon |\overrightarrow{\beta}|^2 + |b|^2}}. \tag{16}$$

Following the calculation presented in Ref.9, we approximate the indistinguishable set with finite-dimensional cylinder sets and reduce the estimation of $\tilde{W}(\mathcal{F}_{\varepsilon,1})$ to the calculation of a Wiener integral with respect to the W measure on trajectories associated with "equidistant stops". We have:

$$\tilde{W}(\mathcal{F}_{\varepsilon,1})$$

$$\leq \int_{\text{quasi-loops},x_0=0\,|\langle \mathbf{x},\overrightarrow{\beta}\rangle|^2<\varepsilon\|\mathbf{x}\|_1^2} d\tilde{W}$$

$$= \lim_{C\to\infty} \int_{\sup_{s\leq l}|x_s|<C\sqrt{l},l\leq N,\,x_0=0|\langle \mathbf{x},\overrightarrow{\beta}\rangle|^2<\varepsilon\|\mathbf{x}\|_1^2} d\tilde{W}$$

$$= \frac{1}{\prod_{l=1}^{\infty}\sqrt{\Delta_l}} \lim_{C\to\infty}\lim_{N\to\infty}$$

$$\frac{\int_{\sup_{s\leq l}|x_s|<C\sqrt{l},l\leq N,\,x_0=0}\int\int\cdots\int_{-\infty}^{\infty} dx_1 dx_2\ldots dx_N \prod_{j=1}^{N} e^{-\frac{|x_j-x_{j-1}|^2}{\Delta_j}}}{\pi^N \int_{|x_N|<C\sqrt{N}} dx_N e^{-\frac{x_N-x_0}{t_N}}}$$

$$\leq \frac{1}{\prod_{l=1}^{\infty}\sqrt{\Delta_l}} \lim_{C\to\infty}\lim_{N\to\infty}$$

$$\frac{\int_{\sup_{s\leq l}|x_s|<C\sqrt{l},l\leq N,\,x_0=0}\int\int\cdots\int_{-\infty}^{\infty} dx_1 dx_2\ldots dx_N e^{-\sum_{s=1}^{N}\frac{|x_s-x_{s-1}|^2}{\delta_s}} e^{\mathcal{B}}}{\pi^N \int_{|x_N|<C\sqrt{N}} dx_N e^{-\frac{|x_N-x_0|^2}{t_N}}}.$$

The integrand of the inner integral in the numerator contains the exponential factor

$$e^{\mathcal{B}} = e^{-(1-\Delta_N)\frac{|x_N-x_{N-1}|^2}{\Delta_N}-(1-\Delta_{N-1})\frac{|x_{N-1}-x_{N-2}|^2}{\Delta_{N-1}}-(1-\Delta_1)\frac{|x_1-x_0|^2}{\Delta_1}},$$

which can be estimated due to (14) by the exponential:

$$e^{-\frac{1}{\varepsilon}|\mathcal{P}_{\overrightarrow{\beta}}\mathbf{x}|^2} = e^{-\frac{1}{\varepsilon|\overrightarrow{\beta}|^2}|\langle \mathbf{x},\,\overrightarrow{\beta}\rangle|^2}.$$

Using this equality, the exponential in the numerator can be estimated from below by the quadratic form

$$-\sum_{m=1}^{N}|x_m-x_{m-1}|^2 - \frac{1}{\varepsilon|\overrightarrow{\beta}|^2}|\langle \mathbf{x},\,\overrightarrow{\beta}\rangle|^2. \tag{17}$$

This quadratic form can be simplified using new vector variables $\xi_m = x_m - x_{m-1}$:

$$\langle \vec{\beta}, \mathbf{x} \rangle = \sum_{m=1}^{\infty} x_m \bar{\vec{\beta}}_m = \sum_{m=1}^{\infty} \xi_m \overline{\sum_{l=m}^{\infty} \vec{\beta}_l}.$$

Recall that the vector $\mathbf{b}$, $b_m = \sum_{l=m}^{\infty} \beta_l$ belongs to l_2. Then the quadratic form in the exponent of the numerator can be presented as a quadratic form of an operator

$$\langle \xi, A_\varepsilon \xi \rangle = |\xi|^2 + \frac{1}{|\vec{\beta}|^2 \varepsilon} |\langle \mathbf{b}, \xi \rangle| = \langle \xi, \left(\mathbf{I} + \frac{|\mathbf{b}|^2}{|\vec{\beta}|^2 \varepsilon} \mathbf{P_b} \right) \xi \rangle,$$

where $\mathbf{P}_b$ is the orthogonal projection onto the one-dimensional subspace in l_2 spanned by the vector $\mathbf{b}$. The ratio of the N-dimensional Gaussian integral in the numerator, normalized by the factor $\pi^{-N/2}$ and the Gaussian integral in the denominator can be expressed as

$$\frac{1}{\pi^{N/2}} \int \int \cdots \int e^{-\langle \xi, A_\varepsilon \xi \rangle} d\xi_1 d\xi_2 \ldots d\xi_N = \frac{1}{\sqrt{\det A_\varepsilon}} = \frac{\sqrt{\varepsilon} \| \vec{\beta} |}{\sqrt{\varepsilon |\vec{\beta}|^2 + |b|^2}}.$$

Finally, we obtain the announced result by taking into account the omitted factor $\prod_l \sqrt{\Delta_l}$.

Acknowledgement

We thank Radu Ionicioiu for his comments on a draft form of this paper. Calude and Pavlov have been supported in part by The Vice-Chancellor's University Development Fund 23124/2002.

References

1. V. A. Adamyan. Scattering matrices for microschemes, *Operator Theory: Adv. & Appl.* 59 (1992), 1–10.
2. V. A. Adamyan, C. S. Calude, B. S. Pavlov. A Quantum Scattering Approach to Undecidable Problems: Preliminary Version, *CDMTCS Research Report* 198, 2002, 29 pp.
3. J. del Alamo, C. Eugster. Quantum field-effect directional coupler, *Appl. Phys. Letters* 56 (1) (1990), 78–83.
4. I. Antoniou, B. Pavlov, A. Yafyasov. Quantum electronic devices based on metal-dielectric transition in low-dimensional quantum structures, in D. S. Bridges, C. Calude, J. Gibbons, S. Reeves, I. Witten (eds.). *Combinatorics, Complexity, Logic, Proceedings of DMTCS'96*, Springer Verlag, Singapore, 1996, 90–104.

5. C. S. Calude. *Information and Randomness: An Algorithmic Perspective*, 2nd Edition, Revised and Extended, Springer Verlag, Berlin, 2002.
6. C. S. Calude, J. L. Casti. Parallel thinking, *Nature* 392, 9 April (1998), 549–551.
7. C. S. Calude, M. J. Dinneen, K. Svozil. Reflections on quantum computing, *Complexity* 6, 1 (2000), 35–37.
8. C. S. Calude, G. Păun. *Computing with Cells and Atoms*, Taylor and Francis Publishers, London, 2001.
9. C. S. Calude, B. Pavlov. Coins, quantum measurements, and Turing's barrier, *Quantum Information Processing* 1, 1–2 (2002), 107–127.
10. J. L. Casti. Computing the uncomputable, *The New Scientist*, 154/2082, 17 May (1997), 34.
11. J. Copeland. Hypercomputation, *Minds and Machines: Journal for Artificial Intelligence, Philosophy and Cognitive Science*, 12, 4 (2002), 461–502.
12. J.-P. Delahaye. Les nombres oméga, *Pour la Science*, 292 May (2002), 98–103.
13. D. Deutsch. Quantum theory, the Church-Turing principle and the universal quantum computer, *Proceedings of the Royal Society London*, A 400 (1985), 97–119.
14. D. Deutsch. *The Fabric of Reality*, Allen Lane, Penguin Press, 1997.
15. G. Etesi, I. Németi. Non-Turing computations via Malament-Hogarth space-times, *International Journal of Theoretical Physics* 41 (2002), 341–370.
16. P. Exner, P. Sheba, M. Tater. Bound states and scattering in quantum waveguides coupled laterally through a boundary window, *J. Math. Phys.* 37 (1996), 4467–4887.
17. R. P. Feynman. Quantum mechanical computers, *Optics News* 11 (1985), 11–20.
18. R. P. Feynman. *Feynman Lectures on Computation*, J. G. Hey and R. W. Allen (Eds.), Addison-Wesley, Reading, Massachusetts, 1996.
19. J. G. Hey (ed.). *Feynman and Computation. Exploring the Limits of Computers*, Perseus Books, Reading, Massachusetts, 1999.
20. J. Gruska. *Quantum Computing*, McGraw-Hill, London, 1999.
21. A. N. Ignatov. Personal communication to B. Pavlov, 1970.
22. T. D. Kieu. Quantum algorithm for the Hilbert's tenth problem, Los Alamos preprint archive `http://arXiv:quant-ph/0110136`, v2, 9 November 2001.
23. R. Landauer. Information is physical, *Physics Today,* 44 (1991), 23–29.
24. R. Landauer. Information is inevitably physical, in Ref.19, 76–92.
25. B. Maclennan. Transcending Turing Computability, *Technical Report UT-CS-01-473*, University of Tennessee, 2001.
26. O. Madelung. *Introduction to Solid-state Theory.* Translated from German by B. C. Taylor. Springer-Verlag, Berlin, New York, 1978.
27. Yu. B. Melnikov, B. S. Pavlov. Two-body scattering on a graph and application to simple nano-electronic devices, *J. Math. Phys.* 36, 6, (1995), 2813-2825.
28. B. Pavlov. A model of zero-range potential with an internal structure, *Teor. Mat. Fizika*, 59 (1984), 345–353. (in Russian) (English Translation: *Theor.*

Math. Physics 59 (1984), 544–550.)
29. B. Pavlov. The theory of extensions and explicitly-solvable models, *Russian Math. Surveys* 42, 6 (1987), 127–168.
30. D. W. Stroock. *Probability Theory. An Analytic View*, Cambridge University Press, Cambridge, 1993.
31. C. Teuscher, M. Sipper. Hypercomputation: Hype or computation?, *Communications ACM* 45,8 (2002), 23–24.
32. K. Svozil. The Church-Turing Thesis as a guiding principle for physics, in C. S. Calude, J. Casti and M. J. Dinneen (eds.). *Unconventional Models of Computation*, Springer, Singapore, 1998, 371–385.

HOMOGENIZATION OF INFINITE DIMENSIONAL DIFFUSION PROCESSES WITH PERIODIC DRIFT COEFFICIENTS

S. ALBEVERIO

Inst. Angewandte Mathematik, Universität Bonn
Wegelerstr. 6, D-53115 Bonn (GERMANY), SFB611; BiBoS;
CERFIM, Locarno; Acc. Architettura USI, Mendrisio;
Dipartimento di Matematica, Università di Trento (ITALY)

M. S. BERNABEI

Inst. Angewandte Mathematik, Universität Bonn
Wegelerstr. 6, D-53115 Bonn; Dipartimento di Matematica e Informatica,
Università di Camerino, 62032 Camerino, Via Madonna delle Carceri, 9
(ITALY)

M. RÖCKNER

Fakultät für Mathematik, Universität Bielefeld
D-33615 Bielefeld (GERMANY); BiBoS

M. W. YOSHIDA

Dept. Systems Engineering The Univ. ELECTRO-COMMUNICATIONS
1-5-1, Chofugaoka, Chofu, Tokyo, 182-8585 (JAPAN)
e-mail yoshida@se.uec.ac.jp fax +81 424 98 0541

A homogenization problem of infinite dimensional diffusion processes indexed by $\mathbf{Z}$ having periodic drift coefficients is considered. By making use of an approximation sequence that converges to the original infinite dimensional diffusion process, the homogenization problem in a generalized sense is discussed. Here, the scaling limit of the original process is not taken directly, but the scaling limit of an approximation sequence, whose running index is controlled by the parameter of the scaling limit, is taken. The existence of a non trivial limiting process which is an infinite dimensional diffusion having constant coefficients is shown.

1. Introduction and setting of the problem

In this paper we discuss the limiting probability distribution of a scaling limit of infinite dimensional diffusion processes on $\mathbf{R}^{\mathbf{Z}}$ with periodic drift coefficients. Here and in the sequel we use $\mathbf{Z}$ and $\mathbf{N}$ to denote the set of integers and natural numbers respectively. By making use of an approximation sequence that converges to the original infinite dimensional diffusion process, we discuss the homogenization problem in a generalized sense, namely, here we do not take the scaling limit of the original process directly, but take the scaling limit of an approximation sequence the index of which is controlled by the parameter of the scaling limit, such that the index diverges to ∞ as the scaling parameter tends to 0 (cf. Problem $\mathrm{Q_G}$ below). We show the existence of a non trivial limiting process for the so modified process which is an infinite dimensional diffusion having constant coefficients.

By the discussion of the modified problem we will be able to reconsider the original homogenization problem (cf. Remark 1.1), however, the detailed discussions of this will be postponed to a future study.

The homogenization problem for diffusion processes with regular coefficients can be best explained by means of a formulation using stochastic differential equations. We begin with the definition of a complete probability space $(\Omega, \mathcal{F}, P; \mathcal{F}_t)$ on which the stochastic differential equations are defined.

Definition 1.1. Let $(\Omega, \mathcal{F}, P; \mathcal{F}_t)$ be a complete probability space with an increasing family of sub σ-field $\mathcal{F}_t \subset \mathcal{F}$ $(t \in \mathbf{R}_+)$. Suppose that on $(\Omega, \mathcal{F}, P; \mathcal{F}_t)$ we are given a family of independent one-dimensional $\mathcal{F}_t$ standard Brownian motion processes $\{B_k(t)\}_{t\in\mathbf{R}_+}$ $(k \in \mathbf{Z})$. □

Suppose that we are given a function $J \in C^\infty(\mathbf{R} \to \mathbf{R}_+)$ such that

(i) there exists $\beta > 0$ and $J(x) \geq \beta$ holds for all $x \in \mathbf{R}$;
(ii) the derivatives of all order of J are bounded functions;
(iii) J is a periodic function with the period 2π.

In Lemma 1.2 of [HS] through an iteration argument it is shown that for each $\epsilon > 0$ and each initial state $\mathbf{x} = (\cdots, x_{-1}, x_0, x_{+1}, \cdots) \in \mathbf{R}^{\mathbf{Z}}$ there exists a unique strong solution $\{\mathbb{X}^\epsilon(t, \mathbf{x})\}_{t\in\mathbf{R}_+}$ on $(\Omega, \mathcal{F}, P; \mathcal{F}_t)$ such that (cf., e.g, [IW] for the concept of strong solution, and for the general infinite dimensional formulation cf. [H], [HKPS])

$$\begin{aligned} X_k^\epsilon(t,\mathbf{x}) &= x_k + \sqrt{2}B_k(t) \\ &+ \frac{1}{\epsilon}\int_0^t \{J'(\frac{X_{k+1}^\epsilon(s,\mathbf{x})}{\epsilon} - \frac{X_k^\epsilon(s,\mathbf{x})}{\epsilon}) \\ &- J'(\frac{X_k^\epsilon(s,\mathbf{x})}{\epsilon} - \frac{X_{k-1}^\epsilon(s,\mathbf{x})}{\epsilon})\}ds, \qquad k \in \mathbf{Z}, \end{aligned} \tag{1}$$

$$\mathbb{X}^\epsilon(t,\mathbf{x}) = (\cdots, X_{-1}^\epsilon(t,\mathbf{x}), X_0^\epsilon(t,\mathbf{x}), X_{+1}^\epsilon(t,\mathbf{x}), \cdots).$$

A formal extension of the homogenization problem of finite dimensional diffusions to the problem of infinite dimensional diffusions will be the following.

Problem Q

Characterize the probability law, if it exists, of the limiting process

$$\lim_{\epsilon\downarrow 0} \mathbb{X}^\epsilon(t,\mathbf{x}), \qquad t \in \mathbf{R}_+. \tag{2a}$$

□

Problem Q has an another equivalent expression Q'. Namely, let $\mathbb{X}^1(\cdot,\mathbf{x})$ be the solution of (1) with $\epsilon = 1$. For the standard Brownian motion process $\{B_k(t)\}_{t\in\mathbf{R}_+}$ the process $\{\epsilon B_k(\frac{t'}{\epsilon^2})\}_{t'\in\mathbf{R}_+}$ is again a standard Brownian motion process. By changing the time variable such that $\frac{t'}{\epsilon^2} = t$ (cf. Remark 2.1-ii)), we see that the probability law of $\{\epsilon\mathbb{X}^1(\frac{t}{\epsilon^2}, \frac{\mathbf{x}}{\epsilon})\}_{t\in\mathbf{R}_+}$ is equivalent to the one of $\{\mathbb{X}^\epsilon(t,\mathbf{x})\}_{t\in\mathbf{R}_+}$. This gives the following equivalent problem:

Problem Q'

Characterize the probability law, if it exists, of the scaling limit such that

$$\lim_{\epsilon\downarrow 0} \epsilon\mathbb{X}^1(\frac{t}{\epsilon^2}, \frac{\mathbf{x}}{\epsilon}), \qquad t \in \mathbf{R}_+. \tag{2b}$$

□

In the formalism of the infinite dimensional diffusions, it does not seem natural to consider the limit (2a) and (2b) for each initial state $\mathbf{x} \in \mathbf{R}^{\mathbf{Z}}$ (cf. Remark 2.2 in the next section), and it may be natural, not only by the limitations caused by the analytic tools at hand but also from the point of view of the physical phenomena of infinite particle systems, that the initial state has a distribution which is not a Dirac point measure.

In addition, to discuss the homogenization problem, the functions χ_k, $k \in \mathbf{Z}$, which are well-known in the literature and defined in Lemma 2.1 play a crucial role, however for the infinite dimensional situation the information concerning the regularities of χ_k ($k \in \mathbf{Z}$) is limited (cf. Lemma 2.1). In this paper we do not discuss above original Problems Q, Q', but rather the following Problem $\mathbf{Q_G}$, where the procedure of taking the scaling limit is modified.

Problem $\mathbf{Q_G}$

For the infinite dimensional diffusion $\{\mathbb{X}^\epsilon(t,\mathbf{x})\}_{t\geq 0}$ in Problem Q, let $\{\mathbb{X}^{\epsilon,N}(t,\mathbf{x})\}_{t\geq 0}$ be an approximation sequence of the original process such that for each $\epsilon > 0$

$$\lim_{N\to\infty} \mathbb{X}^{\epsilon,N}(\cdot,\cdot) = \mathbb{X}^\epsilon(\cdot,\cdot), \qquad \textit{in law.}$$

Characterize the probability law, if it exists, of the limiting process

$$\lim_{\epsilon\downarrow 0} \mathbb{X}^{\epsilon,N(\epsilon)}(t,\mathbf{x}), \qquad t \in \mathbf{R}_+,$$

for some sequence $N(\epsilon) \in \mathbf{N}$ indexed by $\epsilon > 0$, such that

$$\lim_{\epsilon\downarrow 0} N(\epsilon) = +\infty.$$

□

Our intermediate goal is to show that we can find an $N(\epsilon)$ satisfying the assumption of Problem $\mathrm{Q_G}$ and such that $\{\mathbb{X}^{\epsilon,N(\epsilon)}(t,\mathbf{x})\}_{t\in\mathbf{R}_+}$ converges weakly as $\epsilon \downarrow 0$, to a continuous process $\{\mathbb{Y}(t)\}_{t\in\mathbf{R}_+}$, which is an infinite dimensional diffusion with a constant covariance matrix, under the condition that the initial state $\mathbf{x}$ satisfies that $\frac{\mathbf{x}}{\epsilon}$ obeys a probability distribution whose Radon-Nikodym density with respect to the "Gibbs state" of $\{\mathbb{X}^1(t,\mathbf{x})\}_{t\in\mathbf{R}_+}$ is a bounded function.

We give a preliminary formulation of the corresponding result (cf. Theorem 2.2 below). Most of the preparation of the proof is already done in this paper, a complete proof will be presented in a forthcoming paper.

Above we used the terminology "Gibbs state", however the distribution of $\{\mathbb{X}^1(t,\mathbf{x})\}_{t\in\mathbf{R}_+}$ is not a Gibbs state in the strict sense, but rather a Gibbs state of the quotient process $\boldsymbol{\eta}_t$ such that $\mathbb{X}^1(t,\mathbf{x}) = \boldsymbol{\eta}_t \bmod 2\pi$ (cf. Proposition 2.1 in the next section).

Remark 1.1. (scope for the original problem Q)

By making use of $\{\mathbb{X}^{\epsilon,N(\epsilon)}(t,\mathbf{x})\}_{t\in\mathbf{R}_+}$, by which the modified problem $\mathrm{Q_G}$ is solved, we will be able to reconsider the original problem Q. In fact, in our procedures to prove Theorem 2.2 and Lemma 2.2, for each $\epsilon > 0$ we construct the approximation process $\{\mathbb{X}^{\epsilon,N(\epsilon)}(t,\mathbf{x})\}_{t\in\mathbf{R}_+}$ on the same probability space on which the original process $\{\mathbb{X}^{\epsilon}(t,\mathbf{x})\}_{t\in\mathbf{R}_+}$ is defined, and we can then evaluate the distance between these two continuous processes by using a martingale inequality. In this way, we can also discuss the tightness of the probability laws of the original sequence $\{\mathbb{X}^{\epsilon}(t,\mathbf{x})\}_{t\in\mathbf{R}_+}$, $(\epsilon > 0)$. For this discussion we should take a new metric on $\mathbf{R}^{\mathbf{Z}}$ that is weaker than the metric adopted here (cf. Definition 2.1, and section 23 of [I]). This will be discussed in future work.

□

2. Fundamental properties and the main results

As shown in [HS] the process in question really lives on a much smaller space, if the process starts there, than $\mathbf{R}^{\mathbf{Z}}$, more precisely we have that for all $T \geq 0$ there exists $A_T > 0$ such that

$$\begin{aligned} E[\sum_{k\in\mathbf{Z}} \sup_{0\leq t\leq T} \frac{1}{2^{|k|}} |X_k^1(t,\mathbf{x}) - X_k^1(t,\mathbf{x}')|^2] \\ \leq A_T(\sum_{k\in\mathbf{Z}} \frac{1}{2^{|k|}} |x_k - x_k'|^2). \end{aligned} \tag{3}$$

where $\{\mathbb{X}^1(t,\mathbf{x})\}_{t\in\mathbf{R}_+}$ is the process defined by (1) with $\epsilon = 1$ (cf. [HS, Lemma (1.2), estimate (1.2)]). Therefore we define for the sequence $\mathbf{q} = (2^{|k|})_{k\in\mathbf{Z}}$ the Hilbert space

$$l_{\mathbf{q}}^2 \equiv \{\mathbf{x} \in \mathbf{R}^{\mathbf{Z}} \mid \|\mathbf{x}\|_{\mathbf{q}}^2 \equiv \sum_{k\in\mathbf{Z}} \frac{1}{2^{|k|}} x_k^2 < \infty\} \quad \text{with norm } \|\cdot\|_{\mathbf{q}}$$

and consider the processes with path space $C(\mathbf{R}_+ \to l_{\mathbf{q}}^2)$.

Our problem (**Q**), (**Q**') or ($\mathbf{Q_G}$) posed in the previous section consists in proving the weak convergence of the sequence of probability measures on $C(\mathbf{R}_+ \to l_{\mathbf{q}}^2)$, which are the probability laws of the processes $\{\mathbb{X}^{\epsilon}(t,\mathbf{x})\}_{t\in\mathbf{R}_+}$.

Definition 2.1. Let $W^{\mathbf{Z}} \equiv C(\mathbf{R}_+ \to l_{\mathbf{q}}^2)$ be the polish space equipped with the metric

$$d(\mathbf{x}(\cdot),\mathbf{x}'(\cdot)) \equiv \sum_{n=1}^{\infty} \frac{1}{2^n} \Big(d_n(\mathbf{x}(\cdot),\mathbf{x}'(\cdot)) \wedge 1 \Big), \tag{4}$$

with

$$d_n(\mathbf{x}(\cdot),\mathbf{x}'(\cdot)) \equiv \sup_{0\leq s\leq n} \|\mathbf{x}(s) - \mathbf{x}'(s)\|_{\mathbf{q}}, \qquad n \in \mathbf{N}, \quad \mathbf{x}(\cdot),\, \mathbf{x}'(\cdot) \in W^{\mathbf{Z}},$$

$$\mathbf{x}(s) = (\cdots, x_{-1}(s), x_0(s), x_{+1}(s), \cdots), \quad x_k(\cdot) \in C(\mathbf{R}_+ \to \mathbf{R}),\ k \in \mathbf{Z}.$$

Let $\mathcal{B}(W^{\mathbf{Z}})$ be the Borel σ-field of $W^{\mathbf{Z}}$ and $\mathcal{B}_t(W^{\mathbf{Z}})$, $t \in \mathbf{R}_+$, be the sub σ-field of $\mathcal{B}(W^{\mathbf{Z}})$ generated by the open bases $\{\mathbf{x}'(\cdot) \in W^{\mathbf{Z}} \mid d_t(\mathbf{x}(\cdot), \mathbf{x}'(\cdot)) < r\}$, $r \in \mathbf{R}_+$, $\mathbf{x}(\cdot) \in W^{\mathbf{Z}}$, where

$$d_t(\mathbf{x}(\cdot), \mathbf{x}'(\cdot)) \equiv \sup_{0 \le s \le t} \|\mathbf{x}(s) - \mathbf{x}'(s)\|_{\mathbf{q}}.$$

Then $\mathcal{B}(W^{\mathbf{Z}})$ and $\mathcal{B}_t(W^{\mathbf{Z}})$ are identical to the σ-field generated by the cylinder sets of $C(\mathbf{R}_+ \to l^2_{\mathbf{q}})$ and $C([0,t] \to l^2_{\mathbf{q}})$ respectively.

□

Definition 2.2. For each $\epsilon > 0$ and $\mathbf{x} \in \mathbf{R}^{\mathbf{Z}}$, let $\{\mathbb{X}^\epsilon(t, \mathbf{x})\}_{t \in \mathbf{R}_+}$ be the unique strong solution of (1). Let $P^\epsilon_{\mathbf{x}}$ be the probability measure on $(W^{\mathbf{Z}}, \mathcal{B}(W^{\mathbf{Z}}))$ which is the probability law of $\{\mathbb{X}^\epsilon(t, \mathbf{x})\}_{t \in \mathbf{R}_+}$. Then $P^\epsilon_{\mathbf{x}}$ defines a diffusion process taking values in $\mathbf{R}^{\mathbf{Z}}$, and for any $A \in \mathcal{B}(W^{\mathbf{Z}})$

$$P^\epsilon_{\cdot} : \mathbf{R}^{\mathbf{Z}} \mapsto P^\epsilon_{\mathbf{x}}(A)$$

is measurable. Let ν be a Borel probability measure on $\mathbf{R}^{\mathbf{Z}}$, then the probability measure $P^\epsilon_\nu(\cdot) = \int_{\mathbf{R}^{\mathbf{Z}}} P^\epsilon_{\mathbf{x}}(\cdot) d\nu(\mathbf{x})$ on $(W^{\mathbf{Z}}, \mathcal{B}(W^{\mathbf{Z}}))$ is the probability law of the unique strong solution $\{\mathbb{X}^\epsilon(t, \nu)\}_{t \in \mathbf{R}_+}$ of the stochastic differential equation (1) with initial state given by the probability distribution ν.

□

Definition 2.3. Let $T = \{y \in \mathbf{R}^2 : |y| = 1\}$ be the unit circle equipped with the Riemanianian metric. Let $T^{\mathbf{Z}}$ be the product space of T endowed with the product topology, so that $T^{\mathbf{Z}}$ is a Polish space. Let $(W^{\mathbf{Z}}_T, \mathcal{B}(W^{\mathbf{Z}}_T); \mathcal{B}_t(W^{\mathbf{Z}}_T))$ be the measurable space of the Polish space $W^{\mathbf{Z}}_T \equiv C(\mathbf{R}_+ \to T^{\mathbf{Z}})$, such that $\mathcal{B}(W^{\mathbf{Z}}_T)$ is the Borel σ-field of $W^{\mathbf{Z}}_T$ and $\mathcal{B}_t(W^{\mathbf{Z}}_T)$, $t \in \mathbf{R}_+$, is the sub σ-field of $\mathcal{B}(W^{\mathbf{Z}}_T)$ generated by the open sets of $C([0,t] \to T^{\mathbf{Z}})$.

□

Definition 2.4.

(i) We use the symbol $\mathbf{y} = (\cdots, y_{-1}, y_0, y_1, \cdots)$ to denote the point in $T^{\mathbf{Z}}$ as well as the point $(\cdots, \theta_{-1}, \theta_0, \theta_1, \cdots) \in [0, 2\pi)^{\mathbf{Z}}$ such that

$$y_k = (\cos\theta_k, \sin\theta_k) \in T \subset \mathbf{R}^2, \quad k \in \mathbf{Z}.$$

(ii) Let $C_0^\infty(T^{\mathbf{Z}})$ be the set of all finite linear combinations of functions on $T^{\mathbf{Z}}$ that have only a finite number of coordinates, and are infinitely differentiable. We will also interpret the elements of $C_0^\infty(T^{\mathbf{Z}})$ as the smooth functions on $\mathbf{R}^{\mathbf{Z}}$ which have only a finite number of coordinates and are infinitely differentiable periodic functions with the period 2π in each direction.

$\square$

Proposition 2.1. ***(Theorem 1.3 of [HS])***
For each $t \geq 0$, let $\boldsymbol{\eta}_t$ be the measurable function such that

$$\boldsymbol{\eta}_t : W_T^{\mathbf{Z}} \ni \mathbf{y}(\cdot) \longmapsto \mathbf{y}(t) \in T^{\mathbf{Z}}. \tag{5}$$

(i) Then for each $\mathbf{y} \in T^{\mathbf{Z}}$ there exists a unique probability measure $Q_{\mathbf{y}}$ on $(W_T^{\mathbf{Z}}, \mathcal{B}(W_T^{\mathbf{Z}}))$ such that

$$Q_{\mathbf{y}}(\boldsymbol{\eta}_0 = \mathbf{y}) = 1,$$

$$(f(\boldsymbol{\eta}_t) - \int_0^t (Lf)(\boldsymbol{\eta}_s)ds, \mathcal{B}_t(W_T^{\mathbf{Z}}), Q_{\mathbf{y}})$$

is a martingale for each $f \in C_0^\infty(T^{\mathbf{Z}})$, where

$$(L\,f)(\mathbf{y}) = \sum_{k\in\mathbf{Z}} \frac{\partial^2 f}{\partial y_k^2}(\mathbf{y}) + (J'(y_{k+1} - y_k) - J'(y_k - y_{k-1}))\frac{\partial}{\partial y_k} f(\mathbf{y}), \tag{6}$$

(here and in the sequel the periodic function J on $\mathbf{R}$ is interpreted as a function on T (cf. Definition 2.4));

(ii) for each $t \in \mathbf{R}_+$, let $\boldsymbol{\xi}_t$ be the measurable function such that

$$\boldsymbol{\xi}_t : W^{\mathbf{Z}} \ni \mathbf{x}(\cdot) \longmapsto \mathbf{x}(t) \in \mathbf{R}^{\mathbf{Z}}. \tag{7}$$

For $\mathbf{x} \in \mathbf{R}^{\mathbf{Z}}$ let $P_{\mathbf{x}}^1$ be the probability measure on $(W^{\mathbf{Z}}, \mathcal{B}(W^{\mathbf{Z}}))$ defined by Definition 2.2 (by $\epsilon = 1$), which is the probability law of the process $\{\mathbb{X}^1(t,\mathbf{x})\}_{t\in\mathbf{R}_+}$. Then for any bounded measurable function $F : (T^{\mathbf{Z}})^n \to \mathbf{R}$, $n \in \mathbf{N}$, the following holds:

$$\begin{aligned} &E^{Q_{\mathbf{y}}}[F(\boldsymbol{\eta}_{t_1}, \ldots, \boldsymbol{\eta}_{t_n})] \\ &= E^{P_{\mathbf{y}}^1}[F(\boldsymbol{\Phi}(\boldsymbol{\xi}_{t_1}), \ldots, \boldsymbol{\Phi}(\boldsymbol{\xi}_{t_n}))], \qquad \forall \mathbf{y} \in [0, 2\pi)^{\mathbf{Z}}, \end{aligned}$$

where

$$\begin{aligned} \boldsymbol{\Phi} : \mathbf{R}^{\mathbf{Z}} \ni \mathbf{x} &= (\cdots, x_{-1}, x_0, x_1, \cdots) \\ &\longmapsto (\cdots, \Phi(x_{-1}), \Phi(x_0), \Phi(x_1), \cdots) \in T^{\mathbf{Z}} \end{aligned} \tag{8}$$

is the map defined by

$$\Phi(x_k) = y_k \in [0, 2\pi) \quad \textit{if} \quad x_k = y_k \ \textit{mod}\ 2\pi \quad \textit{(cf. Definition 2.4-(i)).}$$

□

Let $\mathcal{B}(T^{\mathbf{Z}})$ be the Borel σ-field of $T^{\mathbf{Z}}$. We denote a sub set of $\mathbf{Z}$ by Λ. For each $\Lambda \subset \mathbf{Z}$, we define the projection

$$\mathbf{y} \in T^{\mathbf{Z}} \longmapsto \mathbf{y}_\Lambda \in T^\Lambda, \tag{9}$$

and define a σ-field $\mathcal{B}_\Lambda$ which is a sub σ-field of $\mathcal{B}(T^{\mathbf{Z}})$ such that $\mathcal{B}_\Lambda$ is the smallest σ-field generated by the functions on $T^{\mathbf{Z}}$ having the support in T^Λ, i.e. these functions admitting the expression such that

$$T^{\mathbf{Z}} \ni \mathbf{y} \longmapsto \varphi(\mathbf{y}_\Lambda) \in \mathbf{R} \tag{10}$$

for some measurable function φ on T^Λ. Let $\mathbb{F}_0$ be the family of all non-empty *finite* subsets $\Lambda \subset\subset \mathbf{Z}$.

Proposition 2.2. ***(Section 3 of [S])***
Let J be the pair potential defined in the previous section. Let $C_0(T^{\mathbf{Z}})$ be the set of continuous functions on $T^{\mathbf{Z}}$ having the support on $T^{\Lambda'}$ for some bounded $\Lambda' \in \mathbb{F}_0$. Then there exists exactly one Gibbs probability measure μ on $(T^{\mathbf{Z}}, \mathcal{B}(T^{\mathbf{Z}}))$ such that

$$< E^\Lambda \varphi, \mu > = < \varphi, \mu >, \qquad \forall \Lambda \in \mathbb{F}_0, \quad \forall \varphi \in C_0(T^{\mathbf{Z}}), \tag{11}$$

where

$$[E^\Lambda \varphi](\mathbf{y}) = \frac{1}{Z_\Lambda(\mathbf{y}_{\Lambda^c})} \int_{T^{\mathbf{Z}}} \varphi(\mathbf{y'}_\Lambda \cdot \mathbf{y}_{\Lambda^c}) \exp(-U^\Lambda(\mathbf{y'}_\Lambda \cdot \mathbf{y}_{\Lambda^c})) d\mathbf{y'}, \tag{12}$$

$$U^\Lambda(\mathbf{y}) \equiv \sum_{k \in \Lambda} \{J(y_k - y_{k-1}) + J(y_{k+1} - y_k)\}, \quad \mathbf{y} \in T^{\mathbf{Z}}, \tag{13}$$

$d\mathbf{y} = (dy)^{\mathbf{Z}}$ is the product measure of the Riemannian measure dy on T. Here we use the notation $\mathbf{y}'_\Lambda \cdot \mathbf{y}_{\Lambda^c} \equiv \mathbf{y}'' \in T^{\mathbf{Z}}$, so that $\mathbf{y}''_\Lambda = \mathbf{y}'_\Lambda$ and $\mathbf{y}''_{\Lambda^c} = \mathbf{y}_{\Lambda^c}$. Moreover $Z_\Lambda(\mathbf{y}_{\Lambda^c})$ in (12) is the normalizing constant of the Gibbs state, i.e.

$$Z_\Lambda(\mathbf{y}_{\Lambda^c}) = \int_{T^{\mathbf{Z}}} \exp[-U^\Lambda(\mathbf{y}'_\Lambda \cdot \mathbf{y}_{\Lambda^c})] d\mathbf{y}'$$

□

Remark 2.1.

(i) By Proposition 2.2 since the Gibbs probability measure corresponding to the pair potential J is unique (for general cases cf. [AKRT], [FU]) , and the martingale problem on $(W_T^{\mathbf{Z}}, \mathcal{B}(W_T^{\mathbf{Z}}), \mathcal{B}_t(W_T^{\mathbf{Z}}))$ considered in Proposition 2.1 is uniquely solved by the system of probability measures $Q_{\mathbf{y}}$, $(\mathbf{y} \in T^{\mathbf{Z}})$, we see that the diffusion process $(\eta_t, Q_{\mathbf{y}})$ is identified with the diffusion processes corresponding to the Dirichlet space $(L^2(T^{\mathbf{Z}}; \mu), \mathcal{E}, \mathcal{D}[\mathcal{E}])$, where $\mathcal{E}$ is the smallest closed extension of (using the same symbol)

$$\mathcal{E}(u,v) = \sum_{j \in \mathbf{Z}} \int_{T^{\mathbf{Z}}} (\frac{\partial}{\partial y_j} u(\mathbf{y}))(\frac{\partial}{\partial y_j} v(\mathbf{y})) d\mu, \quad \text{for} \quad u, v \in C_0^\infty(T^{\mathbf{Z}}), \tag{14}$$

and the domain $\mathcal{D}[\mathcal{E}]$ is the closure of $C_0^\infty(T^{\mathbf{Z}})$ with respect to the norm $|||u||| \equiv \|u\|_{L^2(\mu)} + (\mathcal{E}(u,v))^{\frac{1}{2}}$ (cf. section V of [S], and cf. [F], [MR] for general notions of the Dirichlet space). In the sequel we use this identification.

(ii) Since, for the standard Brownian motion process $\{B_k(t)\}_{t \in \mathbf{R}_+}$ the scaling process $\{\frac{B(\epsilon^2 t)}{\epsilon}\}_{t \in \mathbf{R}_+}$ is also a standard Brownian motion process, in addition since the term

$$\frac{1}{\epsilon} \int_0^t \{J'(\frac{X_{k+1}^\epsilon(s,\mathbf{x})}{\epsilon} - \frac{X_k^\epsilon(s,\mathbf{x})}{\epsilon}) - J'(\frac{X_k^\epsilon(s,\mathbf{x})}{\epsilon} - \frac{X_{k-1}^\epsilon(s,\mathbf{x})}{\epsilon})\} ds$$

in (1) is a Riemann integral of a continuous function of s for a.s. $\omega \in \Omega$, as was briefly mentioned in Problem Q', by changing the time variable t by $\epsilon^2 t$ in (1), then from the stochastic differential equation corresponding to this modification, we see that $\{\epsilon^{-1}\mathbb{X}^\epsilon(\epsilon^2 t, \mathbf{x})\}_{t \in \mathbf{R}_+}$ satisfies the stochastic differential equation which is satisfied by $\{\mathbb{X}^1(t, \frac{\mathbf{x}}{\epsilon})\}_{t \in \mathbf{R}_+}$. Thus in the sequel we use the identification between the processes $\{\epsilon^{-1}\mathbb{X}^\epsilon(\epsilon^2 t, \mathbf{x})\}_{t \in \mathbf{R}_+}$ and $\{\mathbb{X}^1(t, \frac{\mathbf{x}}{\epsilon})\}_{t \in \mathbf{R}_+}$ as well, in the sense that they have the same probability law.

□

Lemma 2.1. *For any $k \in \mathbf{Z}$ the following is well defined as a measurable function of $\mathbf{y} \in T^{\mathbf{Z}}$:*

$$E^{Q_{\mathbf{y}}}[\int_0^\infty \{J'(\eta_s^{k+1}(\cdot) - \eta_s^k(\cdot)) - J'(\eta_s^k(\cdot) - \eta_s^{k-1}(\cdot))\} ds].$$

Denote this function by $\chi_k(\mathbf{y})$, $\mathbf{y} \in T^{\mathbf{Z}}$, then χ_k, $(k \in \mathbf{Z})$, satisfy the following:

$\chi_k(\cdot) \in C(T^{\mathbf{Z}} \to \mathbf{R})$, *(in particular it is bounded) ;* $\quad \chi_k(\cdot) \in L^2(T^{\mathbf{Z}}, \mu)$;

$$\chi_k(\cdot) \in \mathcal{D}[\mathcal{E}], \quad < (\frac{\partial}{\partial y_k}\chi_k)^2, \mu > \leq 1, \quad \sum_{j\neq k} < (\frac{\partial}{\partial y_j}\chi_k)^2, \mu > \leq \frac{1}{4};$$

$$\begin{aligned}
\mathcal{E}(\chi_k(\cdot), v(\cdot)) &\equiv \sum_{j\in\mathbf{Z}} \int_{T^{\mathbf{Z}}} (\frac{\partial}{\partial y_j}\chi_k(\mathbf{y}))(\frac{\partial}{\partial y_j}v(\mathbf{y}))\mu(d\mathbf{y}) \\
&= -\int_{T^{\mathbf{Z}}} \{J'(y_{k+1}-y_k) - J'(y_k - y_{k-1})\} v(\mathbf{y})\mu(d\mathbf{y}) \\
&= \int_{T^{\mathbf{Z}}} \frac{\partial v}{\partial y_k}\mu(d\mathbf{y}) = \mathcal{E}(y_k(\cdot), v(\cdot)) \quad \forall k \in \mathbf{Z}, \quad \forall v \in C_0^\infty(T^{\mathbf{Z}}).
\end{aligned}$$

This in turn implies that $\mathcal{E}(\chi_k(\cdot) - y_k(\cdot), \chi_k(\cdot) - y_k(\cdot)) = 0$, *hence (by [AKR], since* μ *is unique) there exists some constant* $\gamma_k \in \mathbf{R}$ *and*

$$\chi_k(\cdot) - y_k(\cdot) \equiv \gamma_k, \qquad \mu - a.e.,$$

where $y_k(\cdot)$ *is the coordinate function such that* $y_k(\mathbf{y}) = y_k$.

□

Theorem 2.1.
For each $\epsilon > 0$ *let*

$$M_t^{\epsilon,k}(\cdot) = \left(\xi_t^k(\cdot) - \xi_0^k(\cdot)\right) - \left(\epsilon\chi_k(\frac{\boldsymbol{\xi}_t(\cdot)}{\epsilon}) - \epsilon\chi_k(\frac{\boldsymbol{\xi}_0(\cdot)}{\epsilon})\right). \tag{15}$$

Then for μ*-a.s.* $\mathbf{x} \in [0, 2\pi)^{\mathbf{Z}}$, *the stochastic process* $\{M_t^{\epsilon,k}\}_{t\in\mathbf{R}_+}$, $k \in \mathbf{Z}$, *on* $(W^{\mathbf{Z}}, \mathcal{B}(W^{\mathbf{Z}}), P_{\epsilon\mathbf{x}}^{\epsilon})$ *is an* $L^2(P_{\epsilon\mathbf{x}}^{\epsilon})$ *continuous* $\mathcal{B}_t(W^{\mathbf{Z}})$*-martingale. The quadratic variations of* $\{M_t^{\epsilon,k}\}_{t\in\mathbf{R}_+}$, $k \in \mathbf{Z}$, *are given by*

$$< M^{\epsilon,k}(\cdot), M^{\epsilon,l}(\cdot) >_t = \int_0^t a_{k,l}(\frac{\boldsymbol{\xi}_s(\cdot)}{\epsilon})ds, \quad k, l \in \mathbf{Z}, \tag{16}$$

where

$$a_{k,l}(\mathbf{y}) \equiv \sum_{j\in\mathbf{Z}} {\chi'}_{k,j}(\mathbf{y}){\chi'}_{l,j}(\mathbf{y}),$$

with

$${\chi'}_{k,j}(\mathbf{y}) = \begin{cases} \sqrt{2}\dfrac{\partial}{\partial y_j}\chi_k(\mathbf{y}) & j \neq k \\ \sqrt{2}(1 - \dfrac{\partial}{\partial y_k}\chi_k(\mathbf{y})) & j = k. \end{cases}$$

□

Assumption 2.1. *Suppose that we are given a Borel probability measure* ν *on* $T^{\mathbf{Z}}$*, whose Radon-Nikodym derivative with respect to the Gibbs probability measure* μ *satisfies*

$$\|\frac{d\nu}{d\mu}\|_{L^\infty} \equiv \sup_{\mathbf{y}\in T^{\mathbf{Z}}} |\frac{d\nu}{d\mu}(\mathbf{y})| < \infty. \tag{17}$$

Define the probability measure $\bar{\nu}$ *on* $l^2_{\mathbf{q}}$ *by extending* ν *from* $[0,2\pi)^{\mathbf{Z}}$ *to all of* $l^2_{\mathbf{q}}$ *by zero. For each* $\epsilon > 0$*, let* ν_ϵ *be the probability measure on* $l^2_{\mathbf{q}}$ *that is the image of* $\bar{\nu}$ *under the map*

$$l^2 \ni \mathbf{x} \mapsto \epsilon\mathbf{x}.$$

Let $\{\mathbb{X}^\epsilon(t,\nu_\epsilon)\}_{t\in\mathbf{R}_+}$ *be the solution of (1) whose initial distribution is* ν_ϵ*. Let* $P^\epsilon_{\nu_\epsilon}$ *be the probability law of the process* $\{\mathbb{X}^\epsilon(t,\nu_\epsilon)\}_{t\in\mathbf{R}_+}$*.*

Remark 2.2. For the corresponding original homogenization problems of finite dimensional diffusions on $\mathbf{R}^d$ it is possible to take the initial distribution ν of the process $\{\mathbb{X}^\epsilon(t,\mathbf{x})\}_{t\in\mathbf{R}_+}$ to be a Dirac measure on $\mathbf{R}^d$ (cf., e.g., [P] for the case with periodic drift coefficients, cf. also [FNT], [AB] and [A] and references therein for the case with general random drift coefficients), but in the present infinite dimensional formalism this is not allowed. This constraint comes from the fact that to prove the tightness of the sequence of probability measures $\{P^\epsilon_{\nu_\epsilon}\}_{\epsilon>0}$ as well as the convergence of the quadratic variation matrix $(< M^{\epsilon,k}(\cdot), M^{\epsilon,k}(\cdot) >_t)_{k,l}$ to the matrix $\bar{\mathbb{A}}$ with constant elements (cf. next Lemma 2.2), we crucially use in the next section Proposition 3.2 proved by [S], (cf. also Th 4.9 of [AKR] and Proposition 3.3) which is the ergodic theorem for our infinite dimensional diffusions. This ergodic theorem is a crucial result proven by making use of Gross well known *Logarithmic Sobolev Inequality* (cf. [G]).

□

Lemma 2.2. ***(Tightness)***
Under the Assumption 2.1, for each $\epsilon > 0$ *and each* $N \in \mathbf{N}$*, on the probability space* $(W^{\mathbf{Z}}, \mathcal{B}(W^{\mathbf{Z}}), P^\epsilon_{\nu_\epsilon})$ *there exists a sequence of standard* $\mathcal{B}_t(W^{\mathbf{Z}})$*-Brownian motion processes* $\{B^{\epsilon,N}_l(s)\}_{s\in\mathbf{R}_+}$*, (*$l = -N,\ldots,+N$*), and the stochastic process* $(\xi_t(\cdot), P^\epsilon_{\nu_\epsilon})$ *admits the following expression (if necessary*

extending the probability space):

$$\xi_t^k(\cdot) = \xi_0^k(\cdot) - \epsilon\chi_k(\frac{\boldsymbol{\xi}_0(\cdot)}{\epsilon}) + \epsilon\chi_k(\frac{\boldsymbol{\xi}_t(\cdot)}{\epsilon})$$

$$+ \sum_{-N\leq l\leq N} \int_0^t \sigma_{k,l}^N(\boldsymbol{\xi}_s(\cdot))dB_l^{\epsilon,N}(s), \qquad -N \leq k \leq +N, \tag{18}$$

$$\text{where} \quad \sum_{-N\leq j\leq N} \sigma_{k,j}^N(\boldsymbol{\xi}_s(\cdot)) \cdot \sigma_{j,l}^N(\boldsymbol{\xi}_s(\cdot)) = a_{k,l}(\frac{\boldsymbol{\xi}_t(\cdot)}{\epsilon}).$$

By using the family of Brownian motion processes indexed by $N \in \mathbf{N}$ and $\epsilon > 0$, on the probability space $(W^{\mathbf{Z}}, \mathcal{B}(W^{\mathbf{Z}}), P_{\nu_\epsilon}^{\epsilon})$, for each $N \in \mathbf{N}$, define an $R^{\mathbf{Z}}$-valued stochastic process $\{\zeta_t^{\epsilon,N}(\cdot)\}_{t\geq 0}$ as follows

$$\zeta_t^{k,\epsilon,N}(\cdot) = \xi_0^k(\cdot) - \epsilon\chi_k^N(\frac{\boldsymbol{\xi}_0(\cdot)}{\epsilon}) + \epsilon\chi_k^N(\frac{\boldsymbol{\xi}_t(\cdot)}{\epsilon})$$

$$+ \sum_{-N\leq l\leq N} \int_0^t {\chi'}_{k,l}^N(\frac{\boldsymbol{\xi}_s(\cdot)}{\epsilon})dB_l^{\epsilon,N}(s), \qquad -N \leq k \leq +N;$$

$$\zeta_t^{k,\epsilon,N}(\cdot) = \xi_0^k(\cdot) - \epsilon\chi_k^N(\frac{\boldsymbol{\xi}_0(\cdot)}{\epsilon}) + \epsilon\chi_k^N(\frac{\boldsymbol{\xi}_t(\cdot)}{\epsilon}), \qquad |k| > N,$$

where

$${\chi'}_{k,j}^N(\mathbf{y}) = \begin{cases} \sqrt{2}\frac{\partial}{\partial y_j}\chi_k^N(\mathbf{y}), & j \neq k \\ \sqrt{2}(1 - \frac{\partial}{\partial y_k}\chi_k^N(\mathbf{y})), & j = k \end{cases}$$

and $\chi_k^N(\mathbf{y}) = \chi_k(p_N(\mathbf{y}))$, $N \in \mathbf{Z}$, where $p_N : T^{\mathbf{Z}} \to T^{[-N,N]}$ is the canonical projection. Denote the probability law of the process $\{\zeta_t^{\epsilon,N}(\cdot)\}_{t\geq 0}$ by $P_{\nu_\epsilon}^{\epsilon,N}$.

Let $N(\epsilon)$ be a function of $\epsilon > 0$ taking values in $\mathbf{N}$ such that $\lim_{\epsilon\downarrow 0} N(\epsilon) = +\infty$ and there exists some $M < \infty$ such that

$$\sup_{1\geq\epsilon>0} \epsilon\|a_{k,k}^{N(\epsilon)}\|_{L^2(\mu)}^2 < M, \qquad \forall k \in \mathbf{Z},$$

where

$$a_{k,l}^N(\mathbf{y}) = \sum_{-N\leq j\leq +N} {\chi'}_{k,j}^N(\mathbf{y}){\chi'}_{l,j}^N(\mathbf{y}).$$

Then the sequence of probability measures $\{P_{\nu_\epsilon}^{\epsilon,N(\epsilon)}\}_{\epsilon>0}$ on $(W^{\mathbf{Z}}, \mathcal{B}(W^{\mathbf{Z}}))$ is tight.

□

Our aim is to prove the following, for which we give a preliminary formulation:

Theorem 2.2.
Under the Assumption 2.1, the sequence of probability measures $\{P_{\nu_\epsilon}^{\epsilon,N(\epsilon)}\}_{\epsilon>0}$*, defined in Lemma 2.2, converges weakly to a probability measure* P_0^0 *on* $(W^{\mathbf{Z}}, \mathcal{B}(W^{\mathbf{Z}}))$ *which is the probability law of some process* $\{\mathbb{Y}(t)\}_{t\in\mathbf{R}_+}$*, defined on some other complete probability space, such that*

$$\lim_{\epsilon\downarrow 0} E^{P_{\nu_\epsilon}^{\epsilon,N(\epsilon)}}[\varphi(\boldsymbol{\xi}_\cdot)(\cdot)] = \lim_{\epsilon\downarrow 0} E^{P_{\nu_\epsilon}^{\epsilon}}[\varphi(\boldsymbol{\eta}_\cdot^{k,\epsilon,N(\epsilon)}(\cdot))] = E^{P_0^0}[\varphi(\boldsymbol{\xi}_\cdot)(\cdot)]$$
$$= E[\varphi(\mathbb{Y}_\cdot)], \quad \text{for any bounded continuous functional } \varphi \text{ on } W^{\mathbf{Z}}$$

(in particular for any $\varphi \in C_b(C(\mathbf{R}_+ \to l^2_{\mathbf{q}}) \to \mathbf{R})$*,* $(N \in \mathbf{Z})$*).*

□

Remark 2.3. Heuristically one has

$$\mathbb{Y}(t) = \sqrt{2}\int_0^t \bar{\mathbb{A}}^{\frac{1}{2}} d\mathbf{B}_t, \tag{19}$$

where

$$\bar{\mathbb{A}} = \int_{T^{\mathbf{Z}}} (\mathbb{I}-\chi')^t(\mathbb{I}-\chi')d\mu(\mathbf{y}), \quad \text{and} \quad \chi' \equiv (\chi'_{k,j})_{k,j\in\mathbf{Z}}. \tag{20}$$

$\mathbf{B}_t = (\cdots, B_t^{-1}, B_t^0, B_t^1, \cdots)$ is a sequence of independent standard one-dimensional Brownian motion processes. Of course, we have to give rigorous meaning of (19) and (20) since e.g. in (19) convergence of the product of the infinite matrices that have to be settled.

□

3. Proofs of the Theorems

The following are crucially used to prove the Theorems.

Proposition 3.1. *(Theorem (2.23) of [HS])*
Let $p(t, \mathbf{y}, \cdot)$ *be the transition function associated with the diffusion process* $(\boldsymbol{\eta}_t, Q_{\mathbf{y}} : \mathbf{y} \in T^{\mathbf{Z}})$*, i.e.* $E^{Q_{\mathbf{y}}}[f(\boldsymbol{\eta}_{t+s}(\cdot)|\mathcal{B}_s(T^{\mathbf{Z}})] = \int_{T^{\mathbf{Z}}} f(\mathbf{y}')p(t,\mathbf{y},d\mathbf{y}')$*. For* $N \in \mathbf{N}$*, let* $p^{(N)}(t,\mathbf{y},\cdot)$ *be such that*

$$p^{(N)}(t,\mathbf{y},\Gamma) = p(t,\mathbf{y},\tilde{\Gamma}) \qquad \textit{for} \quad \Gamma \in \mathcal{B}(T^{[-N,+N]}),$$

where $\tilde{\Gamma} = \{\mathbf{y} \in T^{\mathbf{Z}} \,|\, \mathbf{y}_{(N)} = (y_{-N}, \cdots, y_N) \in \Gamma\}$.
Then for any multi indices $\mathbf{a}$ *the map*

$$(t, \mathbf{y}, \mathbf{y}_{(N)}) \longrightarrow (\partial^{|\mathbf{a}|}/\partial \mathbf{y}_{(N)}^{|\mathbf{a}|}) p^{(N)}(t, \mathbf{y}, \mathbf{y}_{(N)})$$

is continuous on $(0, \infty) \times T^{\mathbf{Z}} \times T^{[-N,+N]}$.

□

Proposition 3.2. *((6.16) of [S])*
There exists an $\alpha = \alpha(J) > 0$ *which depends only on* J*, and for each* $\Lambda \in \mathbb{F}_0$ *there exists* $K(\Lambda) \in (0, \infty)$ *such that the following holds:*

$$\begin{aligned} &\| \int_{T^{\mathbf{Z}}} \varphi(\mathbf{y}_\Lambda) p(t, \cdot, d\mathbf{y}) - < \varphi, \mu > \|_{L^\infty} \\ &\leq K(\Lambda) e^{-\alpha t} \|\nabla \varphi\|_{L^\infty}, \qquad \forall t > 0, \quad \forall \varphi \in C_\Lambda^\infty(T^{\mathbf{Z}}). \end{aligned} \tag{21}$$

□

Proposition 3.3. *((6.13) of [S])*
There exists an $\alpha = \alpha(J) > 0$ *which depends only on* J*, and the following holds:*

$$\begin{aligned} &\| \int_{T^{\mathbf{Z}}} \varphi(\mathbf{y}) p(t, \cdot, d\mathbf{y}) - < \varphi, \mu > \|_{L^2(\mu)} \\ &\leq e^{-\alpha t} \|\varphi\|_{L^2(\mu)}, \qquad \forall t > 0, \quad \forall \varphi \in C(T^{\mathbf{Z}}). \end{aligned}$$

□

Proof of Lemma 2.1
Let $b_k(y_k; y_{k+1}, y_{k-1}) = J'(y_{k+1} - y_k) - J'(y_k - y_{k-1})$. Since

$$E^{Q_{\mathbf{y}}}[\int_0^t b_k(\eta_{k,s}(\cdot); \eta_{k+1,s}(\cdot), \eta_{k-1,s}(\cdot))ds] = \int_0^t (p_\tau b_k)(\mathbf{y})d\tau,$$

where $(p_t b_k)(\mathbf{y}) = \int_{T^{\mathbf{Z}}} \varphi(\mathbf{y}') p(t, \mathbf{y}, d\mathbf{y}')$, by Proposition 3.2 and the fact that $< b_k, \mu >= 0$ (cf. (11), (12) and (13)), we have

$$\begin{aligned} &| \int_0^t (p_\tau b_k)(\mathbf{y})d\tau - \int_0^s (p_\tau b_k)(\mathbf{y})d\tau| \\ &\leq \int_s^t \|(p_\tau b_k)\|_{L^\infty} d\tau \leq \frac{K(3)}{\alpha} |e^{-\alpha t} - e^{-\alpha s}| \|4J''\|_{L^\infty}. \end{aligned} \tag{22}$$

By Proposition 3.1, since $\int_0^t (p_\tau b_k)(\mathbf{y})d\tau$ is a continuous function of $\mathbf{y}$, combined with the uniform convergence given by (21) we see that there exists $\int_0^\infty (p_\tau b_k)(\cdot)d\tau$ such that

$$\lim_{t\to\infty}\int_0^t (p_\tau b_k)(\cdot)d\tau = \int_0^\infty (p_\tau b_k)(\cdot)d\tau \in C(T^{\mathbf{Z}} \to \mathbf{R}) \subset \bigcap_{1\le p\le\infty} L^p(\mu), \tag{23}$$

where the convergence is uniform in $\mathbf{y} \in T^{\mathbf{Z}}$.

On the other hand, by the identification explained in Remark 2.1-(i) and through Dirichlet form arguments we see that for each $k \in \mathbf{Z}$ and $\lambda > 0$, there exists χ_k^λ such that

$$\chi_k^\lambda(\mathbf{y}) = \int_0^\infty e^{-\lambda\tau}(p_\tau b_k)(\mathbf{y})d\tau \in \mathcal{D}[\mathcal{E}],$$

and the following holds:

$$\mathcal{E}(\chi_k^\lambda, v) + \lambda < \chi_k^\lambda, v >_\mu = - < b_k, v >_\mu, \quad \forall v \in C_0^\infty(T^{\mathbf{Z}}). \tag{24}$$

By (24) we have the following:

$$\|\frac{\partial}{\partial y_k}\chi_k^\lambda\|_{L^2(\mu)} \le 1, \qquad \sum_{j\neq k}\|\frac{\partial}{\partial y_j}\chi_k^\lambda\|_{L^2(\mu)} \le \frac{1}{4}, \quad \forall\lambda > 0. \tag{25}$$

(25) can be seen as follows: Since $\chi_k^\lambda \in \mathcal{D}[\mathcal{E}]$, there exists some sequence $\{u_n^\lambda\}_{n\in\mathbf{N}}$ such that $u_n^\lambda \in C_0^\infty(T^{\mathbf{Z}})$ and

$$\lim_{n\to\infty}\{\mathcal{E}(\chi_k^\lambda - u_n^\lambda, \chi_k^\lambda - u_n^\lambda) + \lambda\|\chi_k^\lambda - u_n^\lambda\|^2_{L^2(\mu)}\} = 0. \tag{26}$$

And, since χ_k^λ is the (unique) solution of (24), if we set $v = u_n^\lambda$ in (24), then

$$\mathcal{E}(\chi_k^\lambda, u_n^\lambda) + \lambda < \chi_k^\lambda, u_n^\lambda >_\mu = - < b_k, u_n^\lambda >_\mu . \tag{27}$$

By (26) and (27) we see that for any $\delta > 0$ there exists $L_\delta^\lambda < \infty$ and

$$|\mathcal{E}(u_n^\lambda, u_n^\lambda) + \lambda < u_n^\lambda, u_n^\lambda >_\mu + < b_k, u_n^\lambda >_\mu | < \delta, \quad \forall n \ge L_\delta^\lambda. \tag{28}$$

On the other hand, by Proposition 3.2, for $u_n^\lambda \in C_0^\infty(T^{\mathbf{Z}})$ (a function having only a finite number of coordinates), which is identified with a C^∞ periodic function (with respect to each coordinates) on $\mathbf{R}^{\mathbf{Z}}$, we can apply the integration by parts formula (cf. (11), (12) and (13)) and then we see that

$$< b_k, u_n^\lambda >_\mu = -\int_{T^{\mathbf{Z}}}(\frac{\partial}{\partial y_k}u_n^\lambda)d\mu(\mathbf{y}). \tag{29}$$

Combining (28) and (29) we have

$$\begin{aligned}&\left(\int_{T^{\mathbf{Z}}}(\frac{\partial}{\partial y_k}u_n^\lambda)^2 d\mu(\mathbf{y}) - \int_{T^{\mathbf{Z}}}(\frac{\partial}{\partial y_k}u_n^\lambda)d\mu(\mathbf{y})\right)\\&+\sum_{j\neq k}\int_{T^{\mathbf{Z}}}(\frac{\partial}{\partial y_j}u_n^\lambda)^2 d\mu(\mathbf{y}) + \lambda\int_{T^{\mathbf{Z}}}(u_n^\lambda)^2 d\mu(\mathbf{y}) \le \delta, \quad \forall\lambda>0,\end{aligned}$$

where the sum which appears in the second term of the left hand side is actually a finite sum. Note that $\delta > 0$ in the above formula can be taken arbitrarily small by taking n large, then by standard argument of quadratic forms and (26) we arrive at (25).

Next, since $b_k \in C_0^\infty(T^{\mathbf{Z}} \to \mathbf{R})$ and $< b_k, \mu >= 0$ from Proposition 3.2, we see that

$$\begin{aligned}&\left|\int_0^\infty e^{-\lambda\tau}(p_\tau b_k)(\mathbf{y})d\tau - \int_0^\infty (p_\tau b_k)(\mathbf{y})d\tau\right|\\&\le \int_0^\infty |e^{-\lambda\tau}-1|\,|(p_\tau b_k)(\mathbf{y})d\tau|\\&\le 4K(3)\|J'\|_{L^\infty}\int_0^\infty |e^{-\lambda\tau}-1|e^{-\alpha\tau}d\tau, \qquad \forall \mathbf{y}\in T^{\mathbf{Z}}, \quad \forall\lambda>0. \quad (30)\end{aligned}$$

We remark that by the Markovian property of p_τ (cf. Theorem 1.4.1 of [F]) since $\|p_\tau b_k\|_{L^\infty} \le \|b_k\|_{L^\infty}$ using Proposition 3.1 the following holds:

$$\chi_k^\lambda(\cdot) = \int_0^\infty e^{-\lambda\tau}(p_\tau b_k)(\cdot)d\tau \in C(T^{\mathbf{Z}} \to \mathbf{R}). \tag{31}$$

Hence, by (22), (23) and (30) we have the following uniform convergence of continuous functions:

$$\lim_{\lambda\downarrow 0}\|\chi_k^\lambda(\cdot) - \int_0^\infty (p_\tau b_k)(\cdot)d\tau\|_{L^\infty} = 0. \tag{32}$$

Consequently, by (25) the family $\{\chi_k^\lambda\}_{\lambda>0}$ forms a weakly relatively compact set of the Hilbert space $\mathcal{D}[\mathcal{E}]$, and then by (23), (24), (31) and (32) we have the conclusion of Lemma 2.1.

□

Lemma 3.1. *(i)* *Let*

$$\chi_k^N \equiv \chi_k \circ p_N \in C(T^{\mathbf{Z}} \to \mathbf{R}), \qquad k \in \mathbf{Z} \tag{33}$$

(cf. Lemma 2.2) and let

$$d_N^k = \int_{T^{\mathbf{Z}}} |\chi_k^N(\mathbf{y}) - \chi_k(\mathbf{y})|^2 d\mu(\mathbf{y}) + \int_{T^{\mathbf{Z}}} \sum_{j\in\mathbf{Z}} |\frac{\partial}{\partial y_j}\chi_k^N(\mathbf{y}) - \frac{\partial}{\partial y_j}\chi_k(\mathbf{y})|^2 d\mu(\mathbf{y}). \tag{34}$$

Then

$$\lim_{N\to\infty} d_N^k = 0. \tag{35}$$

(ii) Let ν be a probability measure on $T^{\mathbf{Z}}$ satisfying the assumption given in Theorem 2.2. Let $\{\chi_k^N\}_{N\in\mathbf{N}}$ be the sequence discussed in (i), then for $N \in \mathbf{N}$, the following holds:

$$E^{Q_\nu}[|\chi_k^N(\eta_\tau(\cdot) - \chi_k(\eta_\tau(\cdot)|^2] + E^{Q_\nu}[\sum_{j\in\mathbf{Z}} |\frac{\partial}{\partial y_j}\chi_k^N(\eta_\tau(\cdot) - \frac{\partial}{\partial y_j}\chi_k(\eta_\tau(\cdot)|^2] \le d_N^k \|\frac{d\nu}{d\mu}\|_{L^\infty}, \ \forall \tau > 0. \tag{36}$$

□

Proof of Lemma 3.1

By Lemma 2.1 (cf. also its proof), since $\chi_k \in \mathcal{D}[\mathcal{E}]$ and $\chi_k \in C(T^{\mathbf{Z}} \to \mathbf{R})$, $\forall k \in \mathbf{Z}$, the sequence $\{\chi_k^N\}_{N\in\mathbf{N}}$, $(k \in \mathbf{Z})$ defined by (33) satisfies (35).

By Proposition 2.2, since μ is an invariant measure of the Markov process $(\boldsymbol{\eta}_t, Q_{\mathbf{y}})$, for any $F \in L^1(\mu)$ the following holds:

$$E^{Q_\mu}[F(\boldsymbol{\eta}_t(\cdot)] = \int_{T^{\mathbf{Z}}} F(\mathbf{y}) d\mu(\mathbf{y}). \tag{37}$$

¿From (35) and (37) the assertion (ii) follows by Assumption 2.1 for ν.

□

Proof of Theorem 2.1

By the identification explained in Remark 2.1-i) and Lemma 2.1, through arguments of Dirichlet space theory on the compact space $T^{\mathbf{Z}}$ (cf. Theorem 5.3.1 of [F]) we see that the processes

$$\Big\{\chi_k(\boldsymbol{\eta}_t(\cdot)) - \int_0^t \{J'(\eta_s^{k+1}(\cdot) - \eta_s^k(\cdot)) - J'(\eta_s^k(\cdot) - \eta_s^{k-1}(\cdot))\} ds\Big\}_{t\in\mathbf{R}_+}, \quad k \in \mathbf{Z},$$

are $L^2(Q_{\mathbf{y}})$, $\mathcal{B}_t(W_T^{\mathbf{Z}})$-martingales on $(W_T^{\mathbf{Z}}, \mathcal{B}(W_T^{\mathbf{Z}}), Q_{\mathbf{y}})$, $\mu - q.e.\mathbf{y} \in T^{\mathbf{Z}}$. In addition from Lemma 2.1, since $\chi_k \in C(T^{\mathbf{Z}} \to \mathbf{R})$, we see that they are continuous martingales.

Hence, by the obvious interpretation of χ_k on $T^{\mathbf{Z}}$ as a periodic function ($\chi_k(\mathbf{\Phi}(\boldsymbol{\xi}_t(\cdot))) = \chi_k(\boldsymbol{\xi}_t(\cdot))$, cf. Prop. 2.1) we see that the processes

$$\Big\{\chi_k(\boldsymbol{\xi}_t(\cdot)) - \int_0^t \{J'(\xi_s^{k+1}(\cdot) - \xi_s^k(\cdot)) - J'(\xi_s^k(\cdot) - \xi_s^{k-1}(\cdot))\}ds\Big\}_{t\in\mathbf{R}_+}, \quad k \in \mathbf{Z},$$

are $L^2(P^1_{\mathbf{x}})$, $\mathcal{B}_t(W^{\mathbf{Z}})$-martingales on $(W^{\mathbf{Z}}, \mathcal{B}(W^{\mathbf{Z}}), P^1_{\mathbf{x}})$, $\mu - q.e.\mathbf{y} \in T^{\mathbf{Z}}, \mathbf{y} = \Phi(\mathbf{x})$. Here we used the fact that by the periodicity of χ_k and J the measurabilities of $\chi_k(\boldsymbol{\xi}_t(\cdot))$ and $J'(\xi_s^{k+1}(\cdot) - \xi_s^k(\cdot)) - J'(\xi_s^k(\cdot) - \xi_s^{k-1}(\cdot))$ with respect to $\mathcal{B}_t(W^{\mathbf{Z}})$ can be replaced by the measurability with respect to $\sigma(\Phi(\boldsymbol{\xi}_s(\cdot)) : s \leq t)$.

For each $\epsilon > 0$ set

$$\begin{aligned} m^{\epsilon,k}_{\epsilon^2 t}(\cdot) &\equiv \epsilon\chi_k(\frac{\boldsymbol{\xi}_{\epsilon^2 t}}{\epsilon}(\cdot)) \\ &-\epsilon\int_0^t \{J'(\frac{\xi^{k+1}_{\epsilon^2 s}}{\epsilon}(\cdot) - \frac{\xi^k_{\epsilon^2 s}}{\epsilon}(\cdot)) - J'(\frac{\xi^k_{\epsilon^2 s}}{\epsilon}(\cdot) - \frac{\xi^{k-1}_{\epsilon^2 s}}{\epsilon}(\cdot))\}ds. \end{aligned}$$

Then, again by the identification explained in Remark 2.1-ii) we see that on $(W^{\mathbf{Z}}, \mathcal{B}(W^{\mathbf{Z}}), P^\epsilon_{\epsilon\mathbf{x}})$ the processes

$$\Big\{m^{\epsilon,k}_{\epsilon^2 t}(\cdot)\Big\}_{t\in\mathbf{R}_+}, \quad k \in \mathbf{Z},$$

are $L^2(P^\epsilon_{\epsilon\mathbf{x}})$, continuous $\mathcal{B}_{\epsilon^2 t}(W^{\mathbf{Z}})$-martingales, $\mu - q.e.$, $\mathbf{y} \in T^{\mathbf{Z}}$, $\mathbf{y} = \Phi(\mathbf{x})$, $\mathbf{x} \in \mathbf{R}^{\mathbf{Z}}$:

$$E^{P^\epsilon_{\epsilon\mathbf{x}}}\Big[m^{\epsilon,k}_{\epsilon^2 t}(\cdot)\Big|\mathcal{B}_{\epsilon^2 s}(W^{\mathbf{Z}})\Big] = m^{\epsilon,k}_{\epsilon^2 s}, \quad 0 \leq s \leq t. \tag{38}$$

Thus, by writting $t' = \epsilon^2 t$, we see that the processes

$$\begin{aligned} &\Big\{\epsilon\chi_k(\frac{\boldsymbol{\xi}_{t'}}{\epsilon}(\cdot)) \\ &-\epsilon\int_0^{\frac{t'}{\epsilon^2}} \{J'(\frac{\xi^{k+1}_{\epsilon^2 s}}{\epsilon}(\cdot) - \frac{\xi^k_{\epsilon^2 s}}{\epsilon}(\cdot)) - J'(\frac{\xi^k_{\epsilon^2 s}}{\epsilon}(\cdot) - \frac{\xi^{k-1}_{\epsilon^2 s}}{\epsilon}(\cdot))\}ds\Big\}_{t'\in\mathbf{R}_+}, \end{aligned} \tag{39}$$

$k \in \mathbf{Z}$, are $L^2(P^\epsilon_{\epsilon\mathbf{x}})$, continuous $\mathcal{B}_{t'}(W^{\mathbf{Z}})$-martingales, $\mu - q.e.$, $\mathbf{y} \in T^{\mathbf{Z}}$, $\mathbf{y} = \Phi(\mathbf{x})$, $\mathbf{x} \in \mathbf{R}^{\mathbf{Z}}$ (for the definition of the function ξ_t on $W^{\mathbf{Z}}$ c.f. Prop. 2.1-(ii)).

In other words, by changing the variable such that $\epsilon^2 s = s'$ in the $P^\epsilon_{\epsilon\mathbf{x}}$-a.s. pathwise representation (39), which involves the Riemann integrals of continuous functions, we see that the processes

$$\begin{aligned} &\Big\{\epsilon\chi_k(\frac{\boldsymbol{\xi}_t}{\epsilon}(\cdot)) \\ &-\frac{1}{\epsilon}\int_0^t \{J'(\frac{\xi^{k+1}_{s'}}{\epsilon}(\cdot) - \frac{\xi^k_{s'}}{\epsilon}(\cdot)) - J'(\frac{\xi^k_{s'}}{\epsilon}(\cdot) - \frac{\xi^{k-1}_{s'}}{\epsilon}(\cdot))\}ds'\Big\}_{t\in\mathbf{R}_+}, \end{aligned} \tag{40}$$

$k \in \mathbf{Z}$, are $L^2(P^\epsilon_{\epsilon\mathbf{x}})$, continuous $\mathcal{B}_t(W^{\mathbf{Z}})$-martingales ($\mu - q.e.$, $\mathbf{y} \in T^{\mathbf{Z}}$, $\mathbf{y} = \Phi(\mathbf{x})$, $\mathbf{x} \in \mathbf{R}^{\mathbf{Z}}$).

On the other hand, on $(W^{\mathbf{Z}}, \mathcal{B}(W^{\mathbf{Z}}), P^\epsilon_{\epsilon\mathbf{x}})$ for any $\mathbf{x} \in \mathbf{R}^{\mathbf{Z}}$, the processes

$$\Big\{\xi^k_t(\cdot) - \frac{1}{\epsilon}\int_0^t \{J'(\frac{\xi^{k+1}_s}{\epsilon}(\cdot) - \frac{\xi^k_s}{\epsilon}(\cdot)) - J'(\frac{\xi^k_s}{\epsilon}(\cdot) - \frac{\xi^{k-1}_s}{\epsilon}(\cdot))\}ds\Big\}_{t\in\mathbf{R}_+}, \quad (41)$$

$k \in \mathbf{Z}$, are $L^2(P^\epsilon_{\epsilon\mathbf{x}})$, continuous $\mathcal{B}_t(W^{\mathbf{Z}})$-martingales. For each $k \in \mathbf{Z}$ by adding (40) and (41), we see that the processes

$$\Big\{\xi^k_t(\cdot) - \epsilon\chi_k(\frac{\boldsymbol{\xi}_t(\cdot)}{\epsilon})\Big\}_{t\in\mathbf{R}_+}, \quad k \in \mathbf{Z}, \quad (42)$$

on $(W^{\mathbf{Z}}, \mathcal{B}(W^{\mathbf{Z}}), P^\epsilon_{\epsilon\mathbf{x}})$ are also $L^2(P^\epsilon_{\epsilon\mathbf{x}})$, continuous $\mathcal{B}_t(W^{\mathbf{Z}})$-martingales ($\mu - q.e.$, $\mathbf{y} \in T^{\mathbf{Z}}$, $\mathbf{y} = \Phi(\mathbf{x})$, $\mathbf{x} \in \mathbf{R}^{\mathbf{Z}}$).

¿From now on, for simplicity, we do not take general $\mathbf{x} \in \mathbf{R}^{\mathbf{Z}}$ but take $\mathbf{x} \in [0,\pi)^{\mathbf{Z}}$ and identify such $\mathbf{x}$ with $\mathbf{y} \in T^{\mathbf{Z}}$ by $\mathbf{y} = \Phi(\mathbf{x})$. The fact that the quadratic variations of $\{M^{\epsilon,k}_t\}_{t\in\mathbf{R}_+}$, $k \in \mathbf{Z}$, are given by (16) is shown as follows: since the probability law of the process $\{\frac{\boldsymbol{\xi}_{\epsilon^2 t}(\cdot)}{\epsilon}\}_{t\in\mathbf{R}_+}$ under $P^\epsilon_{\epsilon\mathbf{x}}$ is identical to $Q_{\mathbf{y}}$, ($\mu - q.e.$ $\mathbf{x} \in [0,2\pi)^{\mathbf{Z}}$), by Lemma 3.1-i) there exist $\chi^N_k \in C^\infty(T^{[-N,+N]} \to \mathbf{R})$, $k \in \mathbf{Z}$ such that

$$\begin{aligned}&\lim_{N\to\infty} E^{P^\epsilon_{\epsilon\mathbf{x}}}\Big[\int_0^t |\chi^N_k(\frac{\boldsymbol{\xi}_{\epsilon^2 s}(\cdot)}{\epsilon}) - \chi_k(\frac{\boldsymbol{\xi}_{\epsilon^2 s}(\cdot)}{\epsilon})|^2 ds \\ &+ \int_0^t \sum_{j\in\mathbf{Z}} |\frac{\partial}{\partial x_j}\chi^N_k(\frac{\boldsymbol{\xi}_{\epsilon^2 s}(\cdot)}{\epsilon}) - \frac{\partial}{\partial x_j}\chi_k(\frac{\boldsymbol{\xi}_{\epsilon^2 s}(\cdot)}{\epsilon})|^2 ds\Big] = 0, \\ &\forall k \in \mathbf{Z}, \quad \forall t \in \mathbf{R}_+, \quad \mu - a.s.\ \mathbf{x} \in [0,2\pi)^{\mathbf{Z}}.\end{aligned}$$

Hence, for $k \in \mathbf{Z}$ and $t \in \mathbf{R}_+$ if we let

$$M^{\epsilon,k,N}_t(\cdot) = \xi^k_t(\cdot) - \epsilon\chi^N_k(\frac{\boldsymbol{\xi}_t(\cdot)}{\epsilon}) - \xi^k_0(\cdot) + \epsilon\chi^N_k(\frac{\boldsymbol{\xi}_0(\cdot)}{\epsilon}),$$

then

$$\begin{aligned}&\lim_{N\to\infty} E^{P^\epsilon_{\epsilon\mathbf{x}}}\Big[\int_0^t |M^{\epsilon,k,N}_s(\cdot) - M^{\epsilon,k}_s(\cdot)|^2 ds\Big] = 0, \\ &\forall k \in \mathbf{Z}, \quad \forall t \in \mathbf{R}_+, \quad \mu - a.s.\ \mathbf{x} \in [0,2\pi)^{\mathbf{Z}}. \end{aligned} \quad (43)$$

By referring to the proof of Doob-Meyer's decomposition theorem for the sub-martingales $\{(M^{\epsilon,k}_t(\cdot))^2\}_{t\in\mathbf{R}_+}$, $k \in \mathbf{Z}$, in the class (DL), we see that their quadratic variations $\{< M^{\epsilon,k} >_t\}_{t\in\mathbf{R}_+}$, $k \in \mathbf{Z}$, are constructed as follows (cf., for e.g., Theorem I-6.12 of [IW]);

$$< M^{\epsilon,k} >_s = (M_s^{\epsilon,k})^2 + E^{P^\epsilon_{\epsilon\mathbf{x}}}\big[A_t \,\big|\, \mathcal{B}_s(W^{\mathbf{Z}})\big] - E^{P^\epsilon_{\epsilon\mathbf{x}}}\big[(M_t^{\epsilon,k})^2 \,\big|\, \mathcal{B}_s(W^{\mathbf{Z}})\big], \quad 0 \le s \le t, \tag{44}$$

$$A_t \equiv \lim_{n\to\infty} A_t^{(n)},$$

$$A_t^{(n)} \equiv \sum_{j=0}^{2^n} \Big\{ E^{P^\epsilon_{\epsilon\mathbf{x}}}[(M_t^{\epsilon,k})^2 \,|\, \mathcal{B}_{t_{n,j}}(W^{\mathbf{Z}})] - (M_{t_{n,j}}^{\epsilon,k})^2 + E^{P^\epsilon_{\epsilon\mathbf{x}}}\Big[(M_{t_{n,j+1}}^{\epsilon,k})^2 - E^{P^\epsilon_{\epsilon\mathbf{x}}}[(M_t^{\epsilon,k})^2 \,|\, \mathcal{B}_{t_{n,j+1}}(W^{\mathbf{Z}})] \,\Big|\, \mathcal{B}_{t_{n,j}}(W^{\mathbf{Z}})\Big] \Big\}, \tag{45}$$

where $t_{n,j} = \frac{jt}{2^n}$. Now, replacing $M^{\epsilon,k}$ by $M^{\epsilon,k,N}$ in the above formulas, we define the process correspondingly by (44). We denote this process by $\{< M^{\epsilon,k,N} >_s\}_{0\le s\le t}$ (We remark that the notation $< M^{\epsilon,k,N} >_s$ for the *semi-martingale* $\{M_t^{\epsilon,k,N}\}_{t\in\mathbf{R}_+}$ used here does not correspond to the usual notation (cf., for e.g., section 2.2 of [K]).). From (43) we have

$$\lim_{N\to\infty} E^{P^\epsilon_{\epsilon\mathbf{x}}}\Big[\int_0^t \big| < M^{\epsilon,k} >_s - < M^{\epsilon,k,N} >_s \big| ds\Big] = 0. \tag{46}$$

By this, in order to characterize the quadratic variation $< M^{\epsilon,k} >_s$, $k \in \mathbf{Z}$, we have only to consider the limit of the process $< M^{\epsilon,k,N} >_s$, $k \in \mathbf{Z}$.

But, since the probability law of $\{M_t^{\epsilon,k,N}\}_{t\in\mathbf{R}_+}$ on $(W^{\mathbf{Z}}, \mathcal{B}(W^{\mathbf{Z}}), P^\epsilon_{\epsilon\mathbf{x}})$ is identical to the probability law of $\{M_t^{\epsilon,k,\mathbf{x},N}(\omega)\}_{t\in\mathbf{R}_+}$, $k \in \mathbf{Z}$, on $(\Omega, \mathcal{F}, P; \mathcal{F}_t)$, where

$$M_t^{\epsilon,k,\mathbf{x},N}(\omega) = X_k^\epsilon(t, \epsilon\mathbf{x})(\omega) - \epsilon\chi_k^N\Big(\frac{\mathbb{X}^\epsilon(t,\epsilon\mathbf{x})(\omega)}{\epsilon}\Big) - \epsilon x_k + \epsilon\chi_k^N(\mathbf{x}),$$

in order to characterize $< M^{\epsilon,k,N} >_{t\in\mathbf{R}_+}$ on $(W^{\mathbf{Z}}, \mathcal{B}(W^{\mathbf{Z}}), P^\epsilon_{\epsilon\mathbf{x}})$ we can make use of $< M_t^{\epsilon,k,\mathbf{x},N}(\omega) >_{t\in\mathbf{R}_+}$ on $(\Omega, \mathcal{F}, P; \mathcal{F}_t)$. Namely, for the semi-martingale $\{M_t^{\epsilon,k,\mathbf{x},N}(\omega)\}_{t\in\mathbf{R}_+}$, $k \in \mathbf{Z}$, which have stochastic integral representations, we can apply Itô's formula and we see that (cf. our definition of $\{< M^{\epsilon,k,\mathbf{x},N}(\omega) >_t\}_{t\in\mathbf{R}_+}$ given above and (44)) $\{< M^{\epsilon,k,\mathbf{x},N}(\omega) >_t\}_{t\in\mathbf{R}_+}$

is the drift term of $\{(M_t^{\epsilon,k,\mathbf{x},N}(\omega))^2\}_{t\in\mathbf{R}_+}$, that is explicitely given by

$$
\begin{aligned}
&< M^{\epsilon,k,\mathbf{x},N}(\omega) >_t = \frac{2}{\epsilon}\int_0^t (M_s^{\epsilon,k,\mathbf{x},N}(\omega)) \\
&\times\Big\{b_\epsilon(X_{k+1}^\epsilon(s,\epsilon\mathbf{x}), X_k^\epsilon(s,\epsilon\mathbf{x}), X_{k-1}^\epsilon(s,\epsilon\mathbf{x})) \\
&-\sum_{j\in\mathbf{Z}}(\frac{\partial}{\partial x_j}\chi_k^N(\frac{\mathbb{X}^\epsilon(s,\epsilon\mathbf{x})}{\epsilon}))b_\epsilon(X_{j+1}^\epsilon(s,\epsilon\mathbf{x}), X_j^\epsilon(s,\epsilon\mathbf{x}), X_{j-1}^\epsilon(s,\epsilon\mathbf{x})) \\
&-\sum_{j\in\mathbf{Z}}\frac{\partial^2}{\partial x_j^2}\chi_k^N(\frac{\mathbb{X}^\epsilon(s,\epsilon\mathbf{x})}{\epsilon}))\Big\}ds \\
&+\int_0^t\Big\{(1-\frac{\partial}{\partial x_k}\chi_k^N(\frac{\mathbb{X}^\epsilon(s,\epsilon\mathbf{x})}{\epsilon}))^2+\sum_{j\neq k}(\frac{\partial}{\partial x_j}\chi_k^N(\frac{\mathbb{X}^\epsilon(s,\epsilon\mathbf{x})}{\epsilon}))^2\Big\}ds,
\end{aligned}
$$

where

$$
b_\epsilon(x_{k+1},x_k,x_{k-1}) = J'(\frac{x_{k+1}}{\epsilon}-\frac{x_k}{\epsilon}) - J'(\frac{x_k}{\epsilon}-\frac{x_{k-1}}{\epsilon}).
$$

In other words, on the probability space $(W^{\mathbf{Z}}, \mathcal{B}(W^{\mathbf{Z}}), P_{\epsilon\mathbf{x}}^\epsilon)$ $(\mu - a.s.\,\mathbf{x}\in[0,2\pi)^{\mathbf{Z}})$ the following holds:

$$
\begin{aligned}
&< M^{\epsilon,k,N}(\cdot) >_t = \frac{2}{\epsilon}\int_0^t (M_s^{\epsilon,k,N}(\cdot)) \\
&\times\Big\{b_\epsilon(\xi_s^{k+1}(\cdot), \xi_s^k(\cdot), \xi_s^{k-1}(\cdot)) \\
&-\sum_{j\in\mathbf{Z}}(\frac{\partial}{\partial x_j}\chi_k^N(\frac{\boldsymbol{\xi}_s(\cdot)}{\epsilon}))b_\epsilon(\xi_s^{j+1}(\cdot), \xi_s^j(\cdot), \xi_s^{j-1}(\cdot)) \\
&-\sum_{j\in\mathbf{Z}}\frac{\partial^2}{\partial x_j^2}\chi_k^N(\frac{\boldsymbol{\xi}_s(\cdot)}{\epsilon}))\Big\}ds \\
&+\int_0^t\Big\{(1-\frac{\partial}{\partial x_k}\chi_k^N(\frac{\boldsymbol{\xi}_s(\cdot)}{\epsilon}))^2+\sum_{j\neq k}(\frac{\partial}{\partial x_j}\chi_k^N(\frac{\boldsymbol{\xi}_s(\cdot)}{\epsilon}))^2\Big\}ds, \qquad (47)
\end{aligned}
$$

But, by the identification explained in Remark 2.1, applying Dirichlet space arguments, Corollary 1 in p.139 of [F], to the functionals $\chi_k \in \mathcal{D}[\mathcal{E}]$ and $\chi_k^N \in \mathcal{D}[\mathcal{E}]$ we see that, if necessary by taking a sub-sequence, for $\mu - q.e.\,\mathbf{x} \in$

$[0,2\pi)^{\mathbf{Z}}$

$$\begin{aligned}&\lim_{N\to\infty}\int_0^t\{\sum_{j\in\mathbf{Z}}(\frac{\partial}{\partial x_j}\chi_k^N(\frac{\boldsymbol{\xi}_s(\cdot)}{\epsilon}))b_\epsilon(\xi_s^{j+1}(\cdot),\xi_s^j(\cdot),\xi_s^{j-1}(\cdot))\\&+\sum_{j\in\mathbf{Z}}\frac{\partial^2}{\partial x_j^2}\chi_k^N(\frac{\boldsymbol{\xi}_s(\cdot)}{\epsilon}))\}ds=b_\epsilon(\xi_s^{k+1}(\cdot),\xi_s^k(\cdot),\xi_s^{k-1}(\cdot)),\qquad(48)\\&P_{\epsilon\mathbf{x}}^\epsilon-a.s.\text{ uniformly in }t\in[0,S],\ \forall S<\infty.\end{aligned}$$

Here we used the following three facts:

(i) the additive functional of finite energy $\{\chi_k(\boldsymbol{\eta}_t(\cdot))\}_{t\in\mathbf{R}_+}$ on $(W_T^{\mathbf{Z}},\mathcal{B}(W_T^{\mathbf{Z}}),Q_{\mathbf{y}})$, $\mu-q.e.\mathbf{y}\in T^{\mathbf{Z}}$, can be decomposed uniquely according to (cf. (2))

$$\begin{aligned}\chi_k(\boldsymbol{\eta}_t(\cdot))&=\int_0^t\{J'(\eta_s^{k+1}(\cdot)-\eta_s^k(\cdot))-J'(\eta_s^k(\cdot)-\eta_s^{k-1}(\cdot))\}ds\\&+\Big\{\chi_k(\boldsymbol{\eta}_t(\cdot))-\int_0^t\{J'(\eta_s^{k+1}(\cdot)-\eta_s^k(\cdot))-J'(\eta_s^k(\cdot)-\eta_s^{k-1}(\cdot))\}ds\Big\},\end{aligned}$$

where the first term of the right hand side is the continuous additive functional of zero-energy and the second term is the martingale additive functional of finite energy;

(ii) the additive functional of finite energy $\{\chi_k^N(\boldsymbol{\eta}_t(\cdot))\}_{t\in\mathbf{R}_+}$ on $(W_T^{\mathbf{Z}},\mathcal{B}(W_T^{\mathbf{Z}}),Q_{\mathbf{y}})$, $\mu-q.e.\mathbf{y}\in T^{\mathbf{Z}}$, also admits a unique decomposition, and the corresponding continuous additive functional of zero-energy is given by

$$\begin{aligned}&\int_0^t\Big\{\sum_{j\in\mathbf{Z}}(\frac{\partial}{\partial x_j}\chi_k^N(\boldsymbol{\eta}_s(\cdot)))b_1(\eta_s^{j+1}(\cdot),\eta_s^j(\cdot),\eta_s^{j-1}(\cdot))\\&+\sum_{j\in\mathbf{Z}}\frac{\partial^2}{\partial x_j^2}\chi_k^N(\boldsymbol{\eta}_s(\cdot)))\Big\}ds;\end{aligned}$$

(iii) $\lim_{N\to\infty}\chi_k^N=\chi_k$ in the sense of Lemma 3.1-i).

Thus by (47) and (48), noting the boundedness of χ_k^N and χ_k (cf. Lemma 2.1 and Lemma 3.1-i)), and if necessary taking a sub-sequence, we see that ($\mu-a.s.\,\mathbf{x}\in[0,2\pi)^{\mathbf{Z}}$)

$$\lim_{N\to\infty}<M^{\epsilon,k,N}(\cdot)>_t=\int_o^t a_{k,k}(\frac{\boldsymbol{\xi}_s(\cdot)}{\epsilon})ds,\quad P_{\epsilon\mathbf{x}}^\epsilon-a.s.,$$

uniformly in $t \in [0, S]$, $\forall S < \infty$.
Since we also have the L^1-convergence (46), we arrive at ($\mu - a.s.\, \mathbf{x} \in [0, 2\pi)^{\mathbf{Z}}$).

$$< M^{\epsilon,k}(\cdot) >_t = < M^{\epsilon,k}(\cdot), M^{\epsilon,k}(\cdot) >_t$$
$$= \int_o^t a_{k,k}(\frac{\boldsymbol{\xi}_s(\cdot)}{\epsilon})ds, \quad P^{\epsilon}_{\epsilon\mathbf{x}} - a.s., \quad \forall t \geq 0.$$

Finally by using the definition of quadratic variation, as an inner product for the space of martingales, we obtain the representation (16) for $< M^{\epsilon,k}(\cdot), M^{\epsilon,l}(\cdot) >_t$, $(k \neq l)$.

□

Proof of Lemma 2.2
For each $\epsilon > 0$, let $\{M_t^{\epsilon,k}(\cdot)\}_{t \in \mathbf{R}_+}$, $(k \in \mathbf{Z})$ be the continuous $\mathcal{B}_t(W^{\mathbf{Z}})$-martingales on $(W^{\mathbf{Z}}, \mathcal{B}(W^{\mathbf{Z}}), P^{\epsilon}_{\nu_\epsilon})$, considered if Theorem 2.1 (cf. (15)). For each $M \in \mathbf{N}$ let $\mathbb{A}|_M$ be a $(2M+1) \times (2M+1)$ matrix such that

$$\mathbb{A}|_M(\frac{\boldsymbol{\xi}_t(\cdot)}{\epsilon}) = \{a_{k,l}(\frac{\boldsymbol{\xi}_t(\cdot)}{\epsilon})\}_{-M \leq k,l \leq +M},$$

where $a_{k,l}$ is defined by Theorem 2.1. Then by Lemma 2.1, the fact that $\|\frac{d\nu}{d\mu}\|_{L^\infty} < \infty$ and Fubini's Lemma for $\mathbf{y} = (y_{-M}, \dots, y_M)$

$$\begin{aligned} 0 &\leq \int_0^T \mathbf{y} \cdot \mathbb{A}|_M(\frac{\boldsymbol{\xi}_s(\cdot)}{\epsilon}) \cdot {}^t\mathbf{y} ds \\ &= 2\int_0^T \sum_{j \in \mathbf{Z}} \Big(\sum_{-M \leq k \leq M} (\chi'_{k,j}(\frac{\boldsymbol{\xi}_s(\cdot)}{\epsilon}))y_k \Big)^2 ds < +\infty, \quad P^{\epsilon}_{\nu_\epsilon} - a.s., \end{aligned} \quad (49)$$

where $\chi'_{k,j}$ is the functional defined in Theorem 2.1. Hence, by the *finite dimensional* continuous L^2-martingale representation theorem by means of standard Brownian motion processes (cf., for e.g., section II-7 of [IW]) we see that for each $\epsilon > 0$ and each $M \in \mathbf{N}$ there exists a sequence of standard $\mathcal{B}_t(W^{\mathbf{Z}})$-Brownian motion processes $\{B_l^{\epsilon,M}(s)\}_{s \in \mathbf{R}_+}$, $(l = -M, \dots, +M)$, and the stochastic process $(\xi_t(\cdot), P^{\epsilon}_{\nu_\epsilon})$ admits the following expression (if necessary extending the probability space):

$$\xi_t^k(\cdot) = \xi_0^k(\cdot) - \epsilon\chi_k(\frac{\boldsymbol{\xi}_0(\cdot)}{\epsilon}) + \epsilon\chi_k(\frac{\boldsymbol{\xi}_t(\cdot)}{\epsilon})$$

$$+ \sum_{-M\le l\le M} \int_0^t \sigma_{k,l}^M(\boldsymbol{\xi}_s(\cdot))dB_l^{\epsilon,M}(s), \qquad -M \le k \le +M, \tag{50}$$

$$\text{where} \qquad \sum_{-M\le j\le M} \sigma_{k,j}^M(\boldsymbol{\xi}_s(\cdot)) \cdot \sigma_{j,l}^M(\boldsymbol{\xi}_s(\cdot)) = a_{k,l}(\frac{\boldsymbol{\xi}_t(\cdot)}{\epsilon}). \tag{51}$$

This shows (18).

Next, for each $\epsilon > 0$ let $\{\zeta_t^{\epsilon,N}(\cdot)\}_{t\ge 0}$ be the process defined in Lemma 2.2. By the identification explained in Remark 2.1 since the probability law of $\{\frac{\boldsymbol{\xi}_{\epsilon^2 t}(\cdot)}{\epsilon}\}_{t\in\mathbf{R}_+}$ on $P_{\nu_\epsilon}^\epsilon$ is identical to P_ν^1, by applying both Proposition 3.2 and Proposition 3.3, we have the estimate on the martingale term of $\{\zeta_t^{\epsilon,N}(\cdot)\}_{t\ge 0}$ as follows

$$E^{P_{\nu_\epsilon}^\epsilon}[| \sum_{-N\le l\le N} \int_0^t {\chi'}_{k,l}^N(\frac{\boldsymbol{\xi}_u(\cdot)}{\epsilon})dB_l^{\epsilon,N}(u)|^4]$$

$$\le \epsilon^4 \int_0^{\frac{t}{\epsilon^2}} \int_0^{\frac{t}{\epsilon^2}} E^{Q_\nu}[a_{k,k}^N(\boldsymbol{\eta}_{u_1})a_{k,k}^N(\boldsymbol{\eta}_{u_2})]du_1du_2$$

$$\le \frac{K_L}{\alpha}(1 - e^{-\frac{\alpha t}{\epsilon^2}})t\epsilon^2\|a_{k,k}^N\|_{L^2(\mu)}\{\|a_{k,k}^N\|_{L^2(\mu)} + \bar{a}_{kk}\}$$

$$+t^2(\bar{a}_{kk})^2 + t^2\big|(\bar{a}_{kk})^2 - (\int_{T^{\mathbf{Z}}} a_{k,k}^N((y))d\mu(\mathbf{y}))^2\big|,$$

where $K_L \equiv \|\frac{d\nu}{d\mu}\|_{L^\infty}$, and $\bar{a}_{kl}$ is the element of the infinite dimensional matrix $\bar{\mathbb{A}}$ with the constant elements defined by (19) in Theorem 2.2.

By an elementary computation for the right hand side of the above formula, and noting that $a_{k,k}$ is defined through χ_k which satisfies $\chi_k(\theta_j\mathbf{y}) = \chi_{k-j}(\mathbf{y})$ (θ_j is the shift operator on the index set $\mathbf{Z}$, by which the l-th component of $\mathbf{R}^{\mathbf{Z}}$ is shifted to the $l+j$-th component) , we see that there exists a constant C and the following holds (as before here and in the sequel we interpret the function ${\chi'}_{k,l}^N$ on $T^{\mathbf{Z}}$ as a periodic function on $\mathbf{R}^{\mathbf{Z}}$ such that $\chi'^{N}_{k,l}(\boldsymbol{\Phi}(\boldsymbol{\xi}_t(\cdot))) = \chi'^{N}_{k,l}(\boldsymbol{\xi}_t(\cdot))$, cf. Prop. 2.1) :

$$\begin{aligned} & E^{P^{\epsilon}_{\nu_\epsilon}}[| \sum_{-N\le l\le N} \int_0^t \chi'^{N}_{k,l}(\boldsymbol{\xi}_u(\cdot))dB^{\epsilon,N}_l(u)|^4] \\ & \le CK_L \|a^N_{k,k}\|^2_{L^2(\mu)}\,\epsilon\, t^{\frac{3}{2}} \\ & +t^2(\bar{a}_{kk})^2 + t^2\Big|(\bar{a}_{kk})^2 - (\int_{T^{\mathbf{Z}}} a^N_{k,k}((y))d\mu(\mathbf{y}))^2\Big|, \\ & \qquad \forall t > 0,\ \forall \epsilon > 0,\ \forall k \in \mathbf{Z},\ \forall N \in \mathbf{N}. \end{aligned} \tag{52}$$

If we replace the term $\int_0^t \chi'^{N}_{k,l}(\boldsymbol{\xi}_u(\cdot))dB^{\epsilon,N}_l(u)$ by $\int_s^t \chi'^{N}_{k,l}(\boldsymbol{\xi}_u(\cdot))dB^{\epsilon,N}_l(u)$ on the left hand side of (52), we have the corresponding evaluation replaced t by $t-s$ on the right hand side of (52) with the same constants.

Now, let us take $N(\epsilon)$ satisfying the assumption of Lemma 2.2, then from (52) (cf. also Lemma 3.1) we have a uniform bound. By this uniform bound (52) and using the Chebyshev's inequality, following a standard procedure (cf. Theorem I-4.2, 4.3 of [IW]), we see that

$$\begin{aligned} & \forall a > 0, \quad \forall \delta > 0, \quad \forall m \in \mathbf{N}, \quad \exists h(a,\delta,m)\ s.t. \\ & P^{\epsilon}_{\nu_\epsilon}\{\max_{t,s\in[0,m],|t-s|\le h'} |\Psi^{\epsilon,N(\epsilon)}_k(t) - \Psi^{\epsilon,N(\epsilon)}_k(s)|^2 > a2^{q'|k|}\} \\ & \le 2^{-2q'|k|}\delta, \qquad \forall h' \le h(a,\delta,m), \quad \forall k \in \mathbf{Z}. \end{aligned} \tag{53}$$

where q' is any non negative number, and

$$\Psi^{\epsilon,N(\epsilon)}_k(t) \equiv \sum_{-N(\epsilon)\le l\le N(\epsilon)} \int_0^t \chi'^{N(\epsilon)}_{k,l}(\boldsymbol{\xi}_u(\cdot))dB^{\epsilon,N(\epsilon)}_l(u).$$

Now, take q' such that $0 < q' < 1$, and define a number $C_{q'} \equiv \sum_{k\in\mathbf{Z}} 2^{-(1-q')|k|}$, then by using the evaluation (53) we see that

$$\begin{aligned} & P^{\epsilon}_{\nu_\epsilon}\{\max_{t,s\in[0,m],|t-s|\le h'} \sum_{k\in\mathbf{Z}} 2^{-|k|}|\Psi^{\epsilon,N(\epsilon)}_k(t) - \Psi^{\epsilon,N(\epsilon)}_k(s)|^2 > aC_{q'}\} \\ & \le \sum_{k\in\mathbf{Z}} P^{\epsilon}_{\nu_\epsilon}\{\max_{t,s\in[0,m],|t-s|\le h'} 2^{-q'|k|}|\Psi^{\epsilon,N(\epsilon)}_k(t) - \Psi^{\epsilon,N(\epsilon)}_k(s)|^2 > a\} \\ & \le \delta \sum_{k\in\mathbf{Z}} 2^{-2q'|k|}, \qquad \forall a > 0,\ \forall \delta > 0,\ \forall m \in \mathbf{N}, \\ & \qquad \text{whenever} \quad h' \le h(a,\delta,m). \end{aligned} \tag{54}$$

Hence, we have (cf. Definition 2.1)

$$\limsup_{h\downarrow 0}\sup_{\epsilon>0} P^{\epsilon}_{\nu_\epsilon}\{\max_{t,s\in[0,m],|t-s|\le h} \|\boldsymbol{\Psi}^{\epsilon,N(\epsilon)}(t) - \boldsymbol{\Psi}^{\epsilon,N(\epsilon)}(s)\|_{\mathbf{q}} > a\} = 0,\ \forall a > 0. \tag{55}$$

Next, by Lemma 2.1 χ_k is a bounded continuous function of $\mathbf{y} \in T^{\mathbf{Z}}$, moreover by construction since $\chi_k(\theta_j \mathbf{y}) = \chi_{k-j}(\mathbf{y})$, for all $\mathbf{y} \in T^{\mathbf{Z}}$, where θ_j is the shift operator on the index $k \in \mathbf{Z}$, we see that there exists a constant M (cf. Lemma 3.1) and

$$\sup_{\mathbf{y}\in T^{\mathbf{Z}}} |\chi_k(\mathbf{y})| = M, \qquad \sup_{\mathbf{y}\in T^{\mathbf{Z}}} |\chi_k^N(\mathbf{y})| \le M \qquad \forall k \in \mathbf{Z}, \quad \forall N \in \mathbf{N}.$$

By using this, and the identification discussed in Remark 2.1 for the process $(\boldsymbol{\eta}_t)_{t\ge 0}$ on $(T^{\mathbf{Z}}, Q_\nu)$, we have the following evaluation for the former term of the process $(\zeta_t^{\epsilon,N(\epsilon)})_{t\ge 0}$:

$$\begin{aligned}&\lim_{h\downarrow 0} \sup_{\epsilon \ge \frac{a}{2M}} P_{\nu_\epsilon}^{\epsilon}\{\max_{t,s\in[0,m],|t-s|\le h} |\epsilon\chi_k^{N(\epsilon)}(\frac{\boldsymbol{\xi}_t(\cdot)}{\epsilon}) - \epsilon\chi_k^{N(\epsilon)}(\frac{\boldsymbol{\xi}_s(\cdot)}{\epsilon})| > a\} \\ &= 0, \qquad\qquad \forall k \in \mathbf{Z}, \quad ,\forall a > 0, \quad ,\forall m \in \mathbf{N}. \qquad (56)\end{aligned}$$

Also, since for $\epsilon < \frac{a}{2M}$,

$$|\epsilon\chi_k^{N(\epsilon)}(\frac{\boldsymbol{\xi}_t(\cdot)}{\epsilon}) - \epsilon\chi_k^{N(\epsilon)}(\frac{\boldsymbol{\xi}_s(\cdot)}{\epsilon})| < a,$$

(56) holds not only for $\epsilon \ge \frac{a}{2M}$ but also for all $\epsilon > 0$. Thus we easily have (cf. (53), (54), (55))

$$\begin{aligned}&\lim_{h\downarrow 0} \sup_{\epsilon > 0} P_{\nu_\epsilon}^{\epsilon}\{\max_{t,s\in[0,m],|t-s|\le h} \|\epsilon\boldsymbol{\chi}^{N(\epsilon)}(\frac{\boldsymbol{\xi}_t(\cdot)}{\epsilon}) - \epsilon\boldsymbol{\chi}^{N(\epsilon)}(\frac{\boldsymbol{\xi}_s(\cdot)}{\epsilon})\|_{\mathbf{q}} > a\} \\ &= 0, \qquad\qquad \forall a > 0. \qquad (57)\end{aligned}$$

Finally, by using the martingale inequality for the L^2-martingales and Lemma 3.1 similar to (52) and (53) we easily see that there exists a constant $M < \infty$ and

$$\begin{aligned}P_{\nu_\epsilon}^{\epsilon}(\sup_{t\in[0,m]} |\Psi_k^{\epsilon,N(\epsilon)}(t)|^2 > \lambda^2) &\le \frac{1}{\lambda^4} E^{P_{\nu_\epsilon}^{\epsilon}}[|\Psi_k^{\epsilon,N(\epsilon)}(t)|^4] \qquad (58)\\ &\le \frac{1}{\lambda^4}\{M(\bar{a}_{kk})^2 m^2 + \epsilon m^{\frac{3}{2}} \|a_{kk}^{N(\epsilon)}\|_{L^2(\mu)}^2\} = \frac{M' m^{\frac{3}{2}}}{\lambda^4}, \quad \forall \epsilon > 0, \ \forall k \in \mathbf{Z},\end{aligned}$$

where the constant $M' < \infty$ does not depend on $k \in \mathbf{Z}$ because of the fact that $\chi_k(\theta_j \mathbf{y}) = \chi_{k-j}(\mathbf{y})$. Hence, for $0 < q' < 1$, as was discussed for (54), we see that there exists a constant $M'' < \infty$ such that

$$P_{\nu_\epsilon}^{\epsilon}\{\max_{t\in[0,m]} \sum_{k\in\mathbf{Z}} 2^{-|k|} |\Psi_k^{\epsilon,N(\epsilon)}(t)|^2 > \lambda^2 C_{q'}\} \le \frac{M'' m^{\frac{3}{2}}}{\lambda^4} \sum_{k\in\mathbf{Z}} 2^{-2q'|k|}, \quad \forall \epsilon > 0,$$

thus for any $\delta > 0$ and $m \in \mathbf{N}$ there exists $\lambda = \lambda(\delta, m) \in \mathbf{N}$ and

$$\sup_{\epsilon > 0} P_{\nu_\epsilon}^{\epsilon}\{\max_{t\in[0,m]} \|\boldsymbol{\Psi}^{\epsilon,N(\epsilon)}(t)\|_{\mathbf{q}} > C_{q'}\lambda(\delta, m)\} \le \delta. \qquad (59)$$

From (55), (57) and (58) we consequently see that for any $\delta > 0$ there exists a relatively compact set $B_\delta \in C(\mathbf{R}_+ \to l^2_{\mathbf{q}})$ such that

$$\sup_{\epsilon>0} P^{\epsilon,N(\epsilon)}_{\nu_\epsilon}(B_\delta) > 1 - \delta.$$

This shows that $\{P^{\epsilon,N(\epsilon)}_{\nu_\epsilon}\}_{\epsilon>0}$ is tight on $C(\mathbf{R}_+ \to l^2_{\mathbf{q}})$.

$\square$

Acknowledgments

The third and the fourth named authours would like to express their great acknowledgements to Professor T. Hida and Professor K. Saito who invited them to the international conference " Quantum Information and Complexity " held at Nagoya Japan, in Jan. 2003, where under the warm hospitality of Meijyo University they fortunately got the chance to present and work on the results of the present paper and to make fruitful discussions.

References

A. S. Albeverio: *Dirichlet forms and applications, Lectures in St. FLOUR 2000,* in Lecture Notes in Mathematics, Springer-Verlag, Berlin (2003).

AB. S. Albeverio and M. S. Bernabei: *Homogenization in random Dirichlet forms,* SFB 611, Bonn preprint (2003).

AKR. S. Albeverio, Y.G. Kondratiev and M. Röckner: *Ergodicity of L^2-semigroups and extremality of Gibbs states, J. Funct. Anal.* 144 (1997), 394-423.

AKRT. S. Albeverio, Y.G. Kondratiev and M. Röckner, T.V. Tsikalenko: *A priori estimates for symmetrizing measures and their applications to Gibbs states, J. Funct. Anal.* 171 (2000), 366-400.

F. M. Fukushima: *Dirichlet forms and Markov processes,* North-Holland, 1980.

FNT. M. Fukushima, S. Nakao and M. Takeda: *On Dirichlet forms with random data- recurrence and homogenization,* in Lecture Notes in Mathematics, 1250, Springer-Verlag, Berlin (1987).

FU. N. Funaki and K. Uchiyama: *From the Micro to the Macro, 1, 2, (in Japanese)* Seriese of Springer contemporary mathematics, (2003) Springer-Verlag, Tokyo.

G. L. Gross: *Logarithmic Sobolev inequalities and contractive properties of semigroups,* in Lecture Notes in Mathematics, Springer-Verlag, Berlin (1992).

H.	T. Hida: *Brownian motion,* Springer-Verlag, Berlin, 1980.
HKPS.	T. Hida, H.-K. Kuo, J. Potthoff, L. Streit: *White Noise: An Infinite Dimensional Calculus.* Kluwer Acadenic Publishers, Dordrecht, 1993.
HS.	R. Holley and D. Stroock: *Diffusions on an infinite dimensional torus, J. Funct. Anal.* 42 (1981), 29-63.
I.	S. Itô: *Introduction to the Lebesgue integral,* Shyoukabou, 1963. (in Japanese)
IW.	N. Ikeda and S. Watanabe: *Stochastic Differential Equations and Diffusion Processes, second edition,* North-Holland, 1989.
K.	H. Kunita: *Stochastic flows and stochastic differential equations,* Cambridge Univ. Press, 1988.
MR.	Z.M. Ma and M. Röckner: *Introduction to the theory of (Non-Symmetric) Dirichlet Forms,* Springer-Verlag, Berlin, 1992.
P.	E. Pardoux: *Homogenization of linear and senilinear second order parabolic PDEs with periodic coefficients: a probabilistic approach, J. Funct. Anal.* 167 (1999), 498-520.
S.	D. Stroock: *Logarithmic Sobolev inequalities for Gibbs states,* in Lecture Notes in Mathematics 1563, Springer-Verlag, Berlin (1993).

CONSTRUCTION OF DIRICHLET FORMS AND SYMMETRIC MARKOV SEMIGROUPS ON $\mathbb{Z}_2$-GRADED VON NEUMANN ALGEBRAS

C. BAHN

Institute of Natural Science, Yonsei University,
Seoul 120-749, Korea
E-mail: bahn@yonsei.ac.kr

C. K. KO

Department of Mathematics, Yonsei University,
Seoul 120-749, Korea
E-mail: kochulki@hotmail.com

Y. M. PARK

Department of Mathematics, Yonsei University,
Seoul 120-749, Korea
E-mail: ympark@yonsei.ac.kr

We give a brief survey on the theory of the Dirichlet forms on standard forms of von Neumann algebras, and then discuss the construction of Dirichlet forms which generate symmetric Markov semigroups on $\mathbb{Z}_2$ graded von Neumann algebras. As an application, we describe the construction of Dirichlet forms and symmetric Markov semigroups on CAR algebras with respect to gauge invariant quasi-free states.

1. Introduction

In this talk we would like to give a brief survey on the theory of non-commutative Dirichlet forms and associated symmetric Markov semigroups on standard forms of von Neumann algebras, and then describe a construction method of Dirichlet forms on $\mathbb{Z}_2$-graded algebras. As an application, we construct Dirichlet forms and symmetric Markov semigroups on CAR algebras with respect to gauge invariant quasi-free states. The main results of this talk are based on Ref. 1.

The theory of commutative Dirichlet forms and associated Markov semigroups, originated by Beurling and Deny, [2] has been developed especially by Fukushima [3] and Silverstein. [4] The theory has been intensively investigated in connection with their important applications to the theory of Markov processes and to various branches of mathematical physics. See Ref. 5 and references therein.

The efforts in constructing and studying of noncommutative (quantum) Markov semigroups are as old as the equilibrium description of quantum systems. [6] The study of noncommutative Dirichlet forms was pioneered by Albeverio and Høegh-Krohn, [7] Sauvageot [8,9,10] and extensively developed by Davies and Lindsay, [11] and Guido, Isola and Scarlatti. [12] All these authors considered Markov property of forms and semigroups only with respect to a tracial state. The abstract theory has been extended to a faithful normal state by Goldstein and Lindsay, [13,14,15] in the setting of the Haagerup's standard forms and by Cipriani [16] in the context of standard forms of von Neumann algebras, respectively.

The need to construct Markov semigroups on von Neumann algebras, which are symmetric with respect to a non-tracial state, is clear for various applications to open systems,[6] quantum statistical mechanics,[17] and quantum probability theory.[18,19,20] Although on the abstract level we have quite well-developed theory,[13,14,15,16] the progress in concrete applications is very slow. We would like to mention a few works in this direction. Matsui considered the completely positive Hamiltonian semigroup for quantum spin chains in the ground state representation. [21] In Refs. 22 and 23, Majewski and Zegarlinski used the generalized conditional expectation to construct generators of spin-flip type dynamics for quantum spin systems. In Ref. 24, Park gave a general construction method of Dirichlet forms on standard forms of von Neumann algebras and applied the method to construct translation invariant Markov semigroups for quantum spin systems. The method of Park has been extended to construct symmetric Markov semigroups on CCR algebras with respect to quasi-free states,[25] on quantum mechanical systems [26] and on CAR algebras with respect to gauge invariant quasi-free states.[1] In Refs. 27 and 28, quantum Ornstein-Uhlenbeck semigroups were constructed by means of noncommutative Dirichlet forms.

In Sec. 2, we give a brief survey on the abstract theory of Dirichlet forms on standard forms of von Neumann algebras. In Sec. 3, we describe a general construction method of Dirichlet forms on $\mathbb{Z}_2$ graded von Neumann algebras proposed in Ref. 1. In section 4, we discuss how we can apply the general construction method to construct symmetric Markov semigroups on CAR algebras with respect to gauge invariant quasi-free states.

2. Review on the Theory of Dirichlet Forms on von Neumann Algebras

In this section, we give a brief review on the theory of noncommutative Dirichlet forms developed by Cipriani. [16]

Let $\mathcal{M}$ be a σ-finite von Neumann algebra acting on a complex Hilbert space $\mathcal{H}$ equipped with an inner product $\langle\cdot,\cdot\rangle$. Let $\xi_0 \in \mathcal{H}$ be a cyclic and separating vector for $\mathcal{M}$. We use Δ and J to denote respectively, the modular operator and the modular conjugation associated to the pair $(\mathcal{M}, \xi_0)$.[17] The associated modular automorphism group is denoted by σ_t: $\sigma_t(A) = \Delta^{it}A\Delta^{-it}, \forall A \in \mathcal{M}, t \in \mathbb{R}$. $j : \mathcal{M} \to \mathcal{M}'$ is the antilinear $*$-isomorphism defined by $j(A) = JAJ$, $A \in \mathcal{M}$.

The natural positive cone $\mathcal{P} \subset \mathcal{H}$ associated with the pair $(\mathcal{M}, \xi_0)$ is the closure of the set

$$\{Aj(A)\xi_0 : A \in \mathcal{M}\}.$$

By a general result, the closed convex cone $\mathcal{P}$ can be obtained by the closure of the set $\{\Delta^{1/4}AA^*\xi_0 : A \in \mathcal{M}\}$ and this cone $\mathcal{P}$ is self-dual in the sense that

$$\Big\{\xi \in \mathcal{H} : \langle \xi, \eta\rangle \geq 0, \ \forall \eta \in \mathcal{P}\Big\} = \mathcal{P}.$$

Then the form $(\mathcal{M}, \mathcal{H}, \mathcal{P}, J)$ is the standard form associated with the pair $(\mathcal{M}, \xi_0)$. For details we refer to Ref. 29 and Sec. 2.5 of Ref. 17. We shall use the fact that $\mathcal{H}$ is the complexification of the real subspace $\mathcal{H}^J = \Big\{\xi \in \mathcal{H} : \langle \xi, \eta\rangle \in \mathbb{R}, \ \forall \eta \in \mathcal{P}\Big\}$, whose elements are called *J-real*: $\mathcal{H} = \mathcal{H}^J \bigoplus i\mathcal{H}^J$. The cone $\mathcal{P}$ gives rise to a structure of ordered Hilbert space on $\mathcal{H}^J$ (denoted by $\leq$) and to an anti-unitary involution J on $\mathcal{H}$, which preserves $\mathcal{P}$ and $\mathcal{H}^J$: $J(\xi + i\eta) = \xi - i\eta$, $\forall \xi, \eta \in \mathcal{H}^J$. Also note that any J-real element $\xi \in \mathcal{H}^J$ can be decomposed uniquely as a difference of two mutually orthogonal, positive elements, called the positive and negative part of ξ, repectively : $\xi = \xi_+ - \xi_-$, $\xi_+, \xi_- \in \mathcal{P}$ and $\langle \xi_+, \xi_-\rangle = 0$. The order interval $\{\eta \in \mathcal{H} : 0 \leq \eta \leq \xi_0\}$ will be denoted by $[0, \xi_0]$. This is a closed convex subset of $\mathcal{H}$, and we shall denote the nearest point projection onto $[0, \xi_0]$ by $\eta \mapsto \eta_I$.

A bounded operator A on $\mathcal{H}$ is *J-real* if $AJ = JA$. A semigroup $\{T_t\}_{t\geq 0}$ on $\mathcal{H}$ is said to be *J-real* if T_t is J-real for any $t \geq 0$. A semigroup $\{T_t\}_{t\geq 0}$ on $\mathcal{H}$ is said to be *symmetric* if each T_t is a self-adjoint operator on $\mathcal{H}$.

Corresponding to any J-real symmetric strongly continuous contraction semigroup T_t on $\mathcal{H}$, there exist a unique closed nonnegative quadratic form

$\mathcal{E}$ given by

$$\mathcal{E}[\eta] := \mathcal{E}(\eta, \eta) = \lim_{t \to 0} t^{-1} \langle \eta, (I - T_t)\eta \rangle,$$

and conversely such a form determines the semigroup through $T_t = e^{-tH}$ where $H = H^* \geq 0$ is determined by

$$\|H^{1/2}\eta\|^2 = \mathcal{E}[\eta], \quad D(H^{1/2}) = \{\eta \in \mathcal{H} : \mathcal{E}[\eta] < \infty\}.$$

A J-real symmetric strongly continuous contraction semigroup T_t on $\mathcal{H}$ is called *sub-Markov* with respect to ξ_0 if

$$T_t([0, \xi_0]) \subset [0, \xi_0], \ \forall t,$$

and *Markov* if also

$$T_t \xi_0 = \xi_0, \ \forall t.$$

A densely defined nonnegative quadratic form $\mathcal{E}$ is called *Markov* with respect to ξ_0 if

$$\eta \in D(\mathcal{E}) \Rightarrow J\eta \in D(\mathcal{E}) \text{ and } \mathcal{E}[J\eta] = \mathcal{E}[\eta] \tag{2.1a}$$

$$\eta = J\eta \in D(\mathcal{E}) \Rightarrow \eta_I \in D(\mathcal{E}) \text{ and } \mathcal{E}[\eta_I] \leq \mathcal{E}[\eta] \tag{2.1b}$$

and is called *Dirichlet* if it is also closed.

The Markov condition (2.1a) and (2.1b) on a closed densely defined nonnegative quadratic form on $\mathcal{H}$ are equivalent to the sub-Markov property of the corresponding semigroup (Theorem 4.11 of Ref. 16, Theorem 5.7 of Ref. 14). This amounts to a noncommutative generalization of the Beurling-Deny characterization of classical symmetric Markov semigroups. As in the commutative case, there is a bijective correspondence between such semigroups and weak *-continuous positive contraction semigroups on $\mathcal{M}$ (Theorem 2.12 of Ref. 16).

If the quadratic form of a J-real contraction semigroup on $\mathcal{H}$ satisfies (2.1a) and $\mathcal{E}[\xi_0] = 0$, then (2.1b) is equivalent to the weaker condition

$$\eta = J\eta \in D(\mathcal{E}) \Rightarrow \eta_\pm \in D(\mathcal{E}) \text{ and } \mathcal{E}(\eta_+, \eta_-) \leq 0. \tag{2.2}$$

See Proposition 4.5(b) and Proposition 4.10 (ii) of Ref. 16.

3. Construction of Dirichlet forms on $\mathbb{Z}_2$-graded von Neumann algebras

In this section, we describe a general construction method of Dirichlet forms on standard forms of $\mathbb{Z}_2$-graded von Neumann algebras proposed in Ref. 1. Let us mention that this method is a modification of that of Ref. 24.

Let $\gamma : \mathcal{M} \to \mathcal{M}$ be a $\mathbb{Z}_2$-grading of $\mathcal{M}$, so that γ is a $*$-automorphism of $\mathcal{M}$ which is involutive: $\gamma^2 = \mathrm{id}$. Thus $\mathcal{M}$ is a direct sum of $\mathcal{M}_e := \{A \in \mathcal{M} : \gamma(A) = A\}$ and $\mathcal{M}_o := \{A \in \mathcal{M} : \gamma(A) = -A\}$. The elements of $\mathcal{M}_e$ are called *even* and those of $\mathcal{M}_o$ *odd*. The identity element $\mathbf{1} \in \mathcal{M}$ is even.

We assume that ξ_0 is invariant under γ in the sense that

$$\langle \xi_0, A\xi_0 \rangle = \langle \xi_0, \gamma(A)\xi_0 \rangle, \quad A \in \mathcal{M}. \tag{3.1}$$

Then there exists a unitary operator U_γ on $\mathcal{H}$ such that

$$\begin{aligned} \gamma(A) &= U_\gamma A U_\gamma^{-1},\ A \in \mathcal{M}, \\ U_\gamma \mathcal{P} &\subset \mathcal{P}, \\ U_\gamma \xi_0 &= \xi_0, \\ U_\gamma^2 &= \mathbf{1}. \end{aligned} \tag{3.2}$$

In fact $U_\gamma A\xi_0 = \gamma(A)\xi_0$, where $A \in \mathcal{M}$. See the proof of Corollary 2.5.32 of Ref. 17. Here we have used the property $\gamma^2 = \mathrm{id}$ for the last relation. In this case, it can be checked that U_γ commutes with Δ and J, which implies for any $t \in \mathbb{R},\ A \in \mathcal{M}$,

$$\begin{aligned} \gamma(\sigma_t(A)) &= \sigma_t(\gamma(A)), \\ \gamma(j(A)) &= j(\gamma(A)). \end{aligned} \tag{3.3}$$

In order to express Dirichlet forms, let us introduce the notion of admissible functions in Ref. 24.

Definition 3.1. An analytic function $f : D \to \mathbb{C}$ on a domain D containing the strip $Im\ z \in [-1/4, 1/4]$ is said to be admissible if the following properties hold:

(a) $f(t) \geq 0 \quad \text{for} \quad \forall t \in \mathbb{R}$,
(b) $f(t + i/4) + f(t - i/4) \geq 0 \quad \text{for} \quad \forall t \in \mathbb{R}$,
(c) there exist $M > 0$ and $p > 1$ such that the bound

$$|f(t + is)| \leq M(1 + |t|)^{-p}$$

holds uniformly in $s \in [-1/4, 1/4]$.

We remark that there exists an admissible function satisfying Definition 3.1. See Lemma 3.1 of Ref. 24.

We are ready to give a construction of Dirichlet forms on the standard form $(\mathcal{M}, \mathcal{H}, \mathcal{P}, J)$ associated with pair $(\mathcal{M}, \xi_0)$, where $\mathcal{M}$ is a $\mathbb{Z}_2$-graded von Neumann algebra by γ. Denote by $\mathcal{M}_{an}$ the dense subset of $\mathcal{M}$ consisting of σ_t-analytic elements on a domain containing the strip

$I_{1/2} := \{z : |\text{Im}z| \leq \frac{1}{2}\}$. By Proposition 2.5.21 of Ref. 17, any element $A \in \mathcal{M}_{an}$ is strongly analytic. For a given admissible function f and odd element $x \in \mathcal{M}_o \cap \mathcal{M}_{an}$, define a sesquilinear form $\mathcal{E} : \mathcal{H} \times \mathcal{H} \longrightarrow \mathbb{C}$ by

$$\begin{aligned}
&\mathcal{E}(\eta, \xi) \qquad\qquad (3.4)\\
&= \int \Big\langle \big(\sigma_{t-i/4}(x) - j(\sigma_{t-i/4}(x^*))U_\gamma\big)\eta, \\
&\qquad\qquad \big(\sigma_{t-i/4}(x) - j(\sigma_{t-i/4}(x^*))U_\gamma\big)\xi\Big\rangle f(t)dt \\
&\quad + \int \Big\langle \big(\sigma_{t-i/4}(x^*) - j(\sigma_{t-i/4}(x))U_\gamma\big)\eta, \\
&\qquad\qquad \big(\sigma_{t-i/4}(x^*) - j(\sigma_{t-i/4}(x))U_\gamma\big)\xi\Big\rangle f(t)dt \\
&\equiv \mathcal{E}^{(1)}(\eta, \xi) + \mathcal{E}^{(2)}(\eta, \xi).
\end{aligned}$$

Here U_γ is the unitary operator satisfying $\gamma(A) = U_\gamma A U_\gamma^{-1}$, $A \in \mathcal{M}$. The associated quadratic form $\mathcal{E}[\cdot]$ is given by

$$\begin{aligned}
\mathcal{E}[\xi] = &\int \|\big(\sigma_{t-i/4}(x) - j(\sigma_{t-i/4}(x^*))U_\gamma\big)\xi\|^2 f(t)dt \\
&+ \int \|\big(\sigma_{t-i/4}(x^*) - j(\sigma_{t-i/4}(x))U_\gamma\big)\xi\|^2 f(t)dt.
\end{aligned} \tag{3.5}$$

The form given in the above is a bounded form since $\|\sigma_{t-i/4}(x)\| = \|\sigma_{-i/4}(x)\|$ for any $x \in \mathcal{M}_{an}$.

The following is the main result which corresponds to Theorem 3.1 of Ref. 24.

Theorem 3.1. *For a given admissible function f and odd element $x \in \mathcal{M}_o \cap \mathcal{M}_{an}$, let $(\mathcal{E}, \mathcal{H})$ be defined as in (3.4). Let H be the self-adjoint operator associated with $(\mathcal{E}, \mathcal{H})$, i.e., $\mathcal{E}[\xi] = \langle \xi, H\xi \rangle$, $\forall \xi \in \mathcal{H}$. Assume that there exists a constant $M > 0$ such that the bound*

$$\sup_{s \in [-1/4, 1/4]} \|\sigma_{t+is}(x)\| \leq M \tag{3.6}$$

holds uniformly in $t \in \mathbb{R}$. Then the following properties hold:
(a) $H\xi_0 = 0$,
(b) $\mathcal{E}[J\xi] = \mathcal{E}[\xi]$, $\forall \xi \in \mathcal{H}$,
(c) $\mathcal{E}(\xi_+, \xi_-) \leq 0$ for $\forall \xi \in \mathcal{H}^J$.
Furthermore $(\mathcal{E}, \mathcal{H})$ is a Dirichlet form.

We will produce the proof of the Theorem 3.1 at the end of this section. The following is a consequence of Theorem 3.1:

Theorem 3.2. *Let H be the self-adjoint operator associated with $(\mathcal{E}, \mathcal{H})$ defined as in (3.4) and let $T_t = e^{-tH}, t \geq 0$. Then $\{T_t\}_{t\geq 0}$ is a J-real, strongly continuous, symmetric Markov semigroup.*

Proof. Clearly $\mathcal{E}[\cdot] \geq 0$. Thus Theorem 3.1 and Theorem 4.11 of Ref. 16 imply that $\{T_t\}_{t\geq 0}$ is J-real, strongly continuous, sub-Markov. It follows from Theorem 3.1 (a) that $T_t\xi_0 = \xi_0$ for any $t \geq 0$. Thus $\{T_t\}_{t\geq 0}$ is Markov. □

We now produce the proof of Theorem 3.1.

Proof of Theorem 3.1 (a) Notice that $U_\gamma \xi_0 = \xi_0$, $JA\xi_0 = \Delta^{\frac{1}{2}} A^* \xi_0$ for any $A \in \mathcal{M}$. We obtain that

$$\begin{aligned}\big(\sigma_{t-i/4}(x) - j(\sigma_{t-i/4}(x^*))U_\gamma\big)\xi_0 &= \Delta^{\frac{1}{4}}\sigma_t(x)\xi_0 - J\Delta^{\frac{1}{4}}\sigma_t(x^*)\Delta^{-\frac{1}{4}}\xi_0 \\ &= \Delta^{\frac{1}{4}}\sigma_t(x)\xi_0 - \Delta^{\frac{1}{4}}\sigma_t(x)\xi_0 \\ &= 0,\end{aligned}$$

and also $\big(\sigma_{t-i/4}(x^*) - j(\sigma_{t-i/4}(x))U_\gamma\big)\xi_0 = 0$. Thus (a) follows from (3.4) and the above facts.

(b) Notice that $U_\gamma x U_\gamma^{-1} = -x$ for any $x \in \mathcal{M}_o$. The first relation in (3.3) implies that

$$\sigma_{t-i/4}(x) = -U_\gamma \sigma_{t-i/4}(x) U_\gamma^{-1}. \tag{3.7}$$

Thus a direct calculation shows that for any $\xi \in \mathcal{H}$,

$$\begin{aligned}&\|\big(\sigma_{t-i/4}(x) - j(\sigma_{t-i/4}(x^*))U_\gamma\big)J\xi\|^2 \\ &= \|U_\gamma J\big(-j(\sigma_{t-i/4}(x))U_\gamma + \sigma_{t-i/4}(x^*)\big)\xi\|^2 \\ &= \|\big(\sigma_{t-i/4}(x^*) - j(\sigma_{t-i/4}(x))U_\gamma\big)\xi\|^2.\end{aligned}$$

Here we have used $U_\gamma^{-1} = U_\gamma$ and U_γ commutes with J. Thus we have $\mathcal{E}^{(1)}[J\xi] = \mathcal{E}^{(2)}[\xi]$. The method used in the above also implies that $\mathcal{E}^{(2)}[J\xi] = \mathcal{E}^{(1)}[\xi]$.

(c) By the expression of $\mathcal{E}(\xi, \eta)$ in (3.4), $\mathcal{E}(\xi_+, \xi_-)$ can be written as

$$\begin{aligned}\mathcal{E}(\xi_+, \xi_-) &= \mathcal{E}^{(1)}(\xi_+, \xi_-) + \mathcal{E}^{(2)}(\xi_+, \xi_-) \\ &= (\mathrm{I}^{(1)} + \mathrm{II}^{(1)}) + (\mathrm{I}^{(2)} + \mathrm{II}^{(2)})\end{aligned} \tag{3.8}$$

where

$$\begin{aligned}\mathrm{I}^{(1)} = \int \Big(&\langle \sigma_{t-i/4}(x)\xi_+, \sigma_{t-i/4}(x)\xi_-\rangle \\ &+\langle \sigma_{t-i/4}(x^*)\xi_-, \sigma_{t-i/4}(x^*)\xi_+\rangle\Big) f(t)\, dt,\end{aligned} \tag{3.9}$$

$$\text{II}^{(1)} = -\int \Big(\langle \sigma_{t-i/4}(x)\xi_+, j\,(\sigma_{t-i/4}(x^*))U_\gamma\xi_- \rangle$$
$$+ \langle j\,(\sigma_{t-i/4}(x^*))U_\gamma\xi_+, \sigma_{t-i/4}(x)\xi_- \rangle \Big) f(t)\,dt.$$

and $\text{I}^{(2)}$ and $\text{II}^{(2)}$ are obtained from $\text{I}^{(1)}$ and $\text{II}^{(1)}$ respectively by replacing x by x^* in the above. Here we have used the facts that $U_\gamma = U_\gamma^{-1}$, the antilinearity of J and (3.7) to obtain the second term of $\text{I}^{(1)}$ in (3.9). As a consequence of Theorem 4(7) of Ref. 29, $\mathcal{M}\xi_+ \perp \mathcal{M}\xi_-$, which implies $\text{I}^{(1)} = 0$. Similarly $\text{I}^{(2)} = 0$. See also the proof of Proposition 5.3(ii) of Ref. 16.

Next, we consider $\text{II}^{(1)}$. Since $\sigma_{t-i/4}(x)^* = \sigma_{t+i/4}(x^*)$, it follows from (3.2) and (3.7) that

$$\begin{aligned} &\langle \sigma_{t-i/4}(x)\xi_+\,,\, j\,(\sigma_{t-i/4}(x^*))U_\gamma\xi_- \rangle \\ &\quad = \langle\, \xi_+\,,\, \sigma_{t+i/4}(x^*) j(\sigma_{t+i/4}(x)^*)U_\gamma\xi_- \rangle \end{aligned} \tag{3.10}$$

and

$$\begin{aligned} &\langle j\,(\sigma_{t-i/4}(x^*))U_\gamma\xi_+\,,\, \sigma_{t-i/4}(x)\xi_- \rangle \\ &\quad = \langle -U_\gamma j\,(\sigma_{t-i/4}(x^*))\xi_+\,,\, -U_\gamma\sigma_{t-i/4}(x)U_\gamma\xi_- \rangle \\ &\quad = \langle \xi_+\,,\, \sigma_{t-i/4}(x) j(\sigma_{t-i/4}(x^*)^*)U_\gamma\xi_- \rangle. \end{aligned} \tag{3.11}$$

Substituting (3.10) and (3.11) into the second expression of (3.9), we get that

$$\text{II}^{(1)} = -\int \langle\, \xi_+, \sigma_{t+i/4}(x^*) j(\sigma_{t+i/4}(x)^*)U_\gamma\xi_- \rangle\, f(t)\,dt$$
$$- \int \langle \xi_+, \sigma_{t-i/4}(x) j(\sigma_{t-i/4}(x^*)^*)U_\gamma\xi_- \,\rangle f(t)\,dt.$$

Notice that for any $x \in \mathcal{M}_{an}$ and $\xi \in \mathcal{H}$, the map

$$z \mapsto j(\sigma_z(x)^*)U_\gamma\xi$$

is analytic on a domain containing the strip $I_{1/2}$. In fact, the analyticity follows from the facts that $\langle \eta, j\big(\sigma_z(x)^*\big)U_\gamma\xi \rangle = \langle \sigma_z(x)^* JU_\gamma\xi, J\eta \rangle = \langle JU_\gamma\xi, \sigma_z(x)J\eta \rangle$ for any $\eta, \xi \in \mathcal{M}$, and that weak analyticity implies strong analyticity (see Theorem VI.4 of Ref. 30).

Using the Cauchy integral theorem, the assumption in the theorem, the property (c) in Definition 3.1 and $\sigma_t(x)^* = \sigma_t(x^*)$, we obtain that

$$\begin{aligned}\mathrm{II}^{(1)} = & -\int \langle \xi_+, \sigma_t(x^*) j(\sigma_t(x^*)) U_\gamma \xi_- \rangle f(t - i/4)\, dt \\ & - \int \langle \xi_+, \sigma_t(x) j(\sigma_t(x)) U_\gamma \xi_- \rangle f(t + i/4)\, dt.\end{aligned}$$

Replacing x by x^* in the above, we obtain the expression of $\mathrm{II}^{(2)}$. Thus we get

$$\begin{aligned}\mathrm{II} = -\int & \langle \xi_+, [\sigma_t(x) j(\sigma_t(x)) + \sigma_t(x^*) j(\sigma_t(x^*))] U_\gamma \xi_- \rangle \\ & \cdot (f(t - i/4) + f(t + i/4))\, dt.\end{aligned}$$

Recall that $U_\gamma \mathcal{P} \subset \mathcal{P}$ in (3.2). Since $\sigma_t(x^\#) j(\sigma_t(x^\#)) U_\gamma \xi_- \in \mathcal{P}$, $\langle \xi_+, \sigma_t(x^\#) j(\sigma_t(x^\#)) U_\gamma \xi_- \rangle \geq 0$ for $\forall t \in \mathbb{R}$, where $x^\#$ is either x or x^*. By the property (b) in Definition 3.1, we conclude that $\mathrm{II} \leq 0$. This proves the part (c) of the theorem.

Clearly $\mathcal{E}[\cdot] \geq 0$. Note that $\mathcal{E}[\xi_0] = 0$. Theorem 3.1 (b),(c) imply that the form $(\mathcal{E}, \mathcal{H})$ satisfies the conditions (2.1a) and (2.2). Thus $(\mathcal{E}, \mathcal{H})$ is a Dirichlet form. □

4. Markov semigroups on CAR algebras with gauge invariant quasi-free states

In this section, we apply the results in Section 3 to construct Dirichlet forms and associated Markov semigroups on CAR algebras with respect to quasi-free states.

We first review the notion of CAR algebras. For the details we refer to Section 5.2.2 of Ref. 17. Let $\mathfrak{h}_0$ be a separable pre-Hilbert space with an inner product $(\cdot, \cdot)$ and $\mathfrak{h}$ the completion of $\mathfrak{h}_0$. Let $\mathcal{A}(\mathfrak{h})$ be the C^*-algebra generated by the identity $\mathbf{1}$ and elements $a(f)$, $f \in \mathfrak{h}$, satisfying

$$\begin{aligned}&(a)\ f \mapsto a(f) \text{ is antilinear,} \\ &(b)\ \{a(f), a(g)\} = 0, \\ &(c)\ \{a(f), a(g)^*\} = (f, g)\mathbf{1}\end{aligned} \tag{4.1}$$

for all f, $g \in \mathfrak{h}$, where $\{A, B\} := AB + BA$. Notice that $\|a(f)\| = \|f\|$ for all $f \in \mathfrak{h}$, and so $\mathcal{A}(\mathfrak{h}_0) = \mathcal{A}(\mathfrak{h})$. Let $\gamma : \mathcal{A}(\mathfrak{h}) \to \mathcal{A}(\mathfrak{h})$ be the $*$-automorphism defined by

$$\gamma(a(f)) = -a(f), \tag{4.2}$$

for all $f \in \mathfrak{h}$. Then $(\mathcal{A}(\mathfrak{h}), \gamma)$ is a $\mathbb{Z}_2$-graded C^*-algebra.

Next, we describe quasi-free states on $\mathcal{A}(\mathfrak{h})$. Let A be a bounded and non-negative operator on $\mathfrak{h}$. In the rest of this section, we assume that A satisfies the following properties:

Assumption 4.1. *(a) There exists $\alpha > 0$ such that $0 < A \leq \alpha \mathbf{1}$.*
(b) The inverse A^{-1} of A exists as a (unbounded) self-adjoint and positive operator on $\mathfrak{h}$.
(c) For any $z \in \mathbb{C}$, A^z leaves $\mathfrak{h}_0$ invariant, i.e., $A^z\mathfrak{h}_0 \subset \mathfrak{h}_0$. Moreover, $z \mapsto A^z f$ is entire analytic for any $f \in \mathfrak{h}_0$.
(d) Any $g \in \mathfrak{h}_0$ is an analytic vector for $A^{-1/2}$.

We remark that a dense subspace $\mathfrak{h}_0$ of $\mathfrak{h}$ satisfying Assumption 4.1 exists by the spectral theorem.

For a given bounded operator A on $\mathfrak{h}$ satisfying Assumption 4.1, the gauge invariant quasi-free state ω on $\mathcal{A}(\mathfrak{h})$ is defined by

$$\omega(a(f_m)^* \cdots a(f_1)^* a(g_1) \cdots a(g_n)) = \delta_{nm} \det((g_i, A(\mathbf{1}+A)^{-1} f_j)) \tag{4.3}$$

for any $f_1, \cdots, f_m, g_1, \cdots, g_n \in \mathfrak{h}$.

Let $(\mathcal{H}_\omega, \pi_\omega, \Omega_\omega)$ be the GNS representation of $(\mathcal{A}(\mathfrak{h}), \omega)$. Define $\mathcal{M}_\omega := \pi_\omega(\mathcal{A}(\mathfrak{h}))''$, $a(f)_\omega := \pi_\omega(a(f)), a(f)^*_\omega := \pi_\omega(a(f)^*), f \in \mathfrak{h}$ and $\gamma_\omega := \pi_\omega \gamma \pi_\omega^{-1}$. From now on we suppress ω from the notations, e.g., $\mathcal{M} := \mathcal{M}_\omega$, $\mathcal{H} := \mathcal{H}_\omega$, $a(f) := a_\omega(f), a(f)^* := a_\omega(f)^*$ and $\gamma := \gamma_\omega$ etc. By continuity, γ extends to $\mathcal{M}$. We also write $\xi_0 = \Omega_\omega$. Let Δ and J denote the modular operator and the modular conjugation associated with the pair $(\mathcal{M}, \xi_0)$ respectively. Thus $\sigma_t(B) = \Delta^{it} B \Delta^{-it}, \forall B \in \mathcal{M}$. It follows from (4.2) and (4.3) that

$$\omega(\gamma(B)) = \omega(B), \ \ \forall B \in \mathcal{M}.$$

Since ω is γ-invariant, that is, $\langle \xi_0, \gamma(B)\xi_0 \rangle = \langle \xi_0, B\xi_0 \rangle, \forall B \in \mathcal{M}$, there exists a unitary operator U_γ on $\mathcal{H}$ satisfying the properties in (3.2).

Denote by $\mathcal{M}_{fin}$ the algebra generated by $\mathbf{1}$ and $a(f)^\#$, $f \in \mathfrak{h}_0$, where $a(f)^\#$ is either $a(f)$ or $a(f)^*$ for $f \in \mathfrak{h}_0$. Since $\mathcal{M}_{fin}$ is norm dense in $\pi_\omega(\mathcal{A}(\mathfrak{h}))$, $\mathcal{M}_{fin}\xi_0$ is also dense in $\mathcal{H}$. For any $f \in \mathfrak{h}_0$, $z \in \mathbb{C}$, we write

$$\begin{aligned} \sigma_z(a(f)) &:= a(A^{i\bar{z}} f), \\ \sigma_z(a(f)^*) &:= a(A^{iz} f)^*. \end{aligned} \tag{4.4}$$

In fact, using (4.3) and Assumption 4.1(c), one can check that for any $f \in \mathfrak{h}_0$ and $\xi \in \mathcal{M}_{fin}\xi_0$, the map $z \mapsto \sigma_z(a(f)^\#)\xi$ is the analytic extension on $\mathbb{C}$ of the map $t \mapsto \sigma_t(a(f)^\#)\xi$ (where $a(f)^\#$ is either $a(f)$ or $a(f)^*$).

Thus for any $g \in \mathfrak{h}_0$, $a(g)$ and $a(g)^*$ are σ_t-entire analytic odd elements of $\mathcal{M}$.

Let us turn to construction of a Dirichlet form which generates the symmetric Markov semigroup on CAR algebra $\mathcal{A}(\mathfrak{h})$ with respect to the quasi-free state ω. An admissible function f is said to be *normalized* if $\int f(t)\,dt = 1$. For a given normalized function f and a complete orthonormal system(CONS) $\{g_j\}_{j=1}^{\infty} \subset \mathfrak{h}_0$ for $\mathfrak{h}$, define a sesquilinear form $\mathcal{E} : D(\mathcal{E}) \times D(\mathcal{E}) \to \mathbb{C}$ as follows:

$$D(\mathcal{E}) = \{\xi \in \mathcal{H} : \sum_{j=1}^{\infty} \mathcal{E}_j(\xi,\xi) < \infty\}, \tag{4.5}$$

$$\mathcal{E}(\eta,\xi) = \sum_{j=1}^{\infty} \mathcal{E}_j(\eta,\xi), \quad \eta,\xi \in D(\mathcal{E})$$

where for each $j \in \mathbb{N}$,

$$\begin{aligned} D(\mathcal{E}_j) &= \mathcal{H}, \\ \mathcal{E}_j(\eta,\xi) &= \int \Big\langle \big(\sigma_{t-i/4}(a(g_j)) - j(\sigma_{t-i/4}(a(g_j)^*))U_\gamma\big)\eta, \\ &\qquad \big(\sigma_{t-i/4}(a(g_j)) - j(\sigma_{t-i/4}(a(g_j)^*))U_\gamma\big)\xi\Big\rangle f(t)dt \\ &\quad + \int \Big\langle \big(\sigma_{t-i/4}(a(g_j)^*) - j(\sigma_{t-i/4}(a(g_j)))U_\gamma\big)\eta, \\ &\qquad \big(\sigma_{t-i/4}(a(g_j)^*) - j(\sigma_{t-i/4}(a(g_j)))U_\gamma\big)\xi\Big\rangle f(t)dt. \end{aligned} \tag{4.6}$$

We also define the associated quadratic forms by

$$\begin{aligned} \mathcal{E}_j[\xi] &= \mathcal{E}_j(\xi,\xi), \quad \xi \in \mathcal{H},\ j \in \mathbb{N}, \\ \mathcal{E}[\xi] &= \sum_{j=1}^{\infty} \mathcal{E}_j[\xi], \quad \xi \in D(\mathcal{E}). \end{aligned} \tag{4.7}$$

We remark that the expression $\mathcal{E}_j(\eta,\xi)$ in (4.6) can be obtained from $\mathcal{E}(\eta,\xi)$ in (3.4) by replacing x by $a(g_j)$. Thus Theorem 3.1 implies that each $\mathcal{E}_j$ is a Dirichlet form.

Indeed we can show that the form $\mathcal{E}$ defined above is a Dirichlet form, and independent of the CONS $\{g_j\} \subset \mathfrak{h}_0$ for $\mathfrak{h}$ and the normalized admissible function f we have chosen. Moreover we can analyze the spectrum of the Dirichlet operator H. But the complete proofs are lengthy and require a decomposition of the GNS space $\mathcal{H}$. So we just state the main results and refer to Ref. 1 for the proofs.

Theorem 4.1. *Let $(\mathcal{E}, D(\mathcal{E}))$ be defined as in (4.5) and (4.6). Then the form $(\mathcal{E}, D(\mathcal{E}))$ is a densely defined Dirichlet form, and independent of the CONS $\{g_j\} \subset \mathfrak{h}_0$ for $\mathfrak{h}$ and the normalized admissible function f we have chosen. Moreover, let $\{T_t\}_{t\geq 0}$ be the semigroup associated to the form $(\mathcal{E}, D(\mathcal{E}))$. Then the semigroup $\{T_t\}_{t\geq 0}$ is J-real, strongly continuous, symmetric and Markov.*

Theorem 4.2. *Let $\{T_t\}_{t\geq 0}$ be the symmetric Markov semigroup associated to the form $(\mathcal{E}, D(\mathcal{E}))$ in Theorem 4.1 and H the Dirichlet operator, i.e., $T_t = e^{-tH}, t \geq 0$. Then the following results hold:*
(a) H is essentially self-adjoint on $\mathcal{M}_{fin}\xi_0$.
(b) The zero is a simple eigenvalue of H with eigenvector ξ_0. Moreover $(0, 2) \cap \sigma(H) = \emptyset$.

Acknowledgments

This work was supported by Korea Research Foundation (KRF-2001-005-D00010), Korean Ministry of Education.

References

1. C. Bahn, C. K. Ko and Y. M. Park, Dirichlet forms and symmetric Markovian semigroups on $\mathbb{Z}_2$-graded von Neumann algebras, preprint mp_arc 03-142.
2. A. Beurling and J. Deny, Dirichlet spaces, *Proc. Nat. Acad. Sci.* **45**, (1959).
3. M. Fukushima, *Dirichlet Forms and Markov Processes*, North-Holland, Kodansha (1980).
4. M. L. Silverstein, *Symmetric Markov Process*, Springer-verlag, (1974).
5. Z. M. Ma and M. Röckner, *An introduction to the thorey of (non-symmetric) Dirichlet forms*, Springer-verlag, (1992).
6. E. B. Davies, *Quantum theory of open systems*, Academic Press, London-New York-San Francisco, (1976).
7. S. Albeverio and R. Høegh-Krohn, Dirichlet forms and Markovian semigroups on C^*-algebras, *Comm. Math. Phys.* **56**, 173-187 (1977).
8. J.L. Sauvageot, Semi-groupe de la chaleur transverse sur la C^* algebre d'un feuilletage riemannien, *C. R. Acad. Sci. Paris. Ser. I* **310**, 531-536 (1990).
9. J.L. Sauvageot, Tangent bimodule and locality for dissipative operators on C^*-algebras, in *Quantum probability and applications IV* (L. Accardi and W. von Waldenfels, Eds.), Lecture Notes in Math. vol. 1396, Springer-Verlag, Berlin-Heidelberg-New York, (1987).
10. J.L. Sauvageot, Quantum Dirichlet forms, differential calculus and semigroups, in *Quantum probability and applications V* (L. Accardi and W. von Waldenfels, Eds.), Lecture Notes in Math. vol. 1442, Springer-Verlag, Berlin-Heidelberg-New York, (1988).

11. E. B. Davies and J.M. Lindsay, Non commutative symmetric Markov semigroups, *Math. Zeit.* **210**, 379-411 (1992).
12. D. Guido, T. Isola and S. Scarlatti, Non-commutative Dirichlet forms on semifinite von Neumann algebras, *J. Funct. Anal.* **135**, 50-75 (1996).
13. S. Goldstein and J. M. Lindsay, Beuring-Deny conditions for KMS-symmetric dynamical semigroups, *C. R. Acad. Sci. Paris. Ser. I* **317**, 1053-1057 (1993).
14. S. Goldstein and J. M. Lindsay, KMS-symmetric Markov semigroups, *Math. Zeit.* **219**, 590-608 (1995).
15. S. Goldstein and J. M. Lindsay, Markov semigroups KMS-symmetric for a weight, *Math. Ann.* **313**, 39-67 (1999)
16. F. Cipriani, Dirichlet forms and Markovian semigroups on standard forms of von Neumann algebras, *J. Funct. Anal.* **147**, 259-300 (1997).
17. O. Bratteli and D. W. Robinson, *Operator algebras and quantum statistical mechanics*, Springer-Verlag, New York-Heidelberg-Berlin, vol I (1979), vol. II (1981).
18. L. Accardi, Topics in quantum probability, *Phys.Rep.* **77**, 169-192 (1981).
19. L. Accaradi, A. Frigerio, and J. T. Lewis, Quantum stochastic processes, *Publ. Res. Inst. Math. Sci.* **18**, 97-133 (1982).
20. K. R. Parthasarathy, *An introduction to quantum stochastic calculus*, Birkhäuser, Basel (1992)
21. T. Matsui, Markov semigroups on UHF algebras, *Rev. Mod. Phys.* **5**, 587-600 (1993).
22. A. W. Majewski and B. Zegarlinski, Quantum stochastic dynamics I: Spin systems on a lattice, *MPEJ* **1**, Paper 2 (1995).
23. A. W. Majewski and B. Zegarlinski, Quantum stochastic dynamics II, *Rev. Math. Phys.* **8** (5), 689-713 (1996).
24. Y. M. Park, Construction of Dirichlet forms on standard forms of von Neumann algebras, *Infinite Dimensional Analysis, Quantum Probability and Related Topics*, Vol. 3, No. 1, 1-14 (2000).
25. C. Bahn, C. K. Ko amd Y. M. Park, Dirichlet forms and symmetric Markovian semigroups on CCR Algebras with quasi-free states, *J. Math. Phys.*, **44**, 723-753 (2003).
26. C. Bahn and C. K. Ko, Construction of unbounded Dirichlet forms on standard forms of von Neumann Algebras, *J. Korean Math. Soc.* **39** (6) 931-951 (2002).
27. F. Cipriani, F. Fagnola and J. M. Lindsay, Spectral Analysis and Feller Properties for Quantum Ornstein-Uhlenbeck Semigroups, *Comm. Math. Phys.* **210**, 85-105 (2000).
28. C. K. Ko and Y. M. Park, Construction of a Family of Quantum Ornstein-Uhlenbeck Semigroups, preprint mp_arc 03-141.
29. H. Araki, Some properties of modular conjugation operator of von Neumann algebras ans noncommutative Radon-Nikodym theorem with chain rule, *Pacific J. Math.* **50** (2), 309-354 (1974).
30. M.Reed and B.Simon, *Method of modern mathmatical physics I, II*, Academic press (1980).

GAUSSIAN WHITE NOISE AND APPLICATIONS TO FINANCE

TAE SU CHUNG
Human Genome Research Institute
College of Medicine
Seoul National University
Seoul, 110-799 Korea
E-mail: epiai@snu.ac.kr

UN CIG JI*
Department of Mathematics
Research Institute of Mathematical Finance
Chungbuk National University
Cheongju, 361-763 Korea
E-mail: uncigji@cbucc.chungbuk.ac.kr

The main purpose of this paper is to develop Gaussian white noise theory which is applied to a unified approach to stochastic calculus derived by several Gaussian Brownian motion containing fractional Brownian motion and generalized Brownian motion. As an application we study a simple model of financial market (Gaussian Black-Scholes market) derived by Gaussian Brownian motion which is no arbitrage and complete.

Keywords: Gaussian Brownian motion, stochastic integral, stochastic differential equation, Gaussian Black-Scholes market, portfolio, arbitrage, complete market.

Mathematics Subject Classification (2000): primary 60H40, 60H05; secondary 90A09, 90A12.

*This work was supported by grant (No. R05-2002-000-00142-0) from the Basic Research Program of the Korea Science & Engineering Foundation.

1. Introduction

The stochastic calculus derived by several processes, which is considered as basic processes, has been widely studied with applications to engineering and mathematical finance, etc. The basic process in mathematical finance describes the uncertainty of a financial market. In the (standard) Black-Scholes model of stock prices, the Brownian motion is used as the basic process and then stochastic calculus for Brownian motion is widely applied to analyze financial market. However, the fractional Brownain motion [5] is used to model a (uncertain) system which does not only depend on the situation at time t but also the whole history of the system up to time t. Also, a generalized Brownian motion [3] which is nonstationary in time is used to explain the position of the Ornstein–Uhlenbeck process in an external force field. Theses basic processes are special types of Gaussian processes.

On the other hand, the white noise theory initiated by Hida [6] has been considerably developed to an infinite dimensional distribution theory on Gaussian space with applications to many fields : stochastic calculus, Feynman path integral, quantum physics and so on (see e.g. Refs. 7, 8, 9, 15, 17). Recently, the (fractional) white noise theory is successfully applied to the study of mathematical finance in Refs. 1, 10.

The main purpose of this paper is to develop Gaussian white noise theory and then we study a unified approach to stochastic calculus derived by several Gaussian Brownian motions containing fractional Brownian motion and generalized Brownian motion. As an application we study a simple model of Gaussian Black-Scholes market derived by Gaussian Brownian motion and then we prove that the Gaussian Black-Scholes market is no arbitrage and complete.

The paper is organized as follows. In Section 2 we recall quickly the Gaussian space with covariance operator. In Section 3 we study Gaussian Brownain motion as a special type of Gaussian process, as examples, fractional Brownian motion and generalized Brownian motion. In Section 3 we revisit the Wiener–Itô–Segal isomorphism. In Section 5 we assemble some basic notions on Gaussian white noise theory. In Section 6 we study Gaussian Itô integral as a stochastic integral and then we prove Gaussian Girsanov formula. In Section 7 we study H–derivative and prove Gaussian Clark–Ocone formula. In Section 8, as an application, we study a simple model of Gaussian Black-Scholes market derived by Gaussian Brownian motion.

2. Gaussian Space

In this paper just for simplicity we start with the special choice of a real Gelfand triple:

$$E \equiv \mathcal{S}(\mathbf{R}) \subset H \equiv L^2(\mathbf{R}, dt) \subset E^* \equiv \mathcal{S}'(\mathbf{R}), \tag{1}$$

where E^* is the strong dual space of E. The norm and the inner product of H are denoted by $|\cdot|_0$ and $\langle \cdot, \cdot \rangle$, respectively. It is known that the Gelfand triple (1) is reconstructed with the harmonic oscillator $A = 1 + t^2 - d^2/dt^2$. For each $p \geq 0$, $E_p = \{\xi \in H \,;\, |\xi|_p \equiv |A^p \xi|_0 < \infty\}$ becomes a Hilbert space and then we have

$$E = \operatorname*{proj\,lim}_{p \to \infty} E_p, \qquad E^* \cong \operatorname*{ind\,lim}_{p \to \infty} E_{-p},$$

where $\cong$ stands the topological isomorphism and E_{-p} is the completion of the Hilbert space H with respect to the Hilbertian norm $|\cdot|_{-p} = |A^{-p} \cdot|_0$. It is noted that for each $t \in \mathbf{R}$ the delta function $\delta_t \in E_{-p}$ for any $p > 5/12$. We also note that there exists a complete orthonormal basis $\{e_j\}_{j=0}^{\infty} \subset E$ of H such that $Ae_j = (2j+2)e_j$. Then

$$\rho \equiv \| A^{-1} \|_{\mathrm{OP}} = \frac{1}{2}, \qquad \| A^{-r} \|_{\mathrm{HS}}^2 = \sum_{j=0}^{\infty} (2j+2)^{-2r} < \infty, \quad r > \frac{1}{2}.$$

For two locally convex spaces $\mathfrak{X}, \mathfrak{Y}$ the space of all continuous linear operators from $\mathfrak{X}$ into $\mathfrak{Y}$ is denoted by $\mathcal{L}(\mathfrak{X}, \mathfrak{Y})$ and is equipped with the topology of uniform convergence on every bounded subset.

Let $K \in \mathcal{L}(E, E^*)$ be a positive selfadjoint operator in the sense of the rigging (1), i.e.,

$$\langle K\xi, \xi \rangle > 0, \qquad \xi \in E \setminus \{0\}; \qquad \langle K\xi, \eta \rangle = \langle K^*\xi, \eta \rangle, \qquad \xi, \eta \in E.$$

Then by the Bochner-Minlos theorem there exists a Gaussian probability measure μ_K on E^* such that the characterization function is given by

$$\exp\left(-\frac{1}{2} \langle K\xi, \xi \rangle \right) = \int_{E^*} e^{i\langle x, \xi \rangle} d\mu_K(x), \qquad \xi \in E, \tag{2}$$

where $\langle \cdot, \cdot \rangle$ is the canonical bilinear form on $E^* \times E$. The probability space (E^*, μ_K) is called a *Gaussian space* with the covariance operator K. For a detail study of Gaussian space we refer to Refs. 11 and 12.

3. Gaussian Processes

Lemma 3.1. *For $\xi \in E$ we put $X_\xi(x) = \langle x, \xi \rangle$. Then the distribution of the random variable X_ξ, i.e., the image measure of μ_K by $x \mapsto X_\xi(x) \in \mathbf{R}$ is the Gaussian measure with mean 0 and variance $|\xi|_K^2$. Moreover, for any pair $\xi, \eta \in E$ we have*

$$\mathbf{E}_{\mu_K}[X_\xi X_\eta] = \int_{E^*} \langle x, \xi \rangle \langle x, \eta \rangle \mu_K(dx) = \langle K\xi, \eta \rangle .$$

Proof. The first assertion follows from the equation (2). For the second assertion we note first that X_ξ and X_η are independent if $\langle K\xi, \eta \rangle = 0$. Then, $\mathbf{E}_{\mu_K}[X_\xi X_\eta]$ is calculated for general $\xi, \eta \in E$ by means of orthogonalization. □

Let $L^2(E^*, \mu_K)$ (resp. $L^2_{\mathbf{R}}(E^*, \mu_K)$) be the Hilbert space of $\mathbf{C}$-valued (resp. $\mathbf{R}$-valued) square integrable functions on E^* with respect to μ_K. It follows from Lemma 3.1 that $\xi \mapsto X_\xi$ is extended to isometric embeddings from H_K into $L^2_{\mathbf{R}}(E^*, \mu_K)$ and from $H_{K,\mathbf{C}}$ into $L^2(E^*, \mu_K)$.

For $t \geq 0$, let $1_{[0,t]}$ be the indicator function of the interval $[0, t]$. From now on we assume that

$$1_{[0,t]} \in H_K, \qquad t \geq 0. \tag{3}$$

For example, if $\tau_K \in E^* \otimes E^*$ is slowly increasing function in $\mathbf{R}^2$, i.e.,

$$\int_0^t \int_0^t \tau_K(u, v) du dv < \infty, \qquad t \geq 0,$$

then $1_{[0,t]} \in H_K$ for any $t \geq 0$.

For each $t \geq 0$, we put

$$B_K(t) = X_{1_{[0,t]}}(\cdot) = \langle \cdot, 1_{[0,t]} \rangle.$$

Then, by the assumption (3), $\{B_K(t)\}_{t \geq 0} \subset L^2(E^*, \mu_K)$ and

(1) $B_K(0) = 0$,
(2) $\mathbf{E}_{\mu_K}[B_K(t)] = 0$ for $t \geq 0$, and
(3) $\mathbf{E}_{\mu_K}[B_K(s) B_K(t)] = \langle K 1_{[0,s]}, 1_{[0,t]} \rangle$ for $s, t \geq 0$,

where $\mathbf{E}_{\mu_K}$ is the expectation with respect to μ_K. The stochastic process $B_K(t)$ is a *Gaussian Brownian motion.*

Example 3.1. (fractional Brownian motion[10]) Now we consider the fractional Brownian motion with Hurst parameter h with $1/2 < h < 1$. Put

$$\mathcal{T}(s, t) = h(2h-1)|s-t|^{2h-2}, \qquad s, t \in \mathbf{R}.$$

Then for any $\xi \in E$ we have

$$\int_{\mathbf{R}} |s-t|^{2h-2}\xi(s)ds = \frac{1}{2h(2h-1)} \int_{\mathbf{R}} |s-t|^{2h} \frac{d^2\xi(s)}{ds^2} ds.$$

Since d^2/ds^2 is the continuous linear operator on E and $|s-t|^{2h}$ is a slowly increasing functions on $\mathbf{R} \times \mathbf{R}$, we define the continuous linear operator $K \in \mathcal{L}(E, E^*)$ with kernel function $\mathcal{T}(s,t)$ by

$$K\xi(t) = \int_{\mathbf{R}} \mathcal{T}(s,t)\xi(s)ds = h(2h-1)\int_{\mathbf{R}} \xi(s)|s-t|^{2h-2}ds.$$

Since $\mathcal{T}$ is slowly increasing function on $\mathbf{R}^2$, we see that $1_{[0,t]} \in H_K$ for $t \geq 0$. The Gaussian Brownian motion $B_K(t)$ associated with the operator K is denoted by $B_h(t)$ and called the *fractional Brownian motion* with Hurst parameter $1/2 < h < 1$. By direct computation, we have

$$\mathbf{E}_{\mu_K}[B_h(s)B_h(t)] = \frac{1}{2}\left(|t|^{2h} + |s|^{2h} - |t-s|^{2h}\right\}, \qquad s,t \geq 0,$$

where μ_K is a Gaussian probability measure on E^* such that

$$\exp\left\{-\frac{1}{2}\int_{\mathbf{R}}\int_{\mathbf{R}} \xi(s)\xi(t)\mathcal{T}(s,t)dsdt\right\} = \int_{E^*} e^{i\langle x,\xi\rangle} d\mu_K(x), \quad \xi \in E.$$

Now we observe that the transform $\mathcal{I}_h$ defined by

$$\mathcal{I}_h\xi(u) = \sqrt{\frac{h(2h-1)\Gamma(3/2-h)}{\Gamma(h-1/2)\Gamma(2-2h)}} \int_u^\infty (t-u)^{h-3/2}\xi(t)dt$$

is an isomorphism from the Hilbert space H_K into H, see Ref. 10. Moreover, a set $\{e_j^{(h)} \equiv \mathcal{I}_K^{-1}(e_j)\}$ forms an orthonormal basis for H_K such that

$$\left|\int_{\mathbf{R}} e_j^{(h)}(s)\mathcal{T}(s,t)ds\right| < C_t(j+1)^{1/6}, \quad j = 0,1,2,\cdots,$$

for some $C_t > 0$, where $\{e_j\}$ is the orthonormal basis for H which is introduced in Section 2.

Example 3.2. (generalized Brownian motion[3]) Now, we consider a special choice of Gelfand triple on $\mathbf{R}_+ = [0,\infty)$:

$$\mathcal{S}(\mathbf{R}_+) \subset L^2(\mathbf{R}_+, dt) \subset \mathcal{S}'(\mathbf{R}_+)$$

which is constructed by standard manner used in Section 2 with the positive selfadjoint operator

$$A = -2\frac{d}{dt}t\frac{d}{dt} + \frac{1}{2}t + 1.$$

For the construction we refer to Ref. 4.

Let $b(t)$ be a strictly increasing continuously differentiable real-valued function on $\mathbf{R}_+$ with $b(0) = 0$ and $b'(t) > 0$ for $t > 0$. We assume that there exists $p \geq 0$ such that $A^{-p}b' \in L^2(\mathbf{R}_+, dt)$. Then we define a continuous linear operator K from $\mathcal{S}(\mathbf{R}_+)$ into $\mathcal{S}'(\mathbf{R}_+)$ by

$$K\xi(t) = b'(t)\xi(t), \quad \xi \in \mathcal{S}(\mathbf{R}_+)$$

and then we can easily verify that for each $t \geq 0$, the indicator function $1_{[0,t]}$ belongs to $H_K = \{f\,;\, \int_{\mathbf{R}_+} b'(t)f(t)^2 dt < \infty\}$. The Gaussian Brownian motion $B_K(t)$ associated with the operator K is denoted by $B_b(t)$ and called the *generalized Brownian motion* with mean 0 and variance $b(\cdot)$. In fact, the covariance of generalized Brownian motion is obtained by

$$\begin{aligned}\mathbf{E}_{\mu_K}[B_b(s)B_b(t)] &= \langle K1_{[0,s]}, 1_{[0,t]}\rangle \\ &= \int_{\mathbf{R}} b'(u)1_{[0,s]}1_{[0,t]}\, du \\ &= \min\{b(s), b(t)\},\end{aligned}$$

for $s, t \geq 0$. Moreover, a set $\{\ell_j^{(b)}(t) \equiv \frac{1}{\sqrt{b'(t)}}\ell_j(t)\}$ forms an orthonormal basis for H_K such that

$$|K\ell_j^{(b)}(t)| = |\sqrt{b'(t)}\ell_j(t)| \leq \sqrt{b'(t)}$$

since $\sup_{t\in\mathbf{R}_+} |\ell_j(t)| \leq 1$, where ℓ_j is the Laguerre function of degree j, see Ref. 4.

4. Wiener-Itô-Segal Isomorphism

For each locally convex space $\mathfrak{X}$ we denote $\mathfrak{X}_{\mathbf{C}}$ the complexification of $\mathfrak{X}$. By the kernel theorem for each $L \in \mathcal{L}(E_{\mathbf{C}}, E_{\mathbf{C}}^*)$ there exists $\tau_L \in (E_{\mathbf{C}} \otimes E_{\mathbf{C}})^*$ such that

$$\langle \tau_L, \xi \otimes \eta\rangle = \langle L\eta, \xi\rangle, \qquad \xi, \eta \in E_{\mathbf{C}}, \tag{4}$$

which is called the *L-trace*. Conversely, for each $\kappa \in (E_{\mathbf{C}} \otimes E_{\mathbf{C}})^*$ there exists $L_\kappa \in \mathcal{L}(E_{\mathbf{C}}, E_{\mathbf{C}}^*)$ such that

$$\langle L_\kappa \eta, \xi\rangle = \langle \kappa, \xi \otimes \eta\rangle, \qquad \xi, \eta \in E_{\mathbf{C}}.$$

Let $K \in \mathcal{L}(E, E^*)$ be a positive operator.

Definition 4.1. For $x \in E^*$ and $n = 0, 1, 2, \cdots$, we define $:x^{\otimes n}:_K \in (E^{\otimes n})^*_{sym}$ inductively as follows:

$$:x^{\otimes 0}:_K = 1;$$
$$:x^{\otimes 1}:_K = x;$$
$$:x^{\otimes n}:_K = x\hat{\otimes} :x^{\otimes(n-1)}:_K -(n-1)\tau_{\widetilde{K}}\widehat{\otimes} :x^{\otimes(n-2)}:_K, \qquad n \geq 2,$$

where $\widetilde{K} = (K + K^*)/2$ is the self-adjoint operator induced by K.

The following lemma is easily verified by induction.

Lemma 4.1. *Let $x \in E^*$. Then*

$$:x^{\otimes n}:_K = \sum_{k=0}^{[n/2]} \frac{(-1)^k n!}{k!(n-2k)!2^k} \tau_{\widetilde{K}}^{\widehat{\otimes} k} \widehat{\otimes} x^{\otimes(n-2k)},$$

$$x^{\otimes n} = \sum_{k=0}^{[n/2]} \frac{n!}{k!(n-2k)!2^k} \tau_{\widetilde{K}}^{\widehat{\otimes} k} \widehat{\otimes} :x^{\otimes(n-2k)}:_K .$$

For each positive operator $K \in \mathcal{L}(E, E^*)$, let H_K be the completion of E with respect to the norm $|\cdot|_K$ induced by the inner product defined as in (4), i.e.,

$$|\xi|_K^2 = \langle K\xi, \xi \rangle, \qquad \xi \in E.$$

Then the inclusion $E \hookrightarrow H_K$ is continuous with dense image. Each positive selfadjoint operator $K \in \mathcal{L}(E, E^*)$ can be extended to $E_{\mathbf{C}}$. For the extension and the Hilbertian norm of $H_{K,\mathbf{C}}$ we use the same notations, i.e.,

$$|\xi|_K^2 = \left\langle K\xi, \overline{\xi} \right\rangle > 0, \qquad \xi \in E_{\mathbf{C}}.$$

From now on we assume that K is a positive self-adjoint operator in $\mathcal{L}(E, E^*)$. Then it is noted that $\widetilde{K} = (K + K^*)/2 = K$.

Theorem 4.1. ***(Wiener–Itô Decomposition)*** *For each $\Phi \in L^2(E^*, \mu_K)$ there exists a unique sequence $f_n \in H_{K,\mathbf{C}}^{\widehat{\otimes} n}$ such that Φ admits the following representation:*

$$\Phi(x) = \sum_{n=0}^{\infty} \left\langle :x^{\otimes n}:_K, f_n \right\rangle,$$

For the proof of Theorem 4.1 we refer to Refs. 2 and 17. The Hilbertian norm on $L^2(E^*, \mu_K)$ is denoted by $\|\cdot\|_K$ and then

$$\|\Phi\|_K^2 = \sum_{n=0}^{\infty} n! |f_n|_K^2.$$

For each $\xi \in E_{\mathbf{C}}$ we put

$$\phi_{K,\xi}(x) = \sum_{n=0}^{\infty} \left\langle :x^{\otimes n}:_K, \frac{\xi^{\otimes n}}{n!} \right\rangle.$$

Then we have

$$\|\phi_{K,\xi}\|_K^2 = \sum_{n=0}^{\infty} n! \left| \frac{\xi^{\otimes n}}{n!} \right|_K^2 = e^{|\xi|_K^2}$$

and $\phi_{K,\xi} \in L^2(E^*, \mu_K)$. The vector $\phi_{K,\xi}(x)$ is called an *exponential vector* or *coherent state*. It is easily proved that $\{\phi_{K,\xi}; \xi \in E_{\mathbf{C}}\}$ spans a dense subspace of $L^2(E^*, \mu_K)$.

Let $\mathcal{H}$ be a Hilbert space with norm $|\cdot|$. For $n \geq 0$ let $\mathcal{H}^{\widehat{\otimes} n}$ denote the n-fold symmetric tensor power of the Hilbert space $\mathcal{H}$. Their norms are denoted by the common symbol $|\cdot|$ for simplicity. Put

$$\Gamma(\mathcal{H}) = \left\{ \phi = (f_n)_{n=0}^{\infty} \,\middle|\, f_n \in \mathcal{H}^{\widehat{\otimes} n}, \| \phi \|^2 \equiv \sum_{n=0}^{\infty} n! \, | f_n |^2 < \infty \right\}.$$

Then $\Gamma(\mathcal{H})$ is called the *(Boson) Fock space* over the Hilbert space $\mathcal{H}$.

Now, we have the following Wiener-Itô-Segal isomorphism between the Boson Fock space $\Gamma(H_{K,\mathbf{C}})$ and $L^2(E^*, \mu_K)$. For the proof we refer to Refs. 2 and 17.

Theorem 4.2. *There exists a unitary isomorphism between* $L^2(E^*, \mu_K)$ *and* $\Gamma(H_{K,\mathbf{C}})$, *which is uniquely determined by the correspondence:*

$$\phi_{K,\xi}(x) = e^{\langle x, \xi \rangle - \langle K\xi, \xi \rangle / 2} \quad \leftrightarrow \quad \left(1, \xi, \frac{\xi^{\otimes 2}}{2!}, \dots, \frac{\xi^{\otimes n}}{n!}, \dots \right), \tag{5}$$

where ξ *runs over* $E_{\mathbf{C}}$.

5. Gaussian White Noise Distributions

In this section, for a positive self-adjoint operator $K \in \mathcal{L}(E, E^*)$, we assume that there exists orthonormal basis $\{e_j^{(K)}\}_{j=0}^{\infty}$ for $H_{K,\mathbf{C}}$ such that

$$|Ke_j^{(K)}(t)| \leq C_t (j+1)^q, \qquad j = 0, 1, 2, \cdots, \quad t \in \mathbf{R}, \tag{6}$$

for some $C_t > 0$ and $q \in \mathbf{R}$. Define a positive (self-adjoint) operator A_K on $H_{K,\mathbf{C}}$ by

$$A_K e_j^{(K)} = (2j+2) e_j^{(K)}, \qquad j = 0, 1, 2, \cdots .$$

Then, for each $p \ge 0$, the dense subspace $D_p \equiv \mathrm{Dom}\,(A_K^p) \subset H_{K,\mathbf{C}}$ becomes a Hilbert space equipped with the norm

$$| \xi |_{K,p} = | A_K^p \xi |_K , \qquad \xi \in \mathrm{Dom}\,(A_K^p).$$

Furthermore, we define D_{-p} to be the completion of $H_{K,\mathbf{C}}$ with respect to the norm $| \xi |_{K,-p} = \left| A_K^{-p} \xi \right|_K$, $\xi \in H_{K,\mathbf{C}}$. Then we have

$$| \xi |_{K,p} \le |||A_K^{-1}|||^{q-p} \, | \xi |_{K,q} , \qquad \xi \in D_q, \quad -\infty < p \le q < \infty,$$

and a chain of Hilbert spaces:

$$\cdots \subset D_p \subset \cdots \subset D_0 = H_{K,\mathbf{C}} \subset \cdots \subset D_{-p} \subset \cdots , \tag{7}$$

where each inclusion is continuous and has a dense image. Moreover, for any $p, q \in \mathbf{R}$ the operator A_K^{p-q} is naturally considered as an isometry from D_p onto D_q. From (7) we obtain a countable Hilbert space and its dual space:

$$D = \operatorname*{proj\,lim}_{p \to \infty} D_p, \qquad D^* = \operatorname*{ind\,lim}_{p \to \infty} D_{-p}.$$

We still denote $\langle F, \xi \rangle_K$ for the dual pairing of $F \in D^*$ and $\xi \in D$. Note that the inductive limit and the strong dual topologies coincide. Obviously,

$$\| A_K^{-q} \|_{\mathrm{HS}}^2 = \sum_{k=0}^{\infty} \frac{1}{(2k+2)^{2q}} < \infty, \quad q > \frac{1}{2}.$$

Therefore, D is a nuclear space.

For simplicity we set

$$\mathcal{W}_p \equiv \Gamma(D_p), \qquad p \ge 0.$$

Then for any $0 \le p \le q$ we naturally come to

$$\begin{aligned} \mathcal{W} \equiv \operatorname*{proj\,lim}_{p \to \infty} \mathcal{W}_p \subset \cdots \subset \mathcal{W}_q \subset \cdots \subset \mathcal{W}_p \subset \cdots \\ \cdots \subset \mathcal{W}_0 = \Gamma(H_{K,\mathbf{C}}) \subset \cdots \subset \mathcal{W}_{-p} \subset \cdots \subset \mathcal{W}_{-q} \subset \cdots \subset \mathcal{W}^*, \end{aligned} \tag{8}$$

where $\mathcal{W}$ becomes a nuclear space, and $\mathcal{W}^*$ is the topological dual space of $\mathcal{W}$, or equivalently $\mathcal{W}^* = \operatorname{ind\,lim}_{p \to \infty} \mathcal{W}_{-p}$. The canonical complex bilinear form on $\mathcal{W}^* \times \mathcal{W}$ is denoted by $\langle\!\langle \cdot, \cdot \rangle\!\rangle_K$. Then for each $\phi \in \mathcal{W}$ there exists a unique sequence (f_n) with $f_n \in D^{\hat{\otimes} n}$ such that

$$\phi(x) = \sum_{n=0}^{\infty} \left\langle : x^{\otimes n} :_K, \, f_n \right\rangle ,$$

and

$$\| \phi \|_{K,p}^2 = \sum_{n=0}^{\infty} n! |f_n|_{K,p}^2 < \infty, \qquad p \geq 0.$$

For simplicity, we write $\phi = (f_n)$. For each $\Phi \in \mathcal{W}^*$, there is a unique sequence (F_n) with $F_n \in (D^{\otimes n})^*_{\mathrm{sym}}$ such that

$$\| \Phi \|_{K,-p}^2 = \sum_{n=0}^{\infty} n! \, | F_n |_{K,-p}^2 < \infty,$$

for some $p \geq 0$, where $| F_n |_{K,-p} = \left| (A_K^{-p})^{\otimes n} F_n \right|_K$. In this case, we write $\Phi = (F_n)$. Moreover, for $\Phi = (F_n)$ and $\phi = (f_n)$ we have

$$\langle\langle \Phi, \phi \rangle\rangle_K = \sum_{n=0}^{\infty} n! \, \langle F_n, f_n \rangle_K ,$$

and it holds that

$$| \langle\langle \Phi, \phi \rangle\rangle_K | \leq \| \Phi \|_{K,-p} \| \phi \|_{K,p} ,$$

where we use the same notation $\langle \cdot, \cdot \rangle_K$ for the dual pairing on $(D^{\otimes n})^*$ and $D^{\otimes n}$ for $n \geq 0$.

For notational convenience, we use same notation for the nuclear rigging corresponding to the Fock rigging (8) under the Wiener-Ito-Segal isomorphism as given in (5). Then we have the following nuclear rigging of functions defined on E^*:

$$\mathcal{W} \subset L^2(E^*, \mu_K) \subset \mathcal{W}^*$$

which is referred to as a *Hida-Kubo-Takenaka space*[14]. An element in $\mathcal{W}$ (resp. $\mathcal{W}^*$) is called an *test (resp. generalized) Gaussian white noise function.*

For $p \in \mathbf{R}$ let $\mathcal{G}_p$ be the space of all sequences $\phi = (f_n)$, $f_n \in H_{K,\mathbf{C}}^{\widehat{\otimes} n}$, such that

$$\| \phi \|_{K,p}^2 \equiv \sum_{n=0}^{\infty} n! e^{2pn} |f_n|_K^2 < \infty.$$

Note that $\mathcal{G}_0 = \Gamma(H_{K,\mathbf{C}})$. Then we have a Fock rigging:

$$\mathcal{G} = \operatorname{proj\,lim}_{p \to \infty} \mathcal{G}_p \subset \Gamma(H) \subset \mathcal{G}^* = \operatorname{ind\,lim}_{p \to \infty} \mathcal{G}_{-p},$$

where $\mathcal{G}$ becomes a countable Hilbert space but not a nuclear space.

Lemma 5.1. *For any $p \geq 0$ we have continuous inclusions with dense images:*

$$\mathcal{W} \subset \mathcal{G}_p \subset \Gamma(H) \subset \mathcal{G}_{-p} \subset \mathcal{W}^*.$$

For the proof, we refer to Ref. 11, thus we have

$$\mathcal{W} \subset \mathcal{G} \subset \Gamma(H) \subset \mathcal{G}^* \subset \mathcal{W}^*.$$

An element in $\mathcal{G}^*$ is said to be *admissible*. The canonical **C**-bilinear form on $\mathcal{G}^* \times \mathcal{G}$ is again denoted by $\langle\!\langle \cdot, \cdot \rangle\!\rangle_K$.

The S-transform of $\Phi \in \mathcal{W}^*$ is a function on D defined by

$$S\Phi(\xi) = \langle\!\langle \Phi, \phi_{K,\xi} \rangle\!\rangle_K, \qquad \xi \in D.$$

Since $\{\phi_{K,\xi} \,|\, \xi \in D\}$ spans a dense linear subspace of $\mathcal{W}$, each $\Phi \in \mathcal{W}^*$ is uniquely determined by its S-transform. Now we state the characterization theorem of S-transform which was first proved in Ref. 18 for the (standard) Hida–Kubo–Takenaka space.

Theorem 5.1. *The S-transform $F = S\Phi$ of $\Phi \in \mathcal{W}^*$ satisfies the following conditions:*

(F1) *For any $\xi, \eta \in D$, the map $z \mapsto F(z\xi + \eta)$ is entire.*
(F2) *There exist $C > 0, a > 0$ and $p \geq 0$ such that*

$$|F(\xi)| \leq C \exp\{a \,|\, \xi \,|_{K,p}^2\}, \qquad \xi \in D.$$

*Conversely, assume that a **C**-valued function F defined on D satisfies the conditions (F1) and (F2). Then there exists a unique $\Phi \in \mathcal{W}^*$ such that $F = S\Phi$.*

For $\Phi, \Psi \in \mathcal{W}^*$, the Wick product $\Phi \diamond_K \Psi$ is defined by the equation

$$S(\Phi \diamond_K \Psi) = (S\Phi)(S\Psi).$$

In fact, the pointwise multiplication of S-transforms of elements in $\mathcal{W}^*$ satisfies the conditions (F1) and (F2). Moreover, we have

$$\phi \diamond_K \psi \in \mathcal{W}, \qquad \phi, \psi \in \mathcal{W},$$

and

$$\phi_{K,\xi} \diamond_K \phi_{K,\eta} = \phi_{K,\xi+\eta}, \qquad \xi, \eta \in D.$$

Here we define the wick exponential $\operatorname{wexp}\Phi$ of $\Phi \in \mathcal{W}^*$ by

$$\operatorname{wexp}_K \Phi = \sum_{n=0}^{\infty} \frac{1}{n!} \Phi^{\diamond_K n},$$

if the series converges in $\mathcal{W}^*$. Then it is easily checked that

$$\mathrm{wexp}_K\{\langle x,\,\xi\rangle\} = e^{\langle x,\,\xi\rangle - \langle \xi,\,\xi\rangle_K/2} = \phi_{K,\xi}(x), \qquad \xi \in D.$$

6. Stochastic Integrals

Lemma 6.1. *Let $\Phi : \mathbf{R} \to \mathcal{W}^*$ be a given function satisfying that for any $\phi \in \mathcal{W}$*

$$\langle\langle \Phi(t),\, \phi \rangle\rangle_K \in L^1(\mathbf{R}, dt).$$

Then there exists a unique element $\Psi \in \mathcal{W}^*$ such that

$$\langle\langle \Psi,\, \phi \rangle\rangle_K = \int_{\mathbf{R}} \langle\langle \Phi(t),\, \phi \rangle\rangle_K \, dt, \qquad \phi \in \mathcal{W}.$$

Proof. The proof is a simple modification of Proposition 8.1 in Ref. 9. □

In this case, Φ is said to be *integrable* in $\mathcal{W}^*$ and Ψ is called a *white noise integral of* Φ, and Ψ is denoted by $\Psi = \int_{\mathbf{R}} \Phi(t)dt$.

Example 6.1. (Gaussian white noise) The Gaussian white noise $W_K(t)$ at time t is defined by the derivative of Gaussian Brownian motion in $\mathcal{W}^*$ as

$$W_K(t) = \frac{d}{dt} B_K(t) = \lim_{\epsilon \to 0} \frac{1}{\epsilon}(B_K(t+\epsilon) - B_K(t)).$$

For the convergence of the limit in $\mathcal{W}^*$, we first need to show that $\delta_t \in D^*$. The fact that $\{(2j+2)^p e_j^{(K)}\}$ forms an orthonormal basis for D_{-p} implies

$$\begin{aligned}
|\,\delta_t\,|_{K,-p}^2 &= \left| \sum_{j=0}^{\infty} \left\langle \delta_t,\, e_j^{(K)} \right\rangle_K e_j^{(K)} \right|_{K,-p}^2 \\
&= \sum_{j=0}^{\infty} (2j+2)^{-2p} |K e_j^{(K)}(t)|^2 \\
&\le C_t^2 2^{-2q} \sum_{j=0}^{\infty} (2j+2)^{-2(p-q)} < \infty,
\end{aligned}$$

for $p > q + 1/2$. Here $C_t > 0$ and $q \in \mathbf{R}$ are taken in the assumption (6). From the fact that $\delta_t \in D^*$ we can easily see that the series

$$\frac{1}{\epsilon}(B_K(t+\epsilon) - B_K(t))$$

converges in $\mathcal{W}^*$ as $\epsilon \to 0$. Actually, for $\phi = (f_n) \in \mathcal{W}$ we have

$$\langle\langle W_K(t), \phi \rangle\rangle_K = \langle \delta_t, f_1 \rangle_K = K f_1(t), \qquad t \in \mathbf{R},$$

and

$$\langle\langle B_K(t), \phi \rangle\rangle_K = \left\langle 1_{[0,t]}, f_1 \right\rangle_K = \int_0^t K f_1(s) ds = \int_0^t \langle\langle W_K(s), \phi \rangle\rangle_K \, ds.$$

Definition 6.1. Let Φ be a $\mathcal{W}^*$-valued function on $\mathbf{R}$ such that $\Phi(t) \diamond_K W_K(t)$ is integrable in $\mathcal{W}^*$. Then we define the *stochastic integral of Itô type* with respect to the Gaussian Brownian motion $B_K(t)$ by

$$\int_{\mathbf{R}} \Phi(t) dB_K(t) \equiv \int_{\mathbf{R}} \Phi(t) \diamond_K W_K(t) dt.$$

Example 6.2. For each $t \geq 0$, we have

$$\begin{aligned} \int_0^t B_K(s) dB_K(s) &= \int_0^t B_K(s) \diamond_K W_K(s) ds \\ &= \frac{1}{2} B_K(t) - \frac{1}{2} \left\langle K 1_{[0,t]}, 1_{[0,t]} \right\rangle . \end{aligned}$$

Example 6.3. (geometric Gaussian Brownian motion) Consider the stochastic differential equation associated with the Gaussian Brownian motion:

$$dX_t = \mu X_t dt + \sigma X_t dB_K(t), \quad X_0 = x,$$

where μ, σ and x are constants. This equation is equivalent to

$$\frac{dX_t}{dt} = (\mu + \sigma W_K(t)) \diamond_K X_t, \quad X_0 = x,$$

in $\mathcal{W}^*$. Since $(\mathcal{W}^*, \diamond_K)$ is a commutative algebra, the formal solution is given by

$$\begin{aligned} X_t &= x \operatorname{wexp}_K \left(\mu t + \sigma \int_0^t W_K(s) ds \right) \\ &= x \operatorname{wexp}_K \left(\mu t + \sigma B_K(t) \right). \end{aligned}$$

Therefore, by a relation between Wick exponential and the usual exponential, we have

$$X_t = x \exp \left(\sigma B_K(t) + \mu t - \frac{1}{2} \sigma^2 \left\langle K 1_{[0,t]}, 1_{[0,t]} \right\rangle \right).$$

From now on we assume that $\|K^{-1}\|_{\mathrm{OP}} < \infty$. Then for $\zeta \in D^*$, the translation operator $\mathcal{T}_\zeta$ on $\mathcal{W}$ is defined by

$$\mathcal{T}_\zeta \phi(x) = \phi(x+\zeta), \qquad \phi \in \mathcal{W}.$$

Then by similar arguments used in Ref. 17 we can show that $\mathcal{T}_\zeta \phi$ belongs to $\mathcal{W}$ and $\mathcal{T}_\zeta \in \mathcal{L}(\mathcal{W}, \mathcal{W})$, and for each $\phi = (f_n) \in \mathcal{W}$

$$\mathcal{T}_\zeta \phi = \left(\sum_{m=0}^{\infty} \frac{(n+m)!}{n!m!} \zeta^{\otimes m} \widehat{\otimes}_m f_{n+m} \right)_{n=0}^{\infty}, \tag{9}$$

where $\widehat{\otimes}_m$ is the right m-contraction of symmetric tensor products, see Ref. 17. Moreover, we have the following

Proposition 6.1. *Let $\zeta \in H_K$ and $\Phi \in \mathcal{G}_{-p}$, where $p \geq 0$. Then $\mathcal{T}_\zeta \Phi \in \mathcal{G}_{-q}$ and $\mathcal{T}_\zeta \in \mathcal{L}(\mathcal{G}_{-p}, \mathcal{G}_{-q})$ for any $q > p + \log \sqrt{2}$.*

Proof. The proof is a simple modification of the proof of Proposition 3.2 in Ref. 11. In the proof, we used the fact that $|\langle \zeta, f \rangle| \leq \|K^{-1}\|_{\mathrm{OP}} |\zeta|_K |f|_K$ for any $\zeta, f \in H_K$. □

Lemma 6.2. *Let $\eta \in H_K$ such that $K\eta \in D^*$ and let $\tilde{\eta} = K\eta$. Then the adjoint operator $\mathcal{T}_{\tilde{\eta}}^*$ of the translation operator $\mathcal{T}_{\tilde{\eta}}$ is represented as*

$$\mathcal{T}_{\tilde{\eta}}^* \Phi = \Phi \diamond_K \phi_{K,\eta}, \qquad \Phi \in \mathcal{W}^*.$$

Proof. Let's calculate S-transforms of both sides. For each $\xi \in D$, we have

$$\begin{aligned} S(\mathcal{T}_{\tilde{\eta}}^* \Phi)(\xi) &= \langle\langle \Phi, \phi_{T,\xi}(\cdot + \tilde{\eta}) \rangle\rangle_K \\ &= \langle\langle \Phi, e^{\langle \cdot + \tilde{\eta}, \xi \rangle - \langle \xi, \xi \rangle_K / 2} \rangle\rangle_K \\ &= e^{\langle \eta, \xi \rangle_K} S\Phi(\xi). \end{aligned}$$

On the other hand,

$$\begin{aligned} S(\Phi \diamond_K \phi_{K,\eta})(\xi) &= S\Phi(\xi) S(\phi_{K,\eta})(\xi) \\ &= e^{\langle \eta, \xi \rangle_K} S\Phi(\xi). \end{aligned}$$

This proves the assertion. □

The following theorem is the Benth–Gjessing version of the Girsanov formula for Gaussian Brownian motion.

Theorem 6.1. *Let $\phi \in L^p(E^*, \mu_K)$ for some $p > 1$ and let $\eta \in H_K$ such that $K\eta \in D^*$, and let $\tilde{\eta} = K\eta$. Assume that the translation $\mathcal{T}_{\tilde{\eta}}\phi$ of ϕ is well–defined. Then $\mathcal{T}_{\tilde{\eta}}\phi \in L^q(E^*, \mu_K)$ for all q with $0 \le q < p$ and we have*

$$\int_{E^*} \phi(x + \tilde{\eta})\mu_K(dx) = \int_{E^*} \phi(x)\, \phi_{K,\eta}(x)\, \mu_K(dx).$$

Proof. The proof is a simple modification of the proof of Theorem 3.16 in Ref. 10. By Lemma 6.2 with $\Phi = 1$, we have $\mathcal{T}^*_{\tilde{\eta}} 1 = \phi_{K,\eta}$ and so for any $\phi \in \mathcal{W}$,

$$\langle\langle 1,\, \mathcal{T}_{\tilde{\eta}}\phi \rangle\rangle_K = \langle\langle \phi_{K,\eta},\, \phi \rangle\rangle_K$$

which is that we want to prove. Since $\phi_{K,\eta}$ belongs to $L^r(E^*, \mu_K)$ for all $r \ge 0$, by applying Hölder's inequality we have $T_{\tilde{\eta}}\phi \in L^q(E^*, \mu_K)$ for all q with $0 \le q < p$. □

From now on, we assume that the kernel function τ_K corresponding to the operator $K \in \mathcal{L}(E, E^*)$ by the kernel theorem is a bona fide function defined on $\mathbf{R} \times \mathbf{R}$.

Corollary 6.1. *Let f be a real-valued bounded function on $\mathbf{R}$ and let $\eta \in H_K$ with $\tilde{\eta} = K\eta \in D^* \cap L^1_{\text{loc}}(\mathbf{R})$. Then we obtain*

$$\boldsymbol{E}_{\mu_K}\left[f(B_K(t) + \int_0^t \tilde{\eta}(s)ds)\right] = \boldsymbol{E}_{\mu_K}[f(B_K(t))\phi_{K,\eta}], \qquad t \ge 0.$$

Proof. For each $t \ge 0$ we define $\phi(x) = f(\langle x,\, 1_{[0,t]}\rangle)$, then $\phi = f(B_K(t)) \in L^p(E^*, \mu_K)$ for any $p \ge 1$. Therefore, by Theorem 6.1 we have

$$\begin{aligned}\mathbf{E}_{\mu_K}\left[f(B_K(t) + \int_0^t \tilde{\eta}(s)ds)\right] &= \int_{E^*} \phi(x + \tilde{\eta})\mu_K(dx)\\ &= \int_{E^*} \phi(x)\phi_{K,\eta}(x)\mu_K(dx)\\ &= \mathbf{E}_{\mu_K}[f(B_K(t))\phi_{K,\eta}].\end{aligned}$$

This implies the proof. □

Theorem 6.2. ***(Gaussian Girsanov formula)*** *Let $T > 0$ be fixed and let η be a integrable function with* supp $\eta \subset [0, T]$. *Suppose that η has the following integral representation:*

$$\eta(t) = \int_{\mathbf{R}} k(s)\tau_K(s,t)ds, \qquad 0 \le t \le T \tag{10}$$

for some function $k \in H_K$ *with* supp $k \subset [0,T]$. *Define a probability measure* $\mu_{K,\eta}$ *on the* σ*-algebra generated by* $\{B_K(s) : 0 \le s \le T\}$ *by*

$$\mu_{K,\eta}(dx) = \phi_{K,-k}(x)\, \mu_K(dx).$$

Then $B_{K,\eta}(t) = B_K(t) + \int_0^t \eta(s)ds$, $0 \le t \le T$, *is also a Gaussian Brownian motion under* $\mu_{K,\eta}$.

Proof. The proof is a simple modification of the proof of Theorem 3.18 in 10. It is sufficient to show that for any $\xi \in E$ with supp$\,\xi \subset [0,T]$,

$$\mathbf{E}_{\mu_{K,\eta}}\left[\psi_\xi(\cdot + \eta)\right] = \mathbf{E}_{\mu_K}\left[\psi_\xi\right],$$

where $\psi_\xi(x) = e^{\langle x,\, \xi\rangle}$. In fact, we have

$$\begin{aligned}\mathbf{E}_{\mu_{K,\eta}}[\psi_\xi(\cdot + \eta)] &= \mathbf{E}_{\mu_K}[\psi_\xi(\cdot + \eta)\phi_{K,-k}(\cdot)] \\ &= \exp\left\{\frac{1}{2}|\xi - k|_K^2 + \langle \eta,\, \xi\rangle - \frac{1}{2}|k|_K^2\right\} \\ &= e^{|\xi|_K^2/2} = \mathbf{E}_{\mu_K}[\psi_\xi],\end{aligned}$$

where we used the property that for any $\xi \in E$ with supp $\xi \subset [0,T]$

$$\langle k,\, \xi\rangle_K = \int_{\mathbf{R}}\int_{\mathbf{R}} k(s)\xi(t)\tau_K(s,t)\, ds\, dt = \langle \eta,\, \xi\rangle$$

which is from the relation between η and k given as in (10). □

7. Integral Representation of Gaussian Functions

7.1. *H-Derivative*

Let $\Phi \in \mathcal{W}^*$ and $\zeta \in D^*$. If $\mathcal{T}_{\epsilon\zeta}\Phi$ is defined for small $|\epsilon| < \epsilon_0$ and if there exists $\Psi \in \mathcal{W}^*$ such that

$$\lim_{\epsilon \to 0} \frac{1}{\epsilon} \langle\!\langle \mathcal{T}_{\epsilon\zeta}\Phi - \Phi,\, \phi\rangle\!\rangle_K = \langle\!\langle \Psi,\, \phi\rangle\!\rangle_K, \qquad \phi \in \mathcal{W},$$

we say that Φ has the *directional derivative* in the direction of ζ and we write $\Psi = D_\zeta\Phi$.

If $\Phi \in \mathcal{W}^*$ has the directional derivatives $D_\zeta\Phi$ for all $\zeta \in H_K$ and if there exists a weakly measurable function $t \mapsto \Psi_t \in \mathcal{W}^*$ such that the function $t \mapsto \langle\!\langle \Psi_t,\, \phi\rangle\!\rangle_K$ belongs to $L^2(\mathbf{R})$ for all $\phi \in \mathcal{W}$ and

$$\langle\!\langle D_\zeta\Phi,\, \phi\rangle\!\rangle_K = \int_{\mathbf{R}} \zeta(t)\, \langle\!\langle \Psi_t,\, \phi\rangle\!\rangle_K\, dt, \qquad \zeta \in H_K, \quad \phi \in \mathcal{W},$$

then we say that Φ is *H-differentiable* and write $\Psi_t = D_t\Phi$. Note that $D_t\Phi$ is determined for almost all $t \in \mathbf{R}$.

Proposition 7.1. *Every $\Phi \in \mathcal{G}^*$ is H-differentiable. Moreover, for $\Phi = (F_n) \in \mathcal{G}^*$, $D_t\Phi = ((n+1)F_{n+1}(t,\cdot))_{n=0}^{\infty} \in \mathcal{G}^*$ for almost all $t \in \mathbf{R}$.*

Proof. This proof is a simple modification of the proof of Proposition 3.3 in Ref. 11. Given $\Phi \in \mathcal{G}^*$ we choose $p \geq 0$ such that $\Phi = (F_n) \in \mathcal{G}_{-p}$. Let $\zeta \in H_K$. By (9) we have

$$\frac{\mathcal{T}_{\epsilon\zeta}\Phi - \Phi}{\epsilon} = \left((n+1)\zeta\widehat{\otimes}_1 F_{n+1} + \sum_{m=2}^{\infty} \frac{(n+m)!}{n!m!} \epsilon^{m-1} \zeta^{\otimes m} \widehat{\otimes}_m F_{n+m} \right)_{n=0}^{\infty}.$$

We put $\Psi = \left((n+1)\zeta\widehat{\otimes}_1 F_{n+1}\right)_{n=0}^{\infty}$ and shall prove that $\Psi = D_\zeta\Phi$. First we see by direct computation that for any $q > p$ there exists a constant C_{q-p} such that

$$|\!|\!| \Psi |\!|\!|_{-q} \leq e^p C_{q-p} \|K^{-1}\|_{\mathrm{OP}} \, | \zeta |_K \, |\!|\!| \Phi |\!|\!|_{-p}.$$

Therefore $\Psi \in \mathcal{G}_{-q}$. Moreover, by a similar estimate as in the proof of Proposition 6.1 we obtain

$$\left|\!\left|\!\left| \frac{\mathcal{T}_{\epsilon\zeta}\Phi - \Phi}{\epsilon} - \Psi \right|\!\right|\!\right|_{-q} \leq \frac{2\epsilon e^{2p}\|K^{-1}\|_{\mathrm{OP}}^2 \, | \zeta |_K^2 \exp(\epsilon^2 e^{2p} \|K^{-1}\|_{\mathrm{OP}}^2 \, | \zeta |_K^2)}{\sqrt{1 - 2e^{-2(q-p)}}} \, |\!|\!| \Phi |\!|\!|_{-p},$$

where $q > p + \log\sqrt{2}$. Hence Φ has the directional derivative in the direction of $\zeta \in H_K$ and $\Psi = D_\zeta\Phi$.

We next prove that Φ is H-differentiable. Consider $t \mapsto \Psi_t = ((n+1)F_{n+1}(t,\cdot))_{n=0}^{\infty}$, which is defined for almost all $t \in \mathbf{R}$. From the estimate

$$\begin{aligned}
\int_{\mathbf{R}} |\!|\!| \Psi_t |\!|\!|_{-q}^2 \, dt &= \|K^{-1}\|_{\mathrm{OP}} \int_{\mathbf{R}} \sum_{n=0}^{\infty} (n+1)^2 n! e^{-2qn} |F_{n+1}(t,\cdot)|_K^2 \, dt \\
&\leq \|K^{-1}\|_{\mathrm{OP}} \sum_{n=0}^{\infty} (n+1) e^{2p} e^{-2(q-p)n} (n+1)! e^{-2p(n+1)} \, | F_{n+1} |_0^2 \\
&\leq e^{2p} C_{q-p}^2 \|K^{-1}\|_{\mathrm{OP}} \, |\!|\!| \Phi |\!|\!|_{-p}^2
\end{aligned}$$

we see that $\Psi_t \in \mathcal{G}_{-q}$ for almost all $t \in \mathbf{R}$. Moreover, it is clear from the above estimate that the function $t \mapsto \langle\!\langle \Psi_t, \phi \rangle\!\rangle_K$ belongs to $H = L^2(\mathbf{R})$ for all $\phi \in \mathcal{W}$. Finally,

$$\langle\!\langle D_\zeta\Phi, \phi \rangle\!\rangle_K = \int_{\mathbf{R}} \zeta(t) \, \langle\!\langle \Psi_t, \phi \rangle\!\rangle_K \, dt, \qquad \phi \in \mathcal{W},$$

is verified easily and we have $D_t\Phi = \Psi_t$ for almost all $t \in \mathbf{R}$. □

Remark 7.1. It is noted that for any $\phi \in \mathcal{W}$, $D_t\phi$ coincides with the Hida derivative $\partial_t\phi$ or $a_t\phi$, see Refs. 8, 9, 15 and 17.

7.2. *Integral Representation*

Let $\mathcal{F}_t^K$ be the σ–algebra generated by $B_K(s)$, $0 \le s \le t$. Then for each $\Phi = (F_n) \in \mathcal{G}^*$, *quasi–conditional expectation* $\widetilde{\mathbf{E}}_t^K(\Phi)$ of Φ with respect to $\mathcal{F}_t^K$ is defined by

$$\widetilde{\mathbf{E}}_t^K(\Phi) \equiv \widetilde{\mathbf{E}}_{\mu_K}[\Phi \,|\, \mathcal{F}_t^K] \equiv (1_{[0,t]}^{\otimes n} F_n).$$

For detail study of the (quasi-) conditional expectation we refer to Refs. 10 and 16. If $\widetilde{\mathbf{E}}_t^K(\Phi) = \Phi$, then we say that Φ is $\mathcal{F}_t^K$–measurable.

Theorem 7.1. ***(Gaussian Clark–Ocone formula)*** *Let $\Phi \in \mathcal{G}^*$ be $\mathcal{F}_T^K$-measurable. Then $\widetilde{\mathbf{E}}_t^K(D_t\Phi) \diamond W_K(t)$ is integrable in $\mathcal{W}^*$ and*

$$\Phi = \mathbf{E}_{\mu_K}(\Phi)\phi_0 + \int_0^T \widetilde{\mathbf{E}}_t^K(D_t\Phi) dB_K(t).$$

Proof. Let $\Phi = (F_n) \in \mathcal{G}^*$ be $\mathcal{F}_T^K$-measurable. Then by Proposition 7.1 and definition of the quasi–conditional expectation we have

$$D_t\Phi = (nF_n(t,\cdot) \in \mathcal{G}^*, \qquad \widetilde{\mathbf{E}}_t^K(D_t\Phi) = \left(n1_{[0,t]}^{\otimes(n-1)} F_n(t,\cdot)\right).$$

Therefore, by direct computation (see the proof of Theorem 3.10 in Ref. 16), for each $\xi \in D$ we have

$$\begin{aligned} S_K\left(\int_0^T \widetilde{\mathbf{E}}_t^K(D_t\Phi) dB_K(t)\right)(\xi) &= S_K\left(\int_0^T \widetilde{\mathbf{E}}_t^K(D_t\Phi) \diamond_K W_K(t) dt\right)(\xi) \\ &= S_K(\Phi)(\xi) - F_0. \end{aligned}$$

Since $\mathbf{E}_{\mu_K}(\Phi) = F_0$, the proof is completed by applying Theorem 5.1. □

Example 7.1. Let $a \in \mathbf{R}$ and let

$$X_t = e^{iaB_K(t) + \frac{1}{2}a^2 \langle K1_{[0,t]}, 1_{[0,t]} \rangle}, \qquad t \ge 0.$$

Then by Example 6.3, we have

$$X_t = 1 + ia \int_0^t X_s dB_K(s), \qquad t \ge 0.$$

8. Application to Finance

Now, we consider two possible investments as a simple model of financial market which is called a *(simple) Gaussian Black-Scholes market*, one is a safe investment and the other one is a risky investment of which the prices F_t and S_t, respectively, satisfy the following equations:

$$dF_t = rF_t dt, \qquad F_0 = 1, \quad 0 \le t \le T; \tag{11}$$

$$dS_t = \mu S_t dt + \sigma S_t dB_K(t), \qquad S_0 = x > 0, \quad 0 \le t \le T, \tag{12}$$

where $r > 0$, μ and $\sigma \neq 0$ are constants. Then by Example 6.3, we have

$$S_t = x \exp\left(\sigma B_K(t) + \mu t - \frac{1}{2}\sigma^2 \left\langle K1_{[0,t]}, 1_{[0,t]}\right\rangle\right), \qquad t \ge 0.$$

A *portfolio* $\theta_t = (u_t, v_t)$ of the investments is an $\mathcal{F}_t^K$–adapted 2–dimensional stochastic process. Then the *value process* V_t of the portfolio θ_t is defined by

$$V_t^\theta = u_t F_t + v_t \diamond_K S_t, \qquad 0 \le t \le T. \tag{13}$$

A crucial assumption is that the portfolio θ_t is called *self–financing*:

$$dV_t^\theta = u_t dF_t + v_t \diamond_K dS_t, \qquad 0 \le t \le T,$$

i.e., changes of wealth is depending only on the changes of prices. Then by (12), we have

$$dV_t^\theta = u_t dF_t + \mu v_t \diamond_K S_t dt + \sigma v_t \diamond_K S_t dB_K(t). \tag{14}$$

From now on, we assume that θ is a self–financing portfolio. Then by (13), we have

$$u_t = \frac{V_t^\theta - v_t \diamond_K S_t}{F_t}$$

and then, by (11) and (14), we have

$$dV_t^\theta = rV_t^\theta dt + \sigma v_t \diamond_K S_t \left(\frac{\mu - r}{\sigma} dt + dB_K(t)\right). \tag{15}$$

Note that, by Theorem 6.2,

$$\widetilde{B}_K(t) = \frac{\mu - r}{\sigma} t + B_K(t)$$

is a Gaussian Brownian motion with respect to the measure $\widetilde{\mu}_K$ defined on $\mathcal{F}_T^K$ by

$$d\widetilde{\mu}_K(x) = \exp\left(- \langle x, k\rangle - \frac{1}{2}|k|_K^2\right) d\mu_K(x),$$

where $k(s) = k(T,s)$ is defined by the following properties: $\operatorname{supp} k \subset [0,T]$ and

$$\int_0^T k(T,s)\tau_K(s,t)ds = \frac{\mu - r}{\sigma}, \qquad 0 \le t \le T \tag{16}$$

if the function k exists. Hence the equation (15) becomes

$$dV_t^\theta - rV_t^\theta dt = \sigma v_t \diamond_K S_t d\widetilde{B}_K(t)$$

which is equivalent to

$$d\left(e^{-rt}V_t^\theta\right) = \sigma e^{-rt} v_t \diamond_K S_t d\widetilde{B}_K(t).$$

Therefore, the value process of the portfolio is given by

$$V_t^\theta = ze^{rt} + \sigma \int_0^t e^{r(t-s)} v_s \diamond_K S_s d\widetilde{B}_K(s), \tag{17}$$

where $z = V_0^\theta$ is the initial fortune.

The portfolio V_t^θ is said to be *admissible* if it is self–financing and

$$\mathbf{E}_{\widetilde{\mu}_K}\left[\left(\int_{\mathbf{R}} v_s \diamond_K S_s d\widetilde{B}_K(s)\right)^2\right] < \infty.$$

An admissible portfolio θ is called an *arbitrage* for this market (F_t, S_t), $t \in [0,T]$ if

$$V_0^\theta \le 0, \quad V_T^\theta \ge 0 \text{ a.s. and } \mu_K\left(\{\omega; V_T^\theta(\omega) > 0\}\right) > 0.$$

It is easily shown that, by (17),

$$\mathbf{E}_{\widetilde{\mu}_K}\left[e^{-rT}V_T^\theta\right] = V_0^\theta.$$

This implies that there is no arbitrage.

Then market (A_t, S_t), $t \in [0,T]$ is called *complete* if for any $\mathcal{F}_T^K$–measurable bounded random variable $G(\omega)$ there exists $z \in \mathbf{R}$ and a portfolio $\theta = (u,v)$ such that

$$G(\omega) = V_T^{\theta,z} \text{ a.s. } \mu_K.$$

From (17) we have

$$e^{-rT}G = z + \sigma \int_0^T e^{-rs} v_s \diamond_K S_s d\widetilde{B}_K(s). \tag{18}$$

By applying Theorem 7.1, we have

$$e^{-rT}G = \mathbf{E}_{\widetilde{\mu}_K}(e^{-rT}G)\phi_0 + \int_0^T \widetilde{\mathbf{E}}_{\widetilde{\mu}_K}[e^{-rT}\widetilde{D}_s G|\widetilde{\mathcal{F}}_s^K]d\widetilde{B}_K(s), \tag{19}$$

where $\widetilde{D}_s$ is the stochastic gradient with respect to $\widetilde{\mu}_K$ and $\widetilde{\mathcal{F}}_s^K$ is the σ–algebra generated by $\{\widetilde{B}_K(s) : s \leq t\}$ which coincides with the σ–algebra $\mathcal{F}_s^K$. By comparing (18) and (19) we see that the market (F_t, S_t), $t \in [0,T]$ is complete. Moreover, there is a unique initial value

$$z = V_0^\theta = \mathbf{E}_{\widetilde{\mu}_K}(e^{-rT}G)$$

and a unique portfolio $\theta_t = (u_t, v_t)$ to hedge the claim G. The initial value is called the *price* of the (European) claim G.

From now on we assume that for each $T > 0$ there exists a function k on $[0,\infty)$ such that supp $k \subset [0,T]$ and (16) is satisfied.

Remark 8.1. The function k satisfying the equation (16) for the fractional Brownian motion is well formulated in Ref. 10. We now consider the case of generalized Brownian motion. By direct computation we have

$$\tau_K(s,t) = b'(t)\delta(s-t).$$

Therefore, we can find $k(T,s)$ satisfying that

$$\int_0^T k(T,s)b'(t)\delta(s-t)ds = \frac{\mu - r}{\sigma}. \tag{20}$$

On the other hand, the left hand side of (20) is $k(T,t)b'(t)$ and hence

$$k(T,t) = \frac{1}{\sqrt{b'(t)}}\frac{\mu - r}{\sigma}.$$

Theorem 8.1. *The Gaussian Black-Scholes market (A_t, S_t) modelled by (11) and (12) has no arbitrage.*

Theorem 8.2. *The Gaussian Black-Scholes market (A_t, S_t) modelled by (11) and (12) is complete and the price of z of a bounded $\mathcal{F}_T^K$–measurable claim G is given by*

$$z = V_0^\theta = \mathbf{E}_{\widetilde{\mu}_K}(e^{-rT}G).$$

Moreover, the hedging portfolio $\theta_t = (u_t, v_t)$ for the claim G is

$$u_t = \frac{V_t^\theta - v_t \diamond_K S_t}{F_t}$$
$$v_t = e^{-r(T-t)}\sigma^{-1}S_t^{\diamond_K -1} \diamond_K \widetilde{\mathbf{E}}_{\widetilde{\mu}_K}[\widetilde{D}_s G|\widetilde{\mathcal{F}}_s^K].$$

Example 8.1. Now we consider the *European call* described by a claim G given by

$$G = (S_T - C)^+,$$

where C is the exercise price. Then the price of the claim G is obtained as following:

$$\begin{aligned} z &= e^{-rT}\mathbf{E}_{\widetilde{\mu}K}[(S_T - C)^+] \\ &= e^{-rT}\mathbf{E}_{\widetilde{\mu}K}\left[\left(x\exp\left(\sigma\widetilde{B}_K(t) + \mu t - \frac{1}{2}\sigma^2\left\langle K1_{[0,t]}, 1_{[0,t]}\right\rangle\right) - C\right)^+\right] \\ &= e^{-rT}\mathbf{E}_{\mu K}\left[\left(x\exp\left(\sigma B_K(t) + \mu t - \frac{1}{2}\sigma^2\left\langle K1_{[0,t]}, 1_{[0,t]}\right\rangle\right) - C\right)^+\right] \end{aligned}$$

Thus by similar arguments used in Ref. 10, we can obtain a explicit formula of the price of the European call option which is called a *Black–Scholes formula.*

References

1. K. Aase, B. Øksendal, N. Privault and J. Ubøe: *White noise generalizations of the Clark-Haussmann-Ocone theorem with application to mathematical finance*, Finance Stoch. **4** (2000), 465–496.
2. Yu. M. Berezansky and Yu. G. Kondratiev: "Spectral Methods in Infinite-Dimensional Analysis," Kluwer Academic, 1995.
3. S.J. Chang, J. G. Choi and D. Skoug: *Integration by Parts Formulas Involving Generalized Fourier-Feynman Transforms on Function Space*, Trans. Amer. Math. Soc., **355** (2003), 2925–2948.
4. Th. Deck, J. Potthoff and G. Vage: *A review of white noise analysis from a probabilistic standpoint*, Acta Appl. Math. **48** (1997), 91–112.
5. L. Decreusefond and A. S. Üstünel: *Stochastic analysis of the fractional Brownain motion*, Potential analysis, **10** (1999), 177–214.
6. T. Hida: "Analysis of Brownian Functionals," Carleton Math. Lect. Notes 13, Carleton University, Ottawa, 1975.
7. T. Hida: "Brownian Motion," Springer–Verlag, 1980.
8. T. Hida, H.-H. Kuo, J. Potthoff and L. Streit (eds.): "White noise: An Infinite Dimensional Calculus", Kluwer Academic Publishers, 1993.
9. Y. Hu, B. Øksendal, J. Ubøe and T. Zhang: "Stochastic Partial Differential Equations," Birkhäuser, 1996.
10. Y. Hu and B. Øksendal: *Fractional white noise calculus and applications to finance*, Infin. Dimen. Anal. Quantum Probab. Rel. Top. **6** (2003) 1-32.
11. U. C. Ji and N. Obata: *Annihilation-derivative, creation-derivative and representation of quantum martingales*, preprint, 2003.
12. U. C. Ji and N. Obata: *Unitarity of Kuo's Fourier-Mehler transform*, preprint, 2003.
13. U. C. Ji and N. Obata: *Generalized Fourier-Gauss transforms as unitary isomorphisms*, in preparation.

14. I. Kubo and S. Takenaka: *Calculus on Gaussian white noise I–IV*, Proc. Japan Acad. **56A** (1980), 376–380; 411–416; **57A** (1981), 433–437; **58A** (1982), 186–189.
15. H.-H. Kuo: "White Noise Distribution Theory," CRC Press, 1996.
16. N. Obata: *Conditional expectation in classical and quantum white noise calculi*, RIMS Kokyuroku **923** (1995), 154–190.
17. N. Obata: "White Noise Calculus and Fock Space," Lect. Notes in Math. Vol. 1577, Springer–Verlag, 1994.
18. J. Potthoff and L. Streit: *A characterization of Hida distributions*, J. Funct. Anal. **101** (1991), 212–229.

ANOTHER NONCANONICAL REPRESENTATION OF A BROWNIAN MOTION RELATED TO SINGULAR VOLTERRA KERNELS

YUJI HIBINO

Faculty of Science and Engineering, Saga University,
840-8502, Saga, JAPAN
E-mail: hibinoy@cc.saga-u.ac.jp

We construct noncanonical representations of a Brownian motion which has an arbitrary finite-dimensional orthogonal complement by using a different method from Ref. 1.

Keywords: Noncanonical representation; Volterra operator.

AMS Subject Classification: 60G15, 60G35

1. Introduction

Let $X = \{X(t); t \geq 0\}$ be a centered Gaussian process. In case that X is represented as

$$X(t) = \int_0^t F(t,u)dB(u) \tag{1}$$

by using a Brownian motion $B = \{B(t); t \geq 0\}$, if $\mathcal{B}_t(X) = \mathcal{B}_t(B)$, for each $t > 0$, then the representation (1) is called the *canonical representation* of X with respect to B. Here $\mathcal{B}_t(X)$ is the σ-field generated by $\{X(s); s \leq t\}$. It should be noted that the function F is nonrandom. The concept of canonical representation may express increase of randomness of X in accordance with that of B. The concept was proposed by Lévy[6] in 1956, and systematically developed by Hida[3] in 1960. In case of a Gaussian process, it is well known that $\mathcal{B}_t(X) = \mathcal{B}_t(B)$, $t > 0$, is equivalent to $H_t(X) = H_t(B)$, $t > 0$. Here $H_t(X)$ is a linear span of $\{X(s); s \leq t\}$. Thus we can express all the statements in terms of linear spans $H_t(\cdot)$ instead of filtration $\mathcal{B}_t(\cdot)$.

In the joint work,[1] we have constructed noncanonical representations of a Brownian motion which has N-dimensional orthogonal complement for any $N \in \mathbf{N}$. In the representation, we used an operator associated with a singular Volterra kernel.

In Sec. 2, we find another method that enables us to construct a noncanonical representation which also has a finite-dimensional orthogonal complement. We can prove the impossibility for construction of a noncanonical representation having an infinite-dimensional orthogonal complement by using the method of iteration.

In Sec. 3, we obtain the condition under which the operators are commutative. As a simple application of the theorem, stationary processes are considered in Sec. 4. There we see that operating N operators corresponds to multiplying N factors in the Blaschke product.

2. Iteration of the operators

For any fixed $N \in \mathbf{N}$, let N functions $g_1, \ldots, g_N \in L^2_{\text{loc}}[0, \infty)$ be linearly independent in $[0, s]$ for any $s > 0$. For the system $\mathbf{g}(s) = (g_1(s), \ldots, g_N(s))$, we define an integral operator $K_{\mathbf{g}}$ on $L^2_{\text{loc}}[0, \infty)$ whose integral kernel is

$$k_{\mathbf{g}}(s,u) = \begin{cases} \mathbf{g}(s)\Gamma^{-1}(s){}^t\mathbf{g}(u) = \sum_{i,j=1}^{N} g_i(s)\Gamma^{ij}(s)g_j(u), & s \geq u, \\ 0, & s < u, \end{cases}$$

where $\left(\Gamma^{ij}(s)\right)$ is the inverse of the Grammian matrix $\Gamma(s)$ of $\mathbf{g}(s)$. It is clear that the inverse of $\Gamma(s)$ exists for any $s > 0$, since g_i's are linearly independent in $[0, s]$ for any $s > 0$. The operator $K_{\mathbf{g}}$ is a Volterra type. We refer to the following theorem which has been proved in the joint work.[1]

Theorem 2.1. *The Gaussian process $\widetilde{B}_{\mathbf{g}} = \{\widetilde{B}_{\mathbf{g}}(t); t \geq 0\}$ defined by*

$$\widetilde{B}_{\mathbf{g}}(t) = \int_0^t (I - K^*_{\mathbf{g}})1_{(0,t)}(u)dB(u) = \int_0^t \left\{1 - \int_u^t k_{\mathbf{g}}(s,u)ds\right\} dB(u) \quad (2)$$

is a Brownian motion and (2) is a noncanonical representation with respect to B satisfying

$$H_t(B) = H_t(\widetilde{B}_{\mathbf{g}}) \oplus LS\left\{\int_0^t g_i(u)dB(u); i = 1, 2, \ldots, N\right\},$$

for each $t > 0$. Here $LS\{\ldots\}$ means the linear span of $\{\ldots\}$.

Remark 2.1. From a symbolical calculation,

$$\int_0^t (I - K^*_{\mathbf{g}})1_{(0,t)}(u)dB(u) = \langle (I - K^*_{\mathbf{g}})1_{(0,t)}, \dot{B}\rangle$$
$$= \langle 1_{(0,t)}, (I - K_{\mathbf{g}})\dot{B}\rangle.$$

We may write the representation (2) as

$$\dot{\widetilde{B}}_{\mathbf{g}}(t) = (I - K_{\mathbf{g}})\dot{B}(t), \quad \widetilde{B}_{\mathbf{g}}(0) = 0.$$

We shall construct, using method of iteration, a noncanonical representation of a Brownian motion which has a two-dimensional orthogonal complement.

Theorem 2.2. *For g, h in $L^2_{\text{loc}}[0, \infty)$, the Gaussian process $\widetilde{B}_{g,h} = \{\widetilde{B}_{g,h}(t); t \geq 0\}$ defined by*

$$\dot{\widetilde{B}}_{g,h}(t) = (I - K_h)(I - K_g)\dot{B}(t), \quad \widetilde{B}_{g,h}(0) = 0, \tag{3}$$

is a Brownian motion and (3) *is a noncanonical representation with respect to B satisfying*

$$H_t(B) =$$
$$H_t(\widetilde{B}_{g,h}) \oplus LS\left\{\int_0^t g(u)dB(u), \int_0^t \left(h(u) - \int_u^t k_g(s,u)h(s)ds\right) dB(u)\right\}$$

for each $t > 0$.

Proof. Put

$$\dot{\widetilde{B}}_1(t) = (I - K_g)\dot{B}(t), \qquad \widetilde{B}_1(0) = 0,$$

then

$$H_t(B) = H_t(\widetilde{B}_1) \oplus LS\left\{\int_0^t g(u)dB(u)\right\}.$$

Since $\dot{\widetilde{B}}_{g,h}(t) = (I - K_h)\dot{\widetilde{B}}_1(t)$,

$$H_t(\widetilde{B}_1) = H_t(\widetilde{B}_{g,h}) \oplus LS\left\{\int_0^t h(u)d\widetilde{B}_1(u)\right\}.$$

Since

$$\int_0^t h(u)d\widetilde{B}_1(u) = \int_0^t (I - K_g^*)h(u)dB(u),$$

$$\begin{aligned}
H_t(B) &= H_t(\widetilde{B}_1) \oplus LS\left\{\int_0^t g(u)dB(u)\right\} \\
&= H_t(\widetilde{B}_{g,h}) \oplus LS\left\{\int_0^t h(u)d\widetilde{B}_1(u)du\right\} \oplus LS\left\{\int_0^t g(u)dB(u)\right\} \\
&= H_t(\widetilde{B}_{g,h}) \oplus LS\left\{\int_0^t (I - K_g^*)h(u)dB(u)\right\} \oplus LS\left\{\int_0^t g(u)dB(u)\right\},
\end{aligned}$$

for each $t > 0$. Thus we have proved the desired result. □

Especially, in the case of $g = h$, we obtain the following result.

Corollary 2.1. *For g in $L^2_{\text{loc}}[0,\infty)$, the Gaussian process $\widetilde{B}_2 = \{\widetilde{B}_2(t); t \geq 0\}$ defined by*

$$\dot{\widetilde{B}}_2(t) = (I - K_g)^2 \dot{B}(t), \quad \widetilde{B}_2(0) = 0, \tag{4}$$

is a Brownian motion and (4) is a noncanonical representation with respect to B satisfying

$$H_t(B) = H_t(\widetilde{B}_2) \oplus LS \left\{ \int_0^t g(u)dB(u), \int_0^t g(u) \log \left(\int_0^u g(v)^2 dv \right) dB(u) \right\},$$

for each $t > 0$.

Proof. The statement is obvious from

$$(I - K_g^*)g(u) = g(u) - g(u) \log \|g\|_t^2 - g(u) \log \left(\int_0^u g(v)^2 dv \right).$$

It is clear that $g(u) \log \left(\int_0^u g(v)^2 dv \right)$ belongs to $L^2_{\text{loc}}[0,\infty)$. We remark that $g(u)$ and $g(u) \log \left(\int_0^u g(v)^2 dv \right)$ are linearly independent. □

Iterating the operator successively, we can construct a Brownian motion $\widetilde{B}_N = \{\widetilde{B}_N(t); t \geq 0\}$ defined by

$$\dot{\widetilde{B}}_N(t) = (I - K_g)^N \dot{B}(t), \quad \widetilde{B}_N(0) = 0.$$

The process has a noncanonical representation having the N-dimensional orthogonal complement such that

$$H_t(B) = H_t(\widetilde{B}_N) \oplus \bigoplus_{k=1}^{N} LS \left\{ \int_0^t (I - K_g^*)^{k-1} g(u) dB(u) \right\}.$$

Thus the method of iteration gives the noncanonical representation having an N-dimensional orthogonal complement for an arbitrary number $N \in \mathbf{N}$. However, this procedure never gives a noncanonical representation having an infinite-dimensional orthogonal complement.

Theorem 2.3. *For any g in $L^2_{\text{loc}}[0,\infty)$,*

$$H_t(B) = \bigoplus_{k=1}^{\infty} LS \left\{ \int_0^t (I - K_g^*)^{k-1} g(u) dB(u) \right\}.$$

Proof. Put $\varphi_k(u) = g(u)(\log \int_0^u g(s)^2 ds)^{k-1}$. It is easily seen that $\{(I - K_g^*)^k g; k \in \mathbf{N}\}$ corresponds to the orthogonalization of $\{\varphi_k; k \in \mathbf{N}\}$. We can see that $\{\varphi_k; k \in \mathbf{N}\}$ is complete in $L^2[0,t]$ for any $t > 0$. This fact leads us to the desired result. □

The above theorem is a generalization of a result obtained in Ref. 5.

The existence of noncanonical representations having an infinite-dimensional orthogonal complement is shown in our previous paper.[2]

3. Commutation of the operators

In this section, we shall consider the condition under which the operators K_g and K_h are commutative. It corresponds to the condition for two Brownian motions obtained in Theo. 2.2 and in Theo. 2.1 for $\mathbf{g} = (g, h)$ to be identical, since it means $(I - K_h)(I - K_g) = I - K_{(g,h)}$.

Theorem 3.1. *For a given g in $L^2_{\mathrm{loc}}[0,\infty)$, the operator K_h, $h \in L^2_{\mathrm{loc}}[0,\infty)$, is commutative with K_g, if and only if*

$$h(t) \in LS\left\{ g(t)\left(\int_0^t g(u)^2 du\right)^{\beta} \right\}, \quad -\frac{1}{2} < \beta.$$

Proof. Thanks to Theo. 2.2, it is sufficient to consider the condition

$$\begin{aligned} & LS\left\{ \int_0^t g(u)dB(u), \int_0^t h(u)dB(u) \right\} \\ &= LS\left\{ \int_0^t g(u)dB(u), \int_0^t \{h(u) - K_g^* h(u)\}\, dB(u) \right\}, \end{aligned}$$

that is to say,

$$LS\{g, h\} = LS\{g,\ h - K_g^* h\}.$$

By putting

$$h - K_g^* h = C_1 g + C_2 h, \quad C_2 \neq 0,$$

we get the integral equation

$$K_g^* h = C_3 g + C_4 h, \quad C_4 \neq 1.$$

Solving this, we obtain

$$h(s) = C g(s)\left(\int_0^s g(u)^2 du\right)^{\beta}, \quad \beta \neq -1, 0, \tag{5}$$

where C is constant.

Thus, the desired h should be in the form (5). In the case of $\beta = 0$, (5) clearly belongs to $L^2_{\rm loc}[0,\infty)$. The function (5) belongs to $L^2[0,t]$ for any $t > 0$, if and only if $2\beta > -1$, because

$$\int_0^t \left\{ g(s) \left(\int_0^s g(u)^2 du \right)^\beta \right\}^2 ds = \int_0^{\|g\|_t^2} x^{2\beta} dx$$

using integration by substitution as $x = \int_0^s g(u)^2 du$.

Therefore the statement is proved. □

4. Application to stationary Gaussian processes

The simple application of the last theorem is the case where g and h are monomials. In the case, the stationary processes are closely connected.

We first review the canonical representation theory for the stationary Gaussian processes. For the precise description, the readers may refer to Refs. 2, 4 and the references cited there.

Suppose a stationary Gaussian process $Y = \{Y(s); s \in \mathbf{R}\}$ to have a *moving average representation*

$$Y(s) = \int_{-\infty}^s G(s-u) dW(u), \tag{6}$$

where $\dot{W} = \{\dot{W}(s); s \in \mathbf{R}\}$ is a Gaussian white noise. Denote by $c(\lambda)$ the inverse Fourier transform of the representation kernel G.

It is well known that the function $c(\lambda)$ belongs to the Hardy class $\mathbf{H}^2$ in the upper half-plane $\mathbf{C}_+$, and is decomposed as

$$c(\lambda) = c_o(\lambda)\Pi(\lambda)D(\lambda)S(\lambda),$$

with notations

$$c_o(\lambda) = \sqrt{2\pi} \exp\left\{ -\frac{1}{2\pi i} \int_{-\infty}^{\infty} \frac{1+w\lambda}{w-\lambda} \cdot \frac{\log f(w)}{1+w^2} dw \right\},$$

$$\Pi(\lambda) = \left(\frac{\lambda - i}{\lambda + i} \right)^k \prod_{n=1}^{\infty} \frac{\lambda - z_n}{\lambda - \overline{z_n}} \cdot \frac{|z_n^2+1|}{z_n^2+1},$$

$$D(\lambda) = e^{ia\lambda},$$

$$S(\lambda) = \exp\left\{ \frac{1}{2\pi i} \int_{-\infty}^{\infty} \frac{1+w\lambda}{w-\lambda} d\beta(w) \right\},$$

where f is the spectral density function of the process Y, $\{z_n\}$ is a complex sequence satisfying $\Im(z_n) > 0$ and $\sum_{n=1}^{\infty} \Im(z_n)/(1+|z_n|^2) < \infty$, a is a non-negative constant and β is a nondecreasing function of bounded variation whose derivative vanishes almost everywhere.

The components $c_o(\lambda)$, $\Pi(\lambda)$, $D(\lambda)$ and $S(\lambda)$ are so-called the *outer function*, the *Blaschke part*, the *delay part* and the *singular part*, respectively.

From the viewpoint of the canonical representation theory, the most important result is that the representation (6) is canonical, if and only if $c(\lambda) = c_o(\lambda)$.

On the other hand, if the representation kernel F of the representation (1) is a homogeneous function of degree α, that is $F(ct, cu) = c^{\alpha} F(t, u)$ for any c, then X can be transformed into a stationary process Y by

$$Y(s) = \frac{1}{\sqrt{2} e^{(2\alpha+1)s}} X(e^{2s}), \quad s \in \mathbf{R}.$$

The process Y is represented as

$$Y(s) = \int_{-\infty}^{s} F(1, e^{-(s-u)}) e^{-(s-u)} dW(u), \tag{7}$$

where $dW(u) = \frac{1}{\sqrt{2}} e^{-u} dB(e^{2u})$.

It should be remarked that this transformation preserves the canonical property. Namely, the representation (1) of X is noncanonical satisfying

$$H_t(X) \perp \int_0^t u^m dB(u), \quad m > -\frac{1}{2},$$

if and only if the representation (7) of Y satisfies

$$H_s(Y) \perp \int_{-\infty}^{s} e^{(2m+1)u} dW(u). \tag{8}$$

Getting back to our subject, we shall consider the case where $g(t) = t^m$ and $h(t) = t^n$ $(m, n > -1/2$, for the sake of convenience, we moreover assume $mn(m-n) \neq 0)$. Then the representation obtained in Theo. 2.2 is

$$\widetilde{B}_{g,h}(t) = \int_0^t \left\{ \frac{(2m+1)(m+n+1)}{m(m-n)} \left(\frac{u}{t}\right)^m + \frac{(m+1)(n+1)}{mn} + \frac{(2n+1)(m+n+1)}{n(n-m)} \left(\frac{u}{t}\right)^n \right\} dB(u). \tag{9}$$

It goes without saying that, in this case, $\widetilde{B}_{(g,h)}$ obtained in Theo. 2.1 for $\mathbf{g} = (g, h)$ is identical with $\widetilde{B}_{g,h}$.

We pay attention to the fact that the representation kernel of (9) is the homogeneous function of degree 0.

By putting

$$\widetilde{Y}_{g,h}(s) = \frac{1}{\sqrt{2}e^s}\widetilde{B}_{g,h}(e^{2s}), \quad s \in \mathbf{R},$$

the Brownian motion $\widetilde{B}_{g,h}$ is transformed into the stationary Brownian motion (the Ornstein-Uhlenbeck process) $\widetilde{Y}_{g,h}(s)$ having the representation

$$\widetilde{Y}_{g,h}(s) = \int_{-\infty}^{s} \left\{ \frac{(2m+1)(m+n+1)}{m(m-n)} e^{-(2m+1)(s-u)} + \frac{(m+1)(n+1)}{mn} e^{-(s-u)} + \frac{(2n+1)(m+n+1)}{n(n-m)} e^{-(2n+1)(s-u)} \right\} dW(u).$$

The inverse Fourier transform of the representation kernel is

$$\frac{1}{1-i\lambda} \cdot \frac{2m+1+i\lambda}{2m+1-i\lambda} \cdot \frac{2n+1+i\lambda}{2n+1-i\lambda},$$

where the first factor corresponds to the outer function, and the rests do to the Blaschke part.

As we have proved in Ref. 2, the noncanonical property (8) comes from the Blaschke part. More precisely, the condition (8) causes the factor $(2m+1+i\lambda)/(2m+1-i\lambda)$ for the inverse Fourier transform of the representation kernel.

Therefore we can understand that operating $(I-K_h)(I-K_g)$ means just successively multiplying the factors $(2m+1+i\lambda)/(2m+1-i\lambda)$ and $(2n+1+i\lambda)/(2n+1-i\lambda)$. Thus as for this case, we are easily convinced that K_g and K_h are commutative.

Acknowledgments

The author would like to express his appreciation to the referees on their suggestions on improving this article.

References

1. Y. Hibino, M. Hitsuda and H. Muraoka; *Construction of noncanonical representations of a Brownian motion.* Hiroshima Math. J. **27** (1997), 439–448.
2. Y. Hibino, M. Hitsuda and H. Muraoka; *Remarks on a noncanonical representation for a stationary Gaussian process.* Recent developments in infinite-dimensional analysis and quantum probability. Acta Appl. Math. **63** (2000), 137–139.

3. T. Hida; *Canonical representations of Gaussian processes and their applications.* Mem. Coll. Sci. Univ. Kyoto **33** (1960), 109–155.
4. T. Hida and M. Hitsuda; Gaussian Processes, Representation and Applications. Amer. Math. Soc. (1993).
5. T. Jeulin and M. Yor; *Filtration des ponts browniens et équations différentielles stochastiques linéaires.* Lect. Notes in Math. **1426**, Springer, Berlin, (1990), 227–265.
6. P. Lévy; *A special problem of Brownian motion and a general theory of Gaussian random functions.* Proc. of 3rd Berkeley Symp. Math. Stat. and Prob. **2** (1956), 133-175.

SOME TOPICS ON WHITE NOISE ANALYSIS

TAKEYUKI HIDA
Meijo University
Nagoya 468-8502, Japan
e-mail: thida@ccfms.meijo-u.ac.jp

SI SI
Faculty of Information Science and Technology
Aichi Prefectural University,
Aichi-ken 480-1198, Japan
e-mail: sisi.ist.aichi-pu.ac.jp

Some recent developments of white noise analysis will be discussed. First, we wish to emphasize the importance of the it innovations in the study of stochastic processes and random fields. It is known that if a Gaussian process has the canonical representation, its innovation can be obtained explicitly, which is worth to be revisited from the viewpoint of innovation theory. Also, we prove the existence of innovations for some stochastic processes which are functions of a Gaussian white noise or of a Poisson noise and even a mixture of them, and discuss their roles. The idea comes from the *reduction* or atomism for random complex systems. White noise theory plays a dominat role in our study.

Math. Subject Classification 2000 : 60H40

1. Introduction

We shall investigate evolutional random complex systems from the view point of the **Reductionism**.

The first step is to form a system of idealized elemental (atomic) random variables (i.e.r.v.) from a given random system. Then, let the given system be expressed as a functional of the i.e.r.v.'s. Finally, analize the functionals, in line with the white noise analysis.

Namely,

Reduction $\longrightarrow$ Synthesis $\longrightarrow$ Analysis

Intuitively speaking, an innovation is a system of i.e.r.v.'s parameterized by t. The basic, and actually standard, innovation involves Gaussian white noise, and Poisson noise (including the case of compound Poisson noise), as well as their combination.

2. A short history on innovation

We ignore the old story and have a quick review of what can be developed to the present research.

1. P. Lévy. Processius stochastiques et mouvement brownien. Gauthier-Villars. 1948, 2nd ed.(with supplementj1964.

   Chapter 2 is essentially devoted to the definition of a stochastic process with special emphasis on the role of innovation, although he does not use this term explicitly.

   Note. The innovation for the discrete parameter case was discussed in Lévy's book "Théorie de l'addition des variables aléatoires. 1937, Chapt. VI.
2. P. Lévy, Random functions: General theory with special reference to Laplacian random functions. Univ. of California Press. 1953.

   To define (or determine) a stochastic process, a formal (intuitive) equation is provided. It is for $\delta X(t)$ F

   $$\delta X(t) = \Phi(X(s), s \leq t; Y(t), t, dt).$$

   It is called a *stochastic infinitesimal equation.*

   The $Y(t)$ contains the information which $X(t)$ gains during an infinitesimal time interval $[t, t+dt)$ and is independent of $X(s)$, $s \leq t$.
3. This idea was realized for Gaussian processes in [3], 1960, in terms of the canonical representation. Section I of this paper is in agreement with the result of the paper by H. Cramér, obtained independently and at almost the same time. In reality, this concept was propsed by P. Lévy in 1955 (his paper was published in 1956: Proceedings of the third Berkeley Symposium).
4. Around 1970, T. Kailath and his collaborators appealed to the innovation theory in mathmatical theory of communication. Let $X(t)$

be the observed data which is of the form

$$X(t) = s(t) + n(t),$$

where $s(t)$ is the signal and $n(t)$ is a noise. The $s(t)$ is independent of the future noise $n(s), s \geq t$.

Set $\hat{s}(t) = E(s(t)|\mathbf{B}_t(X))$. Then,

$$\nu(t) = X(t) - \hat{s}(t)$$

is the innovation of $X(t)$.

It is interesting to note that Shirjaev obtained the similar results around the same time, independently.@@

5. N. Wiener discussed the prediction theory for stationary stochastic process. We now understand the results by using the term innovation or canonical representation. Later he published the book

   Nonlinear problems in random theory, 1958

   where the lectures 12 and 13 on Coding and Decoding are closely related to the innovation theory ([16]).

   G. Kallianpur discussed innovation in N. Wiener's Collected Works Vol 3 ([7]).
6. As for the random field, Doburushin et al and one of the authors have generalized the results on stochastic process to define the innovation. ([11],[15].)
7. The recent work by Accardi-Hida-SiSi (2003)(see [1]) uses the martingale theory developed by Kunita and Watanabe [8] to obtain the innovation.
8. Win Win Htay has reported the innovation approach to linear processes at the IIAS Ohya workshop Jan. 2003 [17].

3. Background

First we consider possible and, in fact, reasonable pattern of the innovation.

A process (depending on continuous parameter) with independdent values at every point is to be a generalized stochastic process. To be homogeneous in time it may be assumed to be stationary. With some other assumptions, like sample function continuity or discontinuity of the first kind, what we can have in mind would be a time derivative of a Lévy process. Let it be denoted by $Z(t)$. The Lévy decomposition or Lévy-Ito decomposition of $Z(t)$ is expressed in the form

$$Z(t) = mt + \sigma B(t) + X(t) \ ,$$

where m is a constant that may be ignored in what follows. The $B(t)$ is a Brownian motion and $X(t)$ is a compind Poisson process. The $X(t)$ can further be decomposed into Poisson processes $X_u(t)$ with jumps of various hights u. To have $X(t)$ as a superposition of the $X_u(t)$'s we need to deal with quasi convergence.

Finally we can conclude that an innovation comes from elemental additive processes $B(t)$ and a system $\{X_u(t), u \in U \subset R-\{0\}\}$. More intuitively speaking, an innovation may be reduced to Gaussian and Poisson type white noises.

4. From data to characteristic functional

Our aim is to obtain the innovation of a given random complex system, say a stochastic process, and the probabilistic structure of the system is, in general, not given in advance. Hence, we must know its probability distribution by observing a sample function. For this purpose we try to find the characteristic functional of the process.

It is reasonable to assume that the information source is a white noise, if not we can apply a whitening filter. Namely, our basic assuption is that the time shift, denoted by T_t, of the information source is *ergodic*.

To fix the idea, consider the case where the input signal is a (Gaussian) white noise denoted by $\dot{B}(t)$. Then T_t acts in such a way that

$$T_t : \dot{B}(s) \longrightarrow \dot{B}(s+t).$$

It is known that $\{T_t\}$ defines a *flow*, that is a one-parameter group of probability measure-preserving transformations. Further it is known that the flow is *ergodic*. Therefore a functional of $\dot{B}(t)$'s which is U_t-invariant is just a constant. Here, U_t is an operator defined by

$$U_t\varphi(x) = \varphi(T_t x).$$

As a result, phase average that is the expectation $E(\varphi)$ is in agreement with the time average $\lim \frac{1}{T}\int_0^T U_t\varphi \ dt$. This property can be applied to the computation of the chatacteristic functional $C(\xi)$.

Let $X = X(t)$ be an observed data. Then,

$$C_X(\xi) = E[e^{i\langle X,\xi\rangle}] = \lim_{T\to\infty} \frac{1}{T}\int_0^T e^{i\langle T_t X,\xi\rangle} dt.$$

This formula is quite useful to obtain charateristic functional of a stochastic process which is a functional of $\dot{B}$ or $\dot{P}$.

5. Innovation

Before we give a definition of innovation, we remind P. Lévy 's *stochastic infinitesimal equation.* For a stochastic process $X(t)$

$$\delta X(t) = \Phi(X(s), s \le t, Y(t), t, dt), \ \ t \in R^1, \tag{5.1}$$

We are thus given an intuitive meaning of innovation $Y(t)$. There it must be assumed that δX_t for an infinitesimal time interval $[t, t+dt)$ means that

$$X(t+dt) - X(t) = \delta X(t) + o(dt).$$

Mathematically rigorous definition is now ready to be given. notationF $\mathbf{B}_t(X)$ is the smallest sigma-field, with respect to which all the $X(s), s \le t$, are measurable.

Definition $Y(t)$ is said to be an innovation of $X(t)$ if

$$\mathbf{B}_{t-}(X) \bigvee \left\{ \bigwedge_{t \in \text{supp } (\eta)} \mathbf{B}(Y(\eta)) \right\} = \bigwedge_{\epsilon > 0} \mathbf{B}_{t+\epsilon}(X). \tag{5.2}$$

holds. Here *supp*(η) stands for the support of η.

Once an innovation $\{Y(t)\}$ is obtained, the given $X(t)$ can be expressed in the form

$$X(t) = \Psi(Y(s), s \le t, t), \tag{5.3}$$

if there is no remote past of the $X(t)$.

Now, it is easy to see that the pair $(\Psi, \{Y(t)\})$ completely determines the probabilistic structure of $X(t)$.

Needless to say, the next step is to analyze functional, the variables of which are $\{Y(t)\}$.

6. Innovation of a Gaussian process

To form an innovatipon os a Gaussian process $X(t)$ is almost the same as the canonical representation theory of the process:

$$X(t) = \int_0^t F(t,u)\dot{B}(u)du. \tag{6.1}$$

For details we refer to the book Hida-Hitsuda, "Gaussian processes" Amer. Math. Soc. 1993. Note that

$$\mathbf{B}_t(X) = \mathbf{B}_t(\dot{B}) \tag{6.2}$$

and hence $\dot{B}(t)$ is the innovation.

Assume that a Gaussian process $X(t)$ has the canonical representation and that $F(t,u)$ is smooth, $F(t,t) \neq 0$. Then, the variation of $X(t)$ exists and $\delta X(t)$ has to be of the form

$$\delta X(t) = dt \int^t \frac{\partial}{\partial t} F(t,u)\dot{B}(u)du + F(t,t)\dot{B}(t)dt,$$

where the second term of the right hand side is of order $\sqrt{dt}$.

The variation in the above sense does not always exists even for the Gaussian process which is interested for us. There is such an example.

Example. Fractional order Brownian motion.

The following example was given by P. Lévy in his Berkely Lecture Notes 1953.

$$X_\alpha(t) = \frac{1}{\Gamma(\alpha+1)} \int_0^t (t-u)^\alpha \dot{B}(u)du,$$

where $\alpha > -1/2$. Let n be the largest integer which is less than $\alpha + 1/2$. Then, $X_\alpha(t)$ is differentiable (in L^2-sense) as many times as n. The k-th derivative is denoted by $X_\alpha^{(k)}(t)$. Then, we have

$$\delta X_\alpha(t) = \sum_1^n X_\alpha^{(k)} \frac{(dt)^k}{k!} + c\dot{B}(t)(dt)^{\alpha+1} + o(dt)^{\alpha+1}.$$

It is proved that $\dot{B}(t)$ is the innovation and $\mathbb{B}_t(X) = \mathbb{B}_t(\dot{B})$ holds for every t.

Since the kernel is a function of (t-u) only, so it is expected to be a stationary Gaussian process with a mild modification. For this purpose the lower limit of the integral should be $-\infty$. Further integrability of the stochastic integral is required. Mandelbrot et al were clever enough to add an integral to give the following formula (1968 SIAM Review)F

$$\frac{1}{\Gamma(\alpha+1)}\int_{\infty}^{0}[(t-u)^{\alpha}-(-u)^{\alpha}]\dot{B}(u)du.$$

The innovation is still $\dot{B}(t)$.

If only stationarity is required, then a fractional order Ornstein Uhlenbeck Brownian motion is given.

$$U_{\alpha}(t)=\frac{1}{\Gamma(\alpha+1)}\int_{-\infty}^{t}(t-u)^{\alpha}e^{-(t-u)}\dot{B}(u)du.$$

7. Examples for which innovation is computed

We shall show some examples.

(1) A process given by a stochastic differential equation.

Let $X(t)$ be given by

$$\delta X(t)=A(X(t))dt+\sigma(X(t))Y(t)dt \tag{7.3}$$

with a suitable initial condition. The $Y(t)$ may be Gaussian noise or compound Poisson noise, even sum of them.

(1.a) Given an equation

$$\delta X(t)=(-aX(t)+f(t))dt+g(t)Y(t)dt\ ,$$

With an initial data, it is proved that the solution exists and that $Y(t)$ is the innovation. The solution $X(t)$ itself is expressed in the form@

$$X(t)=e^{-at}\int_{0}^{t}\{e^{au}f(u)+g(u)Y(u)\}du+C\ .$$

(1.b) Let the function g above be replaced by an affine function of $X(t)$.

$$\delta X(t)=-aX(t)dt+(b+X(t))Y_t dt,\quad Y=\dot{B}\ .$$

Then, it is proved that $Y(t)$ is an innovation and

$$X(t)=\sum_{1}^{\infty}\int_{(-\infty,t]^n}\frac{b^n e^{-at}}{n!}e^{amin(u_1,\ldots u_n)}$$

$$: \dot{B}_{u_1} \ldots \dot{B}_{u_n} : du^n + C \ ,$$

where : : is the Wick product.

(2) Multiple Markov Gaussian process.

Let $X(t)$ be an N-ple Markov Gaussian process and let the canonical representation is given by

$$X(t) = \int_0^t F(t,u)\dot{B}(u)du,$$

where $F(t,u)$ is a Goursat kernel expressed in the form

$$F(t,u) = \sum_1^N f_i(t)g_i(u),$$

where

$$det(f_i(t_j)) \neq 0$$

for different t_j's and where $\{g_i\}$ are linearly independent L^2-functions. Assume now that the covariance $\Gamma(t,s), t \geq s$, is C^N-class as a function of t. Then, there exists an ordinary differential operator L_t of order N such that $\{f_i\}$ is a funndamental system of solutions of the differential equation

$$L_t f = 0.$$

Let L_t be expressed as a Frobenius formula iT. Hida, 1960 or [2;article [8]) to have

$$L_t = \Pi_0^{N-1} \frac{d}{dm_j} \cdot \frac{1}{v_N(t)},$$

where $dm_j(t) = v_j(t)dt$. Now let L_t be factorized as

$$L_t = \frac{d}{dm_0} L_t^1$$

to define the operator L_t^1. If $\{v_j\}$ is suitably chosen, for $t > s$ we obtain with some $g(u)$

$$L_t^1 E(X(t)|\mathbf{B}_s(X)) = \int_0^s g(u)\dot{B}(u)du,$$

The right hand side is an additive process, so that innovation $\dot{B}(s)$ is easily determined (uo to constant).

(3) Linear process (Win Win Htay 2003).

Consider a linear process $X(t)$ (in the sense of P. Lévy), expressed in the form

$$X(t) = \int_0^t F(t,u)\dot{B}(u)du + \int_0^t G(t,u)\dot{P}(u)du.$$

The $\dot{P}$ is a Poisson noiseia compound Poisson process is also acceptablej. Assume that $F(t,u)$ is a canonical kernel and is smooth.AIn addition, $G(t,u)$ is continuous and $G(t,t)$ never vanishes. By observing a sample function (path), $\dot{P}(t)$ and then $\dot{B}(t)$ can be obtained separately. The sum

$$\dot{B}(t) + \dot{P}(t)$$

or the pair

$$(\dot{B}(t), \dot{P}(t)$$

is an innovation.

Remark

There is a new approach to innovation thory by using Martingale theory (See [1]).

References

1. L. Accardi, T. Hida and Si Si, Innovation approach to some stochastic processes. Volterra Center Notes, Nov. 2002.
2. L. Accardi et al eds. Selected papers of Takeyuki Hida, World Sci. Pub. Co. Ltd. 2001.
3. T. Hida, Canonical representation of Gaussian processes and their applications. Mem. College of Sci. Univ. Kyoto, A 33 (1960),109-155.
4. T. Hida and N. Ikeda, Note on linear process. J. of Math. Kyoto Univ. 1 (1961), 75-86.
5. T. Hida, Stationary Stochastic Processes, Mathematical Notes, Princeton University Press, 1970.
6. T. Hida and Si Si, Innovation for random fields. Infinite Dimensional Analysis, Quantum Probability and Related Topics. 1 (1998), 499-509.
7. G. Kallianpur, Some rmifications of Wiener's ideas on nonlinear prediction, N.Wiener's Collected Works. ed. P. Masani. The MIT Press (1981), 402-424.
8. H. Kunita and S. Watanabe, On square integrable martingales. Nagoya Math. J. 30 (1967), 209-245.
9. P. Lévy, Processus stochastiques et mouvement brownien, 1948. 2ème ed. 1965, Gauthier-Villars.

10. P. Lévy, Random functions: General theory with special reference to Laplacian random functions. Univ. California Press. 1953.
11. Si Si, Gaussian processes and Gaussian random fields. Quantum Information II (2000), World Scientific, 195-204.
12. Si Si, Random fields and multiple Markov properties. Proc. of the second conference on Unconventional Models of Computation, 2000. UMC'2K, ed. I.Antoniou et. al, CDMTCS-147CCenter for Discrete Mathematics and Theoretical Computer Science, Solvay Institute, 2000, pp 64-70.
13. Si Si, Random irreversible phenomena. Entropy in subordination, Chaos, Solitons and Fractals 12 (2001), 2873-2876.
14. Si Si, Innovation of the Lévy Brownian motion. Univ. Roma, Volterra Center Notes, N.475, May, 2001.
15. Si Si, Innovation of random complex systems. Tokyo Univ. of Science. Lecture. 2003.
16. N. Wiener, Nonlinear problems in random theory. MIT Press. 1958.
17. Win Win Htay, Note on linear processes. 2003 Jan. IIAS lecture.

AN INNOVATION FOR A GAUSSIAN SEMIMARTINGALE AND THE REPRODUCING KERNEL HILBERT SPACE

MASUYUKI HITSUDA

Faculty of Science, Kumamoto University,
860-8555, Kumamoto, JAPAN

An intrinsic relation between innovation for a Gaussian semimartingale and the reproducing kernel Hilbert space is considered. The problem is essential in the theory of canonical representation for a Gaussian process.

1. Canonical representation and innovation theorem

In the first stage, a simple review is given on the canonical representation and the innovation theorem. In the study of a Gaussian process, especially in the theory of canonical representation, the role of the related reproducing kernel Hilbert space is important, as pointed by Hida.[2] In detail, let $\mathcal{H}(\Gamma_X)$ (resp. $\mathcal{H}_t(\Gamma_X)$) be the reproducing kernel Hilbert space spanned by the covariance function Γ_X for the process $\{X(t); t \in [0,1]\}$ (resp. $\{X(s); s \in [0,t]\}$). On the other hand, let us note the linear span of $\{X(s); s \leq t\}$ by $H_t(X)$ for each $t \in [0,1]$. The following proposition gives a remarkable aspect on the story.[2,3]

Proposition 1.1. *If two independent Gaussian processes* $X_1 = \{X_1(t); t \in [0,1]\}$ *and* $X_2 = \{X_2(t); t \in [0,1]\}$ *are given and satisfy* $\mathcal{H}_t(\Gamma_{X_1}) \cap \mathcal{H}_t(\Gamma_{X_2}) = \{0\}$, $t \in [0,1]$, *then* $H_t(X_1 + X_2) = H_t(X_1) \oplus H_t(X_2)$, $t \in [0,1]$.

The proposition states that the process $X = X_1 + X_2$ generates the equivalent information to the one of generated by X_1 and X_2 time by time, under a condition described by the related reproducing kernel Hilbert space.

In this article the author applies the theory to an innovation theorem for a semimartingale. Throughout the paper the Gaussian process X is assumed to be centered, without loss of generality.

The theory of canonical representation has been developed by Lévy[8,9] and Hida.[2] It will be needed to refer their results as a basement of the present statement. Consider a Gaussian process $X = \{X(t); t \in [0,1]\}$

which is represented as

$$X(t) = \int_0^t F(t,u)dB(u), \quad t \in [0,1], \tag{1}$$

with respect to a Brownian motion $B = \{B(t); t \in [0,1]\}$. If the process X satisfies the condition that

$$H_t(X) = H_t(B), t \in [0,1], \tag{2}$$

then the representation (1) of X is said to be *canonical* with respect to the Brownian motion B. Sometimes a Gaussian process X may not have the canonical representation with respect to a single Brownian motion. Namely, it may have a *generalized canonical representation* with respect to an independent system of Brownian motions such as $(B_1, B_2, \dots, B_N)$,

$$X(t) = \sum_{i=1}^{N} \int_0^t F_i(t,u)dB_i(u), \quad t \in [0,1], \tag{3}$$

satisfying

$$H_t(X) = \bigoplus_{i=1}^{N} H_t(B_i), \quad t \in [0,1], \tag{4}$$

It is called that the process X is of *multiplicity* N in Hida[2]. Sometimes, there is a possibility of $N = \infty$. Such examples are given in Ref. 3. On the other hand, examples of noncanonical representation of the Brownian motion itself are systematically described in the paper.[1]

The following Gaussian semimartingale X will be considered in this paper.

$$X(t) = B(t) + \int_0^t \varphi(s,\omega)ds, \quad t \in [0,1]. \tag{5}$$

Here $B = \{B(t); t \in [0,1]\}$ is a Brownian motion with respect to a given filtration $\mathbf{F} = \{\mathcal{F}_t; t \in [0,1]\}$. The random variable $\varphi(s,\omega)$ is jointly measurable in (s,ω) and $\mathcal{F}_s$-measurable for each $s \geq 0$. The sub-σ-field generated by $\{X(s); s \leq t\}$ is denoted by $\mathcal{B}_t(X)$.

The following innovation theorem gives basic ideas.

Theorem 1.1. *Let X be a stochastic process given by* (5). *If an integrability condition*

$$E\left(\int_0^1 |\varphi(s,\omega)|ds\right) < \infty,$$

is satisfied, then the process $\bar{B} = \{\bar{B}(t); t \in [0,1]\}$*, defined by*

$$\bar{B} : \bar{B}(t) = X(t) - \int_0^t \widehat{\varphi}(s,\omega)ds, \quad t \in [0,1],$$

is a Brownian motion with respect to the filtration $\mathbf{B}(X) = \{\mathcal{B}_t(X); t \in [0,1]\}$*, where* $\widehat{\varphi}(s,\omega) = E(\varphi(s,\omega)|\mathcal{B}_s(X))$*.*

The new Brownian motion $\bar{B}$ is called the *innovated* Brownian motion. The innovation theorem states that we can extract at least one innovation $\bar{B}$. The problem of interest is whether the Brownian motion $\bar{B}$ is the unique innovation or not. In other words, is the process X of multiplicity one in Hida's sense? This paper tries to approach the problem and get a partial result.

Let us introduce some notation related to reproducing kernel Hilbert spaces.

Definition 1.1. Two covariances Γ_{X_1} and Γ_{X_2} are said to be *similar*, if there exist positive constants c_1 and c_2 such that $c_1\Gamma_{X_1} - \Gamma_{X_2}$ and $\Gamma_{X_1} - c_2\Gamma_{X_2}$ are positive definite.

It is well known that $\mathcal{H}(\Gamma_{X_1}) = \mathcal{H}(\Gamma_{X_2})$, if and only if $\Gamma_{X_1} \sim \Gamma_{X_2}$. This means that the similarity of the covariance functions is equivalent to the identity of the reproducing kernel Hilbert spaces as sets of functions. Under the condition, we can get a clear characterization of the linear span $H_t(X)$ as in the following theorem.

Theorem 1.2. *Under the condition* $\Gamma_X \sim \Gamma_B$*,*

$$H_t(X) = \left\{ \int_0^1 P_t\alpha(s)dX(s); \alpha \in L^2[0,1] \right\}, \quad t \in [0,1],$$

where $P_t\alpha$ *means the usual projection defined by* $P_t\alpha(s) = \alpha(s), s \le t; = 0, s > t$*, and the integral is the Wiener-like integral.*[5]

2. Main theorem

Let us consider a Gaussian process X given by

$$X(t) = B(t) + \Phi(t) = B(t) + \int_0^t \varphi(s,\omega)ds. \tag{6}$$

Obviously X is a Gaussian semimartingale. By the use of Jain and Monad's result,[6] the process (B, Φ) is a Gaussian system. The following theorem holds.

Theorem 2.1. *Under the condition* $\Gamma_X \sim \Gamma_B$, *the process* X *defined by* (6) *is expressed in the form*

$$X(t) = X_1(t) + X_2(t), \tag{7}$$

$$X_1(t) = \bar{B}(t) - \int_0^t \int_0^s k(s,u)d\bar{B}(u)ds,$$

$$X_2(t) = \int_0^t \int_0^s m(s,u)dX(u)ds,$$

where k *and* m *are integral kernels which define the operators on* $L^2[0,1]$. *Moreover the last term in* (7) *is an independent process of* $\bar{B}$ *and*

$$H_t(X) = H_t(X_1) \oplus H_t(X_2) = H_t(\bar{B}) \oplus H_t(X_2), \quad t \in [0,1].$$

Remark 2.1. The equation (7) is formally written by

$$\dot{X}(t) = (I - K)\dot{\bar{B}}(t) + M\dot{X}(t),$$

with the initial condition $X(0) = 0$.

For the proof of the theorem, we need some notation and lemmas.

Lemma 2.1. *For each* $\hat{\varphi}(s,\omega)$, $(s,\omega) \in [0,1] \times \Omega$, *there exists an element* $l_s(u) \in P_s L^2[0,1] = \{P_s\alpha; \alpha \in L^2[0,1]\}$, *satisfying*

$$\hat{\varphi}(s,\omega) = \int_0^s l_s(u)dX(u),$$

for almost all $s \in [0,1]$.

Proof. Since $\hat{\varphi}(s,\omega) \in H_s(X)$, we can apply Theo. 1.2 to $\hat{\varphi}(s,\omega)$, noting that it belongs to $H_s(X)$ and equals to $\int_0^s l(s,u)dX(u)$. □

By a simple argument, we can choose an integral kernel $l(s,u) = l_s(u), u \leq s; = 0, u > s$. By using the kernel $l(s,u)$ of Volterra type, the innovated Brownian motion $\bar{B}$ is given as follows:

Lemma 2.2. *The innovated Brownian motion* $\bar{B}$ *is expressed as in the form*

$$\bar{B}(t) = X(t) - \int_0^t \int_0^s l(s,u)dX(u)ds, \tag{8}$$

satisfying

$$E\left(\int_0^1 \left| \int_0^s l(s,u)dX(u) \right| ds\right) < \infty.$$

Proof. The statement is supposed to be clear. □

Remark 2.2. The equation (8) is written formally as

$$\dot{\bar{B}}(t) = (I - L)\dot{X}(t)$$

with $\dot{\bar{B}}(0) = 0$.

By using the adjoint operator L^* of L, the following expression is easily obtained.

Lemma 2.3. *The Brownian motion $\bar{B}$ has another expression*

$$\bar{B}(t) = \int_0^1 (I - L^*)1_t(u)dX(u).$$

More generally,

$$\int_0^1 \alpha(u)d\bar{B}(u) = \int_0^1 (I - L^*)\alpha(u)dX(u),$$

for $\alpha \in L^2[0,1]$.

We get to the contrary

Lemma 2.4. *Let $X_1(t)$ be the projection of $X(t)$ on the subspace $H(\bar{B})$. Then there exists a Volterra kernel $k(s,u)$, X_1 has the canonical representation with respect to $\bar{B}$*

$$\begin{aligned} X_1(t) &= \bar{B}(t) - \int_0^t \int_0^s k(s,u)d\bar{B}(u)ds \\ &= \int_0^1 (I - K^*)1_t(u)d\bar{B}(u). \end{aligned}$$

Lemma 2.5. *The following independent decomposition holds:*

$$\begin{aligned} \int_0^1 \alpha(u)dX(u) &= \int_0^1 (I - M^*)\alpha(u)dX(u) + \int_0^1 M^*\alpha(u)dX(u) \\ &= \int_0^1 (I - K^*)\alpha(u)d\bar{B}(u) + \int_0^1 M^*\alpha(u)dX(u), \quad \alpha \in L^2[0,1], \end{aligned}$$

especially

$$X(t) = \int_0^1 (I - M^*)1_t(u)dX(u) + \int_0^1 M^*1_t(u)dX(u) = X_1(t) + X_2(t),$$

where the operator $M^* = L^* + K^* - L^*K^*$.

Proof. (of Theorem) The results are direct consequence from the lemma above, if we note that the operator M and M^* are projections on the space $L^2[0,1]$. □

Remark 2.3. If the kernel $m = m(s,u) = 0$, the process X is of multiplicity 1.

References

1. Y.Hibino, M.Hitsuda and H.Muraoka; *Construction of noncanonical representations of a Brownian motion,* Hiroshima Math. J. 27 (1997), 439-448
2. T.Hida; *Canonical representations of Gaussian processes and their applications*, Mem. Coll. Sci. Univ. Kyoto 33 (1960), 109-155.
3. T.Hida and M.Hitsuda; Gaussian Processes, Amer. Math. Soc. (1993).
4. M. Hitsuda; *Representation of Gaussian processes equivalent to Wiener process*, Osaka J. Math. 5(1968), 299-312.
5. M. Hitsuda; *Wiener-like integrals for Gaussian processes and the linear estimation problem*, Stochastic Analysis and Applications, Ed. by M.Pinsky(1984; Marcel Dekker), 167-177.
6. N. Jain and D. Monad; *Gaussian quasimartingales*, Z. Wahrs. verw. Gebiete. 59(1982),139-159.
7. T. Kailath; *An innovation approach to least-squares estimation, Part 1, Linear filtering in additive white noise*, IEEE Trans. AC-13(1968),646-665.
8. P.Lévy; *A special problem of Brownian motion and a general theory of Gaussian random functions*, Proc. of 3rd Berkeley Symp. Math. Stat. and Prob. 2 (1956), 133-175.
9. P.Lévy; *Fonctions aléatoires à corrélation linéaire*, Illinois J. Math. 1 (1957), 217-258.
10. R. Liptzer and A. Shiryaev; Statistics of Stochastic Processes (in Russian) NAUKA(1974), English version, Springer(1977).
11. M.Yor; Some aspects of Brownian motion. Part I : Some special functionals, Birkhauser Verlag (1992).

ERROR EXPONENTS FOR DISCRETE-TIME WHITE GAUSSIAN CHANNELS WITH FEEDBACK

SHUNSUKE IHARA
Graduate School of Information Science, Nagoya University, Nagoya 464-8601, Japan
E-mail: ihara@math.human.nagoya-u.ac.jp

and

LIN GUI XIN
Graduate School of Information Science, Nagoya University, Nagoya 464-8601, Japan
E-mail: ksrin@math.human.nagoya-u.ac.jp

We are interested in the asymptotic behavior of the error probability in the information transmission over white Gaussian channels with feedback. It is known that, if the rate of a message is less than the channel capacity, the minumum error probability converges to zero quite rapidly. Shalkwijk and Kailath (1966) investigated the continuous-time white Gaussian channel with feedback and proved that the minimum error probability converges to zero "double exponentially" fast. Then Pinsker (1968) claimed that the minimum error probability decreases more rapidly than the exponential of any order. Zigangirov (1970) also showed the "multiple exponential" decay of the error probability for the discrete-time white Gaussian channel with feedback. However, Pinsker claimed the result without proof and Zigangirov provided only a brief sketch of the proof. Therefore, even though the result itself is known, it is worth giving a detailed proof. The aim of the paper is to give a proof of the "multiple exponential" decay of the minimum error probability in the information transmission over the discrete-time white Gaussian channel with feedback.

1 Introduction

Let us consider a discrete-time white Gaussian channel (WGC)

$$Y_n = X_n + Z_n, \quad n = 1, 2, ..., \tag{1.1}$$

with feedback, where $X = \{X_n\}$ and $Y = \{Y_n\}$ are a channel input and the corresponding output, respectively, and the noise $Z = \{Z_n\}$ is a Gaussian i.i.d. process with distribution $N(0, \sigma^2)$. We assume that the feedback channel is noiseless. We also assume that an average power constraint

$$E[X_n^2] \le A^2, \quad n = 1, 2, ..., \tag{1.2}$$

is imposed on the input signals, where $A > 0$ is a constant.

We are interested in the asymptotic behavior of the error probability to transimit a message over the WGC. Let us consider channel coding and decoding. For each fixed N a message ξ_N, which is independent of Z, is a random variable such that

$$P(\xi_N = m) = \frac{1}{M_N}, \quad m \in \mathcal{M}_N \equiv \{1, 2, ..., M_N\}, \tag{1.3}$$

where M_N is a positive integer. Since the WGC is with noiseless feedback, the input signal X_n at time n is a functional of ξ_N and $Y_1^{n-1} \equiv (Y_1, ..., Y_{n-1})$ and presented in the form

$$X_n \equiv \varphi_N(n, \xi_N, Y_1^{n-1}), \quad n = 1, ..., N.$$

On the other hand, the received message $\widehat{\xi}_N$ is a functional of Y_1^N and written in the form

$$\widehat{\xi}_N \equiv \psi_N(Y_1^N).$$

We call φ_N and ψ_N the encoder and the decoder, respectively. We denote by $\mathcal{C}_N$ the class of all pairs (φ_N, ψ_N) such that the corresponding inputs $X_n = \varphi_N(n, \xi_N, Y_1^{n-1})$, $n = 1, ..., N$, satisfy (1.2). Denote by

$$\pi_N \equiv \pi_N(\varphi_N, \psi_N) = P(\widehat{\xi}_N \neq \xi_N)$$

the error probability. The pair (φ_N, ψ_N) is called an (N, M_N, π_N) code.

Definition 1 A rate $R > 0$ is said to be achievable, if there exists a sequence $(\varphi, \psi) = \{(\varphi_N, \psi_N)\}$ of (N, M_N, π_N) code $(\varphi_N, \psi_N) \in \mathcal{C}_N$ such that

$$\liminf_{N\to\infty} \frac{1}{N} \log M_N \geq R, \tag{1.4}$$

$$\lim_{N\to\infty} \pi_N(\varphi_N, \psi_N) = 0. \tag{1.5}$$

The maximum achievable rate

$$C = \sup\{R;\, R \, \text{is achievable}\} \tag{1.6}$$

is called the capacity of the WGC.

It is well known that, for the WGC subject to (1.2), the capacity is not increased by feedback and is equal to

$$C = \frac{1}{2} \log \frac{A^2 + \sigma^2}{\sigma^2}. \tag{1.7}$$

The capacity can be characterized as follows. Let a message ξ be a random variable, taking values in an arbitrary measurable space, independent of the noise Z. The input X_n is written in the form $X_n = \varphi_N(n, \xi, Y_1^{n-1})$, where

Y_n is the corresponding output. Denote by $I(\xi, Y_1^N)$ the mutual information between ξ and Y_1^N. Then the capacity is equal to the maximum mutual information we can transmit over the WGC;

$$CN = \sup_{(\xi, X_1^N)} I(\xi, Y_1^N),$$

where the supremum is taken for all pairs (ξ, X_1^N) of ξ and X_1^N satisfying (1.2).

In the sequel we may consider only the case where

$$M_N = [e^{RN}].$$

We denote by

$$\pi_N^*(R) = \inf_{(\varphi_N, \psi_N)} \pi_N(\varphi_N, \psi_N)$$

the minimum error probability, where the infimum is taken for all possible pairs (φ_N, ψ_N). By definition, if $R < C$,

$$\lim_{N \to \infty} \pi_N^*(R) = 0. \tag{1.8}$$

On the other hand, (1.8) is not satisfied if $R > C$.

A lot of researches have been doen to study the asymptotic behavior of $\pi_N^*(R)$ as $N \to \infty$. It is known that $\pi_N^*(R)$ converges to zero quite rapidly. Shalkwijk, Kailath [7] first proved that the minimum error probability in the information transmission over the the continuous-time WGC with feedback converges to zero "double exponentially" fast. Then Pinsker [4] claimed that the minimum error probability decreases more rapidly than the exponential of any order (see also [2]). Zigangirov [5] also showed the "multiple exponential" decay of the error probability for the discrete-time WGC with feedback.

To state the result we introduce the notation

$$\exp^{(K)}(x) = \exp\{\exp^{(K-1)}(x)\},$$

where $\exp^{(1)}(x) = \exp(x)$. The result is stated as follows.

Theorem 1 Assume that $R < C$. Then for any positive integer K

$$\lim_{N \to \infty} \pi_N^*(R) \exp^{(K)}(N) = 0. \tag{1.9}$$

The aim of the paper is to give a proof of Theorem 1. Note that Pinsker [4] claimed the result (1.9) but without proof. Zigangirov [5] provided only a brief sketch of the proof of (1.9). Therefore, even though (1.9) is a known result, it is worth giving a detailed proof.

The proof of Theorem 1 will be given in Sec. 3. Here we explain the outline of the proof. In Sec. 2, we study a linear coding scheme which plays important roles to derive an upper bound for the minimum error probability. Let the terminal time N of communication be fixed and denote by ξ_N the message given by (1.3). There exist $\delta > 0$ and an integer n_0 such that $R < (1-\delta)C$ and $(1-\delta)N \leq n_0 < N$. Using the linear coding discussed in Sec. 2, we shall show that at time n_0 the error probability is bounded by $\exp(-e^{2DN})$, where $0 < D < (1-\delta)C - R$ is a constant (Proposition 1). This means that the double exponential decay is achieved at time n_0. Since the WGC is with feedback, the transmitter is able to know whether the received message $\widehat{\xi}_{n_0}$ coincides with the message ξ_N or not. If the message is correctly reproduced at time n_0, then no signal is input at time $n_1 = n_0 + 1$. This enables us to save the average power. This procedure is proposed in Zigangirov [5]. A signal is sent at time n_1 only in the case of $\widehat{\xi}_{n_0} \neq \xi_N$. Then we can show that the triple exponential decay of the error probability is achieved at time n_1. Repeating the same procedure we can show that at time $n_k = n_0 + k$ the $(k+2)$-ple exponential decay of the error probability is achieved (Proposition 2 and Proposition 3).

We should note that the receiver can not know whether the reproduced message coincides with the message to be transmitted or not. This aspect, not taken into account in [2, 5], causes some difficulties. To overcome the difficulties we need to define the decoding scheme in a rather complicated form.

For the continuous-time WGC with feedback we can prove a similar result.

Theorem 2 Assume that an average power constraint is imposed on the continuous-time WGC with feedback. Denote by $\pi_T^*(R)$ the minimum error probability in information transmission over the WGC on the time interval $[0, T]$. If the rate R of a message is less than the channel capacity, then for any positive integer K

$$\lim_{T\to\infty} \pi_T^*(R) \exp^{(K)}(T) = 0. \tag{1.10}$$

Eq. (1.10), given by Pinsker without proof, should be compared with (1.9). The proof of (1.10) will be given elsewhere.

2 Optimal Coding

In this section we investigate a linear coding sheme which plays important roles to prove our main theorem.

Let a message θ be a real valued random variable with $E[\theta^2] = 1$. To transmit θ over the WGC with feedback we consider a linear coding $X_n =$

$\varphi(n, \theta, Y_1^{n-1})$ given by

$$X_n = e^{(n-1)C} \left\{ A\theta - e^C(1 - e^{-2C}) \sum_{j=1}^{n-1} e^{-jC} Y_j \right\}, \tag{2.1}$$

where $X_1 = A\theta$. We define the received message $\widehat{\theta}_n = \psi_n(Y_1^n)$ by

$$\widehat{\theta}_n = A^{-1} e^C (1 - e^{-2C}) \sum_{j=1}^{n} e^{-jC} Y_j. \tag{2.2}$$

Then X_n can be rewritten as

$$X_n = Ae^{(n-1)C}(\theta - \widehat{\theta}_{n-1}), \quad n = 1, 2, ...,$$

where $\widehat{\theta}_0 = 0$.

In the sequel, for notational convenience, for a random variable ζ we denote

$$||\zeta||^2 = E[|\zeta|^2].$$

The following properties are known (see e.g. [2]).

Lemma 1 For the coding scheme given by (2.1) and (2.2) we have

$$||X_n||^2 = A^2, \tag{2.3}$$

$$||\widehat{X}_n||^2 = A^2 ||\widehat{\theta}_n||^2 = A^2(1 - e^{-2Cn}), \tag{2.4}$$

$$||X_1 - \widehat{X}_n||^2 = A^2 ||\theta - \widehat{\theta}_n||^2 = A^2 e^{-2Cn}, \tag{2.5}$$

where $\widehat{X}_n = A\widehat{\theta}_n$ is the orthogonal projection of X_1 on the space spanned by $Y_1, ..., Y_n$. The received message $\widehat{\theta}_n$ can be decomposed in the form

$$\widehat{\theta}_n = (1 - e^{-2Cn})\theta + \widetilde{Z}_n, \tag{2.6}$$

where $\widetilde{Z}_n$ is a linear combination of Z_1, Z_2, ..., Z_n and independent of θ. Moreover, $\widetilde{Z}_n$ is a Gaussian random variable with variance

$$||\widetilde{Z}_n||^2 = e^{-2Cn}(1 - e^{-2Cn}). \tag{2.7}$$

It is also shown that the linear coding scheme given by (2.1) and (2.2) is optimum to transmit a Gaussian message.

Lemma 2 Assume that the message θ is a Gaussian random variable with distribution $N(0, 1)$. Then

$$I(\theta, \widehat{\theta}_n) = I(\theta, Y_1^n) = \frac{n}{2} \log \frac{A^2 + \sigma^2}{\sigma^2} = Cn. \tag{2.8}$$

Moreover, for any $(\widetilde{\varphi}_n, \widetilde{\psi}_n) \in \mathcal{C}_n$,

$$||\theta - \widetilde{\theta}_n||^2 \geq ||\theta - \widehat{\theta}_n||^2 = e^{-2Cn}, \tag{2.9}$$

where $\widetilde{\theta}_n = \widetilde{\psi}_n(\widetilde{Y}_1^n)$ and $\widetilde{Y}_1^n$ is the output coressponding to $(\widetilde{\varphi}_n, \widetilde{\psi}_n)$.

The lemma means that, to transmit the Gaussian message, the coding scheme given by (2.1) and (2.2) is optimum in the sense of, not only maximizing the mututlal information transmitted over the WGC but also minimizing the mean squared error.

3 Proof of Theorem 1

In this section we prove Theorem 1.

Let the terminal time N be fixed. Here we assume that N is large enough. Let the message ξ_N be given by (1.3), where $M_N = [e^{RN}]$. Let $\delta > 0$ be a constant such that $R < (1-\delta)C$ and define n_k $(k = 0, 1, 2, ...)$ by $n_0 = \lceil (1-\delta)N \rceil$ and $n_k = n_0 + k$.

The error probability at time n_0 is discussed in Subsec. 3.1. It is shown that the exponential decay of order 2 is achieved at time n_0. In Subsec. 3.2 we study the error probability at time n_k and prove that the exponential decay of order $k+2$ is achieved at time n_k. Theorem 1 will be proved at the end of this section.

3.1 Double Exponent

On the time interval $\overline{1, n_0}$, we define a coding scheme to transmit the message ξ_N. Let $\{a_m\}$ be a sequence of real values such that

$$\frac{m - x_0}{M_N} = \Phi(a_m), \quad m \in \mathcal{M}_N, \tag{3.1}$$

where $\Phi(x)$ is the probability distribution function of $N(0,1)$;

$$\Phi(x) = 1 - \text{erfc}\,(x) = \frac{1}{\sqrt{2\pi}} \int_{-\infty}^{x} \exp\left(-\frac{y^2}{2}\right) dy,$$

and $0 \leq x_0 < 1$ is a constant. Then we define θ_0 by

$$\theta_0 = a_{\xi_N} \tag{3.2}$$

It is known that we may choose x_0 so that

$$||\theta_0||^2 = 1$$

(see [5]). It is clear that

$$a_{m+1} - a_m > \frac{\sqrt{2\pi}}{M_N} \geq \sqrt{2\pi} e^{-NR}, \quad m \in \mathcal{M}_N. \tag{3.3}$$

We now define the input signal X_n, $1 \leq n \leq n_0$, by (2.1), where θ is replaced by θ_0. In the sequel, for simplicity, we put

$$\gamma_n = 1 - e^{-2Cn}.$$

Note that $\gamma_1 = A^2/(A^2 + \sigma^2)$. To construct the decoder, we define $a_{0,m}$ and $b_{0,m}$ by

$$a_{0,m} = A\gamma_1^{-1}\gamma_{n_0} a_m, \quad b_{0,m} = \frac{1}{2}(a_{0,m} + a_{0,m+1}),$$

and an interval $E_{0,m}$ by

$$E_{0,m} = (b_{0,m-1}, b_{0,m}], \quad m \in \mathcal{M}_N.$$

Receiving the output $Y_1^{n_0}$, the receiver calculates $\widehat{X}_{n_0} = A\widehat{\theta}_{n_0}$ by using (2.2). We now define the reveived message $\widehat{\xi}_{n_0} \equiv \psi(Y_1^{n_0})$ by

$$\widehat{\xi}_{n_0} = m \quad \text{if and only if} \quad \gamma_1^{-1}\widehat{X}_{n_0} \in E_{0,m}. \tag{3.4}$$

To evaluate the error probability

$$\pi_{n_0} = P(\widehat{\xi}_{n_0} \neq \xi_N),$$

we put

$$c_0 = \sqrt{\frac{\pi}{8}} \frac{A\gamma_{n_0}}{\gamma_1} e^{-NR}. \tag{3.5}$$

Proposition 1 The error probability at time n_0 is bounded by

$$\pi_{n_0} \leq 2\,\mathrm{erfc}\left(\frac{2\gamma_1 c_0}{A\sqrt{\gamma_{n_0}(1-\gamma_{n_0})}}\right). \tag{3.6}$$

Moreover, for any $0 < D < (1-\delta)C - R$, if N is large enough,

$$\pi_{n_0} \leq \exp\left(-e^{2DN}\right). \tag{3.7}$$

Although the inequality (3.6) has been obtained by Zigangirov [5], for the completeness we give the proof.

Proof. Suppose that $\xi_N = m$ is sent. Then $X_1 = Aa_m$. It follows from (2.6) and (3.4) that

$$\{\widehat{\xi}_{n_0} = m\} = \left\{\frac{A\widetilde{Z}_{n_0}}{\gamma_1} \in (b_{0,m-1} - a_{0,m}, b_{0,m} - a_{0,m}]\right\}, \tag{3.8}$$

where $\widetilde{Z}_{n_0}$ is a Gaussian random variable with distribution $N(0, \gamma_{n_0}(1-\gamma_{n_0}))$ and independent of ξ_N. It is clear from (3.3) that

$$\begin{cases} b_{0,m} - a_{0,m} = A\gamma_1^{-1}\gamma_{n_0}(a_{m+1} - a_m)/2 > 2c_0, \\ b_{0,m-1} - a_{0,m} = A\gamma_1^{-1}\gamma_{n_0}(a_{m-1} - a_m)/2 < -2c_0. \end{cases} \tag{3.9}$$

As an easy consequence of (3.8) and (3.9) we have

$$\{\widehat{\xi}_{n_0} \neq \xi_N\} \subset \left\{|\widetilde{Z}_{n_0}| > \frac{2\gamma_1 c_0}{A}\right\} \tag{3.10}$$

and

$$\pi_{n_0} \leq P\left(|\widetilde{Z}_{n_0}| > \frac{2\gamma_1 c_0}{A}\right) = 2\,\mathrm{erfc}\left(\frac{2\gamma_1 c_0}{A\sqrt{\gamma_{n_0}(1-\gamma_{n_0})}}\right).$$

Thus (3.6) is obtained. It is true that

$$\mathrm{erfc}(x) \sim \frac{1}{\sqrt{2\pi}x}\, e^{-x^2/2} = o\left(e^{-x^2/2}\right),$$

where $f(x) \sim g(x)$ implies $\lim_{x\to\infty} f(x)/g(x) = 1$. Noting

$$\left(\frac{2\gamma_1 c_0}{A\sqrt{\gamma_{n_0}(1-\gamma_{n_0})}}\right)^2 = \frac{\pi\gamma_{n_0}e^{-2RN}}{2(1-\gamma_{n_0})} < \frac{\pi}{2}e^{2Cn_0}e^{-2RN}$$

and

$$DN < \{(1-\delta)C - R\}N \leq Cn_0 - RN,$$

from (3.6) we have

$$\pi_{n_0} = o\left(\exp\left(-e^{2DN}\right)\right).$$

Therefore (3.7) is true. q.e.d.

3.2 Multiple Exponent

Let us investigate the information transmission on the time interval $\overline{n_1, n_K}$ and evaluate the error probability at time $n_k = n_0 + k$. In a recursive manner, let us define the input signal X_{n_k} and the received message ξ_{n_k}. The signal X_{n_k} is given by

$$X_{n_k} = \begin{cases} A_k W_k, & \text{if } \widehat{\xi}_{n_{k-1}} \neq \xi_N, \\ 0, & \text{if } \widehat{\xi}_{n_{k-1}} = \xi_N, \end{cases} \tag{3.11}$$

where the random variables W_k, $k \geq 1$, are defined by

$$W_k = \begin{cases} \gamma_{n_0} X_1 - \widehat{X}_{n_0}, & k = 1, \\ \gamma_1 (A_{k-1} W_{k-1} - Y_{n_{k-1}}), & k \geq 2, \end{cases} \tag{3.12}$$

$A_k > 0$ is a constant uniquely determined as

$$\|X_{n_k}\|^2 = A^2, \tag{3.13}$$

and $\widehat{\xi}_n$ will be defined by (3.19).

We note here that W_k can be written in the form

$$W_k = \gamma_1 A_{k-1} W_{k-1} 1_{\overline{B}_{k-2}} - \gamma_1 Z_{n_{k-1}}, \tag{3.14}$$

where $\overline{B}_j = \{\widehat{\xi}_{n_j} \neq \xi_N\}$ and 1_G denotes the indicator function of a set G. At first we assume $\{\widehat{\xi}_{n_{k-2}} \neq \xi_N\}$. Then, noting $X_{n_{k-1}} = A_{k-1} W_{k-1}$ we have

$$Y_{n_{k-1}} = A_{k-1} W_{k-1} + Z_{n_{k-1}}.$$

Thus using (3.12) we have

$$W_k = -\gamma_1 Z_{n_{k-1}}.$$

Secondly we assume $\{\widehat{\xi}_{n_{k-2}} = \xi_N\}$. In this case, $X_{n_{k-1}} = 0$ and

$$Y_{n_{k-1}} = Z_{n_{k-1}}.$$

Using (3.12), we have

$$W_k = \gamma_1 A_{k-1} W_{k-1} - \gamma_1 Z_{n_{k-1}}.$$

Thus we have obtained (3.14).

Suppose that we receive the output signals $Y_j = y_j$ $(j < n_k)$ and calculate $\widehat{X}_{n_0} = \widehat{n}_{n_0}$, and define $a_{k,m}$ as

$$a_{k,m} = \begin{cases} \gamma_1 A_1 (a_{0,m} - \gamma_1^{-1} \widehat{x}_{n_0}), & k = 1, \\ \gamma_1 A_k (a_{k-1,m} - y_{n_{k-1}}), & k \geq 2. \end{cases} \tag{3.15}$$

Then, using (3.11) and (3.12), we can easily show that the input signal at time n_k is given in the form

$$X_{n_k} = \begin{cases} a_{k,m}, & \text{if} \quad \widehat{\xi}_{n_{k-1}} \neq \xi_N = m, \\ 0, & \text{if} \quad \widehat{\xi}_{n_{k-1}} = \xi_N. \end{cases} \tag{3.16}$$

To define the decoding, we put

$$b_{k,m} = \frac{1}{2}(a_{k,m} + a_{k,m+1}), \tag{3.17}$$

$$c_k = \gamma_1 A_k c_{k-1}, \tag{3.18}$$

and define intervals E_k^* and $E_{k,m}$ by

$$E_k^* = (-2^{-k}c_k, 2^{-k}c_k), \quad E_{k,m} = (b_{k,m-1}, b_{k,m}] \setminus E_k^*.$$

The reproduced message $\widehat{\xi}_{n_k}$ is now defined by

$$\widehat{\xi}_{n_k} = \begin{cases} m, & \text{if } Y_{n_k} \in E_{k,m}, \\ \widehat{\xi}_{n_{k-1}}, & \text{if } Y_{n_k} \in E_k^*. \end{cases} \tag{3.19}$$

This means that, if the output signal $Y_{n_k} = y_{n_k}$ belongs to E_k^*, then the receiver decides that the message is correctly decoded at time n_{k-1} and accordingly no signal is sent at time n_k.

We shall show that inequalities

$$b_{k,m} - a_{k,m} > 2c_k, \tag{3.20}$$

$$b_{k,m-1} - a_{k,m} < -2c_k, \tag{3.21}$$

hold. The inequalities in (3.9) imply that (3.20) and (3.21) are true for $k = 0$. Suppose that the inequality

$$b_{k-1,m} - a_{k-1,m} > 2c_{k-1} \tag{3.22}$$

is true. Then it follows from (3.15), (3.17), (3.18) and (3.22) that

$$b_{k,m} - a_{k,m} = \gamma_1 A_k (b_{k-1,m} - a_{k-1,m}) > 2\gamma_1 A_k c_{k-1} = 2c_k.$$

Thus we have obtained (3.20). The inequality (3.21) can be proved in the same way.

We are now going to evaluate the error probability

$$\pi_{n_k} = P(\widehat{\xi}_{n_k} \neq \xi_N) = P(\overline{B}_k).$$

By definition

$$\pi_{n_k}^*(R) \leq \pi_{n_k}. \tag{3.23}$$

Proposition 2 The error probability is bounded by

$$\pi_{n_k} < 4 \operatorname{erfc}\left(\frac{c_k}{2^k \sigma}\right) < \sqrt{\frac{8}{\pi}} \exp\left(-\frac{c_k^2}{2^{2k+1}\sigma^2}\right), \quad k \geq 1. \tag{3.24}$$

Proof. Let $k \geq 1$ be fixed. We shall prove (3.24) by deviding into the following three cases.

(1°) $\widehat{\xi}_{n_{k-1}} = \xi_N$;

(2°) $\widehat{\xi}_{n_j} = \xi_N$, $0 \leq \exists j < k-1$, and $\widehat{\xi}_{n_p} \neq \xi_N$, $j < \forall p \leq k-1$;

(3°) $\widehat{\xi}_{n_p} \neq \xi_N$, $0 \leq \forall p \leq k-1$.

Note that, at time n_{k-1}, the message ξ_N is correctly decoded in the case (1°), but incorrectly decoded in (2°) and (3°). Moreover, in (2°) the message has been correctly decoded more than once before n_{k-1}, and never in (3°). Corresponding to each case we introduce the sets

$$\begin{aligned}
G_{1,m} &= \{\widehat{\xi}_{n_{k-1}} = \xi_N = m\}, \\
G_{2,m} &= \{\widehat{\xi}_{n_j} = \xi_N, 0 \le \exists j < k-1, \\
&\qquad\qquad \widehat{\xi}_{n_p} \ne \xi_N, j < \forall p \le k-1, \xi_N = m\}, \\
G_{3,m} &= \{\widehat{\xi}_{n_p} \ne \xi_N, 0 \le \forall p \le k-1, \xi_N = m\},
\end{aligned}$$

and

$$G_i = \bigcup_{m \in \mathcal{M}_N} G_{i,m}, \quad i = 1, 2, 3.$$

We denote the conditional probabilities as

$$\begin{aligned}
q_{i,m} &= P(\widehat{\xi}_{n_k} \ne \xi_N | G_{i,m}), \\
q_i &= P(\widehat{\xi}_{n_k} \ne \xi_N | G_i) = \sum_{m \in \mathcal{M}_N} P(\xi_N = m) q_{i,m}
\end{aligned} \tag{3.25}$$

$(i = 1, 2, 3)$. We shall prove the inequality

$$q_{i,m} < 4 \operatorname{erfc}\left(\frac{c_k}{2^k \sigma}\right), \quad i = 1, 2, 3, \ m \in \mathcal{M}_N. \tag{3.26}$$

Then noting (3.25) and $\pi_{n_k} \le \max(q_1, q_2, q_3)$, we see that the desired inequality (3.24) follows from (3.26).

At first we shall prove (3.26) for $i = 1$. In this case, noting that $X_{n_k} = 0$ and $Y_{n_k} = Z_{n_k}$ is independent of $\widehat{\xi}_{n_{k-1}}$ and ξ_N, we have

$$\begin{aligned}
q_{1,m} &= P(Y_{n_k} \notin E_k^* \cup E_{k,m} | \widehat{\xi}_{n_{k-1}} = \xi_N = m) \\
&= P(Z_{n_k} \notin E_k^* \cup E_{k,m} | \widehat{\xi}_{n_{k-1}} = \xi_N = m) \\
&\le P(|Z_{n_k}| \ge 2^{-k} c_k) \\
&= 2 \operatorname{erfc}\left(\frac{c_k}{2^k \sigma}\right).
\end{aligned}$$

Thus we have obtained (3.26) for $i = 1$.

Secondly we shall prove (3.26) for $i = 2$. It is clear that

$$\begin{aligned}
q_{2,m} &= P(Y_{n_k} \in E_k^* \cup \overline{E}_{k,m} | G_{2,m}) \\
&\le P(Y_{n_k} \in E_k^* | G_{2,m}) + P(Y_{n_k} \notin (b_{k,m-1}, b_{k,m}] | G_{2,m}).
\end{aligned} \tag{3.27}$$

Under the condition $G_{2,m}$, there exists $0 \le j < k-1$ such that $\widehat{\xi}_{n_j} = \xi_N = m$ and $\widehat{\xi}_{n_p} \neq m$ for $j < \forall p \le k$. Therefore

$$y_{n_{j+1}} \notin E^*_{j+1} \cup E_{j+1,m},$$

and

$$y_{n_{j+1}} \notin (b_{j+1,m-1}, b_{j+1,m}].$$

Using (3.20) and (3.21), we know that either

$$y_{n_{j+1}} - a_{j+1,m} > b_{j+1,m} - a_{j+1,m} > 2c_{j+1},$$

or

$$y_{n_{j+1}} - a_{j+1,m} < b_{j+1,m-1} - a_{j+1,m} < -2c_{j+1},$$

holds. Therefore we have

$$|a_{j+1,m} - y_{n_{j+1}}| > 2c_{j+1},$$

and equivalently

$$|a_{j+2,m}| > 2c_{j+2}. \tag{3.28}$$

On the other hand, since $\widehat{\xi}_{n_{j+2}} \neq \xi_N = m$,

$$y_{n_{j+2}} \in E^*_{j+2} \cup \overline{E}_{j+2,m}.$$

This means that

$$|y_{n_{j+2}}| < 2^{-(j+2)} c_{j+2}, \tag{3.29}$$

or

$$y_{n_{j+2}} \notin (b_{j+2,m-1}, b_{j+2,m}]$$

namely

$$|a_{j+2,m} - y_{n_{j+2}}| > 2c_{j+2}$$

is true. In either case we have

$$|a_{j+2,m} - y_{n_{j+2}}| > c_{j+2}$$

(cf. (3.28), (3.29)) and

$$|a_{j+3,m}| > c_{j+3}.$$

We repeat the same argument to obtain

$$|a_{k,m}| > 2^{-(k-j-3)} c_k. \tag{3.30}$$

Noting

$$X_{n_k} = a_{k,m}, \quad Y_{n_k} = a_{k,m} + Z_{n_k}$$

and (3.30), we have

$$\begin{aligned} P(Y_{n_k} \in E_k^* | G_{2,m}) &= P(|a_{k,m} + Z_{n_k}| < 2^{-k} c_k) \\ &\leq P(|Z_{n_k}| > 2^{-k} c_k) \\ &= 2\,\mathrm{erfc}\left(\frac{c_k}{2^k \sigma}\right). \end{aligned} \tag{3.31}$$

We can easily show

$$P(Y_{n_k} \notin (b_{k,m-1}, b_{k,m}] | G_{2,m}) \leq P(|Z_{n_k}| > 2c_k) = 2\,\mathrm{erfc}\left(\frac{2c_k}{\sigma}\right). \tag{3.32}$$

The inequality (3.26) for $i = 2$ follows from (3.27), (3.31) and (3.32).

Finally we shall prove (3.26) for $i = 3$. Under the condition $G_{3,m}$, $\widehat{\xi}_{n_0} \neq m$ and

$$\gamma_1^{-1} \widehat{x}_{n_0} \notin (b_{0,m-1}, b_{0,m}].$$

Therefore

$$|a_{0,m} - \gamma_1^{-1} \widehat{x}_{n_0}| > 2c_0$$

and

$$|a_{1,m}| > 2c_1. \tag{3.33}$$

Since $\widehat{\xi}_{n+1} \neq \xi_N, \ldots, \widehat{\xi}_{n+k-1} \neq \xi_N$, in the same way as in the case of $i = 2$ but replacing (3.28) by (3.33), we can prove (3.26) for $i = 3$. q.e.d.

In the following proposition we shall show that the coefficient A_k appeared in (3.11), (3.13) and (3.18) is large enough.

Proposition 3 For each $k \geq 1$, there exists $D_k > 0$ such that

$$A_k^2 > \exp^{(k+1)}(D_k N) \tag{3.34}$$

and

$$\pi_{n_k} < \exp\{-\exp^{(k+1)}(D_k N)\} \tag{3.35}$$

hold provided N is large enough.

Proof. At first we shall prove (3.34) for $k = 1$. We put

$$\Delta^2 = E[1_{\overline{B}_0} \cdot |\widetilde{Z}_{n_0}|^2].$$

Then

$$A_1^2\Delta^2 = \frac{E[|X_{n_1}|^2]}{A^2} = 1. \tag{3.36}$$

For simplicity we put

$$t = \frac{2\gamma_1 c_0}{A} = \sqrt{\frac{\pi}{2}}\gamma_{n_0} e^{-RN}, \quad \tau_0^2 = \gamma_{n_0}(1-\gamma_{n_0}).$$

Then the distribution of $\widetilde{Z}_{n_0}$ is $N(0,\tau_0^2)$. Using (3.10) we know that $|\widetilde{Z}_{n_0}| > t$ if $\widehat{\xi}_{n_0} \neq \xi_N$. Therefore

$$\Delta^2 < \frac{2}{\sqrt{2\pi}\tau_0}\int_t^\infty x^2 \exp\left(-\frac{x^2}{2\tau_0^2}\right) dx = \sqrt{\frac{2}{\pi}}\tau_0^2 \int_{t/\tau_0}^\infty x^2 \exp\left(-\frac{x^2}{2}\right) dx.$$

Noting the asymptotic behavior

$$\frac{1}{\sqrt{2\pi}}\int_y^\infty x^2 \exp\left(-\frac{x^2}{2}\right) dx \sim \frac{y}{\sqrt{2\pi}} \exp\left(-\frac{y^2}{2}\right), \quad y \to \infty,$$

we have

$$\Delta^2 \lesssim \sqrt{\frac{2}{\pi}} t\tau_0 \exp\left(-\frac{t^2}{2\tau_0^2}\right).$$

Since

$$\frac{t^2}{\tau_0^2} = \frac{\pi\gamma_{n_0}}{2(1-\gamma_{n_0})}e^{-2RN} = \frac{\pi}{2}(1-e^{-2Cn_0})e^{2(Cn_0-RN)},$$

if $0 < D_1 < 2\{(1-\delta)C - R\}$ then $D_1 N < 2\{(1-\delta)C-R\}N < 2(Cn_0 - RN)$ and consequently

$$\Delta^2 < \exp(-e^{D_1 N})$$

hold for large N. Thus, noting (3.36), we have obtained (3.34) for $k=1$.

We note

$$||W_1||^2 = A^2 e^{-2Cn_0}(1-e^{-2Cn_0}) < A^2.$$

Since $(a+b)^2 \le 2(a^2+b^2)$ and $\gamma_1^2||Y_{n_{k-1}}||^2 \le A^2$, it follows from (3.12) that

$$\begin{aligned} ||W_k||^2 &\le 2\gamma_1^2(A_{k-1}^2||W_{k-1}||^2 + ||Y_{n_{k-1}}||^2) \\ &\le 2(\gamma_1^2 A_{k-1}^2||W_{k-1}||^2 + A^2). \end{aligned} \tag{3.37}$$

We may assume $A_j^2\gamma_1^2||W_j||^2 > A^2$ for all j. Then, from (3.37), we obtain the inequality

$$||W_k||^2 \le 4\gamma_1^2 A_{k-1}^2||W_{k-1}||^2. \tag{3.38}$$

Hence we have

$$\|W_k\|^2 \le 4^{k-1}\gamma_1^{2(k-1)} \prod_{j=1}^{k-1} A_j^2 \times \|W_1\|^2 \le 4^{k-1}\gamma_1^{2(k-1)} A^2 \prod_{j=1}^{k-1} A_j^2. \tag{3.39}$$

Noting the definition of A_k (cf. (3.11), (3.13)) and using Hölder inequality, we have

$$A^2 \le A_k^2 \|W_k\|^2 \|1_{\overline{B}_{k-1}}\|^2 = A_k^2 \|W_k\|^2 \pi_{n_{k-1}}. \tag{3.40}$$

Combine (3.39) and (3.40) to get

$$A_k^2 \ge \frac{A^2}{\|W_k\|^2 \times \pi_{n_{k-1}}} \ge \frac{1}{4^{k-1}\gamma_1^{2(k-1)} \pi_{n_{k-1}} \prod_{j=1}^{k-1} A_j^2}. \tag{3.41}$$

We substitute

$$c_k = \gamma_1 A_k c_{k-1} = \cdots = \gamma_1^k c_0 \prod_{j=1}^{k} A_j$$

into (3.24) to obtain

$$\begin{aligned} \pi_{n_k} &< \sqrt{\frac{8}{\pi}} \exp\left\{ -\frac{1}{2\sigma^2} \left(\frac{\gamma_1}{2}\right)^{2k} c_0^2 \prod_{j=1}^{k} A_j^2 \right\} \\ &< \sqrt{\frac{8}{\pi}} \exp\left\{ -\frac{\pi A^2 \gamma_{n_0}^2}{16\sigma^2\gamma_1^2} \left(\frac{\gamma_1}{2}\right)^{2k} e^{-2NR} \prod_{j=1}^{k} A_j^2 \right\}. \end{aligned} \tag{3.42}$$

For simplicity we put

$$\lambda_0 (= \lambda_0(n_0)) = \frac{\pi A^2 \gamma_{n_0}^2}{16\sigma^2\gamma_1^2}.$$

Then combining (3.41) and (3.42) we obtain the inequality

$$A_k^2 > \frac{\sqrt{2\pi}\gamma_1^2}{(2\gamma_1)^{2k} \prod_{j=1}^{k-1} A_j^2} \exp\left\{ \lambda_0 \left(\frac{\gamma_1}{2}\right)^{2(k-1)} e^{-2RN} \prod_{j=1}^{k-1} A_j^2 \right\}. \tag{3.43}$$

We may assume that $\prod_{j=1}^{k-2} A_j^2 > 1$. Then (3.43) can be simplified as

$$A_k^2 > \frac{\sqrt{2\pi}\gamma_1^2}{(2\gamma_1)^{2k} A_{k-1}^2} \exp\left\{ \lambda_0 \left(\frac{\gamma_1}{2}\right)^{2(k-1)} A_{k-1}^2 e^{-2RN} \right\}. \tag{3.44}$$

Here we shall prove the desired inequality (3.34) by induction. We have already proved that the inequality (3.34) holds for $k = 1$. We now suppose that there is a constant $\widetilde{D}_{k-1} > 0$ such that

$$A_{k-1}^2 > \exp^{(k)}(\widetilde{D}_{k-1}N).$$

Then, since the function $f(x) = x^{-2}\exp(cx)$ ($c > 0$ is a constant) is increasing on the interval $(2/c, \infty)$, it follows from (3.44) that

$$\begin{aligned} A_k^2 &> \frac{\sqrt{2\pi}\gamma_1^2}{(2\gamma_1)^{2k}\exp^{(k)}(\widetilde{D}_{k-1}N)} \exp\left\{\lambda_0 \left(\frac{\gamma_1}{2}\right)^{2k} \exp^{(k)}(\widetilde{D}_{k-1}N)e^{-2RN}\right\} \\ &= \frac{\sqrt{2\pi}\gamma_1^2}{(2\gamma_1)^{2k}} \exp\left\{-\exp^{(k-1)}(\widetilde{D}_{k-1}N)\right. \\ &\qquad \left. +\lambda_0 \left(\frac{\gamma_1}{2}\right)^{2k} \exp^{(k)}(\widetilde{D}_{k-1}N)e^{-2RN}\right\}. \end{aligned} \tag{3.45}$$

Therefore, for any $0 < \widetilde{D}_k < \widetilde{D}_{k-1}$, the inequality

$$A_k^2 > \exp^{(k+1)}(\widetilde{D}_k N) \tag{3.46}$$

holds if N is large enough. Thus, by induction, we have proved that the inequality (3.34) holds for any k. We now turn to evaluate the error probability π_{n_k}. Using $\prod_{j=1}^{k-2} A_j^2 > 1$, it follows from (3.42) and (3.46) that

$$\begin{aligned} \pi_{n_k} &< \sqrt{\frac{8}{\pi}} \exp\left\{-\lambda_0 \left(\frac{\gamma_1}{2}\right)^{2k} e^{-2NR} \prod_{j=1}^{k} A_j^2\right\} \\ &< \sqrt{\frac{8}{\pi}} \exp\left\{-\lambda_0 \left(\frac{\gamma_1}{2}\right)^{2k} e^{-2RN} A_k^2\right\} \\ &< \sqrt{\frac{8}{\pi}} \exp\left\{-\lambda_0 \left(\frac{\gamma_1}{2}\right)^{2k} e^{-2RN} \exp^{(k+1)}(\widetilde{D}_k N)\right\}. \end{aligned}$$

Therefore, we can easily show that for any $0 < D_k < \widetilde{D}_k$ the inequality (3.35) holds if N is large enough. q.e.d.

3.3 Proof of Main Theorem

Theorem 1 is straightforward of Proposition 3.

Proof of Theorem 1. Let π_{n_k} be the error probability discussed in Proposition 1 and Proposition 2. We note that, if N is large, $n_k < N$ and

$$\pi_N^*(R) \leq \pi_{n_k}^*(R) \leq \pi_{n_k}.$$

Therefore using (3.35) we have

$$\limsup_{N\to\infty} \frac{\pi_N^*(R)}{\exp\{-\exp^{(k+1)}(D_k N)\}} \le 1.$$

Finally, for any fixed K, by taking k as $k \ge K$ we obtain

$$\begin{aligned}
&\lim_{N\to\infty} \frac{\pi_N^*(R)}{\exp\{-\exp^{(K)}(N)\}} \\
&= \lim_{N\to\infty} \frac{\pi_N^*(R)}{\exp\{-\exp^{(k+1)}(D_k N)\}} \frac{\exp\{-\exp^{(k+1)}(D_k N)\}}{\exp\{-\exp^{(K)}(N)} \\
&= 0.
\end{aligned}$$

The proof of (1.9) is complete. q.e.d.

References

[1] S. Ihara: *Information Theory for Continuous Systems.* World Scientific, Singapore, 1993.

[2] A. J. Kramer: Inproving communication reliability by use of an intermittent feedback channel, Part I: No bandwidth constraints. *IEEE Trans. Inform. Theory*, IT-15 (1969), 52–60.

[3] M. S. Pinsker: *Information and Information Stability of Random Variables and Processes.* Holden-Day, San Francisco, 1964.

[4] M. S. Pinsker: The probability of error in block transmission in a memoryless Gaussian channel with feedback. *Problems Inform. Transmission*, 4 (1968), 1–14.

[5] K. Sh. Zigangirov: Upper bound for the error probability for Gaussian channels with feedback. *Problems Inform. Transmission*, 6 (1970), 159–163.

[6] J. P. M. Schalkwijk and T. Kailath: A coding scheme for additive noise channels with feedback, Part I: No bandwidth constraints. *IEEE Trans. Inform. Theory*, IT-12 (1966), 172–182.

ERROR PROBABILITY OF THE FIXED-LENGTH SOURCE CODING

K. IRIYAMA

Meijo University, Nagoya, 468-8502, Japan

We study the fixed-length lossy source coding for general sources in the sense of Han and Verdú. One of the basic problems in the lossy source coding is to characterize a rate-distortion function which is the infimum of achievable coding rates under the asymptotic constraint of the form $\varepsilon_n(D) \to 0$ as $n \to \infty$, where $\varepsilon_n(D)$ is the probability of the distortion exceeding a lebel D and n is the code-length. In the paper we consider a fidelity criterion of the form $\varepsilon_n(D) \sim \varepsilon$ $(n \to \infty)$, where $\varepsilon \in [0, 1)$ is the prescribed constant. Chen and Alajaji have established a general formula, assuming that the source is with finite alphabet, for the infimum of achievable coding rates subject to the fidelity criterion. The aim of the paper is to show that the minimum achievable rate can be expressed in a different form. We note that, in our case, the alphabet is arbitrary and not necessarily finite.

1. Introduction

In this paper we study the error probability of the fixed-length lossy source coding in a quite general framework. The information source, denoted by $\mathbf{X} = \{X^n\}_{n=1}^{\infty}$, is a general source in the sense of Han and Verdú (cf. [4, 5]) where X^n is a random variable taking values in a set $\mathcal{X}_n$ and $(\mathcal{X}_n, \mathcal{B}_n)$ is a measurable space. Note that the random sequence $\mathbf{X} = \{X^n\}$ is not necessarily stationary nor consistent.

A code $\{\varphi_n\}$ is a sequence of mappings $\varphi_n : \mathcal{X}_n \to \mathcal{X}_n$, $n = 1, 2, \ldots$, where $\varphi_n(\mathcal{X}_n)$ is a finite set. We call $\limsup_{n\to\infty} n^{-1} \log |\varphi_n(\mathcal{X}_n)|$ the rate of the code $\{\varphi_n\}$, where $|\varphi_n(\mathcal{X}_n)|$ denotes the cardinality of $\varphi_n(\mathcal{X}_n)$.

We are given a distortion measure $d_n(\cdot, \cdot)$ on $\mathcal{X}_n \times \mathcal{X}_n$ which is a non-negative valued measurable function. For each fixed distortion level $D \geq 0$, the error probability $\varepsilon_n(D)$ is defined by

$$\varepsilon_n(D) = \Pr\{d_n(X^n, \varphi_n(X^n)) > D\}.$$

In Section 3, we investigate problems in the lossy source coding of general sources. One of the basic problems in the lossy source coding is to characterize the rate-distortion function $R(D|\mathbf{X})$ which is the infimum of achievable coding rates under the asymptotic constraint of the form $\varepsilon_n(D) \to 0$ as $n \to \infty$.

If the coding rate is less than the rate-distortion function $R(D|\mathbf{X})$, then $\varepsilon_n(D)$ does not tend to zero. Recently Chen and Alajaji [1] obtained a formula for the infimum $T(D, \varepsilon|\mathbf{X})$ of achievable rates under the asymptotic constraint of the form $\varepsilon_n(D) \to \varepsilon$ as $n \to \infty$, where the general source is of finite alphabet. The aim of the paper is to show that the rate $T(D, \varepsilon|\mathbf{X})$ can be expressed in a different form (Theorem 3.4). We emphasize that our formula can be applied to general sources of continuous alphabet as well as finite alphabet. The proof is given in section 4. In section 5, some examples are examined.

2. Preliminaries

Let us introduce some notations. For a sequence $\{Z_n\}_{n=1}^{\infty}$ of real random variables,

$$\text{p-}\limsup_{n\to\infty} Z_n = \inf\{R : \lim_{n\to\infty} \Pr\{Z_n > R\} = 0\}$$

and

$$\text{p-}\liminf_{n\to\infty} Z_n = \sup\{R : \lim_{n\to\infty} \Pr\{Z_n < R\} = 0\}$$

denote the limsup in probability and the liminf in probability, respectively, of $\{Z_n\}$ (cf. [4, 5]). Furthermore, we write

$$p_\varepsilon\text{-}\limsup_{n\to\infty} Z_n = \inf\{R : \limsup_{n\to\infty} \Pr\{Z_n > R\} \le \varepsilon\}, \quad 0 \le \varepsilon < 1.$$

Let $\mathbf{Y} = \{Y^n\}$ be a general source, where Y^n is taking values in $\mathcal{X}_n$ and arbitrarily correlated to X^n. Let us denote by P_X the probability distribution of a random variable X. The divergence (relative entropy) of Y^n with respect to X^n is defined by

$$D(Y^n||X^n) = \begin{cases} \int_{\mathcal{X}_n} \log \frac{dP_{Y^n}}{dP_{X^n}}(x) dP_{Y^n}(x), & P_{Y^n} \ll P_{X^n}, \\ \infty, & \text{otherwise,} \end{cases}$$

where $(dP_{Y^n}/dP_{X^n})(x)$ is Radon-Nikodym derivative, and $P_{Y^n} \ll P_{X^n}$ means that P_{Y^n} is absolutely continuous P_{X^n}. We define $D_u(\mathbf{Y}||\mathbf{X})$ and $D_l(\mathbf{Y}||\mathbf{X})$ by

$$D_u(\mathbf{Y}||\mathbf{X}) = \limsup_{n\to\infty} \frac{1}{n} D(Y^n||X^n)$$

$$D_l(\mathbf{Y}||\mathbf{X}) = \liminf_{n\to\infty} \frac{1}{n} D(Y^n||X^n).$$

We define mutual informations as follows:

$$\overline{I}(\mathbf{X};\mathbf{Y}) = p\text{-}\limsup_{n\to\infty} \frac{1}{n} \log \frac{dP_{X^nY^n}}{dP_{X^n} \times P_{Y^n}}(X^n, Y^n)$$

$$\underline{I}(\mathbf{X};\mathbf{Y}) = p\text{-}\liminf_{n\to\infty} \frac{1}{n} \log \frac{dP_{X^nY^n}}{dP_{X^n} \times P_{Y^n}}(X^n, Y^n)$$

$$\overline{I}_\varepsilon(\mathbf{X};\mathbf{Y}) = p_\varepsilon\text{-}\limsup_{n\to\infty} \frac{1}{n} \log \frac{dP_{X^nY^n}}{dP_{X^n} \times P_{Y^n}}(X^n, Y^n).$$

Note that $\lim_{\varepsilon \nearrow 1} \overline{I}_\varepsilon(\mathbf{X};\mathbf{Y}) = \underline{I}(\mathbf{X};\mathbf{Y})$ and $\overline{I}_0(\mathbf{X};\mathbf{Y}) = \overline{I}(\mathbf{X};\mathbf{Y})$.

3. Lossy source coding

In the lossy source coding theorem, the rate-distortion function plays important roles. For the general sources the rate-distortion function has been studied in [2, 4, 12], under a fidelity condition of the form

$$p\text{-}\limsup_{n\to\infty} d_n(X^n, \varphi_n(X^n)) \leq D. \tag{1}$$

On the other hand, we [9] have considered a slightly different condition

$$\lim_{n\to\infty} \varepsilon_n(D) = 0. \tag{2}$$

We now define rate-distortion functions. Denote by $\mathcal{C}_m(R, D|\mathbf{X})$ (resp. $\mathcal{C}(R, D|\mathbf{X})$) the class of all codes $\{\varphi_n\}$ satsfying (1) and

$$\limsup_{n\to\infty} \frac{1}{n} \log M_n \leq R \tag{3}$$

(resp. (2) and (3)).

Definition 3.1. The rate distortion functions of $\mathbf{X}$ are defined as

$$R_m(D|\mathbf{X}) = \inf\{R : \exists \{\varphi_n\} \in \mathcal{C}_m(R, D|\mathbf{X})\}$$
$$R(D|\mathbf{X}) = \inf\{R : \exists \{\varphi_n\} \in \mathcal{C}(R, D|\mathbf{X})\}.$$

Clearly,

$$R_m(D|\mathbf{X}) \leq R(D|\mathbf{X}) \leq R_m(D_0|\mathbf{X}), \quad 0 < D_0 < D. \tag{4}$$

Moreover, if the function $R_m(D|\mathbf{X})$ is continuous at D, then

$$R(D|\mathbf{X}) = R_m(D|\mathbf{X}). \tag{5}$$

Note that the equation (5) does not hold in general. Indeed, if $d_n(x, y) = D + n^{-1}$ $(x \neq y)$, then (5) is not true.

Steinberg and Verdú [12] proved the following theorem.

Theorem 3.1. ([12]) *For any* $D \geq 0$,

$$R_m(D|\mathbf{X}) = \inf_{\mathbf{Y}}\{\overline{I}(\mathbf{X};\mathbf{Y}) : \ \overline{D}(\mathbf{X},\mathbf{Y}) \leq D\}, \tag{6}$$

where $\overline{D}(\mathbf{X},\mathbf{Y}) = p\text{-}\limsup_{n\to\infty} d_n(X^n, Y^n)$.

Chen and Alajaji [1] have investigated the ε-achievable data compression rate at distortion D, where the fidelity criterion is given by

$$\sup\{t : \ \limsup_{n\to\infty} \varepsilon_n(t) \geq \varepsilon\} \leq D. \tag{7}$$

Note that the condition (7) holds if and only if

$$\limsup_{n\to\infty} \varepsilon_n(D_1) < \varepsilon, \quad \text{for all} \quad D_1 > D.$$

In this paper, we consider a simplified condition

$$\limsup_{n\to\infty} \varepsilon_n(D) \leq \varepsilon. \tag{8}$$

Denote by $\mathcal{C}(R, D, \varepsilon|\mathbf{X})$ (resp. $\mathcal{C}_0(R, D, \varepsilon|\mathbf{X})$) the class of all codes $\{\varphi_n\}$ satsfying (3) and (7) (resp. (3) and (8)).

Definition 3.2. The minimum (D, ε)-achievable data compression rates of $\mathbf{X}$ are defined as

$$T(D,\varepsilon|\mathbf{X}) = \inf\{R : \exists\,\{\varphi_n\} \in \mathcal{C}(R, D, \varepsilon|\mathbf{X})\}$$
$$T_0(D,\varepsilon|\mathbf{X}) = \inf\{R : \exists\,\{\varphi_n\} \in \mathcal{C}_0(R, D, \varepsilon|\mathbf{X})\}.$$

Note that $R(D|\mathbf{X}) = T_0(D, 0|\mathbf{X})$. It is clear that if $0 < D_0 < D < D_1$ then

$$T(D_1,\varepsilon|\mathbf{X}) \leq T_0(D,\varepsilon|\mathbf{X}) \leq T(D_0,\varepsilon|\mathbf{X}). \tag{9}$$

Especially, the equation $T(D,\varepsilon|\mathbf{X}) = T_0(D,\varepsilon|\mathbf{X})$ holds, if $T(\cdot,\varepsilon|\mathbf{X})$ is continuous at D.

Theorem 3.2. ([1]) *Suppose that* $\mathcal{X}_n = \mathcal{X}^n$ *and* $\mathcal{X}$ *is a finite set. Then, for* $D > 0$ *and* $0 < \varepsilon < 1$,

$$T(D,\varepsilon|\mathbf{X}) = \inf_{\mathbf{Y}}\{\overline{I}(\mathbf{X};\mathbf{Y}) : \ D_\varepsilon(\mathbf{X},\mathbf{Y}) \leq D\}.$$

where

$$D_\varepsilon(\mathbf{X},\mathbf{Y}) = \sup\{t : \ \limsup_{n\to\infty} \Pr\{d_n(X^n,Y^n) > t\} \geq \varepsilon\}.$$

The main aim of the paper is to give the following expression for $T_0(D, \varepsilon|\mathbf{X})$.

Theorem 3.3. *Suppose that* $d_n(x,x) = 0$ *and* $d_n(x,y) = d_n(y,x)$ *for any* $x, y \in \mathcal{X}_n$. *For any* $\varepsilon \in [0,1)$,

$$T_0(D, \varepsilon|\mathbf{X}) = \inf_{\mathbf{Y}} \{\overline{I}_\varepsilon(\mathbf{X};\mathbf{Y}) : \lim_{n\to\infty} \Pr\{d_n(X^n, Y^n) \le D\} = 1\} \tag{10}$$

where the infimum is taken for all general sources $\mathbf{Y} = \{Y^n\}$ *satisfying* $\Pr\{d_n(X^n, Y^n) \le D\} \to 1$ *as* $n \to \infty$.

The proof will be given in Section 5.

We should emphasize here that, in Theorem 3.3, the alphabet $\mathcal{X}_n$ is not necessarily a finite set. Since Theorem 3.2 was proved under the assumption that $\mathcal{X}_n$ is finite, we need another method to prove Theorem 3.3.

As a corollary of Theorem 3.3, we can easily show the following theorem.

Theorem 3.4. *Suppose that* $d_n(x,x) = 0$ *and* $d_n(x,y) = d_n(y,x)$ *for any* $x, y \in \mathcal{X}_n$. *Then, for any continuous point* $D > 0$,

$$T(D, \varepsilon|\mathbf{X}) = T_0(D, \varepsilon|\mathbf{X}) = \inf_{\mathbf{Y}} \{\overline{I}_\varepsilon(\mathbf{X};\mathbf{Y}) : \overline{D}(\mathbf{X},\mathbf{Y}) \le D\}.$$

Note that we have

$$R(D|\mathbf{X}) = \inf_{\mathbf{Y}} \{\overline{I}(\mathbf{X};\mathbf{Y}) : \lim_{n\to\infty} \Pr\{d_n(X^n, Y^n) \le D\} = 1\} \tag{11}$$

as a special case of (10).

Denote by $\mathcal{C}^*(R, D|\mathbf{X})$ the class of all codes $\{\varphi_n\}$ satsfying

$$\liminf_{n\to\infty} \varepsilon_n(D) < 1$$

and (3). The minimum achievable rate $R^*(D|\mathbf{X})$ is defined by

$$R^*(D|\mathbf{X}) = \inf\{R : \exists \{\varphi_n\} \in \mathcal{C}^*(R, D|\mathbf{X})\}.$$

Since $R^*(D|\mathbf{X}) = \lim_{\varepsilon \nearrow 1} T_0(D, \varepsilon|\mathbf{X})$, as a corollary of Theorem 3.3, we can prove

$$R^*(D|\mathbf{X}) = \inf_{\mathbf{Y}} \{\underline{I}(\mathbf{X};\mathbf{Y}) : \lim_{n\to\infty} \Pr\{d_n(X^n, Y^n) \le D\} = 1\}. \tag{12}$$

Remark 3.1. We put $d = \{d_n\}_{n=1}^\infty$. In the fixed-length lossy source coding, a pair $(\mathbf{X}, d)$ is said to satisfy the strong converse property if $R(D|\mathbf{X}) = R^*(D|\mathbf{X})$. Clearly, the strong converse property is equivalent to $T_0(D, \varepsilon|\mathbf{X}) = \text{const}$, $0 \le \varepsilon < 1$.

It is also an important problem to characterize the minimum achievable rate under the asymptotic condition of the form $\varepsilon_n(D) \sim e^{-nr}$ $(n \to \infty)$. To state the problem precisely, we introduce a condition

$$\liminf_{n\to\infty} \frac{1}{n} \log \frac{1}{\varepsilon_n(D)} \geq r. \tag{13}$$

We denote by $\mathcal{C}_e(R, D, r|\mathbf{X})$ the class of all codes satsfying (3) and (13).

Definition 3.3. The minimum (D, r)-achievable rate of $\mathbf{X}$ is defined as

$$R_e(D, r|\mathbf{X}) = \inf\{R : \exists\{\varphi_n\} \in \mathcal{C}_e(R, D, r|\mathbf{X})\}.$$

In addition, it is interesting to determine

$$R_e^*(D, r|\mathbf{X}) = \inf\{R : \exists\{\varphi_n\} \in \mathcal{C}_e^*(R, D, r|\mathbf{X})\},$$

where $\mathcal{C}_e^*(R, D, r|\mathbf{X})$ is the set of all $\{\varphi_n\}$ satisfying (3) and

$$\limsup_{n\to\infty} \frac{1}{n} \log \frac{1}{1 - \varepsilon_n(D)} \leq r.$$

Clearly,

$$R_e^*(D, r|\mathbf{X}) \leq R^*(D|\mathbf{X}) \leq T_0(D, \varepsilon|\mathbf{X}) \leq R(D|\mathbf{X}) \leq R_e(D, r|\mathbf{X}). \tag{14}$$

For the stationary memoryless source (SMS) with finite alphabet, Marton [11] obtained a formula

$$R_e(D, r|\mathbf{X}) = \sup_{Y:\ D(Y\|X_1)\leq r} R(D|Y) \tag{15}$$

where the distortion measure is additive, namely is given in the form

$$d_n(x^n, y^n) = \frac{1}{n}\sum_{k=1}^{n} d(x_k, y_k) \tag{16}$$

$(x^n = (x_1, \ldots, x_n),\ y^n = (y_1, \ldots, y_n) \in \mathcal{X}^n)$ and

$$R(D|Y) = \inf_{Z:\ E[d(Y,Z)]\leq D} I(Y; Z)$$

is the rate-distortion function of Y. Recently, Ihara and Kubo [6] showed that (15) and

$$R_e^*(D, r|\mathbf{X}) = \inf_{Y:\ D(Y\|X_1)\leq r} R(D|Y)$$

hold for the Gaussian SMS under the mean-squared fidelity criterion. For the general source, Iriyama [9] established the following formulas for the minimum (D, r)-achievable rates.

Theorem 3.5. ([9]) *For any* $r > 0$,

$$\sup_{\mathbf{Y}:\, D_l(\mathbf{Y}\|\mathbf{X})<r} R^*(D|\mathbf{Y}) \le R_e(D,r|\mathbf{X}) \le \sup_{\mathbf{Y}:\, D_l(\mathbf{Y}\|\mathbf{X})\le r} R^*(D|\mathbf{Y}) \qquad (17)$$

and

$$R_e^*(D,r|\mathbf{X}) = \inf_{\mathbf{Y}:\, D_u(\mathbf{Y}\|\mathbf{X})\le r} R(D|\mathbf{Y}). \qquad (18)$$

We should emphasize here that $R_e^*(D,r|\mathbf{X})$ can be expressed in terms of not $R_m(D|\mathbf{X})$ but $R(D|\mathbf{X})$.

If we assume that $d_n(x,y) = 0$ if and only if $x = y$, then the rates $R_e(r|\mathbf{X}) \equiv R_e(0,r|\mathbf{X})$ and $R_e^*(r|\mathbf{X}) \equiv R_e^*(0,r|\mathbf{X})$ are nothing but the minimum r-achievable rates for the lossless coding. When the alphabet $\mathcal{X}$ is a discrete (not necessarily finite) set, the following theorem was proved.

Theorem 3.6. ([7, 8, 10])

$$R_e(r|\mathbf{X}) = \sup_{\mathbf{Y}:\, D_l(\mathbf{Y}\|\mathbf{X})<r} \underline{H}(\mathbf{Y}) \qquad (19)$$

$$R_e^*(r|\mathbf{X}) = \inf_{\mathbf{Y}:\, D_u(\mathbf{Y}\|\mathbf{X})\le r} \overline{H}(\mathbf{Y}). \qquad (20)$$

4. Proof of Theorem 3.3

We denote by $B_n(y,D) = \{x \in \mathcal{X}_n : d_n(x,y) \le D\}$ the D-ball with center y. For each general source $\mathbf{Z} = \{Z^n\}$, we put

$$S(\mathbf{Z}) = p_\varepsilon\text{-}\limsup_{n\to\infty} \frac{1}{n} \log \frac{1}{P_{Z^n}(B_n(X^n,D))}. \qquad (21)$$

In order to prove Theorem 3.3 we need two lemmas.

Lemma 4.1.

$$T_0(D,\varepsilon|\mathbf{X}) = \inf_{\mathbf{Z}=\{Z^n\}} S(\mathbf{Z}), \qquad (22)$$

where the infimum is taken for all general sources $\mathbf{Z}$.

@

Proof. At first we shall prove the inequality

$$T_0(D,\varepsilon|\mathbf{X}) \le \inf_{\mathbf{Z}=\{Z^n\}} S(\mathbf{Z}). \qquad (23)$$

We fix $\delta > 0$. Let $\{Z^{k,n}\}_{k=1,2,\ldots,M_n}$ be an independent and identically distributed (i.i.d.) sequence with distribution P_{Z^n}, where $M_n =$

$\lceil \exp\{n(S(\mathbf{Z}) + 2\delta)\} \rceil$ and $\lceil x \rceil$ denotes the smallest integer not less than x. Then, for any $x \in \mathcal{X}_n$, we have

$$\begin{aligned}\Pr\{d_n(x, Z^{k,n}) > D, k = 1, 2, \ldots, M_n\} &= \prod_{k=1}^{M_n} \Pr\{d_n(x, Z^{k,n}) > D\} \\ &= (1 - P_{Z^n}(B_n(x, D)))^{M_n}. \quad (24)\end{aligned}$$

We define sets T_n and C_n by

$$\begin{aligned}T_n &= \left\{ x \in \mathcal{X}_n : \ \frac{1}{n} \log \frac{1}{P_{Z^n}(B_n(x, D))} \le S(\mathbf{Z}) + \delta \right\} \\ C_n &= \{x \in \mathcal{X}_n : \ d_n(x, Z^{k,n}) > D, \ k = 1, 2, \ldots, M_n\},\end{aligned}$$

respectively. We define the coding function $\widetilde{\varphi}_n$ as follows. If there exists $m \in \{1, \ldots, M_n\}$ such that $d_n(x, Z^{m,n}) \le D$ and $d_n(x, Z^{k,n}) > D$ for all $k \in \{1, \ldots, m-1\}$, then $\widetilde{\varphi}_n(x) = Z^{m,n}$, otherwise $\widetilde{\varphi}_n(x) = Z^{1,n}$. Then

$$\widetilde{\varepsilon}_n(D) \equiv E_Z[\Pr\{X^n \in C_n\}] \quad (25)$$

is the probability of decoding error averaged over the ensemble of random codes, where $E_Z[\cdot]$ denotes the expectation with respect to $Z^{1,n}, \ldots, Z^{M_n,n}$. Since $P_{Z^n}(B_n(x, D)) \ge {M_n}^{-1} e^{n\delta}$ for all $x \in T_n$, from (24) we obtain

$$\Pr\{d_n(x, Z^{k,n}) > D, \ k = 1, 2, \ldots, M_n\} \le \left(1 - \frac{1}{M_n} e^{n\delta}\right)^{M_n}, \quad x \in T_n.$$

Then, noting $(1 - 1/t)^t \to e^{-1}$ as $t \to \infty$, we have

$$\lim_{n\to\infty} \Pr\{d_n(x, Z^{k,n}) > D, \ k = 1, 2, \ldots, M_n\} = 0, \quad x \in T_n,$$

and

$$\lim_{n\to\infty} E_Z[\Pr\{X^n \in T_n \cap C_n\}] = 0. \quad (26)$$

By the definition of $S(\mathbf{Z})$, we have

$$\limsup_{n\to\infty} \Pr\{X^n \in {T_n}^c\} \le \varepsilon. \quad (27)$$

Combining (25), (26) and (27) we have

$$\limsup_{n\to\infty} \widetilde{\varepsilon}_n(D) \le \varepsilon. \quad (28)$$

It follows from (28) that there exists a coding function φ_n, with $|\varphi_n(\mathcal{X}_n)| \le M_n$, such that the error probability $\varepsilon_n(D)$ of φ_n satisfies (8). Therefore we have

$$T_0(D, \varepsilon|\mathbf{X}) \le \limsup_{n\to\infty} \frac{1}{n} \log |\varphi_n(\mathcal{X}_n)| \le S(\mathbf{Z}) + 2\delta.$$

Since $\delta > 0$ is arbitray, we have the inequality (23).

Secondly we shall prove the inequality

$$T_0(D, \varepsilon|\mathbf{X}) \geq \inf_{\mathbf{Z}=\{Z^n\}} S(\mathbf{Z}). \tag{29}$$

By the definition of $T_0(D, \varepsilon|\mathbf{X})$, for each $\delta > 0$, there exists a code $\{\varphi_n\}$ satisfying

$$\limsup_{n\to\infty} \varepsilon_n(D) \leq \varepsilon \tag{30}$$

and

$$\limsup_{n\to\infty} \frac{1}{n} \log |\varphi_n(\mathcal{X}_n)| \leq T_0(D, \varepsilon|\mathbf{X}) + \delta. \tag{31}$$

We define a set A_n by

$$A_n = \{x \in \mathcal{X}_n : \ d_n(x, \varphi_n(x)) \leq D\}.$$

Then

$$\varepsilon_n(D) = P_{X^n}(A_n{}^c).$$

Let $\widehat{\mathbf{Z}} = \{\widehat{Z}^n\}$ be a general source with a probability distribution

$$P_{\widehat{Z}^n}(x) = \frac{1_{\varphi_n(\mathcal{X}_n)}(x)}{|\varphi_n(\mathcal{X}_n)|}, \quad x \in \mathcal{X}_n, \tag{32}$$

where $1_F(x)$ denotes the indicator function of F. Clearly

$$P_{\widehat{Z}^n}(B_n(x, D)) = \frac{|B_n(x, D) \cap \varphi_n(\mathcal{X}_n)|}{|\varphi_n(\mathcal{X}_n)|}.$$

We define a set T_n by

$$T_n = \left\{ x \in \mathcal{X}_n : \ P_{\widehat{Z}^n}(B_n(x, D)) < \frac{1}{|\varphi_n(\mathcal{X}_n)|} \right\}.$$

Then it is clear that

$$T_n = \{x \in \mathcal{X}_n : B_n(x, D) \cap \varphi_n(\mathcal{X}_n) = \phi\} \subset A_n{}^c.$$

Hence we have

$$\limsup_{n\to\infty} P_{X^n}(T_n) \leq \limsup_{n\to\infty} \varepsilon_n(D). \tag{33}$$

It follows from (30) and (33) that

$$\limsup_{n\to\infty} \Pr\left\{ \frac{1}{n} \log \frac{1}{P_{\widehat{Z}^n}(B_n(X^n, D))} > \frac{1}{n} \log |\varphi_n(\mathcal{X}_n)| \right\} \leq \varepsilon,$$

so that

$$\limsup_{n\to\infty} \frac{1}{n} \log |\varphi_n(\mathcal{X}_n)| \geq S(\widehat{\mathbf{Z}}). \tag{34}$$

Using (31) and (34), we have

$$T_0(D, \varepsilon|\mathbf{X}) \geq S(\widehat{\mathbf{Z}}) - \delta.$$

Since $\delta > 0$ is arbitrary, we have the inequality (29). □

Lemma 4.2.

$$\inf_{\mathbf{Z}} S(\mathbf{Z}) = \inf_{\mathbf{Y}} \{\overline{I}_\varepsilon(\mathbf{X};\mathbf{Y}) : \lim_{n\to\infty} \Pr\{d_n(X^n, Y^n) \leq D\} = 1\}. \tag{35}$$

Proof. For general sources $\mathbf{Y} = \{Y^n\}$ and $\mathbf{Z} = \{Z^n\}$,

$$\overline{D}(\mathbf{Y}\|\mathbf{Z}) = p\text{-}\limsup_{n\to\infty} \frac{1}{n} \log \frac{dP_{Y^n}}{dP_{Z^n}}(Y^n)$$

and

$$\underline{D}(\mathbf{Y}\|\mathbf{Z}) = p\text{-}\liminf_{n\to\infty} \frac{1}{n} \log \frac{dP_{Y^n}}{dP_{Z^n}}(Y^n)$$

are called the spectral sup-divergence rate and the spectral inf-divergence rate, respectively. Clearly $\overline{D}(\mathbf{Y}\|\mathbf{Y}) = \underline{D}(\mathbf{Y}\|\mathbf{Y}) = 0$. It is known (cf. [13]) that $\overline{D}(\mathbf{Y}\|\mathbf{Z}) \geq \underline{D}(\mathbf{Y}\|\mathbf{Z}) \geq 0$. Therefore, to prove (35), it is sufficient to show that

$$S(\mathbf{Z}) \geq \inf_{\mathbf{Y}} \{\overline{I}_\varepsilon(\mathbf{X};\mathbf{Y}) + \underline{D}(\mathbf{Y}\|\mathbf{Z}) : \lim_{n\to\infty} \Pr\{d_n(X^n, Y^n) \leq D\} = 1\} \tag{36}$$

and

$$S(\mathbf{Z}) \leq \inf_{\mathbf{Y}} \{\overline{I}_\varepsilon(\mathbf{X};\mathbf{Y}) + \overline{D}(\mathbf{Y}\|\mathbf{Z}) : \lim_{n\to\infty} \Pr\{d_n(X^n, Y^n) \leq D\} = 1\}. \tag{37}$$

We shall first show (36). For simplicity we assume that

$$P_{Z^n}(B_n(X^n, D)) \neq 0, \quad n \in \mathbf{N},$$

almost surely. Then we can define a conditional probability distribution $P_{\widehat{Z}^n|X^n}$ by

$$\frac{dP_{\widehat{Z}^n|X^n}(\cdot|x)}{dP_{Z^n}}(z) = \frac{1_{B_n(x,D)}(z)}{P_{Z^n}(B_n(x,D))}, \quad (x,z) \in \mathcal{X}_n \times \mathcal{X}_n.$$

Clearly

$$\lim_{n\to\infty} \Pr\{d_n(X^n, \widehat{Z}^n) \leq D\} = 1. \tag{38}$$

We can prove that

$$S(\mathbf{Z}) = p_\varepsilon\text{-}\limsup_{n\to\infty} \frac{1}{n} \log \frac{dP_{\widehat{Z}^n|X^n}(\widehat{Z}^n|X^n)}{dP_{Z^n}}(\widehat{Z}^n)$$
$$\geq \overline{I}_\varepsilon(\mathbf{X};\widehat{\mathbf{Z}}) + \underline{D}(\widehat{\mathbf{Z}}\|\mathbf{Z}). \tag{39}$$

Combining (38) and (39), we have the inequality (36). Next we shall prove (37). Let $\mathbf{Y} = \{Y^n\}$ be a general source such that

$$\lim_{n\to\infty} \Pr\{d_n(X^n, Y^n) \leq D\} = 1. \tag{40}$$

Since

$$\Pr\left\{\frac{1}{n}\log\frac{1}{P_{Z^n}(B_n(X^n, D))} > S(\mathbf{Z}) - \delta\right\} > \varepsilon$$

and

$$\Pr\left\{\frac{1}{n}\log\frac{dP_{\widehat{Z}^n|X^n}(\cdot|X^n)}{dP_{Z^n}}(Y^n) > S(\mathbf{Z}) - \delta\right\}$$
$$\geq \Pr\left\{\frac{1}{n}\log\frac{1}{P_{Z^n}(B_n(X^n, D))} > S(\mathbf{Z}) - \delta\right\} - \Pr\{Y^n \in B_n(X^n, D)^c\},$$

it follows from (40) that

$$\limsup_{n\to\infty} \Pr\left\{\frac{1}{n}\log\frac{dP_{\widehat{Z}^n|X^n}(\cdot|X^n)}{dP_{Z^n}}(Y^n) > S(\mathbf{Z}) - \delta\right\} > \varepsilon.$$

Therefore, noting the non-negativity of the divergence, we have

$$S(\mathbf{Z}) \leq p_\varepsilon\text{-}\limsup_{n\to\infty} \frac{1}{n}\log\frac{dP_{Y^n|X^n}(\cdot|X^n)}{dP_{Z^n}}(Y^n)$$
$$-p\text{-}\liminf_{n\to\infty} \frac{1}{n}\log\frac{dP_{Y^n|X^n}(\cdot|X^n)}{dP_{\widehat{Z}^n|X^n}(\cdot|X^n)}(Y^n)$$
$$\leq p_\varepsilon\text{-}\limsup_{n\to\infty} \frac{1}{n}\log\frac{dP_{Y^n|X^n}(\cdot|X^n)}{dP_{Z^n}}(Y^n)$$
$$\leq \overline{I}_\varepsilon(\mathbf{X};\mathbf{Y}) + \overline{D}(\mathbf{Y}\|\mathbf{Z}). \tag{41}$$

The inequality (37) follows from (40) and (41). □

Remark 4.1. In proving Lemma 4.2, we have not used the commutativity of the distortion measure $d_n(\cdot,\cdot)$.

Proof of Theorem 3.3. The equation (10) is clear from (22) and (35).□

5. Examples

This section is devoted to apply our main theorems to SMSs and nonergodic sources.

Example 1 (SMS): We consider a general source $\mathbf{X} = \{X^n\}$ corresponding to an i.i.d. process $\{X_n\}$ taking values in a finite alphabet $\mathcal{X}$, where $X^n = (X_1, \ldots, X_n)$. We assume that a sequence $d = \{d_n\}$ of distortion measures is given by (16). Then, $(\mathbf{X}, d)$ satisfies the strong converse property, i.e,

$$T_0(D, \varepsilon|\mathbf{X}) = T(D, \varepsilon|\mathbf{X}) = R(D|X_1), \quad 0 \leq \varepsilon < 1. \tag{42}$$

We omit the proof.

Example 2 (Nonergodic source): Given general sources $\mathbf{X}_1 = \{X_1^n\}$ and $\mathbf{X}_2 = \{X_2^n\}$, a general source $\mathbf{X} = \{X^n\}$ defined by

$$P_{X^n}(x) = \alpha_1 P_{X_1^n}(x) + \alpha_2 P_{X_2^n}(x), \quad x \in \mathcal{X}_n \tag{43}$$

is called a mixed source of $\mathbf{X}_1$ and $\mathbf{X}_2$, where α_1 and α_2 are constants satisfying $\alpha_1 > 0$, $\alpha_2 > 0$ and $\alpha_1 + \alpha_2 = 1$.

In particular, if $\mathbf{X}_1$ and $\mathbf{X}_2$ are two different SMSs, the mixed source $\mathbf{X}$ can be regarded as a typical and simple example of a stationary nonergodic source.

Theorem 5.1. *Let $\mathbf{X}_i = \{X_i^n\}$ $(i = 1, 2)$ be a general source satisfying the strong converse property and $R(D|\mathbf{X}_1) > R(D|\mathbf{X}_2)$. Then, for the mixed source $\mathbf{X} = \{X^n\}$ defined by (43), we have*

$$T_0(D, \varepsilon|\mathbf{X}) = R(D|\mathbf{X}_1), \quad 0 \leq \varepsilon < \alpha_1, \tag{44}$$

and

$$T_0(D, \varepsilon|\mathbf{X}) = R(D|\mathbf{X}_2), \quad \alpha_1 \leq \varepsilon < 1. \tag{45}$$

We omit the proof.

References

1. P.-N. Chen and F. Alajaji, *A rate-distortion theorem for arbitrary discrete sources, IEEE Trans. Inform. Theory* **44** (1998) 1666–1668.
2. T. S. Han, *An information-spectrum approach to source coding theorems with a fidelity criterion, IEEE Trans. Inform. Theory* **43** (1997) 1145–1164.
3. T. S. Han, *The reliability functions of the general source with fixed-length coding, IEEE Trans. Inform. Theory* **46** (2000) 2117–2132.

4. T. S. Han, *Information-Spectrum Methods in Information Theory* (Springer, 2003).
5. T. S. Han and S. Verdú, *Approximation theory of output statistics, IEEE Trans. Inform. Theory* **39** (1993) 752–772.
6. S. Ihara and M. Kubo, *Error exponent for coding of memoryless Gaussian sources with a fidelity criterion, IEICE Trans.* **E83-A** (2000) 1891–1897.
7. K. Iriyama, *Probability of error for the fixed-length source coding of the general sources, IEEE Trans. Inform. Theory* **47** (2001) 1537–1543.
8. K. Iriyama, *Probability of error for the fixed-length source coding of the general source* (in Japanese). Ph D thesis, Nagoya Univ (2001).
9. K. Iriyama, Probability of error for the fixed-length lossy coding of general sources, submitted.
10. K. Iriyama and S. Ihara, *The error exponent and minimum achievable rates for the fixed-length coding of general sources, IEICE Trans.* **E84-A** (2001) 2466–2473.
11. K. Marton, *Error exponent for source coding with a fidelity criterion, IEEE Trans. Inform. Theory,* **20** (1974) 197–199.
12. Y. Steinberg and S. Verdú, *Simulation of random processes and rate-distortion theory, IEEE Trans. Inform. Theory* **42** (1996) 63–86.
13. S. Verdú and T. S. Han, *General formula for channel capacity, IEEE Trans. Inform. Theory,* **40** (1994) 1147–1157.

ON A DESIGN OF TRANSITION PROBABILITIES AND ESTIMATES OF COVER TIMES

SATOSHI IKEDA

Department of Computer Science,
Tokyo University of Agriculture and Technology,
Naka-cho 2-24-16, Koganei, Tokyo, 184-8588, Japan.
E-mail : bisu@cc.tuat.ac.jp

IZUMI KUBO

Department of Environmental Design, Faculty of Environmental Studies,
Hiroshima Institute of Technology
2-1-1 Miyake, Saeki-ku, Hiroshima 731-5193, Japan
E-mail: kubo@cc.it-hiroshima.ac.jp

MASAFUMI YAMASHITA

Department of Computer Science and Communication Engineering,
Kyushu University,
Hakozaki, Higashi-ku, Fukuoka 812-8581, Japan.
E-mail: mak@csce.kyushu-u.ac.jp

The cover time of the standard random walk on a finite graph with order n is bounded by $2n^3$. When we obtain some local information about topological structure, we are interested in the problem whether cover times are reduced more by suitable design of transition probabilities. We will claim that one of nearly best transition probabilities is inversely proportional to square root of degree of destination vertex. Then, the cover time is extremely improved so that it is less than $3n^2 \log n$.

1. Introduction

Random walks on finite graphs are rich source of attractive researches both in applied mathematics and in computer science. The following is one of its applications. Suppose a computer network having a common resource. To access the resource without any collision, we need an appropriate algorithm. We expect that each computer can determine his strategy without the information of the whole system. For example, he may ask some information only to his adjacent computers.

Israeli and Jalfon's algorithm is that a node having a token may access to the resource, then he sends the token to one of adjacent computers randomly. When we think that the network is an undirected finite graph, token moves as a random walk on the graph.

The cover time has been discussed as a scale of the efficiency of algorithms. Here, the cover time means the expected time for all computers having a chance to access to the resource. In other words, it is the expected number of moves necessary for a random walk on a finite graph to visit all vertices.

We have proposed very good two schemes, which reduce cover times extremely [8,9]. In this paper, we mean by *a scheme* a universal way to determine transition probabilities for any graph G by using only local topological information.

Israeli-Jalfon proposed a scheme such that a computer having the token sends it to adjacent one with equi-probability, that is, the standard random walk on the graph. We call it *Israeli-Jalfon's scheme* and denote it by E. The corresponding cover time is denoted by $C(E,G)$. Blom et al. [3] surveyed works devoted to the cover time for the case (see also [1,4,5,6,12,13,15]). In particular, Aldous [1] showed, for any graph with order n and size m, an upper bound $C(E,G) \leq 4m(n-1) \leq 2n(n-1)^2$, and on the other hand a lower bound $\Omega(n^3)$ is obtained for a lollipop graph L_n, which is a complete graph of order $\lceil n/2 \rceil$ with a tail of length $\lfloor n/2 \rfloor$ (see Fig. 1). Here we use notations; $a_n = O(b_n)$ if $\limsup_{n\to\infty} |a_n|/|b_n| < \infty$, $a_n = \Omega(b_n)$ if $\liminf_{n\to\infty} |a_n|/|b_n| > 0$ and $a_n = \Theta(b_n)$ if both of $a_n = O(b_n)$ and $a_n = \Omega(b_n)$ hold.

The cover time is thus $C(E,G) = \Theta(n^3)$ in this sense, more precisely *the universal cover time* $C(E,n) \equiv \max\{C(E,G) : \text{order}(G) = n\}$ of the standard random walk satisfies $C(E,n) = \Theta(n^3)$.

In this paper, we will show that the bound $O(n^3)$ is reducible to $O(n^2 \log n)$, if the topological information such as degrees of adjacent vertices.

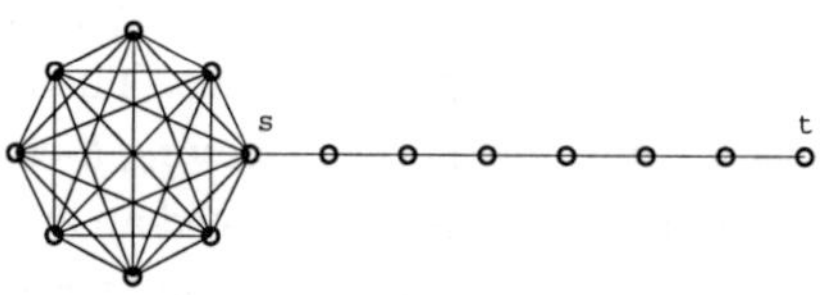

Figure 1. A lollipop graph L_{15}.

2. Preliminaries

Suppose that $G = (V, E)$ is a finite, undirected, simple connected graph. We denote by order(G) the order $|V|$ of G and by size(G) the size $|E|$. For $u \in V$, by $N(u) \equiv \Big\{v : \{u, v\} \in E\Big\}$ we denote the set of vertices adjacent to u. Note that $v \in N(u)$ if and only if $u \in N(v)$. Then $\deg(u) \equiv |N(u)|$ is called *the degree of* $u \in V$, and $\deg(G) \equiv \max_{v \in V} \deg(v)$ is *the degree of* G.

Let $P = \{p_{u,v}, u, v \in V\}$ be a transition matrix on G satisfying $p_{u,v} > 0$ if and only if $v \in N(u)$. Let $\{X_t = X_t(\omega) : t \geq 0\}$ be a Markov chain defined on a suitable probability space (Ω, μ_P) with the transition probability P and an initial distribution $q^{(0)} = \{q_v^{(0)} : v \in V\}$.

For $\omega \in \Omega$, put

$$\tau_v(\omega) \equiv \inf\{t \geq 1 : X_t(\omega) = v\},$$

$$\sigma_V(\omega) \equiv \inf\Big\{i \geq 1 : \{X_0(\omega), X_1(\omega), \cdots, X_i(\omega)\} = V\Big\},$$

which mean *the hitting time* and *the random cover time* for the Markov chain, respectively. *The mean hitting time* $H(P, G; u, v)$ from u to v with respect to P is defined by

$$H(P, G; u, v) \equiv E_{\mu_P}\left[\tau_v(\omega)\Big|X_0(\omega) = u\right]$$

and *the cover time* $C(P, G)$ by

$$C(P, G) \equiv \max_{u \in V} C(P, G, u), \quad C(P, G, u) \equiv \mathbf{E}_{\mu_P}\Big[\sigma_V(\omega)\big|X_0(\omega) = u\Big].$$

In the followings, we discuss some schemes to give transition probabilities to graphs systematically. For a scheme S, $P_S = P_{S,G}$ shows the transition probability for G determined by S. Further, we write $C(S, G) = C(P_{S,G}, G)$ and $H(S, G; u, v) = H(P_{S,G}, G; u, v)$.

Now, we introduce *the universal cover time* $C(S, n)$ *of the scheme* S by

$$C(S, n) \equiv \max\{C(P_{S,G}, G) : \text{order}(G) = n\}.$$

We now introduce a value $H(S, n)$ of the level of mean hitting times by

$$H(S, n) \equiv \max\{\max\{H(P_{S,G}, G; u, v) : u \neq v \in V\} \ : \text{order}(G) = n\}$$

We [8,9] introduced two classes of schemes inspired by Gibbs distributions in statistical mechanics. For a given potential function $U(u, v)$ a transition matrix is defined for $v \in N(u), u \in V$,

$$p_{u,v}^{U,\beta} \equiv \frac{\exp[-\beta U(u, v)]}{\sum_{w \in N(u)} \exp[-\beta U(own)]} \quad \text{for } v \in N(u),$$

and $p_{u,v}^{U,\beta} \equiv 0$ other wise. We discussed two cases

$$U_{\mathrm{I}}(u,v) \equiv \log \deg(v) \quad and \quad U_{\mathrm{II}}(u,v) \equiv \log(\max\{\deg(u), \deg(v)\}). \tag{2.1}$$

Our conclusions were that the order of the universal cover times are $O(n^2 \log n)$ for the both schemes with special values of β ($\beta = 1$ and $\beta = 1/2$, respectively).

The second scheme determined by degrees of neighbors, can be generalized as follows. Let $f(x)$ be a positive function on $\mathbb{N}$ and define a function $F(u)$ on V by $F(u) = F^f(u) \equiv f(\deg(u)),\ u \in V$. A transition matrix P^F is given by

$$p_{u,v}^F \equiv \begin{cases} \dfrac{F(v)}{\sum_{w \in N(u)} F(w)} & \text{if } v \in N(u), \\ 0 & otherwise, \end{cases}$$

that is, each transition probability $p_{u,v}^F$ is proportional to $F(v)$. We call the scheme determined by using F the scheme F^f or F, simply. The mean hitting time, the cover time and the universal cover times corresponding to P^F are denoted by $H(F,G;u,v)$, $C(F,G)$ and $C(F,n)$, respectively.

It is easy to find the unique stationary measure $\pi^F = \{\pi_u^F : u \in V\}$ of P^F:

$$\pi_u^F \equiv \frac{F(u) \sum_{w \in N(u)} F(w)}{\sum_{w \in V} F(w) \sum_{w' \in N(w)} F(w')}.$$

Actually, we can easily check the equality $\pi_v^F = \sum_{u \in V} \pi_u^F p_{u,v}^F$.

Generally, the mean hitting time $H(F,G;u,v)$ satisfies the equation

$$H(F,G;u,v) = \sum_{w \in V} p_{u,w}^F (1 + H(F,G;w,v)) - p_{u,v}^F H(F,G;v,v). \tag{2.2}$$

From this, we get

$$H(F,G;v,v) = \frac{1}{\pi_v^F} \quad \text{for all} \quad v \in V.$$

Appling the equality to Eq. (2.2) for $u = v$, we have

$$\sum_{w \in N(v)} p_{v,w}^F H(F,G;w,v) = \frac{1}{\pi_v^F} - 1,$$

which implies the equality

$$\sum_{w \in N(v)} F(v)F(w)(H(F,G;w,v) + 1) = Z(F,G) \tag{2.3}$$

where $Z(F,G)$ is *the partition function* defined by

$$Z(F,G) \equiv \sum_{w \in V} \sum_{w' \in N(w)} F(w)F(w').$$

A simple application of Eq.(2.3) gives an upper estimate of the mean hitting time

$$H(F,G;u,v) \le \frac{Z(F)}{F(u)F(v)} \quad \text{for} \quad v \in N(u). \tag{2.4}$$

3. Upper estimates

The following theorem and lemma [8,9] give a method to evaluate the cover time. For a given $G = (V,E)$ and a subset $V' \subset V$, we define $C(P,G,V')$ by

$$C(P,G,V') \equiv \max_{u \in V'} C(P,G,V',u), \quad C(P,G,V',u) \equiv E_P[\sigma_{V'}(\omega)|X_0(\omega) = u]$$

for $u \in V'$, where $\sigma_{V'}(\omega)$ is defined by

$$\sigma_{V'}(\omega) \equiv \inf\Big\{i \ge 1 | \ \{X_0(\omega), X_1(\omega), \cdots, X_{i-1}(\omega)\} = V'\Big\}.$$

$C(P,G,V')$ is called *the cover time for the subset V' of V*.

By using the lemma similarly to proofs in our papers [8,9], we can show:

Theorem 3.1. *For any $G = (V,E)$, $V' \subseteq V$ with $|V'| = n'$,*

$$H_{n'-1} \min_{u \ne v \in V'} H(P,G;u,v) \le C(P,G,V') \le H_{n'-1} \max_{u \ne v \in V'} H(P,G;u,v),$$

where H_n denotes the n-th harmonic number $H_n \equiv \sum_{j=1}^{n} \frac{1}{j}$.

Corollary 3.1. *For any scheme S,*

$$C(S,n) \le H(S,n) \log(\mathit{order}(G)).$$

Lemma 3.1. *For any $u, v \in V$, there exists a path $u = v_0, v_1, \cdots, v_\ell = v$ $(v_i \in N(v_i - 1),\ 1 \le i \le \ell)$ satisfying*

$$\Big(N(v_i) \cup \{v_i\}\Big) \bigcap \Big(N(v_j) \cup \{v_j\}\Big) = \emptyset \tag{3.1}$$

for $1 \le i < i+2 < j \le l$. For the shortest path satisfying condition (3.1), we have

$$\sum_{i=0}^{\ell} \deg(v_i) \le 3n - 4.$$

Theorem 3.2. *(i) Put*

$$D_{F,1}(n) \equiv \max_{order(G)=n} \max \left\{ \frac{Z(F,G)}{F(u)F(v)} : v \in N(u),\ u \in V \right\}.$$

Then,

$$H(F,n) \leq nD_{F,1}(n) \quad and \quad C(F,n) \leq 2nD_{F,1}(n).$$

(ii) Put

$$D_{F,2}(n) \equiv \max_{order(G)=n} \max \left\{ \frac{Z(F,G)}{F(u)^2 \deg(u)} : u \in V \right\}.$$

Then,

$$H(F,n) \leq 3nD_{F,2}(n) \quad and \quad C(F,n) \leq 3n \log n D_{F,2}(n).$$

Proof. Let $G = (V,E)$ be a graph with $\mathrm{order}(G) = n$. For any pair $u, v \in V,\ u \neq v$, there exists a path $\{v_i : 0 \leq i \leq \ell\}$ such that $v_0 = u,\ v_\ell = v, v_{i-1} \in N(v_i),\ 1 \leq i \leq \ell\ (\leq n-1)$. Then

$$H(F,G;u,v) \leq \sum_{i=1}^{\ell} H(F,G;v_{i-1},v_i) \leq nD_{F,1}(n)$$

by Eq.(2.4). By Theorem 3.1, we see (i). Next, let $\{v_i : 0 \leq i \leq \ell\}$ be a path in Lemma 3.1. Then we have the enstimation

$$\begin{aligned} H(F,G;u,v) &\leq \sum_{i=1}^{\ell} H(F,G;v_{i-1},v_i) \\ &\leq \sum_{i=1}^{\ell} \frac{Z(F,G)}{F(v_{i-1})F(v_i)} \leq \sum_{i=1}^{\ell} \frac{1}{2} \left(\frac{Z(F,G)}{F(v_{i-1})^2} + \frac{Z(F,G)}{F(v_i)^2} \right) \\ &\leq \sum_{i=1}^{\ell} \frac{\deg(v_{i-1}) + \deg(v_i)}{2} D_{F,2}(n) \leq (3n-4)D_{F,2}(n). \end{aligned}$$ □

We now give a simple application of this theorem. Let $F(u) = \deg(u)^{-1/2}$. In this case, we denote as

$$p^I_{u,v} \equiv p^F_{u,v} = \begin{cases} \dfrac{\deg(v)^{-1/2}}{\deg(w)^{-1/2} \displaystyle\sum_{w \in N(u)} \deg(w)^{-1/2}} & \text{if } v \in N(u), \\ 0 & otherwise. \end{cases}$$

$$\pi_u^I \equiv \pi_u^F = \frac{\deg(u)^{-1/2} \sum_{w \in N(u)} \deg(w)^{-1/2}}{\sum_{w \in V} \deg(w)^{-1/2} \sum_{w' \in N(w)} \deg(w')^{-1/2}}$$

Then, we have

Theorem 3.3. *Suppose that the scheme I is given by $F(u) = \deg(u)^{-1/2}$. Then,*

$$H(I, n) \leq 3n^2 \quad C(I, n) = 3n^2 \log n.$$

Proof. Let us check that $Z(I, G) = Z(F, G)$ for $F(u) = \deg(u)^{-1/2}$.

$$\begin{aligned} Z(I, G) &= \sum_{u \in V, v \in N(u)} \deg(u)^{-1/2} \deg(v)^{-1/2} \\ &\leq \frac{1}{2} \sum_{u \in V, v \in N(u)} \left(\deg(u)^{-1} + \deg(v)^{-1}\right) \\ &= \sum_{u \in V, v \in N(u)} \deg(u)^{-1} = |V| = \text{order}(G) \end{aligned}$$

Therefore $D_{F,2}(n) \leq n$ holds in Theorem 3.2 (ii), since $F(u)^2 \deg(u) = 1$.□

4. Lower estimates

Let us observe a path graph P_n with order n as in Fig. 2.

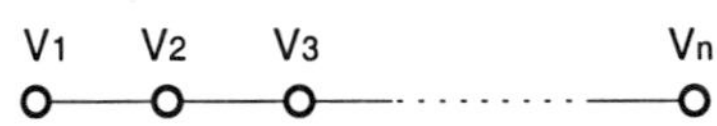

Figure 2. A path graph P_n with order n.

We have shown [9] that the lower bound of path graphs is given by

$$H(S, P_n; u, v) \geq \frac{1}{2}(n^2 - 1) \quad \text{and} \quad C(S, \mathrm{P}_n) \geq \frac{1}{2}(n^2 - 1).$$

for any scheme S. Hence we can claim

Theorem 4.1. *For any scheme S, the following lower estimates hold:*

$$H(S, n) \geq \frac{1}{2}(n^2 - 1) \quad \textit{and} \quad C(S, n) \geq \frac{1}{2}(n^2 - 1).$$

The theorem tells us that we can not reduce more the upper bound given in Theorem in the level of hitting times. Still we have a gap of order of $\log n$ in the level of the cover time. However, it is important to notice that the gap is not so big in practical sense.

As we stated already, the lower bound for Israeli-Jalfon's scheme E is $\Theta(n^3)$, whose critical case is given by lollipop L_n. We have proposed the other scheme S_{II} given by the potential function $U_{II}(u,v) = \log(\max\{\deg(u), \deg(v)\})$ in Eq.(2.1). For both schemes' S_I and S_{II}, $C(S_I, \mathrm{L}_n) = \Theta(n^2)$ and $C(S_{II}, \mathrm{L}_n) = \Theta(n^2)$ hold.

Two critical cases for the scheme S_{II}, are given by a star graph and a cobweb graph as in Fig.3. By applying Theorem 3.1, we can show that $C(S_{II}, \mathrm{Star}_n) = \Theta(n^2 \log n)$ and $C(S_{II}, \mathrm{Cobweb}_n) = \Theta(n^2 \log n)$.

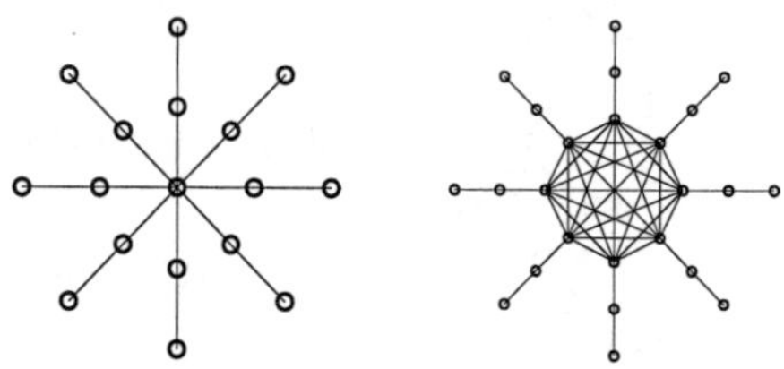

Figure 3. A star graph (left) and a cobweb graph (right).

For scheme S_I, we have $C(S_I, \mathrm{Star}_n) = \Omega(n^{3/2} \log n)$ and $C(S_{II}, \mathrm{Cobweb}_n) = \Omega(n^{3/2} \log n)$. Hence those graphs are not critical for scheme S_I. Actually, we have no example such that

$$C(S_I, G, n) = \Omega(n^2 \log n).$$

Our impression is:

Conjecture. The following universal estimate holds;

$$C(S_I, n) = \Omega(n^2).$$

5. Remarks

Upper estimates of hitting times of the cobweb graph for scheme I is obtained by the exact calculation not hardly. However it is also possible to obatain similar results by a simpler way using Eq. (2.3).

Let us examine the way for the cobweb graph Cobweb_n of order $n = 3r$ and the scheme given by $F(u) = f(\deg(u))$. Here, we denote the hitting times by $H(u,v)$, simply.

There are three kinds of vertices $V_i \equiv \{v \in V : \deg(v) = i\},\ i = 1, 2, r$. Since $|V_i| = r,\ i = 1, 2, r$,

$$Z = Z(f, \text{Cobweb}_n) = 2rf(1)f(2) + 2rf(2)f(r) + r(r-1)f(r)^2.$$

For $v_1, v_1' \in V_1,\ v_1 \neq v_1'$, there exist $v_2, v_2' \in V_2$ and $v_r, v_r' \in V_r$ such that

$$H(v_1, v_1') = H(v_2, v_r) + H(v_r, v_r') + H(v_r', v_2') + H(v_2', v_1'). \tag{5.1}$$

By Eq.(2.3), we have

$$H(v_1, v_1') \leq \frac{Z}{f(1)f(2)} + \frac{Z}{f(2)f(r)} + \frac{Z}{f(r)\min\{f(2), (r-1)f(r)\}}.$$

If $\frac{1}{r-1}f(2) \leq f(r) \leq f(2)$ holds,

$$H(v_1, v_1') \leq 3r\left(\frac{2f(1)f(2) + 3(r-1)f(r)^2}{\min\{f(1), f(2)\}f(r)}\right). \tag{5.2}$$

It is easily seen that the right hand side of Eq.(5.2) takes minimum value when $f(r) = \Theta(r^{-1/2})$. Applying Theorem 3.1 ane Eq.(5.2) to $f(x) = x^{-1/2}$, we have $C(I, \text{Cobweb}_n) = O(n^{3/2})$.

The lower estimate of $C(f, \text{Cobweb}_n)$ can be obtained similarly as follows. Again by Eq.(2.3) and Eq.(5.1),

$$H(v_1, v_1') \geq \frac{Z}{f(2)f(r)} - 4 \geq \frac{2f(1)r}{f(r)} - 4.$$

Again by Theorem 3.1, we have $C(I, \text{Cobweb}_n) = \Omega(n^{3/2}\log n)$.

Next, we try the same argument for the path graph P_n. Number to vertices in V as Fig.2. Then, $V_1 = \{v_1, v_n\}$ and $V_2 = \{v_i : 2 \leq i \leq n-1\}$ and $Z = 4f(1)f(2) + 2(n-3)f(2)^2$ hold. Therefore, we have

$$H(v_1, v_n) + H(v_n, v_1) = \sum_{i=1}^{n-1}(H(v_i, v_i + 1) + H(v_{i+1}, v_i))$$

$$= H(v_{n-}, v_n) + H(v_2, v_1) + \sum_{i=1}^{n-2}(H(v_i, v_{i+1}) + H(v_{i+2}, v_{i+1}))$$

and hence by Eq.(2.3) we have

$$C(I, Path_n) \leq \frac{2Z}{f(1)f(2)} + \frac{2Z}{f(2)\min\{f(1), f(2)\}} + \frac{(n-3)Z}{f(2)^2} - n = O(n^2).$$

Similarly, we see that

$$C(I, Path_n) \geq \frac{Z}{f(1)f(2)} + \frac{Z}{f(2)\max\{f(1), f(2)\}} + \frac{(n-3)Z}{2f(2)^2} - n = \Omega(n^2).$$

This kind of discussion can be performed for graphs with high symmetry. Actually, we can get upper and lower bounds for all examples shown here, similarly. Such observation leads us to the idea that the scheme I is an essentially unique best choice among schemes given by $F(u) = f(\deg(u))$.

References

1. D.J. Aldous, "On the time taken by random walks on finite groups to visit every state", *Z. Wahrsch. verw. Gebiete* 62 361-393, 1983.
2. R. Aleliunas, R.M Karp, R.J. Lipton, L. Lovaász, and C. Rackoff, "Random walks, universal traversal sequences, and the complexity of maze problems", *Proc. 20th Ann. Symposium on Foundations of Computer Science*, 218–223, 1979.
3. G. Blom, L. Holst, and D. Sandell, "Problems and Snapshots from the World of Probability", Springer-Verlag, New York, NY, 1994.
4. G. Brightwell and P. Winkler, "Maximum hitting time for random walks on graphs", *J. Random Structures and Algorithms*, 3, 263-276, 1990.
5. A.Z. Broder and Karlin, "Bounds on covering times", *In 29th Annual Symposium on Foundations of Computer science*, 479-487, 1988.
6. D. Coppersmith, P. Tetali, and P. Winkler, "Collisions among random walks on a graph", *SIAM Journal on Discrete Mathematics*, 6, 3, 363–374, 1993.
7. S. Ikeda, I. Kubo, N. Okumoto and M. Yamashita, "Fair circulation of a token," *IEEE Trans. Parallel and Distributed Systems,* Vol.13, No.4, 367-372, 2002.
8. S. Ikeda, I. Kubo, N. Okumoto and M. Yamashita, "Reducing the hitting time and the cover times of random walks on finite graphs by local topological information," *Proceedings of The 2003 International Conference on VLSI,*, 2003.
9. S. Ikeda, I. Kubo, N. Okumoto and M. Yamashita, "Impact of local topological information on random walks on finite graphs," *Proceedings of Thirtieth International Colloquium on Automata, Languages and Programming,* 2003.
10. L. Isaacson and W. Madsen, "Markov chains: Theory and Application", Wiley series in probability and mathematical statistics, New York, 1976.
11. A. Israeli and M. Jalfon, "Token management schemes and random walks yield self stabilizing mutual exclusion", *Proc. of the 9th ACM Symposium on Principles of Distributed Computing*, 119–131, 1990.
12. P. Matthews, "Covering Problems for Markov Chain", *The Annals of Probability* Vol.16, No.3, 1215-1228, 1988.
13. R. Motowani and P. Raghavan, "Randomized Algorithms", Cambridge University Press, New York, 1995.
14. N. Okumoto, " A study on random walks of tokens on graphs ", M.E.Thesis, Hiroshima Univ., Higashi-Hiroshima, Japan, 1996.
15. J.L. Palacios, "On a result of Aleiliunas et al. concerning random walk on graphs," *Probability in the Engineering and Informational Sciences,* 4, 489–492, 1990.

RECENT PROGRESS ON THE WHITE NOISE APPROACH TO THE LÉVY LAPLACIAN

HUI-HSIUNG KUO

Department of Mathematics
Louisiana State University
Baton Rouge, LA 70808, USA
E-mail: kuo@math.lsu.edu

Let ϕ be a function defined on $L^2(0,1)$. In 1922 P. Lévy introduced the Laplacian $\Delta_L \phi$ of ϕ for harmonic analysis on the space $L^2(0,1)$ by

$$\Delta\phi(x) = \lim_{N\to\infty} \frac{1}{N} \sum_{n=1}^{N} \langle \phi''(x)e_n, e_n \rangle,$$

where $\{e_n; n \geq 1\}$ is an orthonormal basis for $L^2(0,1)$. This Laplacian, called the Lévy Laplacian, has no finite dimensional analogue. It is intrinsically infinite dimensional and possesses peculiar properties. In 1975 T. Hida introduced white noise theory which can be used to study Lévy's functional analysis. In this paper we study recent progress on some properties of the Lévy Laplacian from the white noise viewpoint, e.g., its relationships to the Lévy group, the spherical mean, Wiener processes, Gross Laplacian, and the Fourier transform. More significantly, we discuss K. Saitô's novel idea on the diagonalization of the Lévy Laplacian operator to construct a domain of this operator and to find semigroups and stochastic processes generated by the Lévy Laplacian. We present some of recent results due to Saitô and his collaborators.

1. Introduction

Let H be an infinite dimensional real separable Hilbert space. Suppose a real-valued function f defined on H is twice Fréchet differentiable such that $f''(x)$ is a trace class operator of H. Gross[30] introduced in 1967 the Laplacian $\Delta_G f$ of f defined by

$$\Delta_G f(x) = \text{trace}_H \, f''(x).$$

Thus for any orthonormal basis $\{e_n; n \geq 1\}$ for H, we have

$$\Delta_G f(x) = \sum_{n=1}^{\infty} \langle f''(x)e_n, e_n \rangle. \tag{1}$$

Obviously, this Laplacian operator Δ_G is an infinite dimensional analogue of the finite dimensional Laplacian operator. However, the topology of H is too strong for this Laplacian operator. For example, consider the function $f(x) = |x|^2$, where $|\cdot|$ is the norm on H . The second Fréchet derivative of f is given by $f''(x) = 2I$ with I being the identity operator of H. Hence $\Delta_G f$ is not defined. This was one of the motivations for Gross[29] to introduce the concept of abstract Wiener space in 1965. For example, let K be a Hilbert-Schmidt operator of H and consider the function

$$g(x) = |Kx|^2, \quad x \in H. \tag{2}$$

Then $g''(x) = 2K^*K$ and so $\Delta_G g$ is given by

$$\Delta_G g(x) = 2\,\mathrm{trace}_H\, K^*K.$$

Note that $\Delta_G g(x)$ being defined is reflected by the fact that $|Kx|$ is a measurable semi-norm, see Gross[29] and Kuo[58]. Let $\|\cdot\|$ be a measurable norm on H and let B be the completion of H with respect to $\|\cdot\|$. The pair (H, B) is called an *abstract Wiener space.* Let μ be the standard Gaussian measure on B and define

$$p_t(x, A) = \mu\Big(\frac{A - x}{\sqrt{t}}\Big). \tag{3}$$

Then for any bounded Lipschitz function f on B, the function $u(t, x) = \int_B f(y)\, p_t(x, dy)$ satisfies the heat equation:

$$\frac{\partial u}{\partial t} = \frac{1}{2}\Delta_G u, \quad u(0, x) = f(x). \tag{4}$$

Thus $p_t(x, \cdot)$ is a fundamental solution of the heat equation on the abstract Wiener space (H, B). Moreover, the measures $\{p_t(x, \cdot);\, t > 0, x \in B\}$ are the transition probabilities of a Wiener process on B.

On the other hand, let H be the real Hilbert space $L^2(0, 1)$. Take $h \in L^1(0, 1)$. The function

$$\phi(x) = \int_0^1 h(t)x(t)^2\, dt, \quad x \in L^2(0, 1), \tag{5}$$

can be regarded as the infinite dimensional analogue of the function $f(x_1, \ldots, x_n) = \sum_{k=1}^n a_k x_k^2$ on $\mathbb{R}^n$. Since $\Delta f(x_1, \ldots, x_n) = 2\sum_{k=1}^n a_k$, the corresponding infinite dimensional Laplacian of φ is given by

$$\Delta_L \phi(x) = 2\int_0^1 h(t)\, dt. \tag{6}$$

Obviously, this Laplacian is different from the Gross Laplacian. In fact, we have the second derivative of ϕ given by

$$\langle \phi''(x)\xi, \eta \rangle = \int_0^1 h(t)\xi(t)\eta(t)\, dt, \quad \xi, \eta \in L^2(0,1).$$

If an orthonormal basis $\{e_n; n \geq 1\}$ for $L^2(0,1)$ is uniformly bounded and equally dense (see the next section for definition,) then it is a fact that

$$\Delta_L \phi(x) = \lim_{N\to\infty} \frac{1}{N} \sum_{n=1}^{N} \langle \phi''(x)e_n, e_n \rangle. \tag{7}$$

The Laplacian as defined in Eq. (7) was introduced by Lévy[72] in 1922. Observe that if the Gross Laplacian $\Delta_G\phi(x)$ exists, then the Lévy Laplacian $\Delta_L\phi(x)$ vanishes. Moreover, let $h(\cdot) = 1$, then the function $\phi(x)$ in Eq. (5) is given by $\phi(x) = |x|^2$. In this case, the Gross Laplacian $\Delta_G\phi(x)$ does not exist, while we have the Lévy Laplacian $\Delta_L\phi(x) = 2$.

Recall that the expression $\Delta_L\phi(x)$ in Eq. (6) is motivated by the finite dimensional function $f(x_1, \ldots, x_n) = \sum_{k=1}^n a_k x_k^2$ on $\mathbb{R}^n$, which has the Laplacian $\Delta f(x_1, \ldots, x_n) = 2\sum_{k=1}^n a_k$. However, the Lévy Laplacian as defined by Eq. (7) has no finite dimensional analogue. In this sense, the Lévy Laplacian is an intrinsically infinite dimensional object.

In 1975 Hida[31] introduced the white noise theory. One of the main motivations for this theory was to interpret Lévy's functional analysis by using the white noise language. Within this theory, we can clearly see that the Lévy Laplacian vanishes on ordinary white noise functions and is a rather interesting operator acting on generalized white noise functions.

Another approach to study the Lévy Laplacian has been carried out by M. N. Feller[12−28] in a series of papers. In particular, he has studied the harmonic analysis associated with the Lévy Laplacian. See also the book by Polishchuk[88].

The Lévy Laplacian has been discovered to be closely related to many mathematical objects, e.g., the infinite dimensional rotation group in the work of Hida[41,42] and Obata[78,82], and Yang-Mills equations in the papers by Accardi[1] and his collaborators[2−8] and by Léandre and Volovich[71]. The differential equations involving the Lévy Laplacian have been studied by M. N. Feller as mentioned above, by Chung-Ji-Saitô[10], and by Lávička[70]. The semigroups generated by the Lévy Laplacian and associated stochastic processes have been constructed and studied by Saitô[91,93,94,98−104] and his collaborators[11,69,68,76,77,105−107].

In this paper we will briefly explain the white noise approach to the Lévy Laplacian in Section 2. Some important results will be reviewed in Section 3. In Section 4 we will make some observations which lead to the novel discovery of Saitô on the diagonalization and domains of the Lévy Laplacian in Section 5. Finally we will discuss the semigroups and stochastic processes associated with the Lévy Laplacian in Section 6.

2. White Noise Approach

Let $\mathcal{S}$ denote the Schwartz space of real-valued rapidly decreasing functions on $\mathbb{R}$. The space $\mathcal{S}$ is a nuclear space with the family $\{|\cdot|_p;\, p \geq 0\}$ of norms defined by

$$|f|_p = |A^p f|_0, \quad f \in \mathcal{S}, \tag{8}$$

where $A = -d^2/dx^2 + x^2 + 1$ and $|\cdot|_0$ is the $L^2(\mathbb{R})$-norm. Thus we have a Gel'fand triple

$$\mathcal{S} \subset L^2(\mathbb{R}) \subset \mathcal{S}', \tag{9}$$

where $\mathcal{S}'$ is the dual space of $\mathcal{S}$. Let μ be the standard Gaussian measure on $\mathcal{S}'$. The probability space $(\mathcal{S}', \mu)$ is called a *white noise space.*

Let (L^2) denote the complex Hilbert space $L^2(\mathcal{S}', \mu)$. By the Wiener-Itô theorem, each $\phi \in (L^2)$ has a unique decomposition

$$\phi(x) = \sum_{n=0}^{\infty} I_n(f_n)(x) = \sum_{n=0}^{\infty} \langle :x^{\otimes n}:, f_n \rangle, \quad f_n \in \widehat{L}^2(\mathbb{R}^n), \tag{10}$$

where I_n is the multiple Wiener integral of order n and $:x^{\otimes n}:$ is the n-th order Wick tensor of $x \in \mathcal{S}'$ (see page 33 in the book by Kuo[64].) Moreover, the (L^2)-norm of ϕ is given by

$$\|\phi\|_0 = \left(\sum_{n=0}^{\infty} n!\, |f_n|^2_{L^2(\mathbb{R}^n)} \right)^{1/2}. \tag{11}$$

Note that the Hilbert space (L^2) is unitarily equivalent to the Fock space $\Gamma\big(L^2(\mathbb{R})\big)$ of $L^2(\mathbb{R})$ by the correspondence

$$(f_0, f_1, \ldots, f_n, \ldots) \longleftrightarrow \sum_{n=0}^{\infty} I_n(f_n).$$

For each $p \geq 0$ and ϕ as given in Eq. (10), define $\|\phi\|_p$ by

$$\|\phi\|_p = \left(\sum_{n=0}^{\infty} n!\, |(A^p)^{\otimes n} f_n|^2_{L^2(\mathbb{R}^n)} \right)^{1/2}$$

Note that $\|\phi\|_p = \|\Gamma(A^p)\phi\|_0$ with $\Gamma(A^p)$ being the second quantization of the operator A^p. Let $(\mathcal{S})_p = \{\phi;\ \|\phi\|_p < \infty\}$. Then $(\mathcal{S})_p$ is a complex Hilbert space with norm $\|\cdot\|_p$.

Let $(\mathcal{S}) = \cap_{p\geq 0}(\mathcal{S})_p$ with topology given by the family $\{\|\cdot\|_p;\ p \geq 0\}$ of norms. It can be checked that $(\mathcal{S})$ is a nuclear space and $(\mathcal{S}) \subset (L^2)$. Let $(\mathcal{S})^*$ be the dual space of $(\mathcal{S})$. Then we have a Gel'fand triple

$$(\mathcal{S}) \subset (L^2) \subset (\mathcal{S})^*. \tag{12}$$

Here the identification of (L^2) with its dual space is given by the conjugate, i.e., the bilinear pairing $\langle\langle\cdot,\cdot\rangle\rangle$ of $(\mathcal{S})^*$ and $(\mathcal{S})$ and the inner product $(\cdot,\cdot)$ of (L^2) are related by the equality

$$\langle\langle\phi,\psi\rangle\rangle = (\phi,\bar{\psi}), \quad \phi \in (L^2),\ \psi \in (\mathcal{S}).$$

It turns out that $(\mathcal{S})^* = \cup_{p\geq 0}(\mathcal{S})_{-p}$. Here $(\mathcal{S})_{-p}$ being the completion of (L^2) with respect to the norm

$$\|\phi\|_{-p} = \left(\sum_{n=0}^{\infty} n!\,|(A^{-p})^{\otimes n} f_n|^2_{L^2(\mathbb{R}^n)}\right)^{1/2}. \tag{13}$$

The nuclear space $(\mathcal{S})$ serves as the space of test functions on the white noise space $\mathcal{S}'$. Its dual space $(\mathcal{S})^*$ is the corresponding space of generalized functions on $\mathcal{S}'$.

Let $\mathcal{S}_c$ denote the complexification of $\mathcal{S}$. For each $\xi \in \mathcal{S}_c$, we define the renormalization of $e^{\langle\cdot,\xi\rangle}$ to be the function $:e^{\langle\cdot,\xi\rangle}: = e^{\langle\cdot,\xi\rangle - \frac{1}{2}\langle\xi,\xi\rangle}$. The function $:e^{\langle\cdot,\xi\rangle}:$ is a test function and has the Wiener-Itô decomposition

$$:e^{\langle x,\xi\rangle}: = \sum_{n=0}^{\infty} \frac{1}{n!}\langle :x^{\otimes n}:,\ \xi^{\otimes n}\rangle, \quad x \in \mathcal{S}'. \tag{14}$$

A generalized function Φ in $(\mathcal{S})^*$ can be represented by

$$\Phi = \sum_{n=0}^{\infty}\langle :\cdot^{\otimes n}:,\ F_n\rangle, \quad F_n \in \mathcal{S}'(\mathbb{R}^n), \tag{15}$$

and there exists a positive number p depending on Φ such that $\|\Phi\|_{-p}$ as defined in Eq. (13) is a finite number.

An important tool in white noise theory is the S-transform introduced by Kubo and Takenaka[53−56]. The *S-transform* of a generalized function Φ in $(\mathcal{S})^*$ is defined to be the following function on $\mathcal{S}_c$,

$$S\Phi(\xi) = \langle\langle\Phi,\ :e^{\langle\cdot,\xi\rangle}:\rangle\rangle = e^{-\langle\xi,\xi\rangle/2}\langle\langle\Phi,\ e^{\langle\cdot,\xi\rangle}\rangle\rangle, \quad \xi \in \mathcal{S}_c.$$

If Φ is given by Eq. (15), then in view of Eq. (14) we have

$$S\Phi(\xi) = \sum_{n=0}^{\infty} n!\Big\langle F_n, \frac{1}{n!}\xi^{\otimes n}\Big\rangle = \sum_{n=0}^{\infty}\langle F_n, \xi^{\otimes n}\rangle, \quad \xi \in \mathcal{S}_c. \tag{16}$$

The S-transform uniquely specifies a generalized function, i.e., $S\Phi = S\Psi$ implies that $\Phi = \Psi$. The next fundamental theorem due to Potthoff and Streit[89] gives a characterization of generalized functions in the space $(\mathcal{S})^*$ in terms of their S-transforms. This theorem has been extended to other spaces of generalized functions. See the survey paper by Kuo[66] and references therein.

Theorem 2.1. *A complex-valued function F on $\mathcal{S}_c$ is the S-transform of a generalized function in the space $(\mathcal{S})^*$ if and only if it satisfies the following two conditions:*

(a) For any $\xi, \eta \in \mathcal{S}_c$, the function $F(\xi + z\eta)$ is an entire function of $z \in \mathbb{C}$.

(b) There exist constants $K, a, p \geq 0$ such that

$$|F(\xi)| \leq K \exp\left[a|\xi|_p^2\right], \quad \forall\, \xi \in \mathcal{S}_c.$$

The stochastic process $B(t,x) = \langle x, 1_{[0,t)}\rangle$ for $x \in \mathcal{S}'$ and $t \geq 0$ and $= -\langle x, 1_{[t,0)}\rangle$ for $x \in \mathcal{S}'$ and $t < 0$ is a Brownian motion. Thus an element x in the white noise space $(\mathcal{S}', \mu)$ can be regarded as $\dot{B}$. Moreover, for each $t \in \mathbb{R}$, $\dot{B}(t)$ is a generalized function in the space $(\mathcal{S})^*$ and its S-transform is given by

$$S(\dot{B}(t))(\xi) = \xi(t), \quad \xi \in \mathcal{S}_c.$$

In white noise theory the family $\{\dot{B}(t);\, t \in \mathbb{R}\}$ is taken as a continuum system of coordinates. The renormalization $:\dot{B}(t)^2:$ of $\dot{B}(t)^2$, defined as the limit of $\big(\epsilon^{-1}(B(t+\epsilon) - B(t))\big)^2 - \epsilon^{-1}$ in the space $(\mathcal{S})^*$ as $\epsilon \to 0$, is a generalized function and has the S-transform $S(:\dot{B}(t)^2:)(\xi) = \xi(t)^2, \xi \in \mathcal{S}_c$. For any $h \in L^1(\mathbb{R})$, the function

$$\Phi = \int_{\mathbb{R}} h(t) :\dot{B}(t)^2:\, dt$$

is a generalized function with the S-transform

$$S(\Phi)(\xi) = \int_{\mathbb{R}} h(t)\xi(t)^2\, dt,$$

which is the white noise setup of the function in Eq. (5). On the other hand, the white noise setup of the function in Eq. (2) is given by

$$g(\xi) = \int_{\mathbb{R}^2} k(s,t)\xi(s)\xi(t)\, dsdt, \quad \xi \in \mathcal{S}_c,$$

where $k \in L^2(\mathbb{R}^2)$. This function g is the S-transform of the function

$$\phi = \int_{\mathbb{R}^2} k(s,t)\, dB(s)dB(t) = \int_{\mathbb{R}^2} k(s,t) :\dot{B}(s)\dot{B}(t):\, dsdt,$$

which belongs to the space (L^2) of ordinary Brownian functions.

Now, let $F = S\Phi$ be the S-transform of a generalized function $\Phi \in (\mathcal{S})^*$. By condition (b) in Theorem 2.1 F has functional derivatives of all order. Suppose the second functional derivative is given by

$$\langle F''(\xi)\eta, \zeta\rangle = \int_{\mathbb{R}} F_s''(\xi;t)\eta(t)\zeta(t)\, dt + \int_{\mathbb{R}^2} F_r''(\xi;s,t)\eta(s)\zeta(t)\, dsdt,$$

where $F_s''(\xi;\cdot) \in L^1(\mathbb{R})$ and $F_r''(\xi;\cdot,\cdot) \in L^1(\mathbb{R}^2)$. The functions F_s'' and F_r'' are the *singular* and *regular parts* of F'', respectively.

Let T be a finite interval on $\mathbb{R}$. If the function $\int_T F_s''(\xi;t)\, dt$, $\xi \in \mathcal{S}_c$, satisfies the conditions in Theorem 2.1, then we define the Lévy Laplacians of F and Φ by

$$\Delta_L F(\xi) = \frac{1}{|T|}\int_T F_s''(\xi;t)\, dt, \quad \xi \in \mathcal{S}_c, \tag{17}$$

$$\Delta_L \Phi = S^{-1}\Big(\frac{1}{|T|}\int_T F_s''(\cdot;t)\, dt\Big).$$

Furthermore, if the regular part $F_r''(\xi;s,t)$ is the kernel function of a trace class operator $F_r''(\xi)$ of $L^2(\mathbb{R})$ such that the function $\xi \mapsto \mathrm{trace}_{L^2(\mathbb{R})} F_r''(\xi)$ satisfies the conditions in Theorem 2.1, then we define the Volterra Laplacians of F and Φ by

$$\Delta_V F(\xi) = \mathrm{trace}_{L^2(\mathbb{R})} F_r''(\xi) \quad \xi \in \mathcal{S}_c,$$

$$\Delta_V \Phi = S^{-1}\Big(\mathrm{trace}_{L^2(\mathbb{R})} F_r''(\cdot)\Big).$$

Note that the Volterra Laplacian is an extension of the Gross Laplacian to generalized functions. In this paper we will not discuss the Volterra Laplacian and simply refer to the book by Kuo[64] for more information.

The definition of the Lévy Laplacian in Eq. (17) is related to the one given in Eq. (7) as follows. Suppose an orthonormal basis $\{e_n; n \geq 1\}$ for

$L^2(T)$ is equally dense, i.e.,

$$\frac{1}{N}\sum_{n=1}^{N} e_n(t)^2 \longrightarrow \frac{1}{|T|} \quad \text{in } L^2(T),$$

and uniformly bounded, i.e., $\sup_{n\geq 1}\sup_{t\in T}|e_n(t)| < \infty$. Then for any function F satisfying the above conditions we have

$$\Delta_L F(\xi) = \frac{1}{|T|}\int_T F_s''(\xi;t)\,dt = \lim_{N\to\infty}\frac{1}{N}\sum_{n=1}^{N}\langle F''(x)e_n, e_n\rangle, \qquad \xi \in \mathcal{S}_c.$$

For the proof of this fact, see page 212 in the book by Kuo[64].

3. Properties of the Lévy Laplacian

In this section we will explain some important properties concerning the Lévy Laplacian. In particular, we will state several theorems which show that the Lévy Laplacian is closely related to other objects.

3.1. *The Lévy group*

Let $\mathbb{N}$ denote the set of natural numbers. A permutation σ of $\mathbb{N}$ is called a *Lévy permutation* if it satisfies the condition:

$$\lim_{N\to\infty}\frac{1}{N}\Big|\{n \leq N;\ \sigma(n) > N\}\Big| = 0,$$

where $|A|$ is the cardinality of A. Note that for any two Lévy permutations σ_1 and σ_2, we have

$$\begin{aligned}\Big|\{n \leq N;\ \sigma_1\sigma_2(n) > N\}\Big| \leq \Big|\{n \leq N;\ \sigma_1(n) > N\}\Big| \\ + \Big|\{n \leq N;\ \sigma_2(n) > N\}\Big|.\end{aligned}$$

Thus the set $\mathcal{P}_L(\mathbb{N})$ of all Lévy permutations forms a group, which we call the *Lévy permutation group*. Examples of non-Lévy permutations are easy to construct. For example, consider the following disjoint partition of $\mathbb{N}$,

$$\mathbb{N} = A_1 \cup B_1 \cup A_2 \cup B_2 \cup \cdots \cup A_n \cup B_n \cup \cdots,$$

where A_n's and B_n's are ordered sets given by

$$\begin{aligned} A_1 &= \{1,2\}, & B_1 &= \{3,4\}, \\ A_2 &= \{5,6,7,8\}, & B_2 &= \{9,10,11,12\}, \\ A_3 &= \{13,14,\ldots,24\}, & B_3 &= \{25,26,\ldots,36\}, \\ &\cdots\cdots, & &\cdots\cdots. \end{aligned}$$

Here $|A_n| = |B_n|$ for all $n \geq 1$ and $|A_n| = 2\big(|A_1| + |A_2| + \cdots + |A_{n-1}|\big)$ for $n \geq 2$. Thus the cardinality of A_n is given by

$$|A_1| = 2, \quad |A_n| = 4 \cdot 3^{n-2}, \quad n \geq 2.$$

Let σ be the permutation taking A_n to B_n and B_n to A_n for each n while the order is kept. Then obviously we have

$$\limsup_{N\to\infty} \frac{1}{N}\Big|\{n \leq N;\, \sigma(n) > N\}\Big| = \frac{1}{2}.$$

Thus σ is not a Lévy permutation.

Recall the Gel'fand triple $\mathcal{S} \subset L^2(\mathbb{R}) \subset \mathcal{S}'$ in Eq. (9). Let $\mathcal{O}\big(\mathcal{S}, L^2(\mathbb{R})\big)$ denote the collection of all unitary operators U of $L^2(\mathbb{R})$ such that the restriction of U to $\mathcal{S}$ is a homeomorphism of $\mathcal{S}$. The set $\mathcal{O}\big(\mathcal{S}, L^2(\mathbb{R})\big)$ is a group under the composition of operators. This group is an infinite dimensional rotation group introduced by Yoshizawa[108]. Many interesting subgroups of $\mathcal{O}\big(\mathcal{S}, L^2(\mathbb{R})\big)$ have been discovered by Hida[31,34–37,41,42]. One of them is the Lévy group.

Let $\{e_n; n \geq 1\} \subset \mathcal{S}$ be an orthonormal basis for $L^2(\mathbb{R})$. For each $\sigma \in \mathcal{P}_L(\mathbb{N})$, let g_σ denote the linear operator of $L^2(\mathbb{R})$ defined by

$$x = \sum_{n=1}^{\infty} \langle x, e_n \rangle e_n \longmapsto g_\sigma x = \sum_{n=1}^{\infty} \langle x, e_{\sigma^{-1}(n)} \rangle e_n.$$

Obviously, g_σ is a unitary operator of $L^2(\mathbb{R})$ for each $\sigma \in \mathcal{P}_L(\mathbb{N})$. The *Lévy group* is defined to be

$$\mathcal{G}_L = \Big\{ g_\sigma\,;\, \sigma \in \mathcal{P}_L(\mathbb{N}) \text{ and } g_\sigma : \mathcal{S} \to \mathcal{S} \text{ is a homeomorphism} \Big\}.$$

It is a subgroup of the infinite dimensional rotation group $\mathcal{O}\big(\mathcal{S}, L^2(\mathbb{R})\big)$. For each $g \in \mathcal{G}_L$, its restriction to $\mathcal{S}$ can be extended to the complexification $\mathcal{S}_c$ of $\mathcal{S}$. Thus we can define an operator U_g acting on functions F defined on the space $\mathcal{S}_c$ by

$$U_g F(\xi) = F(g^{-1}\xi), \quad \xi \in \mathcal{S}_c.$$

Theorem 3.1. (Hida and Saitô[50]) *The Lévy Laplacian Δ_L commutes with U_g for all g in the Lévy group $\mathcal{G}_L$, i.e., if F is a function so that $\Delta_L F$ exists, then the Lévy Laplacian of $U_g F$ also exists and the equality holds*

$$\Delta_L U_g F = U_g \Delta_L F.$$

For further properties of the Lévy permutations and the Lévy group, see the papers by Obata[78–80,82].

3.2. *Spherical mean and the Lévy Laplacian*

The concept of spherical mean for functionals defined on the space $L^2(0,1)$ was introduced by P. Lévy[74]. It is closely related to the Lévy Laplacian. We will explain this relationship from the papers by Obata[81], Kuo-Obata-Saitô[67], and the book by Kuo[64]. Functionals defined on $L^2(0,1)$ will be replaced by those functions defined on $\mathcal{S}_c$ which are the S-transforms of generalized functions in $(\mathcal{S})^*$.

Let $\{e_n; n \geq 1\} \subset \mathcal{S}$ be an orthonormal basis for $L^2(\mathbb{R})$. For each $n \geq 2$, let $\lambda^{(n-1)}$ denote the uniform probability measure on the unit sphere S^{n-1} of $\mathbb{R}^n$ and define a mapping θ_n from S^{n-1} into the space $\mathcal{S}$ by

$$\theta_n(x_1, x_2, \ldots, x_n) = x_1 e_1 + x_2 e_2 + x_n e_n.$$

Let F be a function defined on $\mathcal{S}_c$. The *spherical mean* of F at $\xi \in \mathcal{S}_c$ with radius ρ is defined by

$$MF(\xi; \rho) = \lim_{n\to\infty} \int_{S^{n-1}} F(\xi + \rho\theta_n(x))\, d\lambda^{(n-1)}(x),$$

provided that the limit exists. The next theorem has appeared in the papers by Obata[78,81] and Kuo-Obata-Saitô [67]. For the proof, see Theorem 12.27 in the book by Kuo[64].

Theorem 3.2. *Suppose F is a function defined on $\mathcal{S}_c$ and satisfies the conditions:*

(a) $F(\xi + \eta) = F(\xi) + \langle F'(\xi), \eta\rangle + \frac{1}{2}\langle F''(\xi)\eta, \eta\rangle + o(|\eta|_0^2)$. *(Here $|\cdot|_0$ is the $L^2(\mathbb{R})$-norm.)*

(b) F has the spherical mean $MF(\xi;\rho)$ for all small $\rho > 0$.

Then we have

$$\Delta_L F(\xi) = 2 \lim_{\rho\to 0} \frac{MF(\xi;\rho) - F(\xi)}{\rho^2}$$

in the sense that if one side exists, then the other side also exists and the equality holds.

A function F defined on $\mathcal{S}_c$ is said to satisfy the *mean value property* if for any $\xi \in \mathcal{S}_c$ there exists ρ_0 such that

$$MF(\xi;\rho) = F(\xi) \quad \text{for all } \ 0 < \rho < \rho_0.$$

It is shown in Obata[81] that all S-transforms $F = S\phi$ for $\phi \in (L^2)$ satisfy the mean value property. On the other hand, suppose F satisfies the mean value property. Then by the above theorem we have $\Delta_L F = 0$, namely, the function F is Δ_L-harmonic. In particular, all functions in (L^2) and their S-transforms are Δ_L-harmonic. This is an interesting property of the Lévy Laplacian. For comparison, the Gross Laplacian is a densely defined operator on the space (L^2).

3.3. *Associated stochastic processes*

Let μ be the white noise measure on $\mathcal{S}'$. Define $p_t(x, A) = \mu\big((A - x)/\sqrt{t}\,\big)$ as in Eq. (3) for an abstract Wiener space. Then the probability measures $\{p_t(x, \cdot);\, t > 0, x \in \mathcal{S}'\}$ are the transition probabilities of an $\mathcal{S}'$-valued standard Wiener process $W(t)$.

For an orthonormal basis $\{e_n; n \geq 1\} \subset \mathcal{S}$ for $L^2(\mathbb{R})$, the Wiener process $W(t)$ can be represented by

$$W(t) = \sum_{n=1}^{\infty} \langle W(t), e_n \rangle e_n.$$

For fixed t, it follows from Theorem 3.1 in Itô[52] that this series converges almost surely as an $\mathcal{S}'$-valued random variable. For each n, the stochastic process $\langle W(t), e_n \rangle$ is an one-dimensional Brownian motion.

Let $|\cdot|_{-p}$ be the dual norm of $|\cdot|_p$ in Eq. (8), i.e., $|f|_{-p} = |A^{-p} f|_0$ and let $\mathcal{S}_{-p}$ be the completion of $L^2(\mathbb{R})$ with respect to the norm $|\cdot|_{-p}$. Then $\big(L^2(\mathbb{R}), \mathcal{S}_{-p}\big)$ is an abstract Wiener space for any $p > 1/2$. Hence the Wiener process $W(t)$ is actually supported in $\mathcal{S}_{-p}$ for any $p > 1/2$. By Eq. (4) the Gross Laplacian Δ_G is the infinitesimal generator of the Wiener process $W(t)$.

Note that for each N, the stochastic process $\sum_{n=1}^{N} \langle W(t), e_n \rangle e_n$ is a Brownian motion in the space spanned by $\{e_1, e_2, \ldots, e_N\}$. Consider the weighted Brownian motion

$$\widetilde{W}_N(t) = \frac{1}{\sqrt{N}} \sum_{n=1}^{N} \langle W(t), e_n \rangle e_n. \tag{18}$$

The stochastic process $\widetilde{W}_N(t)$ is related to the Lévy Laplacian by the next theorem from Theorem 12.28 in the book by Kuo[64].

Theorem 3.3. *Let F be the S-transform of a generalized function in $(\mathcal{S})^*$. Then*

$$\Delta_L F(\xi) = 2 \lim_{N \to \infty} \lim_{t \to 0} \frac{E\big(F(\xi + \widetilde{W}_N(t)) - F(\xi)\big)}{t}$$

in the sense that if one side exists, then the other side also exists and the equality holds.

In fact, it is easy to check that for any N we have

$$2\lim_{t\to 0}\frac{E\big(F(\xi+\widetilde{W}_N(t)\big)-F(\xi)}{t}=\frac{1}{N}\sum_{n=1}^{N}\langle F''(\xi)e_n,e_n\rangle$$

and so the equality in the above theorem follows from Eq. (7).

The sequence $\{\widetilde{W}_N(t);\, N\geq 1\}$ of weighted Brownian motions has rather peculiar properties as $N\to\infty$. If we regard $\widetilde{W}_N(t), N\geq 1$, as $L^2(\mathbb{R})$-valued stochastic processes, then by the strong law of large numbers we have

$$|\widetilde{W}_N(t)|_0^2=\frac{1}{N}\sum_{n=1}^{N}\langle W(t),e_n\rangle^2\longrightarrow t\quad\text{almost surely,}$$

and

$$\begin{aligned}
&|\widetilde{W}_{2N}(t)-\widetilde{W}_N(t)|_0^2\\
&=\frac{1}{2N}\sum_{n=N+1}^{2N}\langle W(t),e_n\rangle^2+\Big(\frac{1}{\sqrt{2N}}-\frac{1}{\sqrt{N}}\Big)^2\sum_{n=1}^{N}\langle W(t),e_n\rangle^2\\
&=\frac{1}{2}\frac{1}{N}\sum_{n=N+1}^{2N}\langle W(t),e_n\rangle^2+\Big(\frac{1}{\sqrt{2}}-1\Big)^2\frac{1}{N}\sum_{n=1}^{N}\langle W(t),e_n\rangle^2\\
&\longrightarrow(2-\sqrt{2})t\quad\text{almost surely.}
\end{aligned}$$

Therefore, the sequence $\{\widetilde{W}_N(t);\, N\geq 1\}$ does not converge almost surely in $L^2(\mathbb{R})$ as $N\to\infty$.

On the other hand, we can also regard $\widetilde{W}_N(t), N\geq 1$, as $\mathcal{S}_{-p}$-valued stochastic processes. Take e_n to be the Hermite function of order $n-1$, then $Ae_n=2n\,e_n$, $n\geq 1$. Hence for $p>1/2$ we have

$$\begin{aligned}
|\widetilde{W}_N(t)|_{-p}^2&=\frac{1}{N}\sum_{n=1}^{N}\frac{1}{(2n)^{2p}}\langle W(t),e_n\rangle^2\\
&\longrightarrow 0\quad\text{almost surely.}
\end{aligned}$$

Hence $\widetilde{W}_N(t)$ converges to 0 almost surely in $\mathcal{S}_{-p}$ as $N\to\infty$ for any $p>1/2$.

3.4. *Relationship to the Gross Laplacian*

A relationship between the Gross and Lévy Laplacian was discovered by Kuo-Obata-Saitô[67] in 1990. Their idea has been pursued further by Paycha[86,87] and Zhang[109]. In §12.6 of the book by Kuo[64] a more general setup has been discussed, namely, consider the question as to whether it is possible to find operators J_ϵ and positive numbers $c(\epsilon)$ such that

$$\Delta_L F = \lim_{\epsilon \to 0} c(\epsilon) \Delta_G (F \circ J_\epsilon) \text{ ?} \tag{19}$$

Note that the question is to express the Lévy Laplacian in terms of a renormalization of the Gross Laplacian. Consider the function

$$F(\xi) = \int_T f(t) \xi(t)^2 \, dt + \int_{T^2} g(s,t) \xi(s) \xi(t) \, ds dt,$$

where $f \in L^1(T)$ and $g \in L^1(T^2)$. If the equality in Eq. (19) holds for this particular function F, then as derived in §12.6 of the book by Kuo[64] the following two equalities must hold

$$\lim_{\epsilon \to 0} c(\epsilon) \sum_{n=1}^{\infty} \int_T f(t) |(J_\epsilon e_n)(t)|^2 \, dt = \frac{1}{|T|} \int_T f(t) \, dt \tag{20}$$

$$\lim_{\epsilon \to 0} c(\epsilon) \sum_{n=1}^{\infty} \int_{T^2} g(s,t) (J_\epsilon e_n)(s) (J_\epsilon e_n)(t) \, ds dt = 0. \tag{21}$$

It can be easily shown that the equality in Eq. (20) is satisfied if $\{e_n;\, n \geq 1\}$ is uniformly bounded and

$$c(\epsilon) \sum_{n=1}^{\infty} (J_\epsilon e_n)(t)^2 \longrightarrow \frac{1}{|T|} \quad \text{in } L^2(T). \tag{22}$$

In particular, we can integrate t over T to get

$$\lim_{\epsilon \to 0} c(\epsilon) \|J_\epsilon\|_{HS}^2 = \lim_{\epsilon \to 0} c(\epsilon) \sum_{n=1}^{\infty} |J_\epsilon e_n|_0^2 = 1, \tag{23}$$

where $\|\cdot\|_{HS}$ denotes the Hilbert-Schmidt operator norm. On the other hand, the equality in Eq. (21) is satisfied if

$$\lim_{\epsilon \to 0} c(\epsilon) \|J_\epsilon^* J_\epsilon\|_{HS} = 0. \tag{24}$$

We may take $c(\epsilon) = \|J_\epsilon\|_{HS}^{-2}$ so that Eq. (23) is automatically satisfied and Eq. (24) becomes

$$\lim_{\epsilon \to 0} \|J_\epsilon\|_{HS}^{-2} \|J_\epsilon^* J_\epsilon\|_{HS} = 0. \tag{25}$$

In fact, the conditions in Eqs. (22) and (25) together with the condition that $J_\epsilon \to I$ strongly on $L^2(\mathbb{R})$ are also sufficient for the equality in Eq. (19) to hold. This is the next theorem from §12.6 of the book by Kuo[64].

Theorem 3.4. *Let $\{J_\epsilon;\, \epsilon > 0\}$ be a family of continuous linear operators from $\mathcal{S}'$ into $\mathcal{S}$ satisfying the following conditions:*

(a) $J_\epsilon \to I$ *strongly on* $L^2(\mathbb{R})$ *as* $\epsilon \to 0$.

(b) $\lim_{\epsilon\to 0} \|J_\epsilon\|_{HS}^{-2} \|J_\epsilon^* J_\epsilon\|_{HS} = 0$.

(c) There exists a uniformly bounded orthonormal basis $\{e_n;\, n \geq 1\}$ for $L^2(T)$ such that

$$\lim_{\epsilon\to 0} \|J_\epsilon\|_{HS}^{-2} \sum_{n=1}^{\infty} (J_\epsilon e_n)(t)^2 = \frac{1}{|T|} \quad \text{in } L^2(T).$$

Then for any S-transform F with $\Delta_L F$ being defined as in Eq. (17) we have the equality

$$\Delta_L F = \lim_{\epsilon\to 0} \|J_\epsilon\|_{HS}^{-2} \Delta_G (F \circ J_\epsilon).$$

3.5. *Fourier transform and commutators*

The Fourier transform $\mathcal{F}\Phi$ of a generalized function Φ in $(\mathcal{S})^*$ is defined to be the generalized function with S-transform given by

$$(S\mathcal{F}\Phi)(\xi) = (S\Phi)(-i\xi)e^{-\frac{1}{2}\langle \xi,\xi\rangle}, \quad \xi \in \mathcal{S}_c. \tag{26}$$

The Fourier transform $\mathcal{F}$ is a continuous linear operator from $(\mathcal{S})^*$ into itself. It is proved in Hida-Kuo-Obata[47] that $\mathcal{F}$ is the unique (up to a constant multiple) continuous linear operator from $(\mathcal{S})^*$ into itself such that for any $\xi \in \mathcal{S}$ the following equalities hold:

$$\mathcal{F}\widetilde{D}_\xi = i\widetilde{Q}_\xi \mathcal{F}, \quad \mathcal{F}\widetilde{Q}_\xi = i\widetilde{D}_\xi \mathcal{F}, \tag{27}$$

where $\widetilde{D}_\xi$ is the continuous extension to $(\mathcal{S})^*$ of the differential operator D_ξ, in the direction of ξ, acting on $(\mathcal{S})$ and $\widetilde{Q}_\xi$ is the continuous extension to $(\mathcal{S})^*$ of the multiplication operator Q_ξ by $\langle \cdot, \xi\rangle$ acting on $(\mathcal{S})$.

The Dirac delta function δ_x at $x \in \mathcal{S}'$ was introduced in white noise theory by Kubo and Yokoi[57], namely,

$$\langle\langle \delta_x, \phi \rangle\rangle = \phi(x), \quad \phi \in (\mathcal{S}),$$

where the unique continuous version of ϕ is taken. The S-transform of δ_x is given by

$$(S\delta_x)(\xi) = e^{\langle x,\xi\rangle - \frac{1}{2}\langle \xi,\xi\rangle}, \quad \xi \in \mathcal{S}_c. \tag{28}$$

In particular, when $x = 0$, we see from Eqs. (26) and (28) that $\mathcal{F}1 = \delta_0$. Thus $\mathcal{F}$ is the unique linear operator from $(\mathcal{S})^*$ into itself satisfying the equalities in Eq. (27) and $\mathcal{F}1 = \delta_0$.

The next theorem is an interesting relationship between the Fourier transform and the Lévy Laplacian discovered by Hida and Saitô[50].

Theorem 3.5. *The Lévy Laplacian and the Fourier transform are related by the following equality:*

$$\mathcal{F}\big(\Delta_L + \tfrac{1}{2}\big) = -\big(\Delta_L + \tfrac{1}{2}\big)\mathcal{F}.$$

The Lévy Laplacian is related to the adjoint Δ_G^* of the Gross Laplacian and the number operator N by the equalities:

$$\big[\Delta_L, \Delta_G^*\big] = 2I, \quad \big[\Delta_L, N\big] = 2\Delta_L,$$

where $[A, B] = AB - BA$ is the commutator of A and B, and N is the number operator defined by

$$N\bigg(\sum_{n=0}^{\infty}\langle :x^{\otimes n}:, f_n\rangle\bigg) = \sum_{n=1}^{\infty} n\,\langle :x^{\otimes n}:, f_n\rangle$$

The number operator is continuous as an operator from $(\mathcal{S})$ into itself and from $(\mathcal{S})^*$ into itself. For more commutator identities, see the books by Hida-Kuo-Potthoff-Streit[48] and by Kuo[64].

4. Special elementary functions

The Lévy Laplacian is a very peculiar operator. For example, although it is defined from the second derivative, it has the derivation property, namely, for any F and G with $\Delta_L F$ and $\Delta_L G$ being defined, we have

$$\Delta_L(FG) = (\Delta_L F)G + F(\Delta_L G).$$

In terms of generalized functions, let $F = S\Phi$ and $G = S\Psi$. Then

$$\Delta_L(\Phi \diamond \Psi) = (\Delta_L \Phi) \diamond \Psi + \Phi \diamond (\Delta_L \Psi),$$

where $\diamond$ denotes the Wick product, i.e., $S(\Phi \diamond \Psi) = (S\Phi)(S\Psi)$.

There are several classes of special elementary functions on which the Lévy Laplacian operates in particular ways. The third class below is of great importance since it can be used to define various domains of the Lévy Laplacian with nice properties.

4.1. *Normal functionals*

Lévy considered the *normal functionals*, i.e., functions of the form

$$F(\xi) = \int_{\mathbb{R}^n} f(u_1, \ldots, u_n)\xi(u_1)^{j_1} \cdots \xi(u_n)^{j_n}\, du_1 \cdots du_n,$$

where $f \in L^1(\mathbb{R}^n)$ and $j_1, \ldots, j_n \geq 0$ are integers. This function is the S-transform of the generalized function in $(\mathcal{S})^*$

$$\Phi = \int_{\mathbb{R}^n} f(u_1, \ldots, u_n) :\dot{B}(u_1)^{j_1} \cdots \dot{B}(u_n)^{j_n}: du_1 \cdots du_n,$$

where $: \cdot :$ denotes the renormalization. The Lévy Laplacian $\Delta_L F$ exists, but the expression is somewhat complicated to be manageable.

4.2. *Gaussian functionals*

For a complex number $c \neq 1/2$, consider the Gaussian functional Φ_c which is the multiplicative renormalization

$$\Phi_c = \mathcal{N}\exp\left[c\int_{\mathbb{R}} \dot{B}(t)^2\, dt\right]. \tag{29}$$

The S-transform of Φ_c is given by

$$F_c = \exp\left[\frac{c}{1-2c}\int_{\mathbb{R}} \xi(t)^2\, dt\right], \quad \xi \in \mathcal{S}_c. \tag{30}$$

In 1984 we observed in the lecture notes (later published in 1993 as Nagoya University Lecture Notes[62]) a fact about the Lévy Laplacian

$$\Delta_L F_c = \frac{2c}{1-2c} F_c.$$

For any complex number $z \neq -1$, we can choose $c = z/2(1+z)$ so that $\Delta_L F_c = zF_c$. Thus F_c is an eigenfunction of Δ_L with eigenvalue z. On the other hand, as $c \to \infty$ in Eq. (30) we get the function

$$F_\infty = \exp\left[-\frac{1}{2}\int_{\mathbb{R}} \xi(t)^2\, dt\right], \quad \xi \in \mathcal{S}_c.$$

By Theorem 2.1 F_∞ is the S-transform of a generalized function, denoted by Φ_∞, in $(\mathcal{S})^*$. We can check that $\Delta_L F_\infty = -F_\infty$ and so F_∞ is an eigenfunction with eigenvalue -1. Hence every complex number z is an eigenvalue of the Lévy Laplacian. The corresponding eigenfunctions are the Gaussian functionals.

4.3. *Exponential functions*

In recent years Saitô has come up with a novel idea to start from the elementary exponential functions to construct a domain space on which the Lévy Laplacian can operate any number of times. This will produce many interesting objects related to the Lévy Laplacian, e.g., a semigroup whose infinitesimal generator is the L'evy Laplacian.

An *elementary exponential function* is a function of the form

$$F(\xi) = \int_{T^n} f(u_1, \ldots, u_n) e^{ia_1\xi(u_1)+\cdots+ia_n\xi(u_n)}\, du_1 \cdots du_n,$$

where $f \in L^1(T^n)$ and $a_1, \ldots, a_n \in \mathbb{C}$. It is easy to verify that F satisfies the conditions in Theorem 2.1 and so is the S-transform of a generalized function Φ in $(\mathcal{S})^*$ which can be denoted by

$$\Phi = \int_{T^n} f(u_1, \ldots, u_n) :e^{ia_1\dot{B}(u_1)+\cdots+ia_n\dot{B}(u_n)}: du_1 \cdots du_n. \tag{31}$$

A direct computation shows that the singular part F''_s of F'' is given by

$$\begin{aligned} F''(\xi; t) = &-a_1^2 \int_{T^{n-1}} f(t, u_2, \ldots, u_n) e^{ia_1\xi(t)+ia_2\xi(u_2)+\cdots+ia_n\xi(u_n)}\, du_2 \cdots du_n \\ &- \quad \cdots\cdots\cdots\cdots \\ &- a_n^2 \int_{T^{n-1}} f(u_1, \ldots, u_{n-1}, t) e^{ia_1\xi(u_1)+\cdots+ia_{n-1}\xi(u_{n-1})+ia_n\xi(t)}\, du_1 \cdots du_{n-1}. \end{aligned}$$

Integrating the above equation over the t-variable to get

$$\begin{aligned} \Delta_L F(\xi) &= \frac{1}{|T|} \int_T F''_s(\xi; t)\, dt \\ &= \frac{1}{|T|}\big(- a_1^2 F(\xi) - \cdots - a_n^2 F(\xi)\big) \\ &= -\frac{a_1^2 + \cdots + a_n^2}{|T|} F(\xi). \end{aligned}$$

Therefore, the Lévy Laplacians of F and Φ are given by

$$\begin{aligned} \Delta_L F &= -\frac{a_1^2 + \cdots + a_n^2}{|T|} F, \\ \Delta_L \Phi &= -\frac{a_1^2 + \cdots + a_n^2}{|T|} \Phi. \end{aligned} \tag{32}$$

This equality, obtained by Saitô and Tsoi[105], is the starting point of Saitô's idea to construct various domains for the Lévy Laplacian, which we will explain in the next section.

5. Diagonalization and domains of the Lévy Laplacian

In order to study the Lévy Laplacian Δ_L we need to construct a space which can serve as a domain for Δ_L such that the powers Δ_L^k and the exponential $e^{t\Delta_L}$ are all continuous operators on this space. In this section we will explain Saitô's novel idea for constructing various domains for the Lévy Laplacian.

From now on, we fix a *Lévy diagonalization function* h, i.e., a continuous function $h : \mathbb{R} \to \mathbb{C}$ such that there exists a stochastic process $X(t)$ with

$$Ee^{irX(t)} = e^{th(r)}, \quad r \in \mathbb{R}. \tag{33}$$

For example, $h(r) = -|r|^\nu$, $1 \leq \nu \leq 2$, is such a function. In this case, we have the ν-stable process $X(t)$.

The following table shows a comparison of the construction for one- and infinite-dimensional cases. The Schwartz space $\mathcal{S}$ is constructed from the space $L^2(\mathbb{R})$ and the operator $A = -d^2/dx^2 + x^2 + 1$, while the space $\mathbb{D}_{-p,\infty}$ (see Step 4 below) is constructed from the space $(\mathcal{S})_{-p}$ and the Lévy Laplacian Δ_L.

Table 1. Comparison of one- and infinite-dimensional cases.

	space	operator	eigenvalues	powers	domain
one-dim	$L^2(\mathbb{R})$	A	$2k+2$	$A^n, \quad (2k+2)^n$	$\mathcal{S}$
∞-dim	$(\mathcal{S})_{-p}$	Δ_L	$h(\lambda)$	$\Delta_L^n, \quad h(\lambda)^n$	$\mathbb{D}_{-p,\infty}$

Step 1. $\mathbb{E}_\lambda$: *The space of elementary exponential functions*

For each $\lambda \in \mathbb{R}$, let $\mathbb{E}_\lambda$ be the linear space spanned by generalized functions of the form in Eq. (31), i.e,

$$\Phi = \int_{T^n} f(u_1, \ldots, u_n) :e^{ia_1\dot{B}(u_1)+\cdots+ia_n\dot{B}(u_n)}: du_1 \cdots du_n,$$

where $n \geq 1$, $f \in L^1(T^n)$ and $a_1, \ldots, a_n \in \mathbb{C}$ satisfy the conditions:

(a) $a_1 + \cdots + a_n = \sqrt{|T|}\,\lambda$.

(b) $a_1^2 + \cdots + a_n^2 = -|T|h(\lambda)$.

Here condition (b) is motivated by Eq. (32) so that $h(\lambda)$ is an eigenvalue for these functions, while condition (a) is required for the uniqueness of the representation in Lemma 5.3 below.

Lemma 5.1. *For each $\lambda \in \mathbb{R}$, we have*

(1) $\mathbb{E}_\lambda \subset (\mathcal{S})_{-p}$ *for any* $p > 5/12$.

(2) $\mathbb{E}_\lambda \neq \{0\}$.

(3) $\Delta_L \Phi = h(\lambda)\Phi$ *for all* $\Phi \in \mathbb{E}_\lambda$.

Step 2. $\mathbb{E}_{\lambda,-p}$: *The completion of* $\mathbb{E}_\lambda$ *with respect to* $\|\cdot\|_{-p}$

Let $p > 5/12$ be fixed from now on. In view of the fact (1) in Lemma 5.1, we can complete the space $\mathbb{E}_\lambda$ with respect to the $(\mathcal{S})_{-p}$-norm $\|\cdot\|_{-p}$ defined in Eq. (13). Let $\mathbb{E}_{\lambda,-p}$ denote the completion.

Lemma 5.2. *For each $\lambda \in \mathbb{R}$, we have*

(1) $\mathbb{E}_{\lambda,-p} \subset (\mathcal{S})_{-p}$ *for any* $p > 5/12$.

(2) $\mathbb{E}_{\lambda,-p}$ *is a Hilbert space.*

(3) $\Delta_L \Phi = h(\lambda)\Phi$ *for all* $\Phi \in \mathbb{E}_{\lambda,-p}$.

Step 3. $\mathbb{D}_{-p,N}$: *The space of Pettis integrals of* $\mathbb{E}_{\lambda,-p}$ *over* λ

Suppose a weakly measurable function $\Phi : \mathbb{R} \to (\mathcal{S})^*$ is Pettis integrable (see page 243 in the book[64].) Then the Pettis integral $\int_{\mathbb{R}} \Phi(\lambda)\, d\lambda$ exists and is a generalized function in $(\mathcal{S})^*$. The next lemma is from the papers by Kuo-Obata-Saitô[68,69]. The condition (a) in Step 1 is needed for the proof of this lemma (see the papers[68,69].)

Lemma 5.3. *Let* $\Phi, \Psi : \mathbb{R} \to (\mathcal{S})^*$ *be Pettis integrable. Assume that* $\Phi(\lambda), \Psi(\lambda) \in \mathbb{E}_{\lambda,-p}$ *for almost all* $\lambda \in \mathbb{R}$ *and* $\int_{\mathbb{R}} \Phi(\lambda)\, d\lambda = \int_{\mathbb{R}} \Psi(\lambda)\, d\lambda$. *Then* $\Phi(\lambda) = \Psi(\lambda)$ *for almost all* $\lambda \in \mathbb{R}$.

Let $N \geq 0$ be an integer. Consider a Pettis integral Φ given by

$$\Phi = \int_{\mathbb{R}} \Phi(\lambda)\, d\lambda, \qquad \Phi(\lambda) \in \mathbb{E}_{\lambda,-p} \quad \text{a.e. } (\lambda). \tag{34}$$

For such a generalized function Φ, define

$$[\![\Phi]\!]_{-p,N} = \left[\int_{\mathbb{R}} \|\Phi(\lambda)\|_{-p}^2 \Big(\sum_{n=0}^{N} |h(\lambda)|^{2n} \Big)\, d\lambda \right]^{1/2}. \tag{35}$$

Obviously, $0 \leq [\![\Phi]\!]_{-p,N} \leq \infty$. Observe that $[\![\Phi]\!]_{-p,N}$ is well-defined by Lemma 5.3. Moreover, see the infinite-dimensional case in Table 1 for the motivation of the definition for $[\![\Phi]\!]_{-p,N}$.

Now, let $\mathbb{D}_{-p,N}$ denote the space of all Pettis integrals $\Phi = \int_{\mathbb{R}} \Phi(\lambda)\, d\lambda$ with $\Phi(\lambda) \in \mathbb{E}_{\lambda,-p}$, λ-a.e., and $[\![\Phi]\!]_{-p,N} < \infty$. Observe that by Lemma 5.3 each element Φ in $\mathbb{D}_{-p,N}$ can be uniquely represented as a Pettis integral as in Eq. (34).

Lemma 5.4. *For any integer $N \geq 0$, we have*

(1) $\mathbb{D}_{-p,N}$ *is a Hilbert space with norm* $[\![\cdot]\!]_{-p,N}$.
(2) $\mathbb{D}_{-p,N+1} \subset \mathbb{D}_{-p,N}$.
(3) $\Delta_L\big(\int_{\mathbb{R}} \Phi(\lambda)\, d\lambda\big) = \int_{\mathbb{R}} h(\lambda)\Phi(\lambda)\, d\lambda$.
(4) $\Delta_L : \mathbb{D}_{-p,N+1} \longrightarrow \mathbb{D}_{-p,N}$ *is a continuous linear operator.*

Note that the assertion (3) in this lemma gives the diagonalization of the Lévy Laplacian with h being the diagonalization function. The next theorem is from the papers by Kuo-Obata-Saitô[68,69].

Theorem 5.1. *The Lévy Laplacian Δ_L, when restricted to $\mathbb{D}_{-p,N+1}$, is a densely defined self-adjoint operator on $\mathbb{D}_{-p,N}$ for any $p > 5/12$ and any integer $N \geq 0$.*

We point out something related to Eq. (35) and Table 1. For Φ given as in Eq. (34), define

$$[\![\Phi]\!]_{-p} = \left(\int_{\mathbb{R}} \|\Phi(\lambda)\|_{-p}^2\, d\lambda \right)^{1/2}.$$

Then we can use the fact (3) in Lemma 5.4 to express the norm $[\![\Phi]\!]_{-p,N}$ in Eq. (35) as follows:

$$[\![\Phi]\!]_{-p,N} = \left(\sum_{n=0}^{N} \|\Delta_L^n \Phi\|_{-p}^2 \right)^{1/2}.$$

Step 4. $\mathbb{D}_{-p,\infty}$: *The projective limit of* $\{\mathbb{D}_{-p,N};\, N \geq 0\}$

From Step 3 we have the following chain with continuous inclusions:

$$\cdots\cdots \mathbb{D}_{-p,N+1} \subset \mathbb{D}_{-p,N} \subset \cdots \subset \mathbb{D}_{-p,1} \subset \mathbb{D}_{-p,0} \subset (\mathcal{S})_{-p}.$$

Finally, we define $\mathbb{D}_{-p,\infty}$ to be the projective limit of $\{\mathbb{D}_{-p,N};\, N \geq 0\}$. Note that $\mathbb{D}_{-p,\infty} = \cap_{N=0}^{\infty} \mathbb{D}_{-p,N}$ and the topology is given by the family $\{[\![\cdot]\!]_{-p,N};\, N \geq 0\}$ of norms.

The space $\mathbb{D}_{-p,\infty}$ is a domain of the Lévy Laplacian that we mentioned in the beginning of this section. It should be noticed that $\mathbb{D}_{-p,\infty}$ depends not only on $p > 5/12$, but also on the diagonalization function $h(\lambda)$.

Theorem 5.2. *The Lévy Laplacian Δ_L is a continuous linear operator from $\mathbb{D}_{-p,\infty}$ into itself. (Hence all powers Δ_L^n of Δ_L are also continuous linear operators from $\mathbb{D}_{-p,\infty}$ into itself.)*

We summary the above discussion regarding to various spaces and the Lévy Laplacian Δ_L as follows:

(1) We have the spaces $\mathbb{E}_\lambda$, $\mathbb{E}_{\lambda,-p}$, $\mathbb{D}_{-p,N}$, $\mathbb{D}_{-p,\infty}$.
(2) The Lévy Laplacian Δ_L, when regarded as an operator on $\mathbb{E}_\lambda$ and $\mathbb{E}_{\lambda,-p}$, is a multiplication operator by $h(\lambda)$.
(3) The Lévy Laplacian Δ_L as an operator on $\mathbb{D}_{-p,N}$ and $\mathbb{D}_{-p,\infty}$ is a diagonal operator.
(4) On the space $\mathbb{D}_{-p,\infty}$ we can apply the Lévy Laplacian any number of times. Moreover, we can construct a semigroup on $\mathbb{D}_{-p,\infty}$ for Δ_L, which we will do in the next section.

6. Associated Semigroups and Stochastic Processes

Let (H, B) be an abstract Wiener space and let $p_t(x,\cdot)$ be the Gaussian measure defined in Eq. (3). For a bounded continuous function f on B and for $t > 0$, define

$$P_t f(x) = \int_B f(y)\, p_t(x, dy). \tag{36}$$

Let $P_0 = I$. Then the family $\{P_t;\, t \geq 0\}$ is a semigroup and its infinitesimal generator is the Gross Laplacian Δ_G. On the other hand, let W_t be a standard Wiener process with state space B. Then

$$P_t f(x) = E\big[f(W_t)\,\big|\, W_0 = x\big]. \tag{37}$$

Note that $\big(L^2(\mathbb{R}), (\mathcal{S})_{-p}\big)$ is an abstract Wiener space for any $p > 1/2$. Hence we can easily formulate the above two equations in the white noise space for the Gross Laplacian Δ_G with domain being the space $(\mathcal{S})$ of test functions in Eq. (12).

Question: For the Lévy Laplacian, what are the corresponding objects of the semigroup $\{P_t;\, t \geq 0\}$ in Eq. (36), the stochastic process W_t in Eq. (37), and the domain space $(\mathcal{S})$?

First of all, we take $\mathbb{D}_{-p,\infty}$ to be the analogue of $(\mathcal{S})$ for the Lévy Laplacian. As we have seen in the previous section, the Lévy Laplacian Δ_L is a continuous linear operator from $\mathbb{D}_{-p,\infty}$ into itself. Thus we can repeatedly apply Δ_L on $\mathbb{D}_{-p,\infty}$.

6.1. *Semigroups generated by the Lévy Laplacian*

To construct a semigroup $\{G_t;\, t \geq 0\}$ for the Lévy Laplacian, note that G_t is informally given by $G_t = e^{t\Delta_L}$. Since Δ_L is diagonalized along the $\mathbb{E}_{\lambda,-p}$ spaces with eigenvalues $h(\lambda)$, $\lambda \in \mathbb{R}$, we can define G_t as follows. Each $\Phi \in \mathbb{D}_{-p,\infty}$ can be uniquely represented by

$$\Phi = \int_{\mathbb{R}} \Phi(\lambda)\, d\lambda, \quad \Phi(\lambda) \in \mathbb{E}_{\lambda,-p}.$$

For this Φ, we define

$$G_t\Phi = \int_{\mathbb{R}} e^{th(\lambda)} \Phi(\lambda)\, d\lambda.$$

Observe that by Eq. (33) we have $\Re h(\lambda) \leq 0$. Hence by Eq. (35),

$$\begin{aligned} [\![G_t\Phi]\!]_{-p,N}^2 &= \int_{\mathbb{R}} \|e^{th(\lambda)}\Phi(\lambda)\|_{-p}^2 \Big(\sum_{n=0}^{N} |h(\lambda)|^{2n} \Big)\, d\lambda \\ &= \int_{\mathbb{R}} e^{2t\Re h(\lambda)} \|\Phi(\lambda)\|_{-p}^2 \Big(\sum_{n=0}^{N} |h(\lambda)|^{2n} \Big)\, d\lambda \\ &\leq \int_{\mathbb{R}} \|\Phi(\lambda)\|_{-p}^2 \Big(\sum_{n=0}^{N} |h(\lambda)|^{2n} \Big)\, d\lambda \\ &= [\![\Phi]\!]_{-p,N}^2. \end{aligned}$$

Thus the family $\{G_t;\, t \geq 0\}$ of linear operators from $\mathbb{D}_{-p,\infty}$ into itself is equi-continuous. In fact, we have the next theorem from the papers by Kuo-Obata-Saitô[68,69].

Theorem 6.1. *The family $\{G_t;\, t \geq 0\}$ is an equi-continuous semigroup of class (C_0) generated by the Lévy Laplacian Δ_L as a continuous linear operator defined on the space $\mathbb{D}_{-p,\infty}$.*

We should mention that the semigroup $\{G_t;\, t \geq 0\}$ depends on the choice of the diagonalization function $h(\lambda)$ in Eq. (33). For a different choice of the function $h(\lambda)$, we would get a different semigroup. On the other hand, the domain $\mathbb{D}_{-p,\infty}$ of the Lévy Laplacian and the semigroup depends on p. But this dependence is not so essential.

6.2. *Associated stochastic processes*

Unlike the Gross Laplacian, we cannot expect the analogue of Eq. (37) in the following form

$$G_t\Phi(x) = E\big[\Phi(W_t)\,\big|\,W_0 = x\big].$$

This is obviously because $\Phi \in \mathbb{D}_{-p,\infty}$ is a generalized function and so is not defined pointwise. However, we can consider the Lévy Laplacian as an operator acting on S-transforms of generalized functions in $\Phi \in \mathbb{D}_{-p,\infty}$ as defined in Eq. (17). In that case, we will be able to obtain an analogue of Eq. (37) for the Lévy Laplacian.

Let $\widetilde{\mathbb{D}}_{-p,\infty} = S\mathbb{D}_{-p,\infty}$ and endow the topology induced from $\mathbb{D}_{-p,\infty}$ by the S-transform. To avoid confusion, we now use $\widetilde{\Delta}_L$ to denote the Lévy Laplacian acting on the space $\widetilde{\mathbb{D}}_{-p,\infty}$, namely,

$$\widetilde{\Delta}_L = S\Delta_L S^{-1} : \widetilde{\mathbb{D}}_{-p,\infty} \longrightarrow \widetilde{\mathbb{D}}_{-p,\infty}.$$

Thus $\widetilde{\Delta}_L$ is a continuous linear operator from $\widetilde{\mathbb{D}}_{-p,\infty}$ into itself. Define

$$\widetilde{G}_t = SG_tS^{-1} : \widetilde{\mathbb{D}}_{-p,\infty} \longrightarrow \widetilde{\mathbb{D}}_{-p,\infty}.$$

By the above Theorem 6.1, the family $\{\widetilde{G}_t;\, t \geq 0\}$ is an equi-continuous semigroup of class (C_0) generated by the Lévy Laplacian $\widetilde{\Delta}_L$ as a continuous linear operator defined on the space $\widetilde{\mathbb{D}}_{-p,\infty}$.

Recall the stochastic process $X(t)$ in Eq. (33) which has characteristic function $e^{th(r)}$. Take a function $\eta_T \in \mathcal{S}$ such that $\eta_T = 1/\sqrt{|T|}$ on T. By using $X(t)$ and η_T, we define an $\mathcal{S}$-valued stochastic process by

$$\mathbb{X}_t = \xi + X(t)\,\eta_T, \quad \xi \in \mathcal{S}.$$

The next theorem is from the papers by Kuo-Obata-Saitô[68,69].

Theorem 6.2. *For all $F \in \widetilde{\mathbb{D}}_{-p,\infty}$, the following equality holds*

$$\widetilde{G}_tF(\xi) = E\big[F(\mathbb{X}_t)\,\big|\,\mathbb{X}_0 = \xi\big].$$

This theorem is the analogue of Eq. (37) for the Lévy Laplacian. We can interpret $\mathbb{X}_t$ as an $\mathcal{S}$-valued stochastic process generated by the Lévy Laplacian $\widetilde{\Delta}_L$. Note that $\widetilde{\Delta}_L$ and $\mathbb{X}_t$ are intrinsically linked together by the diagonalization function $h(r)$.

Acknowledgments

This research was supported by the Academic Frontier in Science of Meijo University and the Nagoya Winter School, Meijo University, January 6-10, 2003. The author would like to give his deepest appreciation to Professors T. Hida and K. Saitô for the warm hospitality during his numerous visits over many years.

References

1. Accardi, L.: Yang-Mills equations and Lévy-Laplacian; in: *Dirichlet Forms and Stochastic Processes*, de Gruyter (1995) 1–24
2. Accardi, L. and Bogachev, V.: The Ornstein-Uhlenbeck process and the Dirichlet form associated to the Lévy Laplacian; *C. R. Acad. Sci. Paris Ser. I Math.* bf 320 (1995) 597–602
3. Accardi, L. and Bogachev, V.: The Ornstein-Uhlenbeck process associated with the Lévy Laplacian and its Dirichlet form; *Probab. Math. Statist.* **17** (1997) 95–114
4. Accardi, L. and Bogachev, V.: On stochastic analysis associated with the Lévy Laplacian; *Dokl. Akad. Nauk* **358** (1998) 583–587
5. Accardi, L., Gibilisco, P., and Volovich, I. V.: Yang-Mills gauge fields as harmonic functions for the Lévy Laplacian; *Russian J. Math. Phys.* **2** (1994) 235–250
6. Accardi, L., Rozelli, P., and Smolyanov, O. G.: Brownian motion generated by the Lévy Laplacian; *Mat. Zametki* **54** (1993) 144–148
7. Accardi, L. and Smolyanov, O. G.: Trace formulae for Lévy-Gaussian measures and their application; in: *Mathematical Approach to Fluctuations II*, World Scientific (1995) 31–47
8. Accardi, L. and Smolyanov, O. G.: Transformations of Gaussian measures generated by the Lévy Laplacian and generalized traces; *Dokl. Akad. Nauk SSSR* bf 350 (1996) 5–8
9. Blümlinger, M. and Obata, N.: Permutations preserving Cesàro mean, densities of natural numbers and uniform distribution of sequences; *Ann. Inst. Fourier Grenoble* **41** (1991) 665–678
10. Chung, D. M., Ji, U. C., and Saitô, K.: Cauchy problems associated with the Lévy Laplacian in white noise analysis; *Infinite Dimensional Analysis, Quantum Probability and Related Topics* **2** (1999) 131–153
11. Chung, D. M., Ji, U. C., and Saitô, K.: Notes on a C_0-group generated by the Lévy Laplacian; *Proc. American Mathematical Society* (2001)
12. Feller, M. N.: The solvability of infinite-dimensional elliptic equations with constant coefficients; *Dokl. Akad. Nauk SSSR* bf 214 (1974) 59–62
13. Feller, M. N.: Correction: "The solvability of infinite-dimensional elliptic equations with constant coefficients;" *Dokl. Akad. Nauk SSSR* bf 214 (1974) 59–62, *Dokl. Akad. Nauk SSSR* bf 215 (1974) 4–6
14. Feller, M. N.: Solvability of infinite-dimensional elliptic equations with variable coefficients; *Mat. Zametki* bf 25 (1979) 419–424, 477

15. Feller, M. N.: Infinite-dimensional Laplace-Lévy differential operators; *Ukrain. Mat. Zh.* **32** (1980) 69–79, 142
16. Feller, M. N.: Infinite-dimensional selfadjoint Laplace-Lévy differential operators; *Ukrain. Mat. Zh.* **35** (1983) 200–206
17. Feller, M. N.: Infinite-dimensional elliptic equations and operators of P. Lévy type; *Russian Math Surveys* **41** (1986) 119–170
18. Feller, M. N.: Selfadjointness of a nonsymmetrized infinite-dimensional Lévy-Laplace operator; *Ukrainian Math. J.* **41** (1989) 853–856
19. Feller, M. N.: The scope of harmonic functions of an infinite number of variables. I; *Ukrainian Math. J.* **42** (1990) 1419–1423 (1991)
20. Feller, M. N.: The scope of harmonic functions of an infinite number of variables. II; *Ukrainian Math. J.* **42** (1990) 1521–1527 (1991)
21. Feller, M. N.: The scope of harmonic functions of an infinite number of variables. III; *Ukrainian Math. J.* **44** (1992) 362–367
22. Feller, M. N.: Necessary and sufficient conditions of harmonicity of functions of infinitely many variables (Jacobian case); *Ukrainian Math. J.* **46** (1994) 857–862 (1995)
23. Feller, M. N.: Another condition for the harmonicity of functions of an infinite number of variables (the translationally nonpositive case); *Ukrainian Math. J.* **46** (1994) 1777–1781 (1996)
24. Feller, M. N.: On a nonlinear equation not solved with respect to the Lévy Laplacian; *Ukrainian Math. J.* **48** (1996) 805–808 (1997)
25. Feller, M. N.: The Riquier problem for a nonlinear equation solved with respect to the iterated Lévy Laplacian; *Ukrainian Math. J.* **50** (1998) 1796–1800 (1999)
26. Feller, M. N.: The Riquier problem for a nonlinear equation not solved with respect to the iterated Lévy Laplacian; *Ukrainian Math. J.* **51** (1999) 470–475
27. Feller, M. N.: Cauchy problem for non-linear equations involving the Levy Laplacian with separable variables; *Methods Funct. Anal. Topology* **5** (1999) 9–14
28. Feller, M. N.: Notes on infinite-dimensional nonlinear parabolic equations; *Ukrainian Math. J.* **52** (2000) 789–802 (2001)
29. Gross, L.: Abstract Wiener spaces; *Proc. 5th Berkeley Symp. Math. Stat. and Probab.* **2**, part 1 (1965) 31-42, University of California Press, Berkeley
30. Gross, L.: Potential theory on Hilbert space; *J. Funct. Anal.* **1** (1967) 123–181
31. Hida, T.: *Analysis of Brownian Functionals.* Carleton Mathematical Lecture Notes **13**, 1975
32. Hida, T.: *Brownian motion.* Springer-Verlag, 1980
33. Hida, T.: Generalized Brownian functionals; *Proc. IFIP-WG Theory and Application of Random Field, Lecture Notes in Control and Information Sci.* **49** (1983) 89–95, Springer-Verlag
34. Hida, T.: Brownian motion and its functionals; *Ricerche di Matematica* **34** (1985) 183–222
35. Hida, T.: *Analysis of Brownian functionals.* Lecture Notes, IMA, University of Minnesota, 1986

36. Hida, T.: Infinite dimensional rotation group and its applications to quantum dynamics; *Proc. 14th ICGTMP Conf.*, Y. M. Cho (ed.) (1986) 234–237
37. Hida, T.: Infinite-dimensional rotation group and unitary group; *Lecture Notes in Math* **1379** (1989) 125–134, Springer-Verlag
38. Hida, T.: Functionals of Brownian motion; *Lectures in Applied Mathematics and Informatics*, Luigi M. Ricciardi (ed.), Manchester Univ. Press, 1990, 286-329.
39. Hida, T.: A role of the Lévy Laplacian in the causal calculus of generalized white noise functionals; in: *Stochastic Processes, G. Kallianpur Volume*, S. Cambanis (ed.) (1993) 131–139
40. Hida, T.: White noise analysis and applications; *Stochastic Analysis and Applications in Physics*, A. I. Cardoso et al. (eds.) (1994) 119–131, Kluwer Academic Publishers
41. Hida, T.: White noise analysis: An overview and some future directions; *IIAS Reports* 1995-001 (1995)
42. Hida, T.: Infinite dimensional rotation group and white noise analysis; in: *Group Theoretical Methods in Physics, Proc. 20th Colloq. on Group Theoretical Methods in Physics*, A. Arima et al. (eds.) World Scientific (1995) 1–9
43. Hida, T.: White noise approach to fluctuations; *J. Korean Math. Soc.* **35** (1998) 575-581
44. Hida, T.: Harmonic analysis on complex random systems; in: *Infinite Dimensional Harmonic Analysis. Transactions of Japanese-Germany Symp.*, H. Heyer and T. Hirai (eds.) (1999) 160–166
45. Hida, T.: Some methods of computation in white noise calculus; in: *Unconventional Model of Computation UMC'2K*, Solvay Institute, I. Antoniou et al. (eds.) (2001) 85–93
46. Hida, T., Kubo, I., Nomoto, H., and Yoshizawa, H.: On projective invariance of Brownian motion; Publ. RIMS Kyoto University, Series A, **4** (1969) 595–609
47. Hida, T., Kuo, H.-H., and Obata, N.: Transformations for white noise functionals; *J. Funct. Anal.* **111** (1993) 259–277
48. Hida, T., Kuo, H.-H., Potthoff, J., and Streit, L.: *White Noise: An Infinite Dimensional Calculus.* Kluwer Academic Publishers, 1993
49. Hida, T., Obata, N., and Saitô: Infinite dimensional rotations and Laplacians in terms of white noise calculus; *Nagoya Math. J.* **128** (1992) 65–93
50. Hida, T. and Saitô, K.: White noise analysis and the Lévy Laplacian; *Stochastic Processes in Physics and Engineering*, S. Albeverio et al. (eds.) (1988) 177–184, Reidel Publishing Company
51. Ishikawa, A., Saitô, K., and Tsoi, A. H.: Poisson noise analysis based on the Lévy Laplacian; in: *Quantum Information IV*, T. Hida and K. Saitô (eds.) (2002) 103–114, World Scientific
52. Itô, K.: Distribution-valued processes arising from independent Brownian motion; *Math. Z.* **182** (1983) 17–33
53. Kubo, I. and Takenaka, S.: Calculus on Gaussian white noise I; *Proc. Japan Acad.* **56A** (1980) 376–380

54. Kubo, I. and Takenaka, S.: Calculus on Gaussian white noise II; *Proc. Japan Acad.* **56A** (1980) 411–416
55. Kubo, I. and Takenaka, S.: Calculus on Gaussian white noise III; *Proc. Japan Acad.* **57A** (1981) 433–437
56. Kubo, I. and Takenaka, S.: Calculus on Gaussian white noise IV; *Proc. Japan Acad.* **58A** (1982) 186–189
57. Kubo, I. and Yokoi, Y.: A remark on the space of testing random variables in the white noise calculus; *Nagoya Math. J.* **115** (1989) 139-149
58. Kuo, H.-H.: *Gaussian Measures in Banach Spaces.* Lecture Notes in Math. **463**, Springer-Verlag, 1975
59. Kuo, H.-H.: On Laplacian operators of generalized Brownian functionals; *Lecture Notes in Math.* **1203** (1986) 119–128, Springer-Verlag
60. Kuo, H.-H.: Fourier-Mehler transforms in white noise analysis; *Gaussian Random Fields*, K. Itô and T. Hida (eds.) (1991) 257–271, World Scientific
61. Kuo, H.-H.: Lectures on white noise analysis; *Soochow J. Math.* **18** (1992) 229–300
62. Kuo, H.-H.: *Infinite Dimensional Stochastic Analysis.* Nagoya University Lecture Notes in Math. **11** (1993)
63. Kuo, H.-H.: Analysis of white noise functionals; *Soochow J. Math.* **20** (1994) 419–464
64. Kuo, H.-H.: *White Noise Distribution Theory.* CRC Press, 1996
65. Kuo, H.-H.: White noise theory; in *Handbook of Stochastic Analysis and Applications*, D. Kannan and V. Lakshmikantham (eds.) (2002) 107–158, Marcel Dekker
66. Kuo, H.-H.: A quarter century of white noise theory; in: *Quantum Information IV*, T. Hida and K. Saitô (eds.) (2002) 1–37, World Scientific
67. Kuo, H.-H., Obata, N., and Saitô, K.: Lévy Laplacian of generalized functionals on a nuclear space; *J. Funct. Anal.* **94** (1990) 74–92
68. Kuo, H.-H., Obata, N., and Saitô, K.: White noise analysis based on the Lévy Laplacian; *Kokyuroku* **1227** (2001) 170–183, Research Institute of Mathematical Sciences, Kyoto University
69. Kuo, H.-H., Obata, N., and Saitô, K.: Diagonalization of the Lévy Laplacian and related stable processes; *Infinite Dimensional Analysis, Quantum Probability and Related Topics* **45** (2002) 317–331
70. Lávička, R.: The Lévy-Laplacian and differential operators of 2nd order in Hilbert spaces; *Comment. Math. Univ. Carolin.* **39** (1998) 115–135
71. Léandre, R. and Volovich, I. A.: Lévy Laplacian and Yang-Mills equation on manifolds; *Infinite Dimensional Analysis, Quantum Probability and Related Topics* **4** (2001) 161–172
72. Lévy, P.: *Leçons d'analyse fonctionnelle.* Gauthier-Villars, 1922
73. Lévy, P.: *Processus Stochastiques et Mouvement Brownien.* Gauthier-Villars, 1948
74. Lévy, P.: *Problèmes Concrets d'Analyse Fonctionnelle.* Gauthier-Villars, 1951
75. Lévy, P.: Random functions: a Laplacian random function depending on a point of Hilbert space; *Univ. Calif. Publ. Statistics* **2** (1956) 195–206

76. Nishi, K. and Saitô, K.: The Lévy Laplacian and the number operator; *RIMS Kokyuroku* **1193** (2001)
77. Nishi, K., Saitô, K., and Tsoi, A. H.: A stochastic expression of a semi-group generated by the Lévy Laplacian; in: *Quantum Information III*, T. Hida and K. Saitô (eds.) (2001) 105–117, World Scientific
78. Obata, N.: Analysis of the Lévy Laplacian; *Soochow J. Math.* **14** (1988) 105–109
79. Obata, N.: A note on certain permutation groups in the infinite dimensional rotation group; *Nagoya Math. J.* **109** (1988) 91–107
80. Obata, N.: Density of natural numbers and the Lévy group; *J. Number Theory* **30** (1988) 288–297
81. Obata, N.: The Lévy Laplacian and the mean value theorem; *Lecture Notes in Math.* **1379** (1989) 242–253, Springer-Verlag
82. Obata, N.: A characterization of the Lévy Laplacian in terms of infinite dimensional rotation groups; *Nagoya Math. J.* **118** (1990) 111–132
83. Obata, N.: *White Noise Calculus and Fock Space.* Lecture Notes in Math. **1577**, Springer-Verlag, 1994
84. Obata, N.: Harmonic analysis and infinite dimensional Laplacians on Gaussian space; in: *Quantum Probability and Related Topics IX*, L. Accardi et al. (eds.), World Scientific (1994) 327–370
85. Obata, N.: Quadratic quantum white noises and Lévy Laplacian; *Nonlinear Analysis* **47** (2001) 2437–2448
86. Paycha, S.: Renormalized traces as a looking glass into infinite-dimensional geometry; *Infinite Dimensional Analysis, Quantum Probability and Related Topics* **4** (2001) 221–266
87. Paycha, S.: Renormalized traces as a geometric tool; in *Geometric Methods for Quantum Field Theory*, World Scientific (2001) 293–360
88. Polishchuk, E. M.: *Continual Means and Boundary Value Problems in Function Spaces.* Birkhäuser, 1988
89. Potthoff, J. and Streit, L.: A characterization of Hida distributions; *J. Funct. Anal.* **101** (1991) 212–229
90. Saitô, K.: Itô's formula and Lévy's Laplacian; *Nagoya Math. J.* **108** (1987) 67–76
91. Saitô, K.: Lévy's Laplacian in the infinitesimal generator; *Research Report, Meijo Univ.* **28** (1988) 1–5
92. Saitô, K.: Itô's formula and Lévy's Laplacian II; *Nagoya Math. J.* **123** (1991) 153–169
93. Saitô, K.: A group generated by the Lévy Laplacian and the Fourier-Mehler transform; *Pitman Research Notes in Math. Series* **310** (1994) 274–288, Longman Scientific & Technical
94. Saitô, K.: A group generated by the Lévy Laplacian; *RIMS Kokyuroku* **874** (1994) 192–201
95. Saitô, K.: The Lévy Laplacian acting on Hida distributions; *Proc. Applied Math.* **5** (1996) 30–44
96. Saitô, K.: The Fourier-Mehler transform and the Lévy Laplacian; *Infinite Dimensional Harmonic Analysis* **1** D.+M. Gräbner (1996) 195–208

97. Saitô, K.: Transformations approximating a group generated by the Lévy Laplacian; *RIMS Kokyuroku* (1996) 214-224
98. Saitô, K.: A (C_0)-group generated by the Lévy Laplacian; *Journal of Stochastic Analysis and Applications* **16** (1998) 567–584
99. Saitô, K.: A (C_0)-group generated by the Lévy Laplacian II; *Infinite Dimensional Analysis, Quantum Probability and Related Topics* **1** (1998) 425–437
100. Saitô, K.: The Lévy Laplacian and Stochastic Processes; *Infinite Dimensional Harmonic Analysis* **2**, D.+M. Gräbner (1999) 306–318
101. Saitô, K.: Infinite dimensional stochastic processes generated by extensions of the Lévy Laplacian; *Publication of Centro Vito Volterra, Università Degli Studi di Roma "Tor Vergata"* **428** (2000) 1–12
102. Saitô, K.: The Lévy Laplacian and Stochastic Processes; *RIMS Kokyuroku* **1157** (2000) 101–114
103. Saitô, K.: A stochastic process generated by the Lévy Laplacian; *Acta Appl. Math.* **63** (2000) 363–373
104. Saitô, K.: The Lévy Laplacian and stable processes; *Chaos, Solitons & Fractals/ Les Treilles Special Issue*, International Solvay Institute, **12** (2001) 2865–2872
105. Saitô, K., and Tsoi, A. H.: The Lévy Laplacian as a self-adjoint operator; *Quantum Information*, T. Hida and K. Saitô (eds.) (1999) 159–171, World Scientific
106. Saitô, K., and Tsoi, A. H.: The Lévy Laplacian acting on Poisson noise functionals; *Infinite Dimensional Analysis, Quantum Probability and Related Topics* **2** (1999) 503–510
107. Saitô, K., and Tsoi, A. H.: Stochastic processes generated by functions of the Lévy Laplacian; in: *Quantum Information II*, T. Hida and K. Saitô (eds.) (2000) 183–194, World Scientific
108. Yoshizawa, H.: Rotation group of Hilbert space and its application to Brownian motion; in: *Proc. International Conference on Functional Analysis and Related Topics*, 414–423, 1970, University of Tokyo Press
109. Zhang, Y.: The Lévy Laplacian and Brownian particles in Hilbert spaces; *J. Funct. Anal.* **133** (1995) 425–441

On Reversible Computation in Asynchronous Systems

Jia Lee, Ferdinand Peper, Susumu Adachi, Shinro Mashiko
Communications Research Laboratory,
Nanotechnology Group,
588-2 Iwaoka, Iwaoka-cho,
Nishi-ku, Kobe-city, 651-2492
Japan
E-mail: {lijia,peper,sadachi,mashiko}@crl.go.jp

Abstract

Reversible computation has attracted much attention over the years, not only for its promise for computers with radically reduced power consumption, but also for its importance to Quantum Computing. Though studied extensively in a great variety of synchronous computation models, it is virtually unexplored in an asynchronous framework. This paper studies reversibility of asynchronous systems, focussing in particular on delay-insensitive circuits, a type of asynchronous circuit, whose correctness of operation is unaffected by arbitrary delays of signals. We start with Morita's Rotary Element (RE), a delay-insensitive reversible element that can process one signal at a time, and which has four input lines, four output lines, and two states. An RE is universal, because a universal Reversible Turing Machine can be constructed from it, as shown by Morita. We embed an RE on a two-dimensional Self-Timed Cellular Automaton (STCA), a type of cellular automaton that is asynchronous, i.e., that updates its cells at random times that are independent of each other. This proves that the STCA is capable of universal and reversible computation. We then propose a new delay-insensitive reversible element that can process one signal at a time, and which has only three input lines, three output lines, and two states. We prove the universality of this element, by constructing an RE from it. Finally, we investigate reversible delay-insensitive circuits that can process more than one signal at a time. Such circuits are more subtle than their single-signal counterparts, because a simple reordering of signal arrival times may (or may not) destroy their reversibility. We analyze an example in which an irreversible multi-signal circuit is constructed from reversible elements, and speculate about reasons why reversibility is not closed under circuit composition, and about ways to repair this shortcoming.

1 Introduction

Reversibility has played a key role in the quest to develop increasingly powerful and energy-efficient computers. As physical phenomena on the microscale are reversible, computers that test the limits of physics must conduct their computations in a reversible way, lest the microstate information about their history is transformed into some uncontrolled and inaccessible form, such as thermal entropy [11] of the system and its surroundings [4]. Reversibility is also a precondition for Quantum Computation, as the operations of a quantum computer can be expressed in terms of unitary transformations.

Most research on reversible computing [1, 2, 4, 5, 11, 13, 14] implicitly assumes that the underlying system is synchronously timed, i.e., that there is a central clock signal in accordance to which all logic elements switch in lock-step. For example, a Fredkin gate [5] only works correctly if all its input signals arrive roughly at the same time. As asynchronous systems come with a high degree of nondeterminism due to the randomness by which events in them are timed, they are thought to be incompatible with the backward determinism accompanying reversible computing. Still, the fact is that most physical interactions on the microscale are asynchronous. As with reversibility, asynchronicity tends to reduce power consumption and heat dissipation, be it for different reasons: whereas reversible systems reduce power consumption by minimizing entropy loss, asynchronous systems do so by letting logic elements be in a sleeping state if they have no work to do [6, 17]—an approach opposite to that in synchronous systems, where idle logic elements have to engage in dummy switching when receiving clock signals. The study of reversibility in asynchronous systems is also relevant in the context of Quantum Computing, as the absence of a clock signal allows a better isolation of a quantum system from its environment, thus improving its ability to maintain the superpositions of its states.

Synchronous cellular automata are a very popular model to conduct research on reversibility [8, 10, 13, 16, 23]. For a cellular automaton to be reversible, the left-hand side (LHS) of its transition rules usually consists of the same number of elements as on the right-hand side (RHS). This is achieved in the Billiard-Ball Model Cellular Automaton [13] by transition rules that change the states of a block of 2×2 cells all at the same time. Alternatively, in a Partitioned Cellular Automaton (PCA) [10, 16] cells are divided in partitions, and transition rules are applied to the partitions of cells and their neighbors, such that the number of partitions in the LHS of a rule is the same as the number of partitions in the RHS.

In this paper we study reversibility in asynchronous computing systems (see also [12, 18]). In all our constructions we use delay-insensitive circuits, a type of circuit whose correctness of operation is unaffected by arbitrary finite delays of signals. Since one of the reasons to do research on reversible circuit elements is the potential for building circuits consuming low amounts of energy, we shall only consider reversible elements that conserve the number of signals they process, because destroying and creating signals is energy-inefficient. Such elements will be called *conservative*.

In our first construction, we use a delay-insensitive reversible circuit element called a Rotary Element (RE) proposed in [15], which has four input lines, four output lines, and two states, and which can process only one signal at a time. Morita showed in [15] how to build a universal Reversible Turing Machine (RTM) from this element. We embed the RE on an asynchronous cellular automaton, a model in which transitions of cells are timed randomly and independently of each other. This shows that a universal reversible asynchronous cellular automaton exists. Our construction is based on a special type of asynchronous cellular automaton, a so-called two-dimensional Self-Timed Cellular Automaton (STCA) [12, 19, 20], which is basically a PCA with four partitions per cell, each partition carrying one of two possible states. The STCA employs four rotation-symmetric transition rules, each of which takes the four partitions of a cell and one partition of each of the four neighboring cells as arguments, and changes their states.

We then introduce a novel reversible logic element that has three input lines, three output lines, and two states, and which can process one signal at a time. An RE is constructed by building a circuit from these six-line reversible elements. This result demonstrates that an arbitrary RTM can be constructed from the six-line reversible element, signifying the importance of the six-line element for the study of reversibility in asynchronous systems.

We next investigate reversible delay-insensitive circuits that can process more than one signal at a time. To this end we employ an element proposed in [17, 18] that requires two input signals and produces two output signals. We discuss a number of illustrative problems with multi-signal circuits, a major one of which is that multi-signal delay-insensitive reversibility is not strictly closed under composition. In other words, it is possible to construct irreversible circuits from reversible elements, if insufficient care is taken as to how the elements are connected to each other. We also discuss possible solutions to this problem.

This paper is structured as follows. Section 2 gives a short introduction to delay-insensitive circuits. Section 3 describes the RE and shortly discusses Morita's construction to build an RTM from it. Section 4 describes the STCA model in more detail, followed by section 5, which shows how to embed an RE in an STCA, thus proving that an RTM can be simulated on an STCA. A problem with Morita's construction of an RTM from REs is lined out in section 6, and the solution to this problem will be used in our discussion in section 8, where a much more serious problem turns up. In section 7 we describe the six-line reversible delay-insensitive element, and show how to construct an RE from it, proving the element's universality. This is followed by section 8, which discusses problems encountered with multi-signal reversible delay-insensitive circuits, and speculates about possible solutions. This paper finishes with conclusions and a discussion.

2 Delay-Insensitive Circuits

A *Delay-Insensitive circuit* is a circuit whose correctness of operation is robust to arbitrary finite delays of its signals. Operations in delay-insensitive circuits are driven by signals: each circuit element is inactive unless it receives an appropriate set of input signals, after which it operates and becomes inactive again. As delay-insensitive circuits do not require a central clock signal, they are called asynchronous. Unlike in synchronous circuits that use binary valued signals, information in delay-insensitive circuits is usually represented by the presence or absence of one-valued signals on particular wires. To transmit a binary value, for example, two lines are used: a signal on one of the lines denotes the value 0, and a signal on the other line denotes the value 1. Called *dual-rail encoding*, this way of encoding allows a signal on only one of the two lines at a time. The absence of signals on both lines indicates that no information is being transmitted.

Delay-insensitive circuits consist of *modules*, which are elements with a finite number of input and output lines and a finite number of states. When receiving certain signals from the input lines, a module conducts an operation, as a result of which it may change its state and output certain signals to its output lines. Modules can be organized in a hierarchy, ranging from a module constructed from networks of modules, which may be as complex as a complete computer, to modules so simple that they cannot be subdivided any further, which we shall call *elements*. Modules are connected to each other by their input and output lines, such that the input of one module is the output of another module, or, in case self-connections are allowed, of the same module. It is assumed that a module can detect the presence of signals on each of its input and output lines, independently of the position of a signal on a line. A line is thus considered as a unit on which either a signal is present or not, and a signal is considered as an entity that fits exactly on a line. In practice this boils down to a model, in which a signal, after a possible delay to enter a line, fills the line instantaneously, rather than being a kind of particle moving from one end of a line to the other. Figure 1 gives examples of three delay-insensitive elements.

The *Merge* element merges two input streams of signals into a single output stream. It does so by redirecting a signal at either of its two input lines to its output line. The Merge element is not reversible, because from its output signal it cannot be deduced which of its input lines received the corresponding input signal. The Merge element keeps the number of signals in a circuit unchanged, and is therefore conservative. The *Fork* element produces one signal at each of its two output lines upon receiving a signal at its input line. A kind of fan-out element, the fork increases the number of

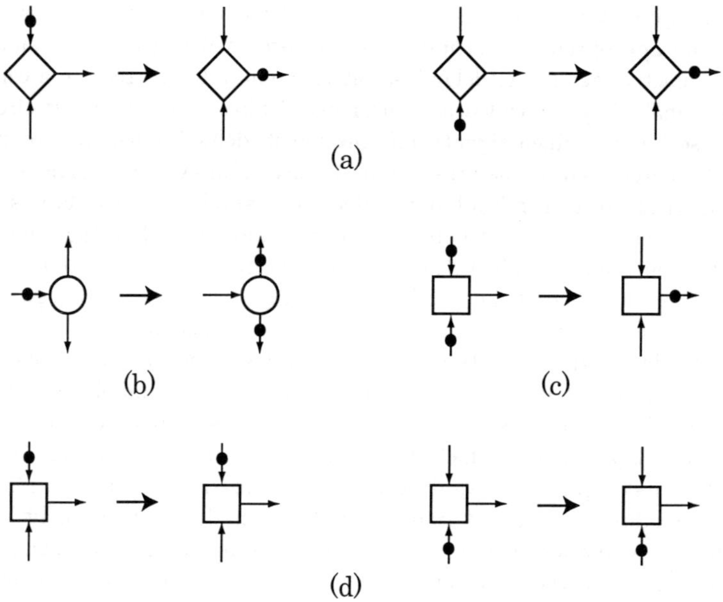

Figure 1: Functionality of three delay-insensitive elements, described by transition rules. The LHS (part left of the central arrow) of each rule denotes the state before a transition takes place, and the RHS (part right of the arrow) denotes the state after the transition. A blob on a line denotes a signal. (a) Merge element, which redirects a signal at any of its two input lines to its output line, (b) Fork element, which produces a signal at each of its two output lines upon receiving a signal at its input line, and (c) Join element, which produces a signal at its output line upon receiving a signal to each of its two input lines. If a Join element receives only one input signal, like in (d), the signal stays pending until the second input signal arrives.

signals in a circuit, and is thus not conservative. The *Join* element produces a signal on its output line whenever it receives one signal at each of its two input lines. When only one of the input lines receives a signal, it stays pending until a signal is received at the other input line. The first signal thus waits for the second signal to be input before they are joined into an output signal by the Join element. The Join decreases the number of signals in a circuit, and is thus not conservative.

3 An Eight-Line Reversible Element: Morita's Rotary Element

The *Rotary Element* (RE) introduced by Morita [15] is a reversible delay-insensitive element that can process one signal at a time. It has four input lines $\{n, e, s, w\}$, four output lines $\{n', e', s', w'\}$, and two states, the H-state and the V-state (see Fig. 2), which are displayed as horizontal and vertical bars respectively. When a signal arrives at one of the input lines of an RE, the RE operates on the signal, by possibly changing its state and transmitting the signal to an appropriate output line. If a signal comes from a direction parallel to the rotating bar of an RE, then it passes straight through to the opposite output line, without changing the direction of the bar (the state of the RE), as in Fig. 3(a); if a signal comes from a direction orthogonal to the rotating bar, then it turns right and rotates the bar by 90 degrees (Fig. 3(b)). Simultaneous signals on two or more input or output lines of an RE are not allowed. Since there is a one-to-one correspondence between the LHS and the RHS of each transition rule in Fig. 3, the RE is reversible. Obviously an RE conserves the number of signals.

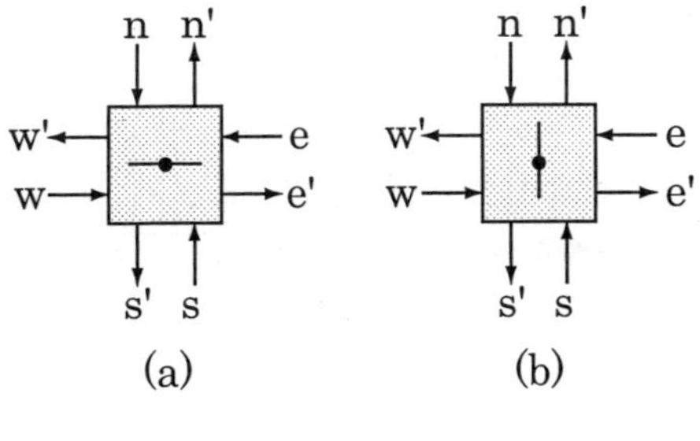

Figure 2: (a) An RE in the H-state, and (b) an RE in the V-state.

The RE is capable of universal computation, in the sense that any RTM can be realized by a network of an infinite number of RE modules, whereby at most one signal moves around in the entire network at any time [15]. The circuit can operate in an asynchronous mode, obliterating the need of a central clock signal, and in such a mode, delays in modules or lines do not

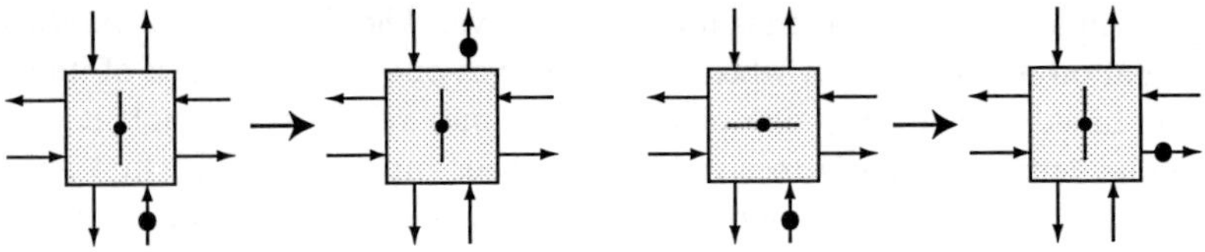

Figure 3: Transition rules describing an RE operating on an input signal in (a) parallel case, and (b) orthogonal case. A signal is denoted by a blob on a line.

affect the correct operation of the circuit. An example of a circuit composed of RE modules that simulates a particular RTM is given in Fig. 4, whereby the partial network of RE modules on the left is used for finite-state control, and the infinite array of blocks on the right represents the infinite number of tape cells of the Turing machine. This circuit computes the function $g(n) = 2n$, starting with a signal input to the "Begin" line. The computation ends when a signal emerges from the "End" line. A universal RTM can be constructed by equiping the circuit in Fig. 4 with an appropriate finite-state control unit.

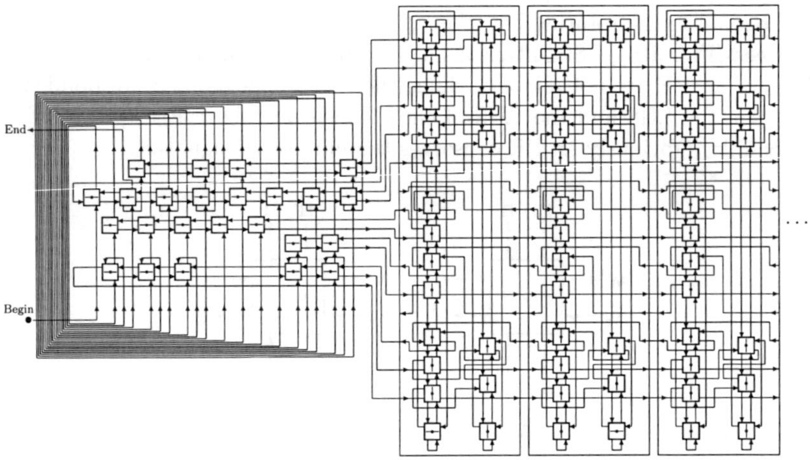

Figure 4: Realization of a Reversible Turing Machine (RTM) by REs. Each RE in the construction has its state set before the operation of the machine starts, such as to ensure a proper initial configuration of the Turing machine. (Reprinted from [15] copyright©2001, Springer)

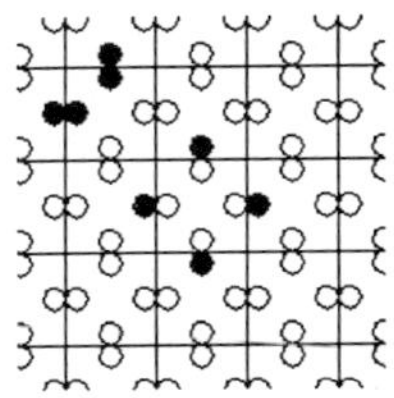

Figure 5: Cellular space of a Self-Timed Cellular Automaton (STCA).

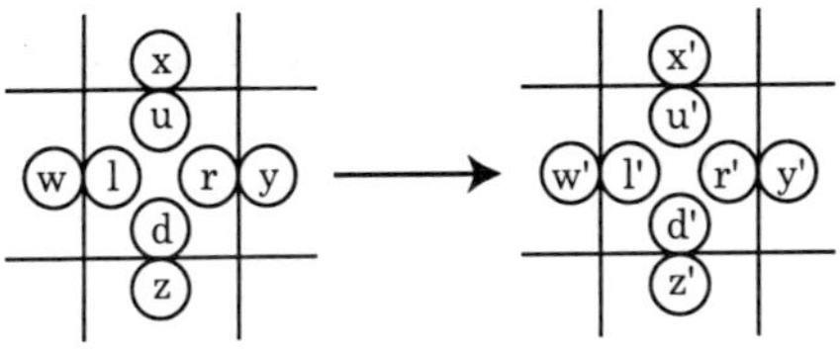

Figure 6: Transition rule used in an STCA. The variables in the circles denote states of the corresponding bits, which may be 0 or 1. The LHS of the rule (the part left of the arrow) gives the states of a cell's bits before a transition, and the RHS gives the states after the transition.

4 Self-Timed Cellular Automata

A *Self-Timed Cellular Automaton* (STCA) [19] is a two-dimensional array of identical cells, each of which can be in one of 16 states. Each cell's state is partitioned into four bits that are positioned along the cell's sides (see Fig. 5), where a filled circle denotes a 1-bit, and a blank circle denotes a 0-bit.

Each cell undergoes state transitions via transition rules that operate on the four bits of the cell and the nearest bit of each of its four neighbors. A transition rule

$$f(u, r, d, l, x, y, z, w) = (u', r', d', l', x', y', z', w')$$

can be depicted as in Fig. 6. where $u, u', r, r', d, d', l, l', x, x', y, y', z, z', w, w' \in \{0, 1\}$, and $(u, r, d, l, x, y, z, w) \neq (u', r', d', l', x', y', z', w')$. In STCA, transitions of cells are timed independently of each other and occur at arbitrary times, with the condition that no two neighboring cells may undergo transitions simultaneously. This is to avoid situations where neighboring cells attempt to simultaneously update the two bits common to the cells—a *write* conflict, which could lead to undefined values of bits. If no transition rule applies to the bits associated with a cell, the bits are left unchanged, even if the cell was randomly selected by the timing mechanism to undergo a transition. In this case, the cell is considered to undergo a *dummy* transition. Due to their timing model, STCA belong to the class of asynchronous cellular automata.

Let Conf(c) denote the set of all configurations over an STCA. For any $c_i, c_j \in \text{Conf}(c)$, there exists a global transition from c_i to c_j written by $c_i \rightarrow c_j$, iff c_j can be obtained by random selection and update of at least one cell in c_i at a time, and no neighboring cells in c_i can change a bit common to them to different values at the same time [20]. Moreover, global transitions in an STCA are reversible iff

$$\forall c_i, c_j, c_h \in \text{Conf}(c),\ c_i \rightarrow c_j \wedge c_h \rightarrow c_j \Longrightarrow c_i = c_h.$$

5 Embedding Rotary Elements in Self-Timed Cellular Automata

In this section, we introduce an STCA that can embed a network of REs (see also [12]). The interactions between the cells are described by the transition rules in Fig. 7. Since there is a one-to-one correspondence between the LHS and the RHS of each rule, the rules are locally reversible. The rules are also bit-conserving, since the numbers of 1-bits in the LHS and RHS of each rule are the same.

An interconnection line is implemented in the STCA by a straight continuous path of cells, all bits of which are 0. A signal is represented as a pattern of one 1-bit and seven 0-bits, as in Fig. 8(a). Applying transition rule 1 in Fig. 7 twice to this pattern moves the signal along the path of cells towards the east in Fig. 8(b). We adopt the convention that the rotation-symmetric equivalents of transition rules may also serve as transition rules. This allows, for example, the use of transition rule 1 for transmitting signals in directions towards the north, south, and west as well.

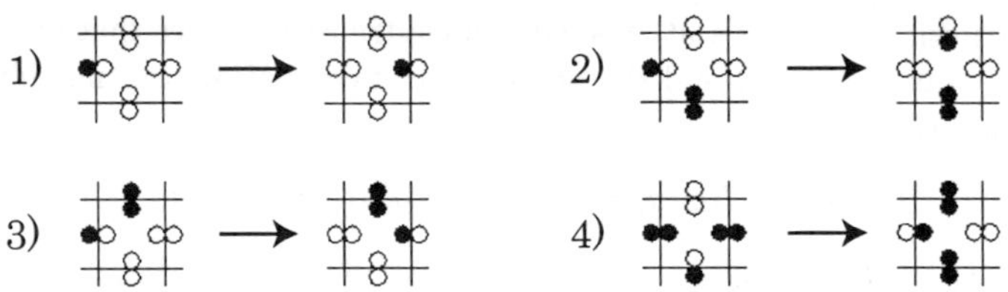

Figure 7: Transition rules used for implementing an RE on an STCA.

Figure 9(a) shows a pattern that is used as a left-turn element for signals. The sequence of configurations in Fig. 9(b) denotes a signal turning left. A signal coming from the opposite direction just passes through the left-turn element, as in Fig. 9(c). Using three left-turn elements, we can construct a right-turn element.

The pattern in Fig. 10(a) represents a rotation bar. A signal coming from a direction orthogonal to the bar will turn left and rotate the bar by 90 degrees as in Fig. 10(b).

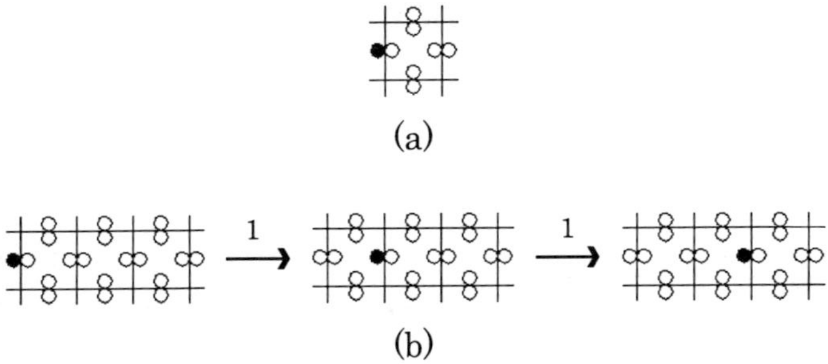

Figure 8: (a) A signal in the STCA. (b) Transmitting a signal to the east. The integer above each arrow denotes the label in Fig. 7 of the transition rule (or its rotation-symmetric equivalent) used to update the states of the cells in the configurations.

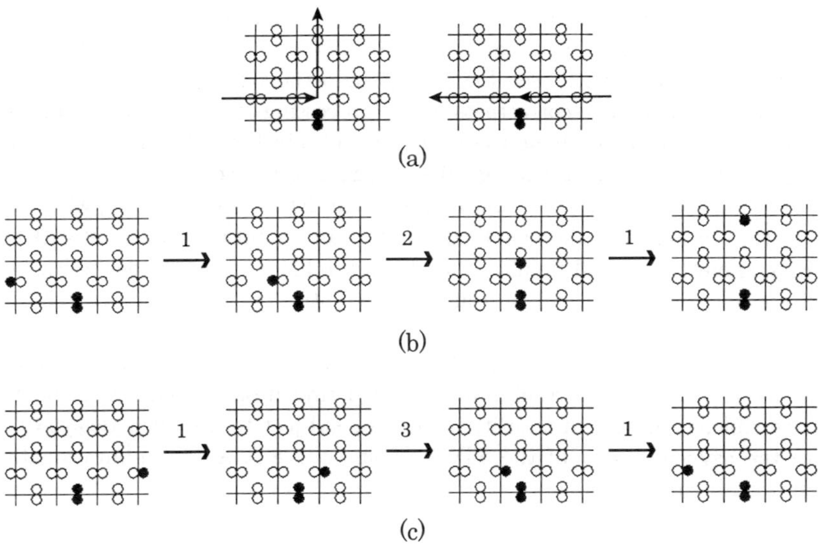

Figure 9: (a) The pattern on the left represents a left-turn element for signals coming from the west; the pattern at the right is the same, but it has no effect on signals coming from the opposite direction. The arrow indicates the direction in which a signal propagates. (b) A signal from the west turning left to the north. (c) A signal from the east passing through to the west.

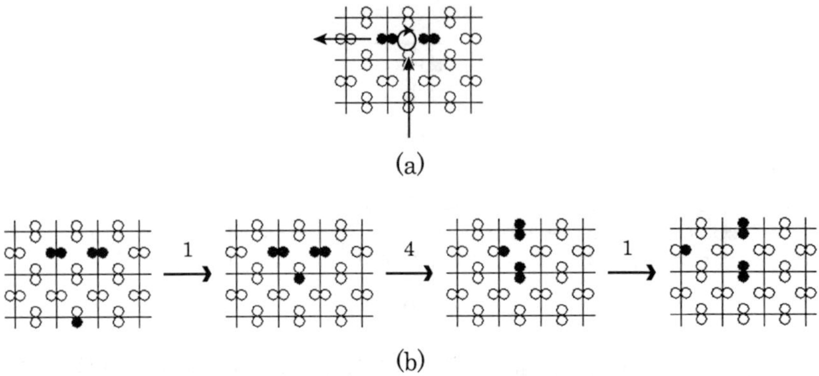

Figure 10: (a) A rotation bar in horizontal position (H-state). (b) A signal from the south turning to the west, while setting the bar to a vertical position (V-state).

A rotary element embedded in the STCA is a composition of the rotation bar and left-turn elements, as in Fig. 11. Figure 12 illustrates the operations of an RE operating on a signal that arrives at one of its input lines at directions parallel with (see Fig. 12(a)) or orthogonal to (see Fig. 12(b)) the rotation bar. The correctness of the processing in Fig. 12 can be easily verified, by locally applying the transition rules in Fig. 7 to the patterns in Fig. 12, tracing a signal through the configuration of cells.

Using the configuration that represents an RE in Fig. 11, along with the left-turn pattern and the signal pattern, we can lay out an infinite network of REs on our STCA and simulate an arbitrary RTM, such as the one in Fig. 4. Since only one signal moves around in the entire circuit in Fig. 4, at most one cell in the entire configuration undergoes a transition at a time, and this cell is the one at which the signal arrives. In this case, the local reversibility of the transition rules results in the global transitions of our STCA being reversible, even though transitions may be subject to arbitrary delays due to the asynchronicity of the STCA. Thus, our STCA is computationally universal and reversible.

6 A Problem with Self-Connections of Rotary Elements

In this section we discuss a problem with the construction in Fig. 4 of the RTM from REs, a problem that, though relatively benign, will come to haunt us in a more serious form in delay-insensitive reversible circuits that allow multiple signals to occur simultaneously (see section 8). The problem occurs when an RE has a self-connection, i.e., a line connecting an output

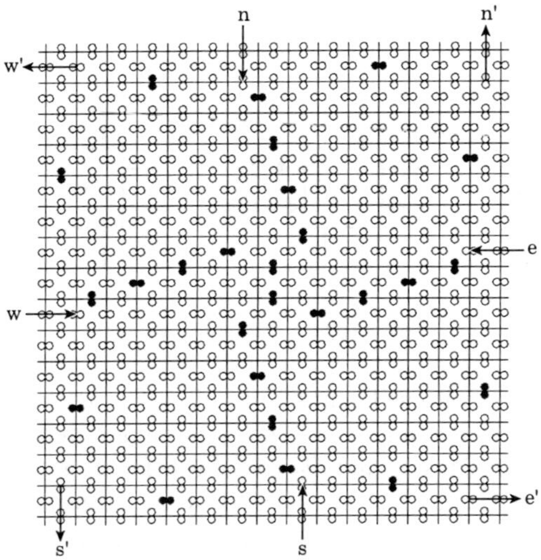

Figure 11: An RE in V-state, embedded in an STCA. Positions of the input and output lines in this RE slightly differ from those in the RE in Fig. 2.

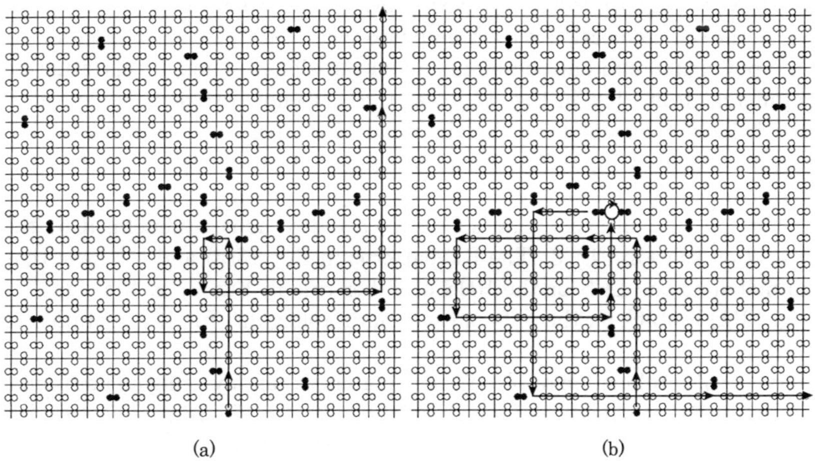

Figure 12: Traces of signals moving in the RE configuration when the rotation bar is (a) parallel with, or (b) orthogonal to a signal's direction.

to an input of the RE. An example where this occurs in Fig. 4 is the upper right RE of a tape cell of the RTM, or the lower left or lower right RE of a tape cell. Let's consider an RE that contains a signal on its self-connection (see Fig. 13(a)). This signal has arrived into this position due to the arrival

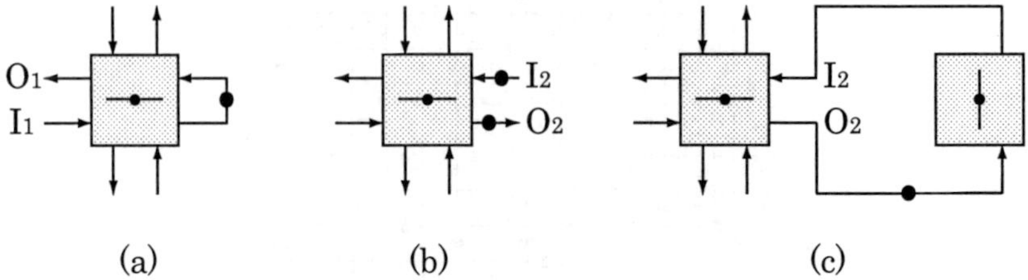

Figure 13: Problem with self-connections of an RE and a possible solution. (a) Signal stuck on self-connection. The signal cannot move, because, viewed from the RE, both lines I_2 and O_2 contain a signal, a situation similar to that in (b). This situation is not defined by the transition rules in Fig. 3. (c) Insertion of a dummy RE in the self-connection shields either the input line I_2 or the output line O_2 from the signal, as a result of which the originally self-connected RE now only sees one signal, guaranteeing its correct operation.

of a signal at input line I_1, and the signal is supposed to reenter the RE and leave it via output line O_1. Now, consider the situation in Fig. 13(b), in which the self-connection is broken up into two separate lines, one containing an input signal, and one containing an output signal. Though this situation is different from that in Fig. 13(a), as there are now two signals instead of one, it looks the same from the viewpoint of the RE. In both cases the RE sees one signal at input line I_2 and one signal at output line O_2. From the viewpoint of the RE there are thus two signals at its input and output lines in Fig. 13(a), a situation that should be avoided, as it is undefined by the transition rules in Fig. 3. The RE patterns in Fig. 13(a) and (b) have thus no match to the LHS or the RHS of either rule in Fig. 3, so the signal in Fig. 13(a) is stuck on the self-connection, disabling any further operation of the RE.

Can this problem be resolved? Fortunately, it can, by using an additional (dummy) RE to shield the input from the output of the self-connecting RE (see Fig. 13(c)). Now, the signal is either at the input line of the originally self-connecting RE or at the output line, but not at both simultaneously, so, our problem disappears. Another way to look at this solution is that the inserted dummy RE is a delay element, which makes it impossible that the signal is at both ends of the self-connection at the same time. By substituting the construction in Fig. 13(c) for each self-connecting RE in Fig. 4, the resulting RTM construction becomes correct.

When simulating the RTM in Fig. 4 on a cellular automaton, the above problem would not manifest itself, because in the implementation of the RTM on the cellular automaton, interconnection lines are represented as a path of cells, as pointed out in section 5. A path of cells does not exactly behave like an interconnection line, but is rather a string of dummy modules that do nothing else than transferring a signal, like the shielding module in Fig. 13(c). Also, a signal on a cellular automaton behaves more like a particle, rather than filling a line completely, so our assumption in section 2 that a module can check the presence or absence of a signal on a line does not hold. In an implementation of the RTM in Fig. 4 on a cellular automaton, shielding modules are thus implicitly used in self-connections. Shielding modules represented by cells are also used in other interconnection lines, but there they have no effect other than delaying signals.

7 A Six-Line Reversible Element

Although the RE can be used to realize universal reversible computers that operate asynchronously, it is somewhat complex, because it has four input lines and four output lines. Intuitively, the less complex a logic element is, the easier it is to implement physically. Is it possible to construct a universal Turing machine from an element with less lines? The answer is yes. In this section, we introduce a reversible delay-insensitive element which has three input lines, three output lines, and two states. This six-line element accepts one signal as input and produces one signal as output, so it is conservative. We will construct an RE from this element, thus proving its universality.[1] Let the three input lines of our element be $\{T, T', S\}$, the three output lines be $\{T_A, T_B, S'\}$, and the two states be A and B (see Fig. 14). A signal arriving on input line T is transferred to output line T_A (resp. T_B) when the element is in state A (resp. B), and the state is changed to B (resp. A) (Fig. 15(a)). A signal arriving on input line S is transferred to output line S' (resp. T_A) when the element is in state A (resp. B), and the state is always set to A (Fig. 15(b)). When the element is in state B, a signal that arrives on input line T' is merely transferred to output line T_B without changing the state of the element (Fig. 15(c)). The arrival of a signal on input line T' when the element is in state A is not allowed.[2] Moreover, the simultaneous existence

[1]Independently, Morita's group has systematically listed up all conservative reversible delay-insensitive elements accepting one input signal, with three input lines, three output lines, and two states [22]. The universality of some of the elements different from our element are proven in [22], whereas the universality of our element is proven in the current paper.

[2]It is possible to define an operation in which a signal arrives on input line T' while the element is in state A—for example such an operation may transfer the signal to output line S', and change the state to B—but it is unnecessary for our purposes and it merely makes the element more complicated, so we omit this possibility.

of more than one signal on the input or output lines is not allowed. Our element is reversible, because there is a one-to-one correspondence between the LHS and the RHS of each of the transition rules in Fig. 15.

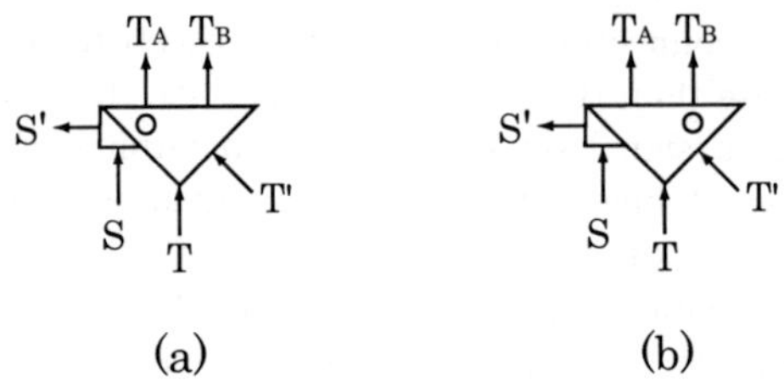

Figure 14: A six-line reversible element in (a) the A-state, and (b) the B-state

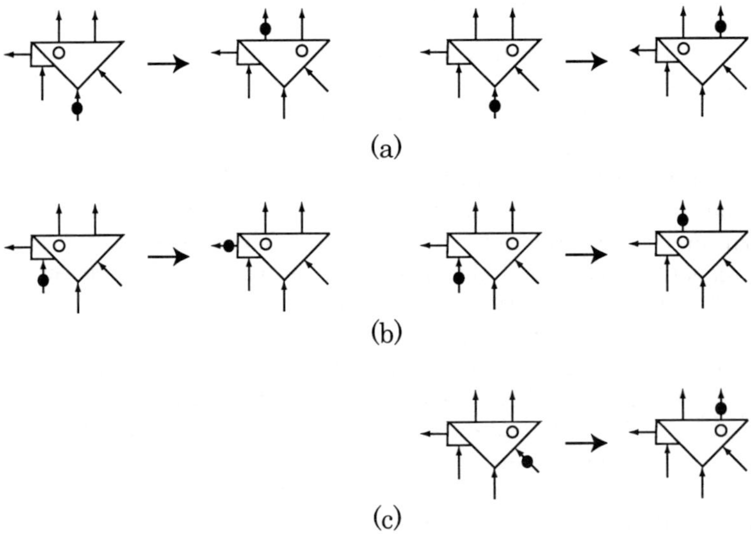

Figure 15: Transition rules describing the behavior of the six-line reversible element when a signal arrives on input line (a) T, (b) S, or (c) T'

Before constructing an RE using the six-line reversible element, we construct another useful module. Called a *Coding-Decoding* (C-D) module, it has four input lines $\{C_0, C_1, C_2, D\}$, four output lines $\{D_0, D_1, D_2, C\}$, and three states $\{0, 1, 2\}$ with 0 being the initial state (see Fig. 16(a)). Signals cannot arrive on input line C_0, C_1, or C_2 unless the C-D module is in state 0. A signal arriving on input line C_i ($i \in \{0, 1, 2\}$) changes the state of the C-D module from 0 to i, and is transferred to output line C. Subsequently, a signal coming from input line D is transferred to output line D_i if the C-D

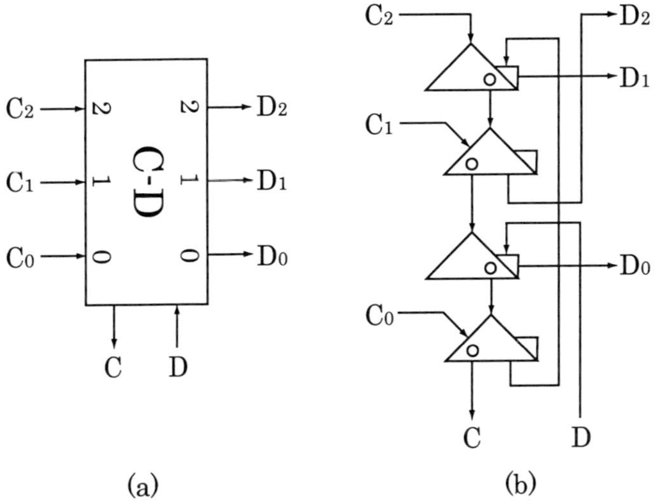

Figure 16: (a) A C-D module, and (b) its realization using our six-line reversible element.

module is in state i, and the state is always reset to 0. At most one input signal can be processed by a C-D module at any time. The realization of the C-D module from our six-line reversible element is given in Fig. 16(b), the validity of which can be easily verified by assuming a signal at one of the input lines and applying the transition rules in Fig. 15, tracing the route of the signal through the circuit.

Figure 17 shows the realization of an RE using C-D modules and six-line reversible elements. The four elements N_s, N_e, N_n, and N_w being in states A, B, A, and B respectively represent the RE being in state V, whereas N_s, N_e, N_n, and N_w being in states B, A, B, and A respectively represent the RE being in state H. We can thus construct a circuit from C-D modules and six-line reversible modules that simulates an arbitrary RTM, whereby there is at most one signal moving around, like in the RTM in [15]. Consequently, our delay-insensitive six-line reversible element is universal.

8 Multi-Signal Reversible Circuits

Up to this point we have only considered reversible delay-insensitive elements that can process a single signal at a time. In this section we investigate some of the issues arising when reversible delay-insensitive elements can process more than one signal at a time. Again we require that circuit elements are conservative. The simplest possible element satisfying these conditions is the so-called *Conservative Join*, abbreviated as *CJoin*, an element introduced

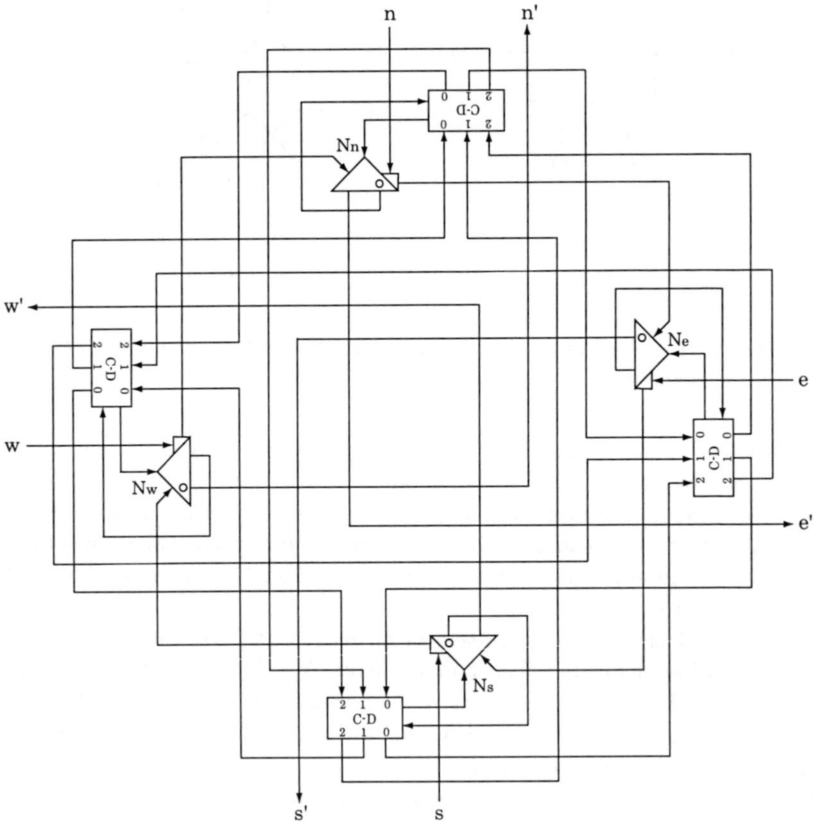

Figure 17: Realization of an RE using six-line reversible elements and C-D modules, the latter of which are composed of six-line reversible elements. All C-D modules in the circuit are initially in state 0 and the RE in state V.

in [17] that is like the Join element in section 2, but then with two output lines instead of one (see Fig. 18(a)). The transition rules of the CJoin in

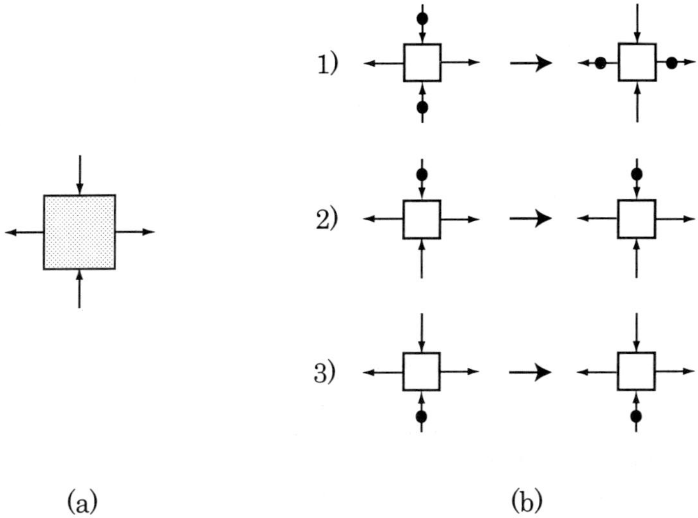

Figure 18: (a) A Conservative Join element (CJoin). (b) Behavior of the CJoin, defined by transition rules. Upon receiving one signal at each of its input lines, the element outputs one signal at each of its output lines, by applying transition rule (1). If only one of the input lines has received a signal, the element keeps the signal pending like in rule (2) or (3), until a signal from the other input line is received, after which transition rule (1) applies.

Fig. 18(b) appear reversible, because there is a one-to-one correspondence between their respective LHSs and RHSs. We now encounter the first potential problem that may jeopardize reversibility. Consider the two sequences of states of the CJoin and its interconnection lines in Fig. 19. The sequences differ in that the input signals to the CJoin arrive in a different order. As the two sequences converge towards the same state, it is impossible to derive from the latter part of each sequence what the former part is. We have thus lost the property of reversibility due to the different orders at which signals can arrive at the CJoin. In Delay-Insensitive systems, however, the order at which input signals arrive at an element is unimportant as to the resulting output, because a different order can be assumed to be the result of delays experienced by the input signals, delays that should not affect the

(a) (b)

Figure 19: Two sequences in which input signals arrive at a CJoin. In (a) the input signal from the north arrives first, whereas in (b) the input signal from the south arrives first. Both sequences are different, but they converge towards the same state of two input signals being input to the CJoin. In a strict sense, this destroys reversibility, because from the latter parts of the sequences the former parts cannot be uniquely determined.

output result.[3] Due to the delay-insensitive character of the circuits, we shall therefore assume that sequences corresponding to different orders at which signals arrive at elements are identical if these sequences can be transformed into each other by imposing delays on signals. The two sequences in Fig. 19 can thus be considered as identical, thereby implying that reversibility is not lost. In this interpretation we thus consider the CJoin to be reversible.

Is every circuit built up from reversible delay-insensitive elements itself reversible and delay-insensitive? To see that the answer to this question is more subtle than may be expected in first instance, we construct a Merge (see section 2) from REs and CJoins. As observed in section 2, a Merge is irreversible, since from its output signal it is impossible to derive from which line the corresponding input signal was received. REs and CJoins, on the other hand, are reversible. Consider the construction in Fig. 20. When a signal is input to line I_1, the subsequent actions of the elements give rise to the output of a signal on line O. It can be easily checked by applying the elements' transition rules, following their local actions, that an input signal to line I_2 also gives rise to the output of a signal on line O. Now, let's investigate in more detail what happens when a signal is input to line I_1. Two signals will then be output from CJ_1, i.e., one to RE_1, and one to RE_2, bringing about the situation of RE_2 in Fig. 21. Examining the transition rules in Fig. 3, we conclude that the situation in Fig. 21 is undefined. In other words, the circuit in Fig. 20 is incorrect, because it leads to a situation at RE_2 that does not match the LHS or RHS of any of the transition rules for the RE in Fig. 3. This is very similar to the problem lined out in section 6.

Before discarding the circuit in Fig. 20, we try to repair it in a way

[3]There is one exception in delay-insensitive circuits in which the order of arrival of input signals does affect the output of an element, and this is if the element conducts *arbitration*, i.e., when it must choose between competing input signals, each choice resulting in a different output. We shall not consider arbitration in this paper, since its nondeterminism seems incompatible with the backward determinism of reversibility.

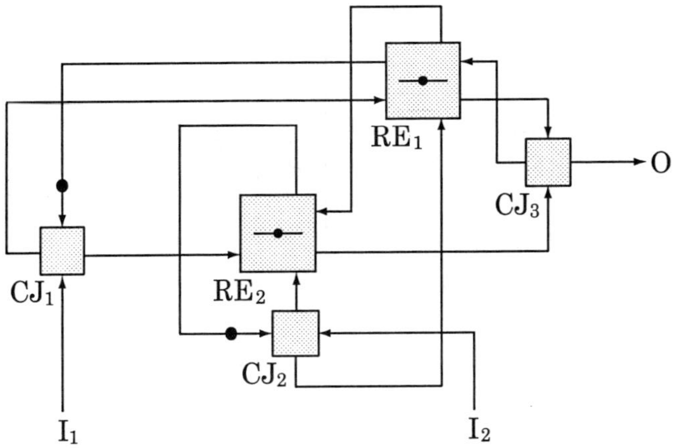

Figure 20: Circuit that is build up from REs and CJoins and that behaves as a Merge element. Even though the REs and the CJoins are reversible, the resulting Merge circuit is irreversible. Why does this happen?

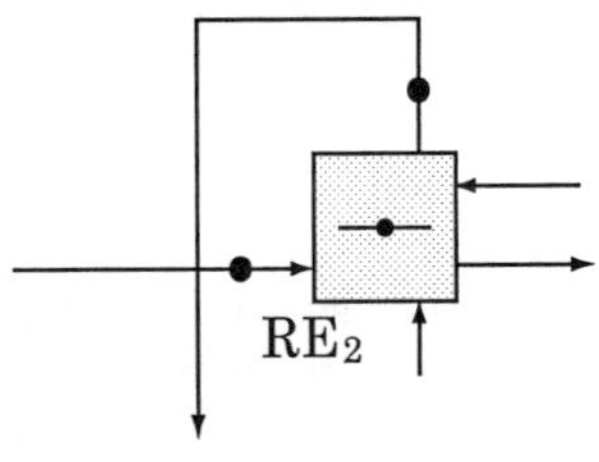

Figure 21: A situation undefined by the RE transition rules that occurs when the circuit in Fig. 20 receives an input signal from line I_1. Note that the signal that was pending at the input of CJ_2 in Fig. 20 has been depicted closer to RE_2 for illustrative reasons. This has no influence on the circuit behavior, because the signal is assumed in section 2 to fill the entire interconnection line from RE_2 to CJ_2.

similar to that of section 6. To this end we add Rotary Element RE_3 to the circuit such as to shield the pending signal on the line from RE_2 to CJ_2 from the output side of RE_2 (see Fig. 22). Now, the situation in Fig. 21

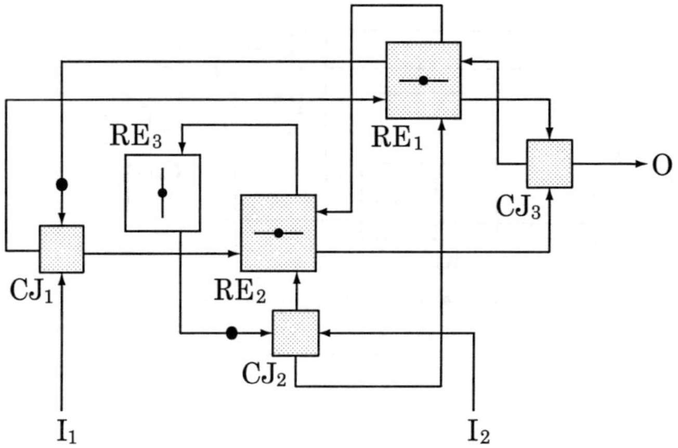

Figure 22: The same circuit as in Fig. 20, but then Rotary Element RE_3 has been added to shield RE_2 from the pending signal at CJ_2. The rotary function of RE_3 is not used, but rather RE_3's ability to transfer a signal from one of its input lines to one of its output lines. RE_3 can thus be considered as just an identity element, or, in other words, a piece of wire along which a signal can be transferred. This construction still causes undefined situations at RE_2, because the signal pending at the input line of CJ_2 can move backwards through RE_3 due to the reversibility of RE_3, thus interfering with the operation of RE_2 when a signal is input to the circuit at line I_1.

may not occur if the signal pending at CJ_2 stays effectively between RE_3 and CJ_2. However, the impossibility to guarantee this is exactly where the next problem occurs: due to the reversibility of RE_3, the signal pending at CJ_2 may travel in the reverse direction beyond RE_3, thus interfering with the operation of RE_2. The undefined situation in this circuit shown in Fig. 21 can thus not be solved by the insertion of element RE_3 with the aim of shielding element RE_2 from an output signal it has produced previously. Unlike in section 6, here shielding doesn't work, because there are two independent signals to deal with.

The circuit in Fig. 20 has more problems. Due to the reversibility of the circuit elements, the signal pending between RE_2 and CJ_2 may be moved back through RE_2 and RE_1, until it is on the line between RE_1 and CJ_3. In this state the circuit is unable to accept a signal from line I_1, because that would give undefined situations.

We conclude that the example in Fig. 20 of an irreversible circuit built from reversible delay-insensitive elements, though illustrative, is incorrect, and that the circuit seems not to be repairable in a straightforward way. It may thus still be possible that only reversible circuits can be built from reversible elements, provided the elements are connected correctly to each other, i.e., in such a way that no situations occur in the circuit that are undefined with respect to the transition rules of the elements. Finding the conditions in accordance to which the elements should be connected to each other is still an open problem.

9 Conclusions and Discussion

In this paper, we discuss reversible delay-insensitive models that are universal in the sense that they can simulate a universal Reversible Turing Machine (RTM). All our constructions are based on Morita's use of his Rotary Element (RE) for realizing an RTM. In our first construction we implement the RE on a Self-Timed Cellular Automaton (STCA) that requires only four rotation-symmetric transition rules to express the interactions between its cells. For our second construction we propose a novel reversible delay-insensitive element with three input and three output lines—less than the four input and four output lines of the RE. This element is universal, like the RE, which follows from the fact that it is possible to construct an RE from the element. Since only a single signal is used at any time to conduct computations with these models, the ordering of the operations is strictly defined. Together with the local reversibility of the transition rules, this is sufficient to guarantee reversibility of the models on a global level.

The models are not reversible in the traditional sense, as a signal may undergo arbitrary delays at any time. In synchronous reversible systems, such behavior is not permitted, because signals that are static in two successive clock cycles, should remain so indefinitely, due to the reversibility condition. However, a clock is an artificial construct that plays no role in most reversible processes in the physical world. It may thus be unnecessary to require for a process to be reversible that it behaves in accordance to a synchronous timing scheme. Delay-insensitivity is a more natural state of affairs: the molecule cascades in [7] is an example of a system in nature that exhibits it. We thus believe it justified to call our single-signal delay-insensitive models reversible. Whether such models indeed guarantee a decreased energy-consumption due to the preservation of entropy, as is shown theoretically for synchronous reversible models (but apparently never demonstrated experimentally), remains to be proven.

If more than one signal at a time is allowed in delay-insensitive circuits, a whole new situation occurs. As section 8 shows, the occurrence of signal delays may change the order at which signals arrive at an element, and in a strict sense this destroys reversibility. Following a less strict interpretation, however, we assume that reversibility is maintained under different orderings of signal arrival times, because such an interpretation is better compatible with delay-insensitivity.

A potentially more serious problem with multiple-signal reversible delay-insensitive circuits is illustrated by an example in which an irreversible circuit is composed from reversible circuit elements. This example, though illustrating some typical pitfalls in designing reversible delay-insensitive circuits, is shown to be incorrect: it leads to undefined situations in which there are multiple signals on the input and output lines of an RE—a problem similar to the problem in section 6, where a single signal on a self-connection of the RE was incorrectly interpreted by the RE as two signals, one signal on the input and one on the output to which the self-connection connects. Solving this problem by placing dummy REs in interconnection lines to shield other REs from signals—a solution working well for the single-signal case—does not work for the multiple-signal case, because due to reversibility, a signal may move backwards through a dummy RE, nullifying the shielding function.

What can we learn from the example in Fig. 20? The lesson is that when building a reversible delay-insensitive circuit, care should be taken as to how the circuit elements are connected. While this also holds for irreversible delay-insensitive circuits, it is especially important to keep it in mind for reversible circuits. More specifically:

- Reversible circuits are reversible! That is, they can compute backwards, and interfere with the operations of subcircuits at their input side. This is even more of an issue if there is no clock to drive the computation forward, like in delay-insensitive circuits.
- Not only should attention be given to input signals of a circuit element, but also to output signals, because they can still interfere with the element's operation. This point becomes especially clear in a Quantum Computation framework (e.g. [3]), as the inclusion of a complex conjugate in each operation suggests that input and output should be treated at the same level.

There is a very bright side to our failure to repair the example in Fig. 20: it leaves open the possibility that exact conditions can be formulated, that, if satisfied, guarantee that circuits composed of reversible elements are reversible themselves. Finding such conditions is an important research topic for reversible delay-insensitive systems.

We finally wish to mention a probablistic approach [21] (see also [9]) to the study of reversibility that we believe may be fruitful for the analysis on a microscale level of asynchronous systems in terms of Brownian motion. In this approach, reversibility is defined as the symmetry in time of the probability distribution of a Brownian Bridge, which is a Markov Gaussian

Process characterized by a certain covariance function. Complementary to our program, this approach may lead to a better characterization in terms of the entropy associated with computation in asynchronous reversible systems.

Acknowledgements

We would like to thank Prof. Kenichi Morita at Hiroshima University, Japan, for the helpful discussions. This research is financed by the Ministry of Public Management and Home Affairs in Japan.

References

[1] C. Bennett (1973) Logic reversibility of computation, *IBM J. Research and Development* **17**, 525–532

[2] A. De Vos, B. Raa, and L. Storme (2002) Generating the group of reversible logic gates, *J. Physics A: Mathematical and General* **35**, 7063–7078

[3] R.P. Feynman (1985) Quantum Mechanical Computers, *Optics News* **11**, 11-20

[4] M. Frank, C. Vieri, M.J. Ammer, N. Love, N.H. Margolus, and T. Knight Jr. (1998) A scalable reversible computer in silicon, In: Calude, C.S., Casti, J., Dinneen, M.J. (eds.): *Unconventional Models of Computation*, Springer-Verlag, Singapore, 183–200

[5] E. Fredkin and T. Toffoli (1982) Conservative logic, *International J. Theoretical Physics* **21**, 219–253

[6] S. Hauck (1995) Asynchronous design methodologies: an overview, *Proc. IEEE* **83**(1), 69–93

[7] A.J. Heinrich, C.P. Lutz, J.A. Gupta, and D.M. Eigler (2002) Molecule cascades, *Science* **298**, 1381-1387

[8] P. Hertling (1998) Embedding cellular automata into reversible ones, In: C.S. Calude, J. Casti, M.J. Dinneen (eds.), *Proc. 1st Conf. Unconventional Models of Computation (UMC'98)*, Springer-Verlag, Singapore, 243–256

[9] T. Hida (2000) Complexity in White Noise Analysis, In: *Quantum Information II*, T. Hida and K. Saitô (Eds.), World Scientific Publishing, Singapore, 61–70

[10] K. Imai and K. Morita (2000) A computation-universal two-dimensional 8-state triangular reversible cellular automaton, *Theoretical Computer Science* **231**(2), 181–191

[11] R. Landauer (1961) Irreversibility and heat generation in the computing process, *IBM J. Research and Development* **5**, 183–191

[12] J. Lee, F. Peper, S. Adachi, K. Morita, and S. Mashiko (2002) Reversible computation in asynchronous cellular automata, *Proc. 3rd Conf. on Unconventional Models of Computation (UMC'02), Lecture Notes in Computer Science (LNCS)* **2509**, Springer-Verlag, Berlin Heidelberg, 220–229

[13] N. Margolus (1984) Physics-like models of computation, *Physica D* **10**, 81–95

[14] R.C. Merkle (1993) Reversible electronic logic using switches, *Nanotechnology* **4**, 21–40

[15] K. Morita (2001) A simple universal logic element and cellular automata for reversible computing, *Lecture Notes in Computer Science (LNCS)* **2055**, 102–113

[16] K. Morita, Y. Tojima, and K. Imai (2001) A simple computer embedded in a reversible and number-conserving two-dimensional cellular space, *Multi. Val. Logic* **6** 483–514

[17] P. Patra (1995) Approaches to design of circuits for low-power computation, *Ph.D. Thesis* University of Texas at Austin

[18] P. Patra and D.S. Fussell (1996) Conservative Delay-Insensitive Circuits, *Workshop on Physics and Computation (PhysComp)*, pages 248-259

[19] F. Peper, T. Isokawa, N. Kouda, and N. Matsui (2002) Self-timed cellular automata and their computational ability, *Future Generation Computer Systems* **18**(7), 893–904

[20] F. Peper, J. Lee, S. Adachi, and S. Mashiko (2003) Laying out circuits on asynchronous cellular arrays: a step towards feasible nanocomputers?, *Nanotechnology* **14**, 469-485.

[21] Si Si (1999) Reversibility of random fields, Paper presented at *Nagoya Workshop on Infinite Dimensional Analysis and Quantum Probability*, Nov., organized by N. Obata.

[22] K. Tanaka (2003) Universal Reversible Logic Elements with 3 Inputs, 3 Outputs, and 2 States, *Master Thesis*, in Japanese, Hiroshima University, Japan.

[23] T. Toffoli (1977) Computation and construction universality of reversible cellular automata, *J. Computer and System Sciences* **15**, 213–231

A CHARACTERIZATION OF GENERALIZED LÉVY WHITE NOISE FUNCTIONALS *

YUH-JIA LEE

Department of Applied Mathematics
National University of Kaohsiung
Kaohsiung, TAIWAN 811
E-mail: yjlee@nuk.edu.tw

HSIN-HUNG SHIH

Center of General Education
Kun Shan University of Technology
Tainan, TAIWAN 710

The generalized Lévy functionals are characterized in terms of their Segal-Bargmann transforms (S-transforms, for abbreviation). In the course of the proof the locality of the functionals are studied. The results extend the Potthoff-Streit's characterization theorem and Lee's characterization theorem for Gaussian case to non-Gaussian cases.

1. Introduction

In the investigation of Gaussian white noise analysis, the Segal-Bargmann transform (S-transform, for abbrev.) plays a key role. In order to identify if a functional is the S-transform of certain generalized Gaussian white noise functionals (also known as U-functionals), it is necessary to characterize the generalized functionals in terms of their U-functionals. Such a characterization theorems were done by Potthoff-Streit[34] and Lee[22] and later by many other authors. In [22,24], the locality of generalized functionals was also given.

Recently the generalization of Hida's theory to non-Gaussian cases, such as generalized Poisson functionals or the more general cases of generalized

*Rsearch supported by the National Science Council of ROC(Taiwan)

Lévy functionals, have received great attention. Thanks to the integral representation of the S-transform of Lévy functionals [25], we are able to extend Hida's distribution theory to Lévy functionals [29]. In this paper we characterize generalized Lévy functionals in terms of their U-functionals. Using the growth estimate of a given U-functionals, we can also identify the locality of the generalized Lévy functional associated with the given U-functionals. The theorem extends both the Potthoff-Streit's characterization theorem [34] and also the characterization theorem (with locality) in Lee[22,24].

2. Preliminaries

By a Lévy process we mean an additive process $X = \{X(t) : t \in \mathbb{R}\}$ on a probability space $(\Omega, \mathcal{M}, \mathcal{P})$, which is a right continuous process having left limits and the characteristic function $\mathbb{E}[e^{ir(X(t)-X(s))}]$ of $X(t) - X(s)$, $t > s$, is given by $\mathbb{E}[e^{ir(X(t)-X(s))}] = \exp[(t-s)f_X(r)]$, where $f_X(r)$ is the Lévy function associated with X, which is uniquely represented as

$$f_X(r) = i\,\mu\, r - \frac{\sigma^2 r^2}{2} + \int_{|u|>0} \left(e^{iru} - 1 - \frac{iru}{1+u^2}\right) \frac{1+u^2}{u^2}\, d\beta(u), \quad r \in \mathbb{R}, \tag{1}$$

where $i = \sqrt{-1}$, μ is a real constant, β is a positive finite measure on $\mathbb{R}$ with $\sigma^2 = \beta(\{0\})$ (see [4,37]). In this paper, we denote the Lévy measure of X by β_0, which is a positive measure on $\mathbb{R}_*(= \mathbb{R} \setminus \{0\})$ defined by

$$\beta_0(E) = \int_E \frac{1+u^2}{u^2}\, d\beta(u)$$

for $E \in \mathcal{B}(\mathbb{R}_*)$.

Let $\mathcal{S}'$ be the space of tempered distributions on $\mathbb{R}$, which is the dual of the Schwartz space $\mathcal{S}$. As we have shown in [25] that under the condition that

$$\int_{-\infty}^{+\infty} |u|\, d\beta(u) < \infty$$

the Minlos theorem (see e.g. [5,6]) guarantees that there exists a unique probability measure Λ on $(\mathcal{S}', \mathcal{B}(\mathcal{S}'))$ with the characteristic functional $\mathcal{C}$ on $\mathcal{S}$ given by

$$\mathcal{C}(\eta) = \exp\left[\int_{-\infty}^{+\infty} f_X(\eta(t))\, dt\right], \quad \eta \in \mathcal{S}, \tag{2}$$

where $(\cdot,\cdot)$ is the $\mathcal{S}'$-$\mathcal{S}$ pairing. Λ is then called the Lévy white noise measure and $(\mathcal{S}',\mathcal{B}(\mathcal{S}'),\Lambda)$ which will serve as the underlying probability space in our study is called the Lévy white noise space. Further, if β satisfies the following absolute moment condition

$$\int_{-\infty}^{+\infty} |u|^n \, d\beta(u) < \infty \quad \forall n \in \mathbb{N} \tag{3}$$

then the space $L^2(\mathcal{S}',\Lambda)$ of quadratic integrable functionals would include all the cylinder polynomials. Many important cases, such as Gaussian, Poisson, Gamma white noises measure, etc. satisfy the condition (3).

2.1 *The chaos decomposition of Lévy white noise functionals*

In this paper, we consider only the class of measures β in (1) associated with Λ that satisfy the absolute moment condition (3). The condition(3) is equivalent to the condition that the class of all cylinder polynomials is contained in $L^2(\mathcal{S}',\Lambda)$, the complex L^2-space over $\mathcal{S}'$ with respect to Λ (see [25]). For $\eta \in \mathcal{S}$, it is easy to see that the mean and variance of the random variables $(\cdot,\,\eta)$ are given by

$$\mathbb{E}[(\cdot,\,\eta)] = \tau_1 \int_{-\infty}^{+\infty} \eta(t)\, dt \quad \text{and} \quad \mathbb{V}[(\cdot,\,\eta)] = \tau_2 \int_{-\infty}^{+\infty} \eta(t)^2\, dt, \tag{4}$$

where $\tau_1 = \mu + \int_{-\infty}^{+\infty} u\, d\beta(u)$ and $\tau_2 = \int_{-\infty}^{+\infty} (1+u^2)\, d\beta(u)$. Now, for each $\rho \in L^1 \cap L^2(\mathbb{R}, dt)$, choose a sequence $\{\eta_n\} \subset \mathcal{S}$ so that $\eta_n \to \rho$ in $L^1 \cap L^2(\mathbb{R}, dt)$ under the norm $|\cdot|_{L^1(\mathbb{R},dt)} + |\cdot|_{L^2(\mathbb{R},dt)}$. Then it follows from (4) that $\{(\cdot,\,\eta_n)\}$ forms a Cauchy sequence in $L^2(\mathcal{S}',\Lambda)$. Denote by $\langle\cdot,\,\rho\rangle$ the L^2-limit of $\{(\cdot,\,\eta_n)\}$. When $\rho = 1_{(s,\,t]}$, the indicator of $(s,\,t]$, the characteristic function of $\langle\cdot,\rho\rangle$ is exactly the same as (1). Thus the Lévy process X on $(\mathcal{S}',\mathcal{B}(\mathcal{S}'),\Lambda)$ can be represented by

$$X(t\,;\,x) = \begin{cases} \langle x,\, 1_{[0,\,t]}\rangle, & \text{if } t \geq 0 \\ -\langle x,\, 1_{[t,\,0]}\rangle, & \text{if } t < 0, \ x \in \mathcal{S}'. \end{cases}$$

Taking the time derivative formally, we get $\dot{X}(t\,;\,x) = x(t)$, $x \in \mathcal{S}'$. From this observation we may regard the elements of $\mathcal{S}'$ as the sample paths of Lévy white noise and we call the members of $L^2(\mathcal{S}',\Lambda)$ quadratic integrable Lévy white noise functionals. For $\eta = \eta_1 + i\,\eta_2 \in L^1_c \cap L^2_c(\mathbb{R}, dt)$ with η_1, $\eta_2 \in L^1 \cap L^2(\mathbb{R}, dt)$, define $\langle\cdot,\,\eta\rangle$ by $\langle\cdot,\,\eta_1\rangle + i\,\langle\cdot,\,\eta_2\rangle$, where V_c denotes the complexification of a real locally convex space V, then the identities in (4) also hold for the complex-valued random variable $\langle\cdot,\,\eta\rangle$.

The following three important Lévy white noise measures all satisfy the Condition (3):

- the Gaussian white noise measure Λ_W: $\mu = 0$, $\beta = \delta_0$.
- the Poisson white noise measure Λ_P: $\mu = 1/2$, $\beta = (1/2)\delta_1$.
- the Gamma white noise measure Λ_G: $\mu = \int_0^\infty e^{-u}/(1+u^2)\,du$, $\beta(E) = \int_{E\cap[0,+\infty)} u\,e^{-u}/(1+u^2)\,du$ for $E \in \mathcal{B}(\mathbb{R})$,

where δ_a is the Dirac measure concentrated on the point a.

Let $\mathcal{B}_b(\mathbb{R}^2_*)$ be the class of all bounded Borel subsets E of $\mathbb{R}^2_* = \mathbb{R}^2 \setminus \{(t,0) : t \in \mathbb{R}\}$. Define the product measure $d\nu(t,u) = d\beta_0(u)\,dt$ on $\mathcal{B}(\mathbb{R}^2_*)$. For $E \in \mathcal{B}_b(\mathbb{R}^2_*)$, let $N(E\,;\,\cdot)$ be a random variable on $(\mathcal{S}', \mathcal{B}(\mathcal{S}'))$ defined by

$$N(E\,;\,x) = \left|\{\,(t,u) \in E : X(t\,;\,x) - X(t-\,;\,x) = u\,\}\right|.$$

Then $N(E\,;\cdot)$ is Poisson distributed with the intensity measure ν, and the system of $\{\,N(E\,;\,x) - \nu(E) : E \in \mathcal{B}_b(\mathbb{R}^2_*)\,;\, x \in \mathcal{S}'\,\}$ forms an independent random measure with zero mean. We denote "$N(E\,;\,x) - \nu(E)$" by "$N_0(E\,;\,x)$". Then $\mathbb{E}\big[N_0(E)\,N_0(F)\big] = \nu(E\cap F)$. Let $f = \sum_{i=1}^m a_i\,1_{E_i}$ be a complex-valued simple function on $\mathbb{R}^2_*$, where a_i's $\in \mathbb{C}$ and E_i's $\in \mathcal{B}_b(\mathbb{R}^2_*)$ are disjoint. The stochastic integral $\int_{\mathbb{R}^2_*} f(s)\,dN_0(s)$ with respect to N_0 is defined by

$$\int_{\mathbb{R}^2_*} f(s)\,dN_0(s) := \sum_{i=1}^m a_i\,N_0(E_i).$$

Then we have the isometry

$$\left\| \int_{\mathbb{R}^2_*} f(s)\,dN_0(s) \right\|^2_{L^2(\mathcal{S}',\Lambda)} = \int_{\mathbb{R}^2_*} |f(s)|^2\,d\nu(s).$$

Extend by continuity the integral $\int_{\mathbb{R}^2_*} f(s)\,dN_0(s)$ is then defined for any $f \in L^2_c(\mathbb{R}^2_*, \nu)$.

The Lévy-Itô decomposition (see [10,37]) ensures that for $b > a$,

$$X(b) - X(a) = \tau_1\,(b-a) + \sigma(B(b) - B(a)) + \int_{\mathbb{R}^2_*} u 1_{(a,b]\times\mathbb{R}_*}(t,u)dN_0(t,u), \tag{5}$$

where $B = \{B(t) : t \in \mathbb{R}\}$ is an 1-dimensional Wiener process, independent of the system of $\{N(E) : E \in \mathcal{B}_b(\mathbb{R}^2_*)\}$. Let λ be a positive measure on

$\mathcal{B}(\mathbb{R}^2)$ defined by $d\lambda(t,u) = (1+u^2)\,d\beta(u)\,dt$ and let M be a $L^2(\mathcal{S}',\Lambda)$-valued function on $\{E \in \mathcal{B}(\mathbb{R}^2) : \lambda(E) < +\infty\}$ defined by

$$M(E) = \sigma \int_{-\infty}^{+\infty} 1_E(t,0)\,dB(t) + \int_{\mathbb{R}^2_*} u\,1_E(t,u)\,dN_0(t,u).$$

Then the system of $\{\,M(E\,;x) \,:\, E \in \mathcal{B}(\mathbb{R}^2)$ with $\lambda(E) < +\infty;\ x \in \mathcal{S}'\,\}$ forms an independent random measure with zero mean and $\mathbb{E}[M(E)\,M(F)] = \lambda(E\cap F)$ for all $E,\,F \in \mathcal{B}(\mathbb{R}^2)$. The multiple stochastic integral with respect to M, introduced by Itô in [11], is described as follows.

Let V_n denote the class of all complex-valued symmetric simple functions on $(\mathbb{R}^2)^n$ of the form

$$\sum_{i=1}^{m} a_i\,\widehat{1}_{E_1^i\times\cdots\times E_n^i} \tag{6}$$

in which $\lambda(E_j^i) < +\infty$ for $j=1,\dots,n$, $E_1^i < \cdots < E_n^i$, and $\widehat{1}_{E_1^i\times\cdots\times E_n^i}$ is the symmetrization of $1_{E_1^i\times\cdots\times E_n^i}$. Here, "$A < B$" means that $t_1 < t_2$ for $(t_1,u_1) \in A$ and $(t_2,u_2) \in B$. For any $g \in V_n$ of the form (6), the multiple stochastic integral $I_n(g)$ of g with respect to M is defined by $\sum_{i=1}^m a_i \prod_{j=1}^n M(E_j^i)$ which satisfies the isometry

$$\big\|I_n(g)\big\|^2_{L^2(\mathcal{S}',\Lambda)} = n!\int_{\mathbb{R}^2}\cdots\int_{\mathbb{R}^2} \big|g(s_1,\dots,s_n)\big|^2\,d\lambda(s_1)\cdots d\lambda(s_n).$$

Since V_n is dense in $\widehat{L^2_c}((\mathbb{R}^2)^n,\lambda^{\otimes n})$, the closed subspace of $L^2_c((\mathbb{R}^2)^n,\lambda^{\otimes n})$ consisting of all symmetric complex-valued functions in $L^2_c((\mathbb{R}^2)^n,\lambda^{\otimes n})$, $I_n(g)$ can be extended by continuity to all $g \in \widehat{L^2_c}((\mathbb{R}^2)^n,\lambda^{\otimes n})$, called the multiple Lévy-Itô integral of order n with respect to M. In notation, we write

$$I_n(g) = \int_{\mathbb{R}^2}\cdots\int_{\mathbb{R}^2} g(s_1,\dots,s_n)\,dM(s_1)\cdots dM(s_n).$$

From the definition of multiple Lévy-Itô integrals, it is obvious that as $n \neq m$, $I_n(g)$ and $I_m(h)$ are orthogonal to each other in $L^2(\mathcal{S}',\Lambda)$ for $g \in \widehat{L^2_c}((\mathbb{R}^2)^n,\lambda^{\otimes n})$ and $h \in \widehat{L^2_c}((\mathbb{R}^2)^m,\lambda^{\otimes m})$. In particular, for any $g \in L^2_c(\mathbb{R}^2,\lambda)$,

$$I_1(g) = \sigma\int_{-\infty}^{+\infty} g(t,0)\,dB(t) + \int_{\mathbb{R}^2_*} u\,g(t,u)\,dN_0(t,u). \tag{7}$$

For $\eta \in L^2_c(\mathbb{R},dt)$, we identify η with $\eta\otimes 1$, where $\eta\otimes h(t,u) = \eta(t)\,h(u)$ for any $(t,u) \in \mathbb{R}^2$. Then the random variable $\langle\cdot,\eta\rangle$ with $\eta \in L^1_c \cap L^2_c(\mathbb{R},dt)$

can be expressed in terms of the Lévy-Itô integral by

$$\langle \cdot, \eta \rangle = \tau_1 \int_{-\infty}^{+\infty} \eta(t)\, dt + I_1(\eta). \tag{8}$$

Note that the centered Lévy process $\widetilde{X} = \{X(t) - \tau_1 t : t \in \mathbb{R}\}$ is a martingale so that the stochastic integral with respect to $\widetilde{X}$ is defined. The following proposition show the connection between $I_n(\phi)$ the n-the multiple Lévy integral with integrand ϕ.

Proposition 2.1. [29] *Let $\phi \in \widehat{L^2_c}(\mathbb{R}^n, (dt)^{\otimes n})$ be given. Then*

$$I_n(\phi) = \int_{-\infty}^{+\infty} \cdots \int_{-\infty}^{+\infty} \phi(t_1, \ldots, t_n)\, d\widetilde{X}(t_1) \cdots d\widetilde{X}(t_n). \tag{9}$$

Theorem 2.1. *(Itô [11]) Let φ be given in $L^2(\mathcal{S}', \Lambda)$. Then there exist uniquely a sequence of kernel functions $\phi_n \in \widehat{L^2_c}((\mathbb{R}^2)^n, \lambda^{\otimes n})$, $n \in \mathbb{N} \cup \{0\}$, such that φ is equal to the orthogonal direct sum $\sum_{n=0}^{\infty} \oplus I_n(\phi_n)$. In notation, we write $\varphi \sim (\phi_n)$.*

2.2 Analyticity of Lévy white noise measures

Let $U \subset \mathcal{S}_c$ be open. A complex-valued functional φ on $\mathcal{S}_c$ is said to be holomorphic on U if φ is locally bounded on U and for all $x \in U$ and $y \in \mathcal{S}_c$, the mapping $z \mapsto \varphi(x + z\, y)$ from $\mathbb{C}$ to $\mathbb{C}$ is holomorphic in some neighborhood of zero in $\mathbb{C}$. An probability measure ν is called analytic (in the sense of Kondratiev *et al.*) if its characteristic functional $\mathcal{C}_\nu$ is holomorphic on an open neighborhood of zero on $\mathcal{S}_c$ (see [15,16,17,38]). ν is said to be *completely analytic* if its characteristic function $\mathcal{C}_\nu$ is holomorphic on the whole space $\mathcal{S}_c$, . It is worth noting that the Gaussian, Poisson, and Gamma white noise measures are all analytic measures, but in general Lévy white noise measures Λ are not necessarily analytic.

In the existing literature, the authors used to consider only the Lévy white noise functionals with a further assumptions either on the analyticity of the measure Λ or on the measure β. It is desirable to make clear the relationship between the analyticity of Λ and the associated measure β before continuing our further discussions.

Let $A = -d^2/dt^2 + (1 + t^2)$ be a densely defined self-adjoint operator on $L^2(\mathbb{R}, dt)$ and $\{e_n : n \in \mathbb{N}_0\}$ be the CONS of $L^2(\mathbb{R}^2, dt)$, consisting of all Hermite functions on $\mathbb{R}$, formed by the eigenfunctions of A with corresponding eigenvalues $2n + 2$, $n \in \mathbb{N} \cup \{0\}$. For any $p \in \mathbb{R}$ and $\eta \in L^2(\mathbb{R}, dt)$, define $|\eta|_p := |A^p \eta|_{L^2(\mathbb{R},dt)}$ and let $\mathcal{S}_p$ be the completion of the class $\{\eta \in L^2(\mathbb{R}, dt) : |\eta|_p < +\infty\}$ with respect to $|\cdot|_p$-norm. Then $\mathcal{S}_p$ is a real separable Hilbert space and we have the continuous inclusions:

$$\mathcal{S} \subset \mathcal{S}_p \subset \mathcal{S}_q \subset L^2(\mathbb{R}, dt) \subset \mathcal{S}_{-q} \subset \mathcal{S}_{-p} \subset \mathcal{S}', \quad p > q \geq 0.$$

Next let

$$K_\beta = \sup\left\{m \geq 0 : \int_{-\infty}^{+\infty} e^{m|u|}\, d\beta(u) < +\infty\right\},$$

and for any $p \geq 1$, let

$$K_{\nu,p} = \sup_{\eta\in\mathcal{S}, |\eta|_p \leq m'} \left\{m' \geq 0 : \int_{\mathbb{R}^2_*} (e^{|u\,\eta(t)|} - 1 - |u\eta(t)|) d\nu(t,u) < +\infty \right\};$$

$$K_{\Lambda,p} = \sup_{\eta\in\mathcal{S}, |\eta|_p \leq m''} \left\{m'' \geq 0 : \int_{\mathcal{S}'} e^{(x,\,\eta)}\, \Lambda(dx) < +\infty \right\}.$$

Then we have

Lemma 2.1. [29] *For $p \geq 1$,*

$$2^p \sqrt[4]{e^2\pi}\, K_\beta \geq K_{\Lambda,p} \geq K_{\nu,p} \geq \frac{1}{2\, c_p}\, K_\beta,$$

where $c_p = \sup\{|\delta_t|_{-p} : t \in \mathbb{R}\}$.

As a consequence of Lemma 2.3, we have

Theorem 2.2.

(i) The Lévy white noise measure Λ is analytic if and only if $K_\beta > 0$.
(ii) The Lévy white noise measure Λ is completely analytic if and only if $K_\beta = +\infty$.

We remark that Λ_W and Λ_P are both completely analytic. For the Gamma white noise measure Λ_G, it is easily seen that $K_\beta = 1$ by a direct calculation. So, by Theorem 2.4, Λ_G is analytic but not completely analytic.

We recall that an equivalent description of analytic measures given in [Lemma 3.9[17]] which is stated as a lemma below.

Lemma 2.2. [17] *Let m be a probability measure on $(\mathcal{S}', \mathcal{B}(\mathcal{S}'))$. The following statements are equivalent.*

(i) m is an analytic measure.
(ii) there exist $p_m \in \mathbb{N}$ and $C > 0$ such that $\left|\int_{\mathcal{S}'} (x,\, \eta)^n\, dm(x)\right| \leq n!\, C^n\, |\eta|^n_{p_m}$, $\eta \in \mathcal{S}_{p_m,c}$.

(iii) *there exist* $p'_m \in \mathbb{N}$ *and* $\epsilon_m > 0$ *such that* $\int_{\mathcal{S}'} \exp[\epsilon_m |x|_{-p'_m}]\, dm(x) < +\infty$.

It follows from Lemma 2.3, Theorem 2.4, Lemma 2.5, and [25] , we characterize the analyticity of Λ as follows.

Theorem 2.3. *The following statements are equivalent.*

(i) *The Lévy white noise measure* Λ *is analytic.*
(ii) *There exists* $m > 0$ *such that* $\int_{-\infty}^{+\infty} e^{m|u|}\, d\beta(u) < +\infty$.
(iii) *There exists* $p \geq 1$ *and* $\epsilon > 0$ *such that* $\int_{\mathcal{S}'} e^{\epsilon|x|_{-p}}\, \Lambda(dx) < +\infty$.
(iv) *For all* $q \geq 1$, *there exists* $\epsilon > 0$ *such that* $\mathbb{E}[e^{\langle \cdot, \eta \rangle}] < +\infty$ *for all* $\eta \in \mathcal{S}$ *with* $|\eta|_q \leq \epsilon$.

In this case, $\mathbb{E}[e^{\langle \cdot, \eta \rangle}] = \exp\left[\int_{-\infty}^{+\infty} f_X(-i\, \eta(t))\, dt\right]$, *where* f_X *is the Lévy function defined in (1).*

To end this subsection we sum up the properties of the complete analyticity of the measure Λ into a characterization theorem.

Theorem 2.4. *Assume that*

$$\int_{|u|>0} |u|\, d\beta_0(u) < +\infty.$$

Then the following statements are equivalent.

(i) *The Lévy white noise measure* Λ *is completely analytic.*
(ii) *For any* $m > 0$, $\int_{-\infty}^{+\infty} e^{m|u|}\, d\beta(u) < +\infty$.
(iii) *For* $p > 5/4$ *and for all* $\epsilon > 0$, $\int_{\mathcal{S}'} e^{\epsilon|x|_{-p}}\, \Lambda(dx) < +\infty$.
(iv) $\mathbb{E}[e^{\langle \cdot, \eta \rangle}] < +\infty$ *for all* $\eta \in \mathcal{S}$.

2.3 The ***S****-transform*

For $\varphi \in L^2(\mathcal{S}', \Lambda)$, the well-known Itô theorem (see Itô [11]) asserts that φ can be written in a unique manner as an orthogonal direct sum of multiple Lévy-Itô integrals $I_n(\phi_n)$ with kernel functions $\phi_n \in \widehat{L^2}((\mathbb{R}^2)^n, \lambda^{\otimes n})$, where λ is the product measure $dt \otimes (1 + u^2)\, d\beta(u)$ (see Theorem 2.2).

Define the S-transform of φ on $L^2(\mathbb{R}^2, \lambda)$ by

$$S\varphi(g) = \sum_{n=0}^{\infty} \int_{\mathbb{R}^2} \cdots \int_{\mathbb{R}^2} \phi_n(x_1, \ldots, x_n)\, g(x_1) \cdots g(x_n)\, d\lambda(x_1) \cdots d\lambda(x_n).$$

Then S-transform is a unitary operator maps from $L^2(\mathcal{S}', \Lambda)$ onto the Bargmann-Segal-Dwyer space $\mathcal{F}^1(L^2_c(\mathbb{R}^2, \lambda))$ over $L^2_c(\mathbb{R}^2, \lambda)$ (see Theorem 3.8). Moreover, we have

$$\|\varphi\|^2_{L^2(\mathcal{S}',\Lambda)} = \sum_{n=0}^{\infty} \frac{1}{n!} \|D^n S\varphi(0)\|^2_{\mathcal{L}^{(n)}_{(2)}(L^2_c(\mathbb{R}^2,\lambda))},$$

where D is the Fréchet derivative of $S\varphi$ and $\|\cdot\|_{\mathcal{L}^{(n)}_{(2)}(H)}$ denotes the Hilbert-Schmidt operator norm of a n-linear functional on a Hilbert space H.

Equivalently the S-transform $S\varphi$ of φ may also be defined as an analytic function in $L^2_c(\mathbb{R}^2, \lambda)$ given by

$$S\varphi(g) = \int_{\mathcal{S}'} \varphi(x)\, \mathcal{E}_M(g)(x)\, \Lambda(dx) \quad \text{for all } g \in L^2_c(\mathbb{R}^2, \lambda), \tag{10}$$

where $\mathcal{E}_M(g) = \sum_{n=0}^{\infty} \frac{1}{n!} I_n(g^{\otimes n})$, also known as the coherent state functional associated with g. It has been shown in [25] that $\mathcal{E}_M(g)$ has a stochastic representation given as follows.

Let $\mathcal{O} = \{g \in L^2_c(\mathbb{R}^2, \lambda) : g^* \in L^1_c \cap L^\infty_c(\mathbb{R}^2_*, \nu)\}$, where $g^*(t,u) = u\, g(t,u)$. Then

$$\mathcal{E}_M(g) = \gamma_g\, \vartheta_g$$

for $g \in L^2_c(\mathbb{R}^2, \lambda)$, where γ_g is the L^2-limit of $\{\gamma_{g_m}\}$, $\{g_m\} \subset \mathcal{O}$, $g_m \to g$ in $L^2_c(\mathbb{R}^2, \lambda)$ and

$$\gamma_{g_m}(x) = \exp\left[-\int_{\mathbb{R}^2_*} g^*_m(x)\, d\nu(x)\right] \prod_{t \in J_X(x)} (1 + g^*_m(t,\, X(t;\, x) - X(t-;\, x))), \tag{11}$$

$J_X(x) = \{t \in \mathbb{R} : X(t;\, x) - X(t-;\, x) \neq 0\}$; and

$$\vartheta_g(x) = \exp\left[-\frac{\sigma^2}{2} \int_{-\infty}^{+\infty} g(t,0)^2\, dt + \sigma \int_{-\infty}^{+\infty} g(t,0)\, dB(t;\, x)\right], \quad x \in \mathcal{S}'. \tag{12}$$

The other important transform $\mathcal{T}\varphi$, called the $\mathcal{T}$-transform introduced by Hida in white noise analysis, is defined as a complex-valued functional on the real locally convex space $L^1 \cap L^2(\mathbb{R}, dt)$ so that

$$\mathcal{T}\varphi(\eta) = \int_{\mathcal{S}'} \varphi(x)\, e^{i\langle x,\, \eta\rangle}\, \Lambda(dx), \quad \eta \in L^1 \cap L^2(\mathbb{R}, dt). \tag{13}$$

Theorem 2.5. [25] *Let $\xi \in L^2_c(\mathbb{R}^2, \lambda)$ so that*

$$\int_{\mathbb{R}^2_*} \left(e^{2\,|\xi^*(t,u)|} - 1 - 2\,|\xi^*(t,u)|\right) d\nu(t,u) < +\infty, \tag{14}$$

where $\xi^*(t,u) = u\xi(t,u)$. Then

$$\int_{\mathcal{S}'} \varphi(x)\, e^{I_1(\xi)(x)}\, \Lambda(dx) = \mathbb{E}[e^{I_1(\xi)}] \cdot S\varphi(\phi_\xi) \quad \textit{for any } \varphi \in L^2(\mathcal{S}', \Lambda), \tag{15}$$

where $\phi_\xi(t,u) = (e^{\xi^*(t,u)} - 1)/u$ if $u \neq 0$; otherwise, $\phi_\xi(t,u) = \xi(t,u)$.

For $\eta \in L^1 \cap L^\infty(\mathbb{R}, dt)$, put

$$\xi_m(t,u) = i\,\eta(t)\, 1_{\{u:1/m \le |u| \le m\}}(u), \quad m \in \mathbb{N}.$$

Then ξ_m's satisfy the condition (14). Replacing ξ by ξ_m in (15) and letting $m \to \infty$ one obtains the connection between the S-transform and $\mathcal{T}$-transform by the following equality

$$\mathcal{T}\varphi(\eta) = \mathbb{E}[e^{i\langle\cdot,\,\eta\rangle}] \cdot S\varphi(\phi_{i\eta}). \tag{16}$$

Since the class $\{e^{i\langle\cdot,\eta\rangle} : \eta \in \mathcal{S}\}$ is total in $L^2(\mathcal{S}', \Lambda)$, (13) and (16) imply that the S- and $\mathcal{T}$-transforms are both injective maps on $L^2(\mathcal{S}', \Lambda)$.

Example 2.1.

(1) When Λ is the Gaussian white noise measure Λ_W, let $\xi = \eta$ for $\eta \in L^2_c(\mathbb{R}, dt)$. Then $\phi_\xi = \eta$ since $\beta = \delta_0$. We have

$$\mathcal{E}_M(\phi_\xi) = \mathcal{E}_M(\eta) = \exp\left[\langle\cdot,\,\eta\rangle - \frac{1}{2}\int_{-\infty}^{+\infty} \eta(t)^2\, dt\right]. \tag{17}$$

(2) When Λ is the Poisson white noise measure Λ_P, let $\xi = \mathrm{Log}(1+\eta)$ for $\eta \in L^1_c(\mathbb{R}, dt)$ with $|\eta|_\infty < 1$. Then $\phi_\xi = \eta$ since $\beta = (1/2)\,\delta_1$. We have

$$\mathcal{E}_M(\phi_\xi) = \mathcal{E}_M(\eta) = \exp\left[\langle\cdot,\,\mathrm{Log}(1+\eta)\rangle - \int_{-\infty}^{+\infty} \eta(t)\, dt\right]. \tag{18}$$

(3) When Λ is the Gamma white noise measure Λ_G, let $\xi = \eta/(\eta+1)$ for $\eta \in L^1_c \cap L^\infty_c(\mathbb{R}, dt)$ with $|\eta|_\infty < 1$. Then $\xi \in L^1_c \cap L^\infty_c(\mathbb{R}, dt)$ and it also satisfies the condition (14). By Theorem 2.9 and a direct calculation, we obtain for $x \in \mathcal{S}'$

$$\mathcal{E}_M(\phi_\xi)(x) = \frac{e^{I_1(\xi)(x)}}{\mathbb{E}[e^{I_1(\xi)}]} = \exp\left[\langle x,\,\eta/(1+\eta)\rangle - \int_{-\infty}^{+\infty} \mathrm{Log}(1+\eta(t))\, dt\right].$$

3. The Spaces of Test and Generalized Functions

In order to set up the calculus of Lévy white noise functionals, we need to construct the spaces of test and generalized functions. Follow the idea of Gaussian white noise calculus, we start with the construction of a Gel'fand triple on $L^2(\mathbb{R}^2, \lambda)$.

3.1 A Gel'fand triple associated with $\mathbf{L^2(\mathbb{R}^2, \lambda)}$

Consider the real Hilbert space $L^2(\mathbb{R}, \gamma)$, where $d\gamma(u) = (1 + u^2)\, d\beta(u)$ is a finite measure on $(\mathbb{R}, \mathcal{B}(\mathbb{R}))$. Let $\zeta_0 \equiv 1/\sqrt{\tau_2}$ and choose a complete orthonormal basis (CONS, in abbreviation) $\{\zeta_0, \zeta_1, \dots\}$ of $L^2(\mathbb{R}, \gamma)$, consisting of continuous functions on $\mathbb{R}$. It can be obtained by applying the Gram-Schmidt orthogonalization process to $\{1, u, u^2, \dots\}$.

Define a linear operator A_β densely on $L^2(\mathbb{R}, \gamma)$ by $A_\beta\, \zeta_n = r_n\, \zeta_n$ for $n = 0, 1, \dots$, where the eigenvalues r_n's satisfy the assumption that $1 < r_0 \leq r_1 \leq \cdots$ and $\|A_\beta^{-\alpha}\|_{HS}^2 = \sum_n 1/r_n^{2\alpha} < +\infty$ for some $\alpha > 0$. For example, when Λ is the Gamma white noise measure Λ_G, $\{\zeta_n\}$ can be chosen so that $\zeta_n = L_n^{(1)}/\sqrt{n+1}$ for $n \in \mathbb{N} \cup \{0\}$, where $\{L_n^{(1)}\}$ is the system of Laguerre polynomials on $[0, +\infty)$ with parameter 1. In this case, if we take $A_\beta = -u\,(d^2/du^2) + (u-2)(d/du) + 2$, then $A_\beta\, \zeta_n = (n+2)\, \zeta_n$, i.e., $r_n = n + 2$, for $n \in \mathbb{N} \cup \{0\}$.

In the sequel, let

$$\boldsymbol{\alpha_\beta = \inf\{\alpha > 0 : \|A_\beta^{-\alpha}\|_{HS} < +\infty\}.}$$

For each $p \geq 0$, let $\mathcal{E}_p$ be the set of all $\zeta \in L^2(\mathbb{R}, \gamma)$ with $|A_\beta^p\, \zeta|_{L^2(\mathbb{R},\gamma)} < +\infty$. Then $\mathcal{E}_p$ is a real separable Hilbert space with the inner product $\langle \cdot, \cdot \rangle_{p,\beta}$ given by

$$\langle \zeta,\, \psi \rangle_{p,\beta} = \int_{-\infty}^{+\infty} A_\beta^p\, \zeta(u)\, A_\beta^p\, \psi(u)\, d\gamma(u) \qquad \text{for all } \zeta, \psi \in \mathcal{E}_p.$$

and induced norm by $|\cdot|_{p,\beta} := \sqrt{\langle \cdot, \cdot \rangle}$.

Denote by $\mathcal{E}_{-p}$ the dual of $\mathcal{E}_p$. Then $\mathcal{E}_{-p}$ is isometrically isomorphic to the completion of $L^2(\mathbb{R}, \gamma)$ with the inner product $\langle \cdot, \cdot \rangle_{-p,\beta}$ and $|\cdot|_{-p,\beta}$-norm by

$$\langle \zeta,\, \psi \rangle_{-p,\beta} = \int_{-\infty}^{+\infty} A_\beta^{-p}\, \zeta(u)\, A_\beta^{-p}\, \psi(u)\, d\gamma(u) \qquad \text{for all } \zeta, \psi \in L^2(\mathbb{R}, \gamma),$$

where we identify $x \in \mathcal{E}_{-p}$ with $\sum_n (x, \zeta_n)\,\zeta_n$ in which $(\cdot,\cdot)$ is the $\mathcal{E}_{-p}$-$\mathcal{E}_p$ pairing. Let $\mathcal{E} = \varprojlim_{p>0} \mathcal{E}_p$. Then $\mathcal{E}$ is a nuclear space with the dual $\mathcal{E}' = \varinjlim_{p>0} \mathcal{E}_{-p}$. $\mathcal{E} \subset L^2(\mathbb{R}, \gamma) \subset \mathcal{E}'$ forms a Gel'fand triple.

For $p \in \mathbb{R}$, denote by $\mathcal{N}_p$ the Hilbert space $\mathcal{S}_p \otimes \mathcal{E}_p$ with $|\!|\cdot|\!|_p$-norm defined by $|\!| e_i \otimes \zeta_j |\!|_p = |e_i|_p\,|\zeta_j|_{p,\beta}$. So, $\mathcal{N}_0 = L^2_c(\mathbb{R}^2, \lambda)$. Let $\mathcal{N} = \mathcal{S} \otimes \mathcal{E}$ with the dual $\mathcal{N}' = \mathcal{S}' \otimes \mathcal{E}'$.

Proposition 3.1.

(1) For any $p, q \in \mathbb{R}$ with $q - p > \max\{\alpha_\beta, 1/2\}$, the embedding $\mathcal{N}_q \hookrightarrow \mathcal{N}_p$ is Hilbert-Schmidt.

(2) $\mathcal{N} = \varprojlim_{p>0} \mathcal{N}_p$ is a nuclear space induced by the family $\{(\mathcal{N}_p, |\!|\cdot|\!|_p) : p \geq 0\}$.

(3) The dual $\mathcal{N}_p'$ of $\mathcal{N}$, $p \in \mathbb{R}$, is isometrically isomorphic to $\mathcal{N}_{-p}$ by identifying $x \in \mathcal{N}_p'$ with $\sum_{n,m} (x, e_n \otimes \zeta_m)\, e_n \otimes \zeta_m$, where $(\cdot,\cdot)$ is the $\mathcal{N}_p'$-$\mathcal{N}_p$ pairing.

(4) $\mathcal{N}' = \varinjlim_{p>0} \mathcal{N}_{-p}$ and $\mathcal{N} \subset L^2(\mathbb{R}^2, \lambda) \subset \mathcal{N}'$ forms a Gel'fand triple. Moreover, the inclusions $\mathcal{N} \subset \mathcal{N}_p \subset \mathcal{N}_q \subset L^2(\mathbb{R}^2, \lambda) \subset \mathcal{N}_{-q} \subset \mathcal{N}_{-p} \subset \mathcal{N}'$, $p > q \geq 0$, are all continuous.

To simplify notation, from now on we denote

$$|\!|f|\!|_p := |f|_{\mathcal{N}_{p,c}^{\widehat{\otimes} n}}, \quad \textit{whenever } f \in \mathcal{N}_{p,c}^{\widehat{\otimes} n} \tag{19}$$

where $\widehat{\otimes}$ denotes the symmetric tensor product. In addition, we relabel the CONS

$$\{e_n \otimes \zeta_m : n, m = 0, 1, \dots\} \quad \textit{by} \quad \{f_0, f_1, \dots\}, \tag{20}$$

in such a way that $f_0 = e_0 \otimes \zeta_0$ and let

$$\lambda_j = |\!| A e_j \otimes A_\beta \zeta_j |\!|_0 = |\!| f_j |\!|_1 \tag{21}$$

for $j \in \mathbb{N} \cup \{0\}$.

3.2 Test and generalized functionals

For $p \in \mathbb{R}$ and $\varphi \in L^2(\mathcal{S}', \Lambda)$, define

$$\|\varphi\|_p^2 = \sum_{n=0}^{\infty} \frac{1}{n!} \|D^n S\varphi(0)\|^2_{\mathcal{L}^{(n)}_{(2)}(\mathcal{N}_{-p})}$$

and let $\mathcal{L}_p$ be the completion of the class $\{\varphi \in L^2(\mathcal{S}', \Lambda) : \|\varphi\|_p < +\infty\}$ with respect to $\|\cdot\|_p$-norm. Then $\mathcal{L}_p$, $p \in \mathbb{R}$, is a Hilbert space with the inner product $\langle\!\langle \cdot, \cdot \rangle\!\rangle_p$ induced by $\| \cdot \|_p$-norm. For $p, q \in \mathbb{R}$ with $q \geq p$, $\mathcal{L}_q \subset \mathcal{L}_p$ and the embedding $\mathcal{L}_q \hookrightarrow \mathcal{L}_p$ is of Hilbert-Schmidt type, whenever $q - p > \max\{\alpha_\beta, 1/2\}$. Let $\mathcal{L} = \varprojlim_{p>0} \mathcal{L}_p$. Then $\mathcal{L}$ is a nuclear space. $\mathcal{L}$ will serve as the space of test functions and the dual space $\mathcal{L}'$ of $\mathcal{L}$ the space of generalized functions. In our investigation, members of $\mathcal{L}'$ are called generalized Lévy white noise functionals. In this way, we obtain a Gel'fand triple $\mathcal{L} \subset L^2(\mathcal{S}', \Lambda) \subset \mathcal{L}'$ and have the continuous inclusion:

$$\mathcal{L} \subset \mathcal{L}_p \subset \mathcal{L}_q \subset L^2(\mathcal{S}', \Lambda) \subset \mathcal{L}'_q \subset \mathcal{L}'_p \subset \mathcal{L}' = \varinjlim_{p>0} \mathcal{L}'_p, \quad p \geq q > 0,$$

where $L^2(\mathcal{S}', \Lambda)$ is identified with its dual space by the Riesz representation theorem. In what follows, the dual pairing of $\mathcal{L}'$ and $\mathcal{L}$ will be denoted by $\langle\!\langle \cdot, \cdot \rangle\!\rangle$. We note that for $\varphi \in L^2(\mathcal{S}', \Lambda)$ and $\psi \in \mathcal{L}$, $\langle\!\langle \varphi, \psi \rangle\!\rangle = \langle\!\langle \varphi, \overline{\psi} \rangle\!\rangle_0$ ("$\bar{\ }$" denotes the complex-conjugate).

Then the set, consisting of all functionals

$$\frac{1}{\sqrt{n_1! \cdots n_m!}} I_n(((A^{-p} \otimes A_\beta^{-p}) f_{j_1})^{\otimes n_1} \widehat{\otimes} \cdots \widehat{\otimes} ((A^{-p} \otimes A_\beta^{-p}) f_{j_m})^{\otimes n_m})$$

for $n_1 + \cdots + n_m = n$, $n \in \mathbb{N} \cup \{0\}$, and $j_1 < \cdots < j_m$, forms a CONS of $\mathcal{L}_p$ for any $p \in \mathbb{R}$. By identifying an $F \in \mathcal{L}'_p$ with the function

$$\sum_{n=0}^{\infty} \sum_{\substack{n_1+\cdots+n_m=n \\ j_1<\cdots<j_m}} \frac{\langle\!\langle F, I_n(f_{j_1}^{\otimes n_1} \widehat{\otimes} \cdots \widehat{\otimes} f_{j_m}^{\otimes n_m}) \rangle\!\rangle}{n_1! \cdots n_m!} I_n(f_{j_1}^{\otimes n_1} \widehat{\otimes} \cdots \widehat{\otimes} f_{j_m}^{\otimes n_m}), \tag{22}$$

it is easy to see that $\mathcal{L}'_p$ is isometrically isomorphic to $\mathcal{L}_{-p}$. Thus $\mathcal{L}' = \varinjlim_{p>0} \mathcal{L}_{-p}$.

For $F \in \mathcal{L}'$ and $n \in \mathbb{N}_0$, let

$$F_n = \sum_{\substack{n_1+\cdots+n_m=n \\ j_1<\cdots<j_m}} \frac{\langle\!\langle F, I_n(f_{j_1}^{\otimes n_1} \widehat{\otimes} \cdots \widehat{\otimes} f_{j_m}^{\otimes n_m}) \rangle\!\rangle}{n_1! \cdots n_m!} f_{j_1}^{\otimes n_1} \widehat{\otimes} \cdots \widehat{\otimes} f_{j_m}^{\otimes n_m}. \tag{23}$$

Then each F_n is in $\in \mathcal{N}_{-p,c}^{\widehat{\otimes} n}$. Moreover,

$$\langle\!\langle F, \varphi \rangle\!\rangle = \sum_{n=0}^{\infty} n!\,(F_n, \phi_n) \quad \text{and} \quad \|F\|_{-p}^2 = \sum_{n=0}^{\infty} n! \, | F_n |_{-p}^2, \tag{24}$$

where $(\cdot, \cdot)$ is the $\mathcal{N}_{-p,c}^{\widehat{\otimes} n}$-$\mathcal{N}_{p,c}^{\widehat{\otimes} n}$ pairing. Being motivated by (24), we define a generalized multiple Lévy-Itô integral $I_n(F_n)$ with the kernel function $F_n \in \mathcal{N}_c'^{\widehat{\otimes} n}$ as a generalized Lévy white noise functional by

$$\langle\!\langle I_n(F_n), \varphi \rangle\!\rangle = n!\,(F_n, \phi_n) \quad \text{for any } \varphi \sim (\phi_n) \in \mathcal{L},$$

where $(\cdot,\cdot)$ is the $\mathcal{N}_c'^{\widehat{\otimes}n}$-$\mathcal{N}_c^{\widehat{\otimes}n}$ pairing. Consequently, each F in $\mathcal{L}'$ can be expressed as an orthogonal direct sum $\sum_{n=0}^{\infty} \oplus I_n(F_n)$, written by $F \sim (F_n)$, of generalized multiple Lévy-Itô integrals.

For $g \in \mathcal{N}_{p,c}$, it is easy to see that $\|\mathcal{E}_M(g)\|_p = e^{(1/2)|g|_p^2}$ for any $p > 0$. Hence $\mathcal{E}_M(g) \in \mathcal{L}_p$ if and only if $g \in \mathcal{N}_{p,c}$ for $p > 0$. Then we define the S-transform for all generalized Lévy white noise functionals as follows.

Definition 3.1. The S-transform SF of a generalized Lévy white noise functional $F \in \mathcal{L}_{-p}$, $p > 0$, is defined to be a complex-valued function on $\mathcal{N}_{p,c}$ such that

$$SF(g) = \langle\langle F, \mathcal{E}_M(g)\rangle\rangle \quad \text{for } g \in \mathcal{N}_c.$$

The functional $U = SF$ is called the U-functional of F.

Let $F \sim (F_n) \in \mathcal{L}'$. Then, for $g \in \mathcal{N}_c$, $SF(g) = \sum_{n=0}^{\infty} (F_n, g^{\otimes n})$, where $(\cdot,\cdot)$ is the $\mathcal{N}_c'^{\widehat{\otimes}n}$-$\mathcal{N}_c^{\widehat{\otimes}n}$ pairing. From this formula, it is easy to verify the following results.

Proposition 3.2. *Let $F \sim (F_n)$ be in $\mathcal{L}_{-p}$, $p \in \mathbb{R}$. Then*

(i) *SF is analytic on $\mathcal{N}_{p,c}$, satisfying the exponential growth condition:*

$$|SF(h)| \le \|F\|_{-p}\, e^{(1/2)|h|_p^2} \quad \textit{for any } h \in \mathcal{N}_{p,c}.$$

(ii) *$D^n SF(0)$ is a Hilbert-Schmidt type n-linear functional on $\mathcal{N}_{p,c}^{\widehat{\otimes}n}$. Moreover*

$$\|F\|_{-p}^2 = \sum_{n=0}^{\infty} \frac{1}{n!} \|D^n SF(0)\|^2_{\mathcal{L}_{(2)}^{(n)}[\mathcal{N}_{p,c}]}.$$

(iii) *For any $\varphi \in \mathcal{L}_p$,*

$$\langle\langle F, \varphi\rangle\rangle = \sum_{n=0}^{\infty} \frac{1}{n!} \langle\langle D^n SF(0), D^n S\varphi(0)\rangle\rangle_{HS}.$$

where, for any two n-linear operators T and S

$$\langle\langle T, S\rangle\rangle_{HS} := \sum_{i_1,\dots,i_n=0}^{\infty} T(f_{i_1},\dots,f_{i_n})\, S(f_{i_1},\dots,f_{i_n}).$$

Let H be a complex Hilbert space. For $r > 0$, denote by $\mathcal{F}^r(H)$ the class of analytic functions on H with norm $\|\cdot\|_{\mathcal{F}^r(H)}$ such that

$$\|f\|^2_{\mathcal{F}^r(H)} = \sum_{n=0}^{\infty} \frac{r^n}{n!} \|D^n f(0)\|^2_{\mathcal{L}_{(2)}^{(n)}[H]} < +\infty,$$

$\mathcal{F}^r(H)$ is called the Bargmann-Segal-Dwyer space.

Follow the same proof as in [22] we have

Theorem 3.1.

(*i*) *The S-transform is an isomorphism from $\mathcal{L}_{-p}$ onto the Bargmann-Segal-Dwyer space $\mathcal{F}^1(\mathcal{N}_{p,c})$ for $p \in \mathbb{R}$. More precisely, for $f \in \mathcal{F}^1(\mathcal{N}_{p,c})$ and $\varphi \in \mathcal{L}_p$,*

$$\langle\langle S^{-1}f, \varphi\rangle\rangle = \sum_{n=0}^{\infty} \frac{1}{n!} \langle\langle D^n f(0), D^n S\varphi(0)\rangle\rangle_{HS}.$$

(*ii*) *Let $\mathcal{D}_p$ be the closed subspace of $\mathcal{L}_p$, $p \in \mathbb{R}$, consisiting of elements $\varphi = \sum_{n=0}^{\infty} I_n(\phi_n)$ such that $\sum_{n=0}^{\infty} n!\, \tau_2^n\, |\phi_n|_p^2 < +\infty$.*
Then the S-transform is an isomorphism from $\mathcal{D}_{-p}$ onto $\mathcal{F}^{1/\tau_2}(\mathcal{S}_{p,c})$ for $p \in \mathbb{R}$.

Now we are ready to state the characterization theorem for generalized Lévy white noise functionals.

Theorem 3.2. [29]

(*i*) *Suppose that a complex-valued function G defined on $\mathcal{N}_c$ is analytic and satisfies the following growth condition:*

$$\exists\, p \in \mathbb{R},\ \exists\, c > 0\ \ni\ |G(\xi)| \leq c\, e^{(1/2)|\xi|_{-p}^2},\quad \forall\, \xi \in \mathcal{N}_c. \tag{25}$$

Then there exists a unique $F \in \mathcal{L}_{p-\ell_0}$ such that $SF = G$ with $\|F\|_{p-\ell_0} \leq K c$, where ℓ_0 is the least positive integer such that $e^2\, \|A^{-\ell_0} \otimes A_\beta^{-\ell_0}\|_{HS}^2 < 1$, and K is a constant, independent of the choice of F.

(*ii*) *Suppose that a complex-valued function G defined on $\mathcal{N}_c$ is analytic and satisfies the condition:*

$$\forall\, p \geq 0,\ \exists\, c > 0\ \ni\ |G(\xi)| \leq c\, e^{(1/2)|\xi|_{-p}^2},\quad \forall\, \xi \in \mathcal{N}_c. \tag{26}$$

Then there exists a unique $\varphi \in \mathcal{L}$ such that $S\varphi = G$.

Proof. (i) Let f_k, $k \in \mathbb{N}_0$, be defined as given in (20) and $\lambda_k = |f_k|_1$. Then $\lambda_k^q = |f_k|_q$ for any $q \in \mathbb{R}$. Let $\widetilde{f}_k = \lambda_k^p\, f_k$, $k \in \mathbb{N}_0$. By Cauchy integral formula, we have

$$
\begin{aligned}
&|\, D^nG(0)(\widetilde{f}_{i_1},\dots,\widetilde{f}_{i_n})\,|^2 \\
=\ & \left| \left(\frac{1}{2\pi i}\right)^n \int_{|z_1|=r}\cdots\int_{|z_n|=r} \frac{G(\,z_1\widetilde{f}_{i_1}+\cdots+z_n\widetilde{f}_{i_n}\,)}{z_1^2\cdots z_n^2}\,dz_1\cdots dz_n \right|^2 \qquad (27)\\
\le\ & n^n\cdot c^2\cdot e^n,\ \ \forall\ i_1,\dots,i_n\in\mathbb{N}_0,\ \ (\text{by (25) and setting } r=n^{-\frac{1}{2}}).
\end{aligned}
$$

Observe that for $n\in\mathbb{N}$,

$$
\begin{aligned}
&\sum_{\substack{n_1+\cdots+n_m=n\\ j_1<\cdots<j_m}} \frac{\lambda_{j_1}^{2(p-\ell_0)n_1}\cdots\lambda_{j_m}^{2(p-\ell_0)n_m}}{n_1!\cdots n_m!}\,|\,D^nG(0)(f_{j_1}^{n_1},\dots,f_{j_m}^{n_m})\,|^2\\
=\ & \sum_{\substack{n_1+\cdots+n_m=n\\ j_1<\cdots<j_m}} \frac{\lambda_{j_1}^{-2n_1\ell_0}\cdots\lambda_{j_m}^{-2n_m\ell_0}}{n_1!\cdots n_m!}\,|\,D^nG(0)(\widetilde{f}_{j_1}^{n_1},\dots,\widetilde{f}_{j_m}^{n_m})\,|^2\\
\le\ & \alpha\, c^2\, e^{2n}\cdot\left(\sum_{k=0}^{\infty}\lambda_k^{-2\ell_0}\right)^n,\quad (\text{by (27) and Stirling's formula}),
\end{aligned}
$$

where $\alpha=\sup\{\frac{n^n}{e^n\cdot n!}:n\in\mathbb{N}_0\}$ and "$f_{j_i}^{n_i}$"(resp. "$\widetilde{f}_{j_i}^{n_i}$") is an abbreviation of "$\underbrace{f_{j_i},\dots,f_{j_i}}_{n_i\text{ times}}$"(resp. "$\underbrace{\widetilde{f}_{j_i},\dots,\widetilde{f}_{j_i}}_{n_i\text{ times}}$"). Define $F=\sum_{n=0}^{\infty}I_n(F_n)$, where

$$
F_n=\sum_{\substack{n_1+\cdots+n_m=n\\ j_1<\cdots<j_m}} \frac{D^nG(0)(f_{j_1}^{n_1},\dots,f_{j_m}^{n_m})}{n_1!\cdots n_m!}\,f_1^{\otimes n_1}\widehat{\otimes}\cdots\widehat{\otimes}f_{j_m}^{\otimes n_m},\ n\in\mathbb{N}_0.
$$

Then $F\in\mathcal{L}_{p-1}$ with $\|\,F\,\|_{p-1}\le(\frac{\alpha}{1-r})^{\frac{1}{2}}\,c$ in which $r=e^2\sum_{k=0}^{\infty}\lambda_k^{-2\ell_0}<1$. Since

$$
D^nSF(0)(f_{j_1}^{n_1},\dots,f_{j_m}^{n_m})=\ D^nG(0)(f_{j_1}^{n_1},\dots,f_{j_m}^{n_m}),\quad n\in\mathbb{N},
$$

where $n_1+\cdots+n_m=n$; $j_1<\cdots<j_m$, F is the only choice such that $SF=G$.

(ii) is an immediate consequence of (i). □

Remark 3.1.

(1) By (24) and Proposition 3.5(ii), we see that for any $p\in\mathbb{R}$, $\|SF\|_{\mathcal{F}^1(\mathcal{N}_{p,c})}=\|F\|_{-p}$. Let $\mathcal{A}_p$, $p\in\mathbb{R}$, be the space of all analytic functions f on $\mathcal{N}_{-p,c}$ satisfying the condition: $\|f\|_{\mathcal{A}_p}=\sup\{|f(z)|\,e^{-(1/2)|z|_{-p}^2}:z\in\mathcal{N}_{-p,c}\}<+\infty$. Then $\{(\mathcal{A}_p,\|\cdot\|_{\mathcal{A}_p}):p\in\mathbb{R}\}$ forms a system of compatible and comparable Banach

spaces. By Proposition 3.5(i), $SF \in \mathcal{A}_{-p}$ and $\|SF\|_{\mathcal{A}_{-p}} \leq \|F\|_{-p}$ for $F \in \mathcal{L}_{-p}$, $p \in \mathbb{R}$. Let $\mathcal{A}_\infty = \varprojlim_{p>0} \mathcal{A}_p$. Then $\mathcal{A}_\infty$ is the space consisting of all analytic functions on $\mathcal{N}_c'$ satisfying the condition (26). By Theorem 3.2 and [23], $\mathcal{A}_\infty = S(\mathcal{L}) = \cap_{p>0} \mathcal{F}^1(\mathcal{N}_{-p,c})$ as topological algebras.

(2) In the work of Kondratiev *et al.* [15,16,17] for the study of the non-Gaussian analytic measures, the space $\mathcal{E}^1_{\text{min}}(\mathcal{S}_c')$ plays the role of the test functions space, where $\mathcal{E}^1_{\text{min}}(\mathcal{S}_c')$ is the projective limit of $\{\mathcal{E}^1_{p,k}(\mathcal{S}_c')\}$ for which $\mathcal{E}^1_{p,k}(\mathcal{S}_c')$ consists of all analytic functionals φ on $\mathcal{S}_c'$ such that $\sup\{|\varphi(z)|\, e^{-(1/k)\,|z|_{-p}} : z \in \mathcal{S}_{-p,c}\} < +\infty$. However, in view of Theorem 2.7, $\mathcal{E}^1_{\text{min}}(\mathcal{S}_c')$ is not suitable to serve as the test functions space unless the Lévy white noise measure Λ is analytic. This shows that there is a close connection between the regularity properties of the test functions space and the Lévy measure β_0. However, it is not clear to us so far.

(3) For the Poisson white noise measure Λ_P, $\mathcal{N}_c \equiv \mathcal{S}_c$ and $\mathcal{E}^1_{\text{min}}(\mathcal{S}_c')$ is a dense subspace of $\mathcal{L}$ (see [16,38]). In this case, $\mathcal{E}^1_{\text{min}}(\mathcal{S}_c')$ is the analytic version of the subspace of $L^2(\mathcal{S}', \Lambda_P)$ which is the projective limit of $\{\mathcal{L}_{k,p}\}$, where $\mathcal{L}_{k,p} \subset L^2(\mathcal{S}', \Lambda_P)$ is the Hilbert space with the $\|\cdot\|_{k.p}$-norm given by

$$\|f\|^2_{k,p} = \sum_{n=0}^{\infty} \frac{1}{2^{kn}} \|D^n Sf(0)\|^2_{\mathcal{L}^{(n)}_{(2)}(\mathcal{S}_{-p,c})}, \quad k, p > 0.$$

Then $\mathcal{E}^1_{\text{min}}(\mathcal{S}_c') \subset \mathcal{L} \subset L^2(\mathcal{S}', \Lambda_P)$, in which the set inclusions are continuous.

References

1. Albeverio, S., Daletsky, Yu. L., Kondratiev, Yu. G. and Streit, L.: Non-Gaussian Infinite Dimensional Analysis, *J. Funct. Anal.138 (1996) 311-350.*
2. *Berezansky, Y. M. and Kondratiev, Yu. G. : Spectral Methods in Infinite Dimensional Analysis, (in Russian), Naukova Dumka, Kiev, 1988. English translation, Kluwer Academic Publishers, Dordrecht, 1995.*
3. *Dineen, S. : Complex Analysis on Infinite Dimensional Spaces, Springer-Verlag, 1999.*
4. *Doob, J. L. : Stochastic Processes, Wiley, New York, 1953.*
5. *Gel'fand, I. M. and Vilenkin, N. Y. : Generalized Functions, Vol 4. Academic Press, 1964.*
6. *Hida, T. : Analysis of Brownian Functionals, Carleton Mathematical Lecture Notes 13 (1975).*

7. *Hida, T. and Ikeda, N. : Analysis on Hilbert Space with Reproducing Kernel Arising from Multiple Wiener Integral, In Proc. 5th Berkeley Symp. on Math. Stat. and Prob., Vol. II, part 1 (1967) 117-143.*
8. *Hida, T., Kuo, H.-H., Potthoff, J., and Streit, L. : White Noise: An Infinite Dimensional Calculus, Kluwer Academic Publishers, 1993.*
9. *Itô, K. : Multiple Wiener Integral, Jap. J. Math. 22 (1952) 157-169.*
10. *Itô, K. : On Stochastic Processes, Jap. J. Math. 18 (1942) 261-301.*
11. *Itô, K. : Spectral Type of Shift Transformations of Differential Process with Stationary Increments, Trans. Amer. Math. Soc. 81 (1956) 253-263.*
12. *Ito, Y. : Generalized Poisson functionals, Prob. Theory Relat. Fields 77 (1988) 1-28.*
13. *Ito, Y. and Kubo, I. : Calculus on Gaussian and Poisson White Noises, Nagoya Math. J. 111 (1988) 41-84.*
14. *Kallianpur, G. : Stochastic Filtering Theory, Springer-Verlag, Berlin, 1980.*
15. *Kondratiev, Yu. G. and Lytvynov, E. W. : Operators of Gamma White Noise Spaces, Infinite Dimensional Anal. Quantum Prob. Related Topics 3(3) (2000) 303-335.*
16. *Kondratiev, Yu. G. , Silva, J. L. , Streit, L. , and Us, G. F. : Analysis on Poisson and Gamma Spaces, Infinite Dimensional Anal. Quantum Prob. Related Topics 1(1) (1998) 91-118.*
17. *Kondratiev, Yu. G. , Streit, L. , Westerkamp, W. , and Yan, J.-A. : Generalized Functions in Infinite Dimensional Analysis, Hiroshima Math. J. 28 (1998) 213-260.*
18. *Kubo, I. and Takenaka, S. : Calculus on Gaussian White Noises I, Proc. Japan Acad. Ser. A Math. Sci. 56 (1980) 376-380; Calculus on Gaussian White Noises II, Proc. Japan Acad. Ser. A Math. Sci. 56 (1980) 411-416; Calculus on Gaussian White Noises III, Proc. Japan Acad. Ser. A Math. Sci. 57 (1981) 433-437; Calculus on Gaussian White Noise IV, Proc. Japan Acad. Ser. A Math. Sci. 58 (1982) 186-189.*
19. *Kuo, H.-H.: White Noise Distribution Theory, CRC Press, 1996.*
20. *Kuo, H.-H. and Russek, A. Russek: White Noise Approach to Stochastic Integration, J. Multivariate Anal., 24 (1988) 218-236.*
21. *Lee, Y.-J. : Generalized Functions on Infinite Dimensional Spaces and its Application to White Noise Calculus, J. Funct. Anal. 82 (1989) 429-464.*
22. *Lee, Y.-J. : Analytic Version of Test Functionals, Fourier Transform and a Characterization of Measures in White Noise Calculus, J. Funct. Anal. 100 (1991) 359-380.*
23. *Lee, Y.-J. : Integral Representation of Second Quantization and its Application to White Noise Analysis, J. Funct. Anal. 100 (1991) 359-380.*
24. *Lee, Y.-J. : A characterization of generalized functions on infinite dimensional spaces and Bargmann-Segal Analytic functions, In Gaussian Random Fields, Series on Probability and Statistics Vol.1 (1991)272-284*
25. *Lee, Y.-J. and Shih, H.-H. : The Segal-Bargmann Transform for Lévy Functionals, J. Funct. Anal. 168 (1999) 46-83.*
26. *Lee, Y.-J. and Shih, H.-H. : Donsker's Delta Function of Lévy Processes, Acta Appl. Math. 63 (2000) 219-231.*

27. Lee, Y.-J. and Shih, H.-H. : Itô Formula for Generalized Lévy Functionals, In: Quantum Information II, Eds. Hida, T. and K. Saitô, World Scientific (2000) 87-105.
28. Lee, Y.-J. and Shih, H.-H. : The Product Formula of Multiple Lévy-Itô Integrals, to appear in Bull. Inst. Math. Academia Sinica, 2003.
29. Lee, Y.-J. and Shih, H.-H. : A Characterization of Generalized Lévy Functionals, preprint, 2003
30. Lytvynov, E. W. , Rebenko, A. L., and Schepa'nuk, G. V. : Wick Calculus on Spaces of Generalized Functions of Compound Poisson White Noise, Rep. Math. Phys. 39 (1997) 219-248.
31. Meyer, P. A. : Quantum Probability for Probabilistss, Lecture Notes in Math. 1538, Spring-Verlag, 1993.
32. Obata, N. : White Noise Calculus and Fock Space, Lecture Notes in Math. 1577, Springer-Verlag, 1994
33. Parthasarathy, K. R. : An Introduction to Quantum Stochastic Calculus, Basel/ Boston/ Berlin, Birkauser (1992).
34. Potthoff, J. and Streit, L. : A Characterization of Hida Distributions, J. Funct. Anal. 101, 212-229
35. Privault, N. : Chaotic and Variational Calculus in Discrete and Continuous Time for the Poisson Process, Stochast. Stochast. Rep. 51 Nos. 1-2 (1994) 83-109.
36. Privault, N. : A Different Quantum Stochastic Calculus for the Poisson Process, Prob. Theory Relat. Fields. 105 (1996) 225-278.
37. Sato, K. : Lévy Processes and Infinitely Divisible Distributions, Cambridge University Press, 1999.
38. Silva, J. L. : Studies in Non-Gaussian Analysis , Ph.D. Dissertation, University of Maderira, 1998.
39. Szegö, G. : Orthogonal Polynomials, AMS Colloquium Publication, Vol 23, 1939.

A NEW MEASUREMENT OF TIME SERIAL CORRELATIONS IN STOCK PRICE MOVEMENTS AND ITS APPLICATION

T. MATSUOKA

Faculty of Mangement of Administration and Information
Tokyo University of Science, Suwa
Chino City, Nagano 391-0292, Japan
E-mail: matsuoka@rs.suwa.tus.ac.jp

M.OHYA

Department of Information Science
Tokyo University of Science
Noda City, Chiba 278-8510, Japan
E-mail: Ohya@is.noda.tus.ac.jp

We shall introduce a new method in order to find a time serial correlation in actual price movements of stock. Some information quantities are used in our approach. It will be shown by our method that such a time serial correlation has a hierarchical structure. Moreover actual price movements can be characterized by such a hierarchical structure. For example the Bubble Break in Japanese Market is characterized by this structure. The classification for several securities are given by the hierarchical structure, by which we are led to solve a portfolio selection problem.

1. Introduction

In the simple Black-Sholes model it is assumed that a price of risky stock evolves according to a geometric Brownian motion. However, from a lot of research characterizing the stochastic process of real assets[1-6], it is widely recognized that actual price movements of stock have a certain correlation in several time points. It is important to analyze empirical financial data in order to clear statistical properties of those data.

In this paper we shall introduce a new method to analyze empirical financial data using the entropy and the fractal dimension of a state. The fractal dimension of a state, which was introduced by one of the present authors (MO) [7, 8], is a new criterion to measure a complexity of phenomena. We will discuss, by our method, the following basic question of finance: Is a random walk assumption consistent with empirical financial data in Japanese market? It will be answered that the consequence obtained by that assumption is not consistent with the price of real market. This answer means that there is a time serial correlation in actual price movements, and such a time serial correlation has a hierarchical structure. For example the crash of Japanese financial market in 1990, so called "Bubble Break", is characterized by the hierarchical structure. Specifically it will be shown that the dynamics of Japanese market is not stable.

Moreover, if such a hierarchical structure is useful to characterize the dynamics of each security, then a finer clustering of securities can be done in order to solve the portfolio selection problem. H.Markowitz says, in his paper [9]: "The process of selecting a portfolio may be divided into two stages. The first stage starts with observation and experience and ends with beliefs about the future performance of available securities. The second stage starts with the relevant beliefs about future performances and ends with the choice of portfolio." Then he also mentions that it is necessary to avoid investing in securities with high covariance among themselves.

In our analysis the fractal dimension of a state plays more important role than the entropy. So we simply review the notion and the definition of the fractal dimension of a state given in [7,8]. Then a random walk assumption for an empirical data in real market is explained. In the following sections it will be shown that there exists a time serial correlation with a hierarchical structure in evolution of Japanese stock index Nikkei 225. Finally we shall show some results obtained by applying our method to the portfolio selection problem [10].

2. Fractal dimensions of a state

The normal fractal theory usually treats only geometrical sets [11]. It is desirable to extend its application to other fields. For this purpose, Ohya introduced the concept of fractal dimensions for general states[a] [7,8].

The discrete classical system is represented by a pair (X, P), where X is a set composed of n events (i.e., $X = \{x_1, x_2, \cdots, x_n\}$) and P is a probability distribution of X (i.e., $P = (p_1, p_2, \cdots, p_n)$, $\sum p_i = 1$, $p_i \geq 0$), (X, P) is called a complete event system and P is called a state of X. Let us consider two event systems (X, P) and (Y, Q) as an input system and

[a]A general state means a state on a general quantum system. In a classical system a state means a probability measure.

an output system, respectively. In the following discussion both systems are composed of n events for simplicity. Then a channel Λ^* transmitting information from the input system (X, P) to the output system (Y, Q) is represented by a transition probability matrix $(p(j|i))$, namely $q_j = (\Lambda^* P)_j = \sum p(j|i) p_i$. For discrete classical system, the entropy $S(P)$ and the mutual entropy with respect to the initial state P and the channel Λ^* are given by

$$S(P) = -\sum_{i=1}^{n} p_i \log p_i, \tag{1}$$

$$I(P; \Lambda^*) = \sum_{i,j=1}^{n} p(j|i) p_i \log \frac{p(j|i)}{q_j}. \tag{2}$$

This mutual entropy means how much information contained in P is correctly transmitted to Q through the channel $\Lambda^* : Q = \Lambda^* P$.

The fractal dimensions of a state are defined by means of the above entropies and the ε-entropy defined below:

Let C be the set of all channels. The ε-entropy of a state[b] is given by

$$S(P; \varepsilon) \equiv \inf \{J(P; \Lambda^*); \Lambda^* \in C, \|P - \Lambda^* P\| \leq \varepsilon\} \tag{3}$$

where $\|P - Q\| = \sum |p_i - q_i|$ and

$$J(P; \Lambda^*) \equiv \sup \{I(P; \Gamma^*); \Gamma^* \in C, \Gamma^* P = \Lambda^* P\} \tag{4}$$

Two fractal dimensions of a state are defined as follows:

Definition 2.1. The capacity dimension of order ε :

$$d_C(P; \varepsilon) \equiv \frac{S(P; \varepsilon)}{\log \frac{1}{\varepsilon}} \tag{5}$$

Definition 2.2. The information dimension of order ε :

[b]It is well-known that the ε-entropy of a random variable was introduced by Kolmogorov, however Ohya's ε-entoropy is a little different from Kolmogorov's. We can fefer to [12] concernig the difference between these definitions of ε-entropy.

$$d_I(P;\varepsilon) \equiv \frac{S(P;\varepsilon)}{S(P)} \tag{6}$$

When the distance of P and Q is given by ε (i.e., $\|P-Q\| = \|P-\Lambda^* P\| = \varepsilon$), we consider the following transition probability matrix $(p(j|\,i))$ for a chosen k, l $(1 \le k,\ l \le n\ )$

$$q_j = \sum p(j|\,i)\,p_i \quad \text{where} \quad \begin{cases} q_k = p_k + \frac{\varepsilon}{2} \\ q_l = p_l - \frac{\varepsilon}{2} \\ q_j = p_j \quad (j \neq k,\ l) \end{cases}. \tag{7}$$

The following theorem holds and its proof is given by [13].

Theorem 2.1. *If the set C is the set of channels constructed by (7) and ε satisfies the condition $0 \le \frac{\varepsilon}{2} \le \min(p_1, ..., p_n)$, then*

$$S(P;\varepsilon) = S(P) - \left(p_{\max} + \frac{\varepsilon}{2}\right)\log\left(p_{\max} + \frac{\varepsilon}{2}\right) + p_{\max}\log p_{\max} + \frac{\varepsilon}{2}\log\frac{\varepsilon}{2}$$

where $p_{\max} = \max(p_1, ..., p_n)$.

3. A random walk assumption for an empirical stock data in real market

In standard stochastic financial models it is often assumed that the model of stock prices follows a geometrical Brownian motion. Let $\{P_t\}_{t\in Z}$ be a set of empirical prices of an asset where P_t represents a price of the asset at time t. Let us call $\{P_t\}$ an empirical price process. The empirical price process $\{P_t\}$ can be rewritten as follows by taking the logarithm of $\{P_t\}$ over a time interval $[0, T]$

$$\log P_T = \log P_{T-1} + \Delta P_T = \log P_0 + \sum_{t\,=\,1}^{T} \Delta P_t \tag{8}$$

where $\Delta P_t = \log \frac{P_t}{P_{t-1}}$. If $\{P_t\}$ follows a geometrical Brownian motion, then the incremental process $\{\Delta P_t\}$ will be recognized as an independent identical distribution process (i.i.d. process for short) with a normal distribution $N(\mu, \sigma)$ where μ is the mean and σ is the standard deviation of ΔP_t. Using a symbol $\Delta\tilde{P}_t$ given by $\Delta\tilde{P}_t = \frac{(\Delta P_t - \mu)}{\sigma}$, then

$$\log P_T = \log P_0 + \mu T + \sigma \sum_{t=1}^{T} \Delta\tilde{P}_t. \tag{9}$$

Namely, the following proposition holds by a random walk assumption.

Proposition 3.1. *If an empirical price process $\{P_t\}$ follows a geometrical Brownian motion, then the frequency distribution of the normalized increment $\Delta\tilde{P}_t$, gained from $\{P_t\}$, will go to the normal distribution $N(0,1)$ approximately as the amount of data increases.*

In this paper the proposition 3.1 is checked critically for an empirical price process $\{P_t\}$ by comparing a computer simulation of prices with an i.i.d. process. We study the property of the distribution of $\Delta\tilde{P}_t$ by calculating its information quantities.

We introduce some notations:

(1)A price process of an asset over the period $[T,T']$;

$$\{P_t\}_{T\le t\le T'} \text{ where } t,T,T' \text{ represent dates.}$$

(2)A "x-day log-return" at date t ;

$$\Delta P_t^x = \log \frac{P_t}{P_{t-x}}.$$

(3)A frequency distribution of normalized "x-day log-returns" obtained from the empirical data in the periods $[T,T']$ or y, where y corresponds to the years of period $[T,T']$;

$$\Pr(\Delta\tilde{P}_t^x)_{T\le t\le T'} \textit{ or } \Pr(\Delta\tilde{P}_t^x)_y.$$

With $\Pr(\Delta\tilde{P}_t^x)_y$, from the proposition 3.1, we can claim that the following approximation holds for any x:

$$\lim_{y\to\infty} \Pr(\Delta\tilde{P}_t^x)_y \cong N(0,1). \tag{10}$$

In the following sections we will show the results of analyzing empirical data of Nikkei 225, which is the popular index of Japanese market, and those of other securities like SONY. We calculate both the entropy and the information dimension of their frequency distributions, namely $S\left(\Pr(\Delta\tilde{P}_t^x)_{T\le t\le T'}\right)$, $d_I\left(\Pr(\Delta\tilde{P}_t^x)_{T\le t\le T'};\varepsilon\right)$, and ε is fixed and given as e.q., 0.00229, satisfying the condition $\frac{\varepsilon}{2} \le \min\{p_i\}$ for all distributions analyzed in this paper.

4. A time serial correlation in stock price movements in Japan market

4.1. *Some results of a scramble test for Nikkei 225*

At first we analyze the computer simulation of incremental process $\{\Delta B_t\}$ according to an i.i.d. process with normal distribution $N(0,1)$. Notice that there are about 260 data obtained from the prices of real asset over one year, because the number of business days over one year is also about 260 days. So the frequency distribution of the computer simulation corresponding to y-years $\Pr(\Delta B_t)_y$ is composed of $y \times 260$ data. In Figure 1 and Figure 2 the values of the entropy $S(\Pr(\Delta B_t)_y)$ and the information dimension $d_I(\Pr(\Delta B_t)_y;\varepsilon)$ are plotted from one-year distribution $\Pr(\Delta B_t)_1$ to thirteen-years distribution $\Pr(\Delta B_t)_{13}$.

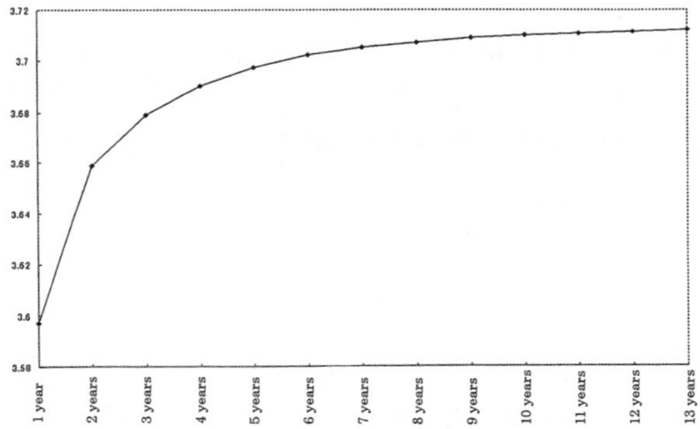

Figure 1. Change of the entropy for computer simulation of i.i.d proces

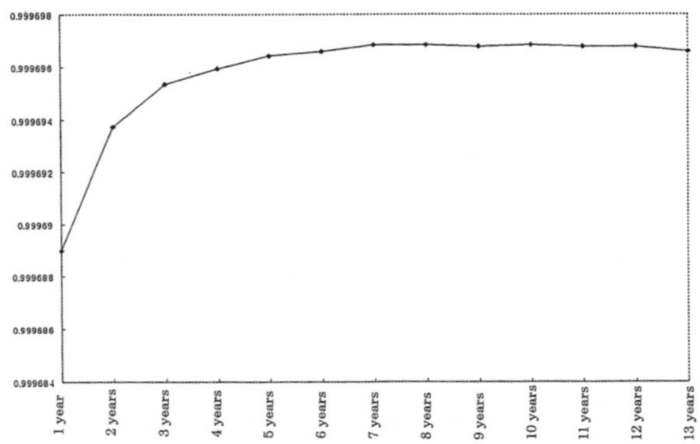

Figure 2. Change of the information dimension for computer simulation of i.i.d. process

Both of the criteria increase monotonously as the amount of data increases, namely according to the change of y, and they also seem to converge to a fixed value, because $\Pr(\Delta B_t)_y$ goes to $N(0,1)$ as the amount of data increases. From this result and the proposition 3.1, the following proposition holds.

Proposition 4.1. *If an empirical price process* $\{P_t\}$ *follows a geometrical Brownian motion, then for any* x *both of the entropy and the information dimension of the distribution* $\Pr(\Delta\tilde{P}_t^x)_y$ *increase monotonously as the amount of data increases, and they converge to fixed values.*

However we can show that the evolution of the stock index Nikkei 225 is not recognized as a geometrical Brownian motion. Figure 3 and Figure 4 show the results of calculating the information quantities for the evolution of Nikkei 225 over thirteen years from April 1983 to December 1995. We analyzed the frequency distribution of "20-day log-returns" (so called "monthly returns") of Nikkei 225. We plotted the values of the entropy and the information dimensions of the distribution[c] $\left\{S\left(\Pr(\Delta\tilde{P}_t^{20})_y^{\mathrm{Nikkei225}}\right)\right\}_{1\le y\le 13}$, $\left\{d_I\left(\Pr(\Delta\tilde{P}_t^{20})_y^{\mathrm{Nikkei225}};\varepsilon\right)\right\}_{1\le y\le 13}$.

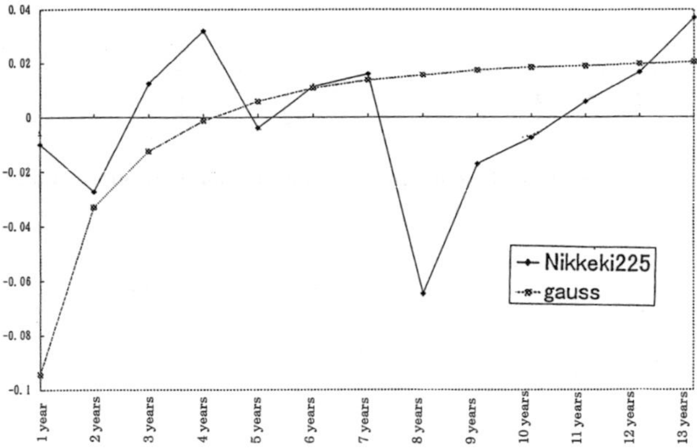

Figure 3. Change of the entropy for Nikkei 225

[c]In order to compare Nikkei 225 with computer simulation, the value of axis of ordinates is shifted by a mean value; $d_I^*(\Pr_y;\varepsilon) = d_I(\Pr_y;\varepsilon) - mean_{1\le y\le 13}(d_I)$ in Figure 3 and Figure 4.

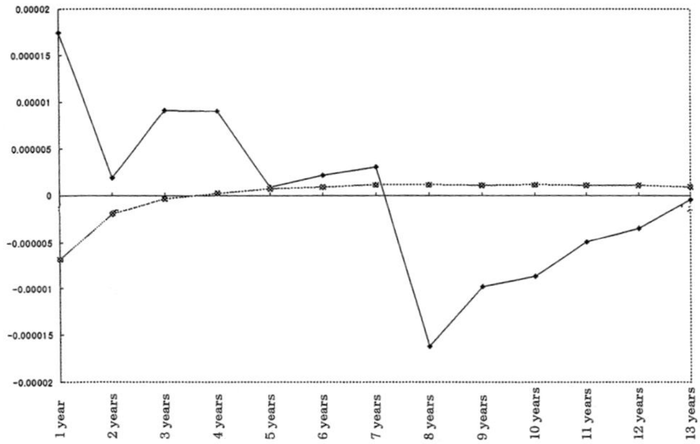

Figure 4. Change of the information dimension for Nikkei 225

It is clear that the changes of the information criteria of Nikkei 225 are different from those of computer simulation. Both of the criteria of Nikkei 225 do not increase monotonously, nor converge to a fixed value. So the following assertion holds by taking the contraposition of the proposition 4.1.

Assertion 4.1. The evolution of Nikkei 225 does not follow a geometrical Brownian motion. In other words, a geometrical Brownian motion model is not consistent with the evolution of Nikkei 225.

Let us explain these difference. We take a random permutation of the time series for the price data of Nikkei 225, then the time order of the price data is broken. Let us call such a new price process obtained after a random permutation the "scrambled price process"[d]. We calculate the entropy and the information dimension of the scrambled price process. Figure 5 and Figure 6 show the changes of the entropy and the information dimension of the scrambled price process of Nikkei 225.

After a random permutation it seems that both the entropy and the information dimension of Nikkei 225 increase monotonously and converge to fixed values. These results mean that the scrambled price process approximately follows a geometrical Brownian motion. From this result and the assertion 4.1 the following statement holds.

[d]E. E. Peaters estimated the performance of S&P 500 monthly returns using a scrambled test for R/S (rescaled range) analysis[6].

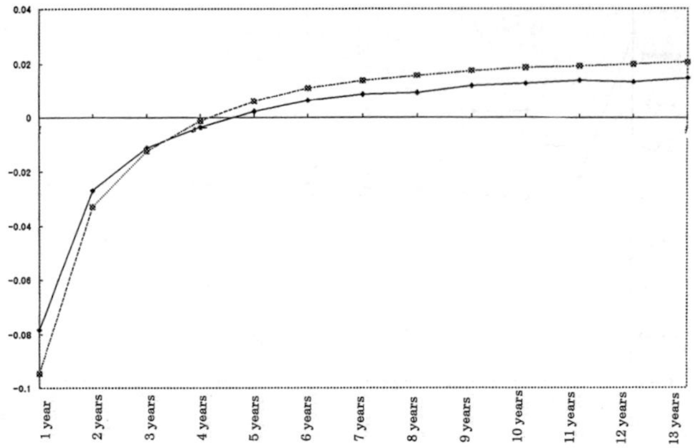

Figure 5. Change of the entropy for scrambled Nikkei 225

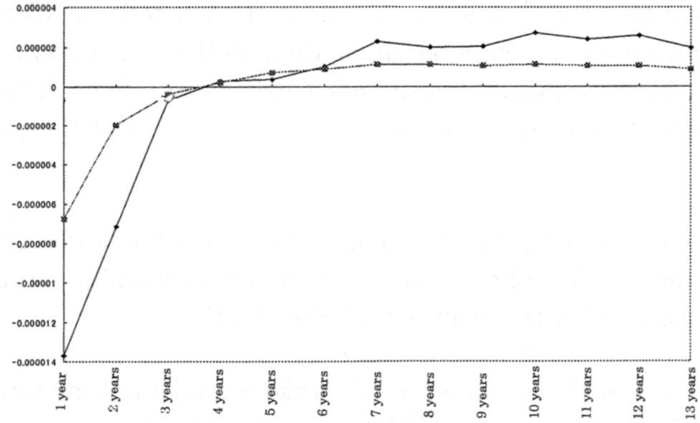

Figure 6. Change of the information dimension for scrambled Nikkei 225

Assertion 4.2. There is a time serial correlation in the evolution of the index Nikkei 225.

4.2. *What is the difference between the entropy and the information dimension?*

In the previous sub-section it was shown that both the entropy and the information dimension could be useful to show the existence of a time serial correlation in actual price movements. Before showing other results we reconfirm an important property of the information dimension mentioned in [8]. The information dimension d_I provides a measure of the complexity of a probability P. In particular, when $S(P) = S(Q)$, where Q is another probability, d_I gives us a new order of the complexity between P and Q, namely $d_I(P) \neq d_I(Q)$.

Figure 7 shows the change of the entropy for the average prices of TOYOTA, NEC and SONY (TNS for short). We calculated the information dimensions of the frequency distributions of "30-day log-returns" of the average price process and we plotted $\left\{S\left(\Pr(\Delta\tilde{P}_t^{30})_y^{\mathrm{TNS}}\right)\right\}_{1\le y\le 13}$.

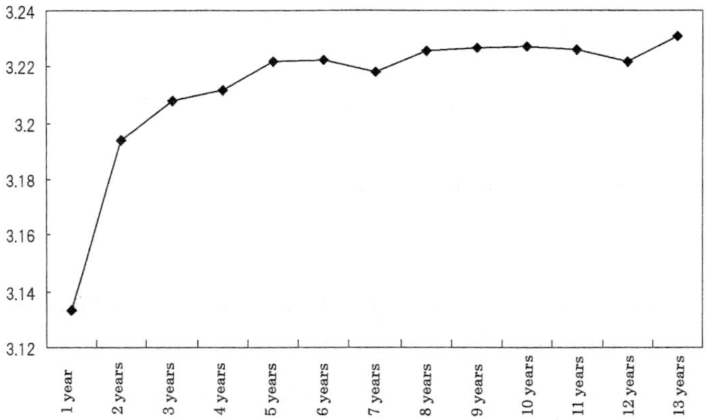

Figure 7. Change of the entropy for TNS average

From Figure 7 the entropy change seems to increase monotonously, so we may claim that "30-day log-return" of the price process approximately follows a Brownian motion from the entropy point of view. What about the information dimension? The change of the information dimensions $\left\{d_I\left(\Pr(\Delta\tilde{P}_t^{30})_y^{\mathrm{TNS}};\varepsilon\right)\right\}_{1\le y\le 13}$ are plotted in Figure 8.

The change of the information dimension can not be considered monotone increasing. Thus the average price process seems to be different from a geometrical Brownian motion on the view of the information dimension. That is, the following assertion holds.

Assertion 4.3. The information dimmension can be more useful than the entropy itself, becuase the information dimension can show the existence of time serial correlation in the evolution of real asset even when the entropy can not.

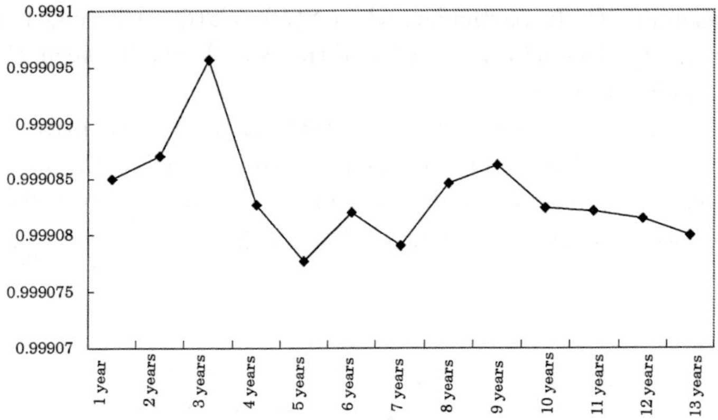

Figure 8. Change of the information dimension for TNS average

In the following section we show the results of our analysis by using only the information dimension.

4.3. *A hierarchical structure of a time serial correlation and the characterization of the Bubble Break*

In the sub-section 4.1 we calculated the information dimension of monthly return distributions $\Pr(\Delta\tilde{P}_t^{20})_y^{\text{Nikkei225}}$. Now we can also calculate the information dimension of any "x-day log-return" distributions $\Pr(\Delta\tilde{P}_t^{x})_y^{\text{Nikkei225}}$. We prepare some distributions, namely distributions $\Pr(\Delta\tilde{P}_t^{x})_y^{\text{Nikkei225}}$, where $x = 1, 5, 10, 15, 20, 40$ and $y = 6, 7, 8, 9, 10, 11, 12, 13$ years. Then we compute the information dimensions of these distributions and the average information dimensions of "x-day log-return" distributions over the period from 6-years to 13-years[e]. In Figure 9 these average values for each "x-day log-return" distribution are plotted.

If the evolution of Nikkei 225 follows a geometrical Brownian motion, then the information dimension will be the same for any "x-day log-return". However it seems that there is an order for the information dimensions ac-

---

[e]The average value is given as follows:

$$\overline{d}_I\left(\Pr(\Delta\tilde{P}_t^{x})^{\text{Nikkei225}};\varepsilon\right) = \frac{1}{8}\sum_{y=6}^{13} d_I\left(\Pr(\Delta\tilde{P}_t^{x})_y^{\text{Nikkei225}};\varepsilon\right).$$

Why do we average from 6-years? Because at 6-years, as shown in Fig.1 and Fig.2, there is enough data to show a distribution approximating a normal distribution, in order to later use it on actual prices of Nikkei 225.

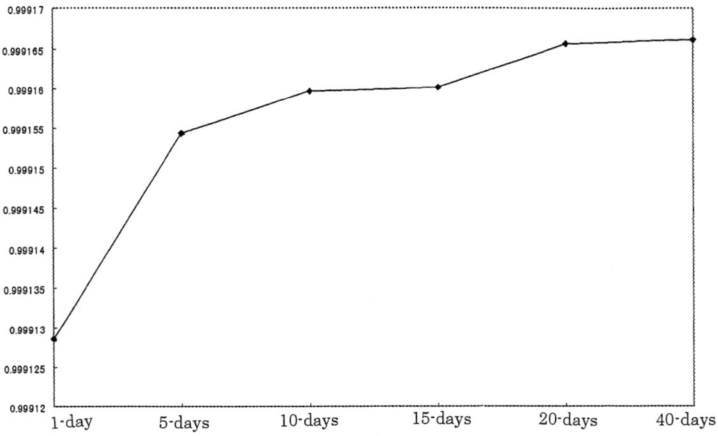

Figure 9. Hierarchical structure of time serial correlations in Nikkei 225

cording to the order of time interval x. It can be said that, for example, the randomness of "20-day log-returns" is bigger than that of "1-day log-returns" in the prices of Nikkei 225. In addition to the assertion 4.2 we can also claim that the randomness of an incremental process depends on interval x. We call this order "hierarchical structure of time serial correlations". Using such a hierarchical structure, it is shown that the crash of Japanese market in 1990, so-called "Bubble Break", can be characterized. We prepare a series of distributions $\left\{\Pr(\Delta\tilde{P}_t^x)^{\text{Nikkei225}}_{1983.1.1\le t\le T'}\right\}$ obtained from the empirical prices of Nikkei 225 where T' increases by one week from $T' = 15/12/1986$ to $T' = 31/1/1995$ and $x = 1, 5, 10, 15, 20, 40$. Then we compute their information dimension. In Figure 10 the series for the information dimension

$$\left\{d_I\left(\Pr(\Delta\tilde{P}_t^x)^{\text{Nikkei225}}_{1983.1.1\le t\le T'};\varepsilon\right)\right\}_{1986.12.15\le T'\le 1995.01.7}$$

is plotted.

We recognize that the pattern of the hierarchical structure changes over the period [1986.12.15, 1995.01.31]. The total period may be divided into five short periods corresponding to the change of such hierarchical structures. In Figure 10 these five periods are denoted by [A], [B], [C], [D], [E]. Figure 11 represents the evolution of Nikkei 225 divided into five periods corresponding to the classification done in Figure 10.

Comparing Figure 10 with Figure 11 we can see that the empirical price process of Nikkei 225 is characterized by the change of its hierarchical structure of time serial correlations. For example, the period [C] corresponds

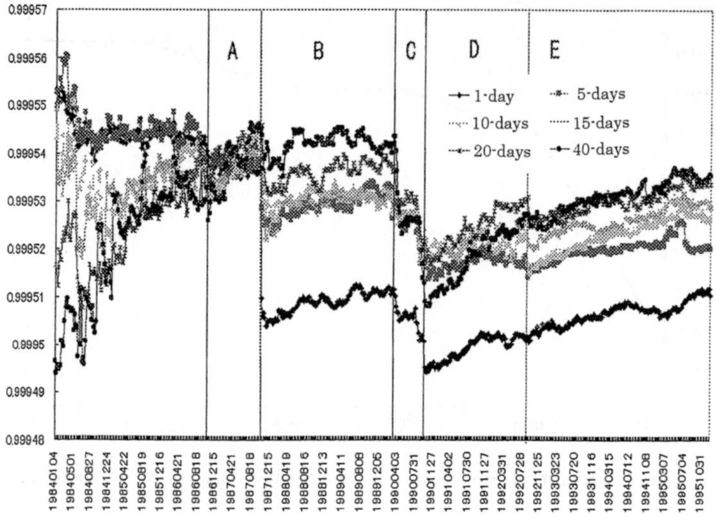

Figure 10. Change of hierarchical structure in Nikkkei 225

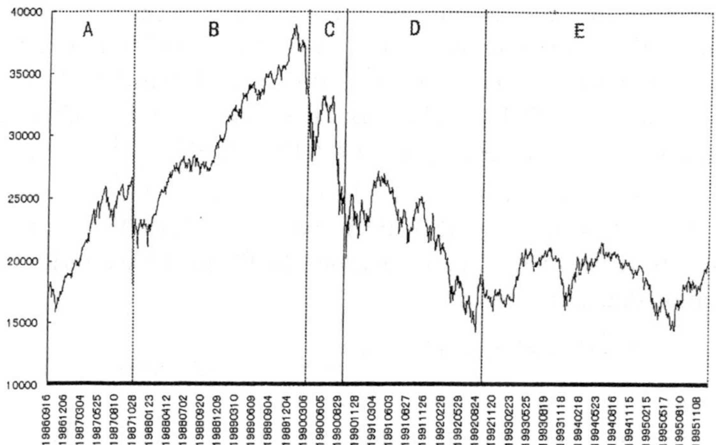

Figure 11. Price process of Nikkei 225

to the market crash called the "Bubble Break". The tendency of price movements of Nikkei 225 at each period may be explained as follows:

Period [A] ; Price movements begin to increase before the bubble.

Period [B] ; Price movements increase during the bubble.

Period [C] ; Price of Nikkei 225 falls after the bubble.

Period [D] ; Price movements fall more slowly after the bubble.

Period [E] ; Price movements are stable after the bubble.

The change of tendency of price movements can be characterized by the change of hierarchical structure of time serial correlations represented by the information dimensions. It could be suggested that the change of hierarchical structure corresponds to the change of dynamics of Nikkei 225. We summarize the above discussion as the following assertion.

Assertion 4.4. If there exists a hierarchical structure in an evolution of real asset, then the hierarchical structure will be closely related to the dynamics of the asset.

As an example, the Bubbl break in the evolution of Nikkei 225 is characterized by the change of its hierarchical structure.

It is natural to think that each individual security has its own dynamics depending on its internal structure. If the dynamics of each individual security is characterized by its hierarchical structure of time serial correlation, then we have some possibility to classify many securities by the difference of such hierarchical structures.

In the next section, our classification is applied to the portfolio selection problem.

5. Application to the portfolio selection problem [10]

In this section some results from the analysis of 80 security companies in Japan market are shown. The 80 securities are divided into two groups, group I and II shown in Table 1 and Table 2.

Here the evolution of the above securities from April 1987 to November 1997 are analyzed. For every security we calculate the information dimensions of probability distributions from "1-day log-return" to "200-day log-return" with 1-day increment, and also from 6-years to 8-years, yearly. Namely we calculate the information dimensions of the distributions $\left\{\Pr(\Delta\tilde{P}_t^x)_y\right\}_{1\leq x\leq 200,\ 6\leq y\leq 8}$. Finally we average the information dimensions from 6-years to 8years for every x. For example, in Figure 12 the computations of the information dimensions for SONY, $\left\{\overline{d_I}\left(\Pr(\Delta\tilde{P}_t^x)^{\mathrm{SONY}};\varepsilon\right)\right\}_{1\leq x\leq 200}$ are plotted.

As mentioned in the previous section, it could be said that such a structure illustrated in Figure 12 corresponds to the hierarchical structure of the time serial correlation in the stock price process of SONY. It seems that SONY's hierarchical structure has something like a cyclic property,

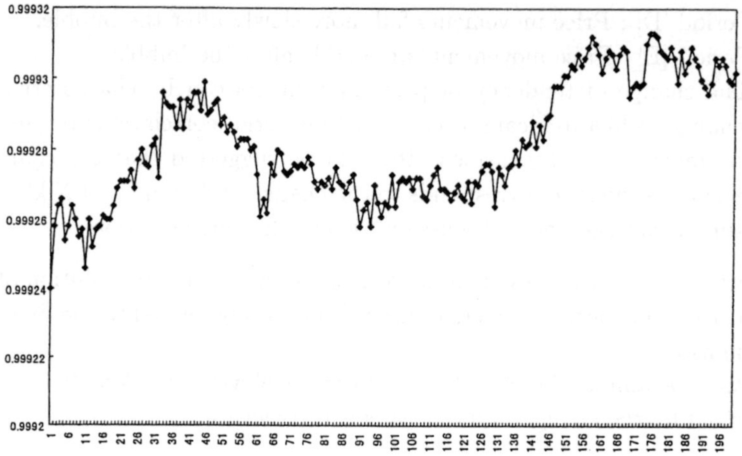

Figure 12. Hierarchical structure of time serial correlations in Sony

the degree of time serial correlation becomes stronger again over the time interval 45-days. It can be shown that every security has its own hierarchical structure. Comparing the hierarchical structures of many securities, we find that some of them are similar and some of them are different. In order to classify 40 securities of group I and II, we apply a normal clustering method to the series of the information dimensions like in Figure 12. We calculate the covariance matrix of those series of 40 securities, and applying the "furthest neighbor method" to such a covariance matrix we can get a classification of 40 securities. Our classification is done by the similarity of hierarchical structures of time serial correlation in the price process, or by the similarity of dynamics of every security. The results of such a classification for Group I and II are shown in Table 3 and Table 4. We can also get another classification of 40 securities by applying the same clustering method to the actual price processes of the 40 securities so called "normal clustering". The results of normal clustering for Group I and II are shown in Table 5 and Table 6[f].

Two types of classifications give different results of clustering. For example, TOSHIBA (6502) and SONY(6758) belong to the same category by our classification, however these two securities belong to different categories by normal classification.

We consider the merits of the two types of classification by analyzing the performance of portfolios that we construct as follows:

[f]Here we used the clustering method on the monthly price processes of 40 securities.

Step 1: We choose one security from each category A, B, C, looking at its price at the end of March 1995.

$$\mathrm{a} \in \mathrm{A},\ \ \mathrm{b} \in \mathrm{B},\ \ \mathrm{c} \in \mathrm{C}$$

Step 2: We construct a equal-weighted portfolio as follows;

$$R = R^{\mathrm{a}} + R^{\mathrm{b}} + R^{\mathrm{c}}$$

where $R^{\mathrm{a}} = R^{\mathrm{b}} = R^{\mathrm{c}}$, namely we invest the same asset to each security "a", "b", "c".[g].

Step 3: We investigate the performance of the portfolio return at date t from the end of May 1995 to the end of November. 1997 as follows;

$$\begin{aligned} R_{05.95} &= R^{\mathrm{a}}_{05.95} + R^{\mathrm{b}}_{05.95} + R^{\mathrm{c}}_{05.95} \\ R_{06.95} &= R^{\mathrm{a}}_{06.95} + R^{\mathrm{b}}_{06.95} + R^{\mathrm{c}}_{06.95} \\ &\ \vdots \\ R_{11.97} &= R^{\mathrm{a}}_{11.97} + R^{\mathrm{b}}_{11.97} + R^{\mathrm{c}}_{11.97} \end{aligned}$$

where R^{a}_t is the return of the security "a" at date t.

Step 4: We do this investigation for all portfolios, namely all combinations of securities chosen according to step 1.

Step 5: We calculate the statistics of the portfolio returns as follows:

Standard deviation of return: $D(R_{05.95}), D(R_{06.95}), \cdots, D(R_{11.97})$

Kurtosis of return: $K(R_{05.95}), K(R_{06.95}), \cdots, K(R_{11.97})$

The result of comparing our classification with normal classification for Group I is shown in Table 6.6 and that of Group II is shown in Table 6.7.

Let us denote the return of a portfolio constructed using our classification by $R_{\text{information}}$, and that of normal classification by R_{price}. From Table 6.6 and 6.7, it can be shown that the following inequalities hold very well in the short term corresponding to the period [May 1995, October 1996];

$$D(R_{\text{information}}) < D(R_{\text{price}}), K(R_{\text{information}}) > K(R_{\text{price}})$$

However, in the long term corresponding to the period [November 1996, November 1997], the opposite inequalities hold. In the short term our classification gives us more stable portfolio returns than the normal one. In

[g]Here we calculated the return of the portfolio based on a total asset eqaul to 1.

the long term our classification seems to be not so good. Why does the estimation for the two types of classifications depend on the length of term? If the dynamics of the price process of a security change according to how that security's performance is influenced by the market due to its internal structure, then it is natural to think that in the near future its dynamics will not be so different from those in the present. The market does not change quickly except for anomalies like the Bubble Break. However in the long future a security's dynamics will likely change because of the change in market conditions. Our classification holds based on the similarity of dynamics of return in the near future. However, in the long future the dynamics of return of some securities may change. Therefore our classification does not correspond to new dynamics of return. Namely the hierarchical structure of each security may be also changed. If we renew the classification using the information dimensions for new data, then it may be possible to get better results.

6. Summary and Discussion

In this paper we showed that the evolution of Nikkei 225 may be different from a geometrical Brownian motion, for an incremental price process of Nikkei 225 could not be recognized as an i.i.d process. In other words, it can be said that there exists a time serial correlation in the evolution of Nikkei 225. In addition we showed that there exists a hierarchical structure of time serial correlations and such a structure may be closely related to the dynamics of price movements of Nikkei 225. In fact the Bubble Break can be characterized by the change of hierarchical structure of time serial correlations.

Under the assumption that each security has its own hierarchical structure, we classified forty securities according to the difference of their hierarchical structures. Then it was shown that the portfolios constructed from our classification performed better than that from a normal classification in a short period. The fact suggests that the hierarchical structure of each security may be changed in a long period.

From our characterization of Bubble Break and the performance of our portfolios, it seems that there exits a significant change in the evolution of actual security. In the previous sections we call this significant change a "change of dynamics". The fact means that, for example, it is difficult to recognize the evolution of Nikkei 225 as a stationary process. Therefore we claim the following assertion:

Assertion 6.1. If an evolution of an asset is a non-staionaly process, then it will be useful to analyze such an evolution by means of the change of hierarchical structure.

Finally we mention future direction of our research.

(1) To clear why the information dimension of a state is more useful than the entropy in our analysis.

(2) To clear which mathematical model is consistent with the results of our analysis.

(3) To construct a mathematical model for a real asset recognized as a non-stationary process.

Acknowledgments

The authors would like to thank Prof. Allanus H. Tsoi for his useful comments. He suggests that the fractional Brownian motion will be related to the results of our analysis.

References

1. P. Cootner, ed. *The random character of stock market price*, MIT press, 1964.
2. B. B. Mandelbrot, J.W. Van Ness, *SIAM Rev*, **10**, 422 (1968).
3. B. B. Mandelbrot, *Fractals and Scaling in Finance, Risk*, Springer-Verlag, 1977.
4. U. A. Miller, M. M. Dacorogna, R. B. Olsen, O.V. Pictet, M. Schwarz, C. Morgenegg, *Journal of Banking and Finance*, **14**, 1189 (1990).
5. M. Larrain, *Empirical tests of chaotic behavior in a nonlinear interest rate model*, Financial Analysts Journal, in press, 1991.
6. E. E. Peters, *Chaos and Order in The Capital Markets*, John Wiley and Sons, 1991.
7. M. Ohy, *Proc. Symp. Appl. Func. Anal.*, **11**, 45 (1988).
8. M. Ohya, *Quantum Probability and Related Topics*, **6**, World Sientific, 359 (1991).
9. H. Markowitz, *The Journal of Finance*, 77 (1952).
10. A. Isogai, T. Matsuoka, *Frontier of Financial Technology*, 311 (1998) (in Japanese).
11. B. B. Mandelbrot, *The Fractal Geometry of Nature*, W. H. Freeman and Company, 1982.
12. K. Inoue, T. Matsuoka, M. Ohya, *Open Sys. and Information Dyn.*, **7**, 41 (2000).
13. T. Matsuoka, M. Ohya, *Chubu Forum for Mathematical Sciences*, **4**, 15 (1998).

Table 1. 40 securities in Group I

code	security	code	security
1812	Shimizu	1803	Kajima
1886	Aoki	2502	Asahi Breweries
3102	Kanebo	3401	Teijin
3407	Asahi Chmeical Ind.	4063	Shin-etu Chemical
4151	Kyouwa Hakko Kogyo	4507	Shionogi
4901	Fuji Photo Film	5001	Nippon Oil
5108	Bridgestone	5401	Nippon Steel
5403	Kawasaki Steel	5406	Kobe Steel
5711	Mitsubishi Material	6301	Komatsu
6502	Toshiba	6701	NEC
6758	Sony	7201	Nissan Motor
7203	Toyota Motor	7267	Honda Motor
7751	Cannon	7911	Toppan Printing
8058	Mitsubishi	8231	Mistukoshi
8317	Fuji Bank	8402	Mitsubishi Trust & Banking
8604	Nomura Securities	8751	Tokio Marine & Fire
8802	Mitsubishi Estate	9001	Tobu Railway
9008	Keisei Electric Railway	9062	Nippon Express
9101	Nippon Yusen	9202	All Nippon Airways
9501	Tokyo Electric Power	9531	Tokyo Gas

Table 2. 40 securities in Group II

code	security	code	security
1801	Taisei	2202	Meiji Seika
2501	Sapporo Breweries	2531	Takara Shuzo
2602	Nisshin Oil Miles	2802	Ajinomoto
3402	Toray Industrial	3404	Mitsubishi Raiyon
3863	Nippon Paper Ind.	4010	Mitsubishi Chemical
4403	NOF	4503	Yamanouchi Pharm
5201	Asahi Glass	5713	Sumitomo Metal Mining
5801	Furukawa Electric	5803	Fujikura
6361	Ebara	6501	Hitachi
6503	Mitsubishi Electric	6504	Fuji Electric
6702	Fujitsu	6752	Matsusita Elect. Ind.
6753	SHRAP	6764	Sanyo Electric
6773	Pioneer Electronic	6841	Yokokawa Electric
7003	Mitsui Shipbuilding	7011	Mitsubishi Heavy Ind.
7012	Kawasaki Heavy Ind.	7013	Ishikawa jima-Harima
7261	Mazda Motor	7762	Citzen Watch
7912	Dai Nippon Printing	7951	Yamaha
8031	Mitui & Co.	8063	Nissho Iwai
8320	Sanwa Bank	8755	Yasuda Fire & Marine
9007	Odakyu Electric Railway	9503	Kansai Electric Power

Table 3. A classification for Group I using the informataion dimensions

Category A	1803	5108	5406	7911	8402	8751
	8802	9062				
Category B	4901	1812	9008			
Category C	1886	2502	3102	3401	3407	4063
	4597	4151	5108	5001	5401	5711
	6301	6502	6701	6758	7201	7203
	7267	7751	8058	8231	8604	8317
	9001	9101	9062	9501	9531	

Table 4. A classification for Group II using the informataion dimensions

Category A	1801	2802	3402	4010	7003	7012
	7013	8755				
Category B	2531	2602	3404	3863	4503	6361
	6501	6503	6504	6702	6752	6753
	6764	6773	6841	7261	7762	7951
	8031	8063	8320	9007		
Category C	2202	2501	4403	5201	5713	5801
	5803	7011	7912	9503		

Table 5. A classification for Group I using the normal clustering

Category A	1803	1812	1886	2502	3102	3401
	3407	5001	5711	6301	8058	8231
	8317	8402	8604	8751	8802	9001
	9008	9062	9101	9202	9501	9531
Category B	5401	5403	5406	6502	7201	
Category C	4063	4151	4507	4901	5108	6701
	6758	7203	7267	7751	7911	

Table 6. A classification for Group II using the normal clustering

Category A	1801	2501	3402	3404	4010	4403
	5201	5713	6361	6503	6504	6764
	7003	7011	7012	7013	7261	8031
	8063	8320	8755	9007	9503	
Category B	2202	2531	2602	2802	3863	4503
	7912	7951				
Category C	5801	5803	6501	6702	6752	6753
	6773	6841	7262			

Table 7. Comparisons the performances of our portfolios with normal portfolios on Group I (Note: "1" includes that S.D.(or Kurtois) of our portfolio is smaller (or bigger) than normal one. "0" is the opposite.)

Date	Normal portfolios		Our portfolios		Comparrisons	
	S.D.	Kurtos	S.D.	Kurtos	S.D.	Kurtos
05.95	0.0195	4.2353	0.0201	4.2392	0	1
06.95	0.04	3.3606	0.0452	3.1337	0	0
07.95	0.0435	2.92	0.0451	2.9439	0	1
08.95	0.0529	2.4291	0.0409	2.4852	1	1
09.95	0.0522	2.8928	0.0419	3.3357	1	1
10.95	0.0575	2.9202	0.0496	3.2355	1	1
11.95	0.0523	2.7471	0.0427	2.7831	1	1
12.95	0.0667	2.6071	0.057	2.6714	1	1
01.96	0.0697	2.6021	0.0563	2.6999	1	1
02.96	0.0729	2.6794	0.0558	2.8453	1	1
03.96	0.0724	2.8948	0.0688	2.7151	1	0
04.96	0.0702	2.7317	0.0616	2.7061	1	0
05.96	0.076	2.8317	0.066	2.9157	1	1
06.96	0.0882	2.8345	0.0777	2.9193	1	1
07.96	0.0827	2.8473	0.0768	2.97	1	1
08.96	0.0816	2.6767	0.0792	2.9128	1	1
09.96	0.0897	2.8655	0.0846	3.043	1	1
10.96	0.0948	2.8637	0.0916	2.9696	1	1
11.96	0.1050	2.9734	0.1143	3.1573	0	1
12.96	0.1588	2.9093	0.1389	2.9162	0	1
01.96	0.139	2.8679	0.1637	2.8226	0	0
02.97	0.1536	3.1552	0.1847	2.9353	0	0
03.97	0.162	3.1519	0.2032	2.8327	0	0
04.97	0.1777	3.1784	0.2444	2.6822	0	0
05.97	0.1765	3.3807	0.2183	2.8875	0	0
06.97	0.1843	3.5383	0.2246	2.9448	0	0
07.97	0.2356	4.0506	0.2767	3.2345	0	0
08.97	0.24	4.1851	0.269	3.3943	0	0
09.97	0.2842	4.2517	0.322	3.3745	0	0
10.97	0.2528	3.8579	0.2939	3.1453	0	0
11.97	0.3154	4.3351	0.3655	3.4379	0	0

Table 8. Comparisons the performances of our portfolios with normal portfolios on Group II

Date	Normal Portfolios		Our Portfolios		Comparrisons	
	S.D.	Kurtos	S.D.	Kurtos	S.D.	Kurtos
05.95	0.031	2.6308	0.0364	2.6868	0	1
06.95	0.0493	2.5832	0.0487	2.6411	1	1
07.95	0.0355	2.556	0.0354	2.7604	1	1
08.95	0.043	2.5756	0.0457	2.8974	0	1
09.95	0.0568	3.018	0.0537	3.0811	1	1
10.95	0.0743	2.8035	0.0673	2.7812	1	0
11.95	0.0674	2.919	0.0609	2.9611	1	1
12.95	0.0587	3.195	0.0549	3.1888	1	0
01.96	0.0691	3.2995	0.0592	3.7989	1	1
02.96	0.0616	2.9759	0.0562	3.263	1	1
03.96	0.0619	2.8949	0.0613	3.057	1	1
04.96	0.0608	2.8101	0.0591	2.8808	1	1
05.96	0.0605	2.8026	0.0625	2.9075	0	1
06.96	0.0673	2.7274	0.0681	2.9242	0	1
07.96	0.0586	2.7839	0.0623	2.8096	0	1
08.96	0.0648	2.8225	0.0662	2.9624	0	1
09.96	0.0671	2.7634	0.0732	2.9212	0	1
10.96	0.0618	2.8239	0.0674	2.9041	0	1
11.96	0.0755	2.9064	0.086	2.8459	0	0
12.96	0.0915	2.9662	0.1019	2.8507	0	0
01.96	0.1058	3.0616	0.1107	2.9255	0	0
02.97	0.1115	3.1711	0.1166	3.0244	0	0
03.97	0.1202	2.9608	0.1213	2.9411	0	0
04.97	0.1272	2.9408	0.1289	2.9148	0	0
05.97	0.1145	2.7714	0.1296	2.6773	0	0
06.97	0.1117	2.676	0.13	2.6739	0	0
07.97	0.1344	2.7526	0.162	2.6785	0	0
08.97	0.1395	2.7692	0.1561	2.7168	0	0
09.97	0.1649	2.812	0.1811	2.8638	0	0
10.97	0.1551	2.8006	0.167	2.7862	0	0
11.97	0.1776	2.8023	0.1946	2.6953	0	0

A STOCHASTIC HEISENBERG INEQUALITY

C. MUELLER*

Department of Mathematics,
University of Rochester,
Rochester, NY 14627, USA
E-mail: cmlr@math.rochester.edu

A. STAN

Department of Mathematics,
University of Rochester,
Rochester, NY 14627, USA
E-mail: astan@math.rochester.edu

An analogue of the Fourier transform will be introduced for all square integrable continuous martingale processes whose quadratic variation is deterministic. Using this transform we will formulate and prove a stochastic Heisenberg inequality.

1. Introduction

The Heisenberg inequality in $\mathbb{R}$ (see Folland and Sitaram[3] and Strichartz[5]) says that there exists a positive constant c such that for any $f \in L^2(\mathbb{R})$ and a and b real numbers, we have:

$$\int_{\mathbb{R}} (x-a)^2 |f(x)|^2 dx \cdot \int_{\mathbb{R}} (\gamma - b)^2 |\hat{f}(\gamma)|^2 d\gamma \geq c \parallel f \parallel_2^4 . \qquad (1)$$

In the above inequality, $\hat{f}$ denotes the Fourier transform of f:

$$\hat{f}(\gamma) = \int_{\mathbf{R}} e^{-2\pi i \gamma x} f(x) dx$$

while $\parallel f \parallel_2$ represents the L^2-norm of f. Besides, both integrals from the left-hand side are assumed to be finite.

2000 Mathematics Subject Classifications: 60G44, 60G15.
Key words and phrases: martingale, Itô's formula, commutation relationship, unitary operator, Schwarz inequality, Heisenberg inequality.
*Supported by an NSA grant.

The mathematical interpretation of inequality (1) is that an L^2-function and its Fourier transform cannot both be localized. See Folland and Sitaram[3].

In this paper the function f will be replaced by a complex valued stochastic process $\{Y_t\}_{t\in I}$, where I is an interval of the form $[0,T]$, for some $T>0$, or $I=[0,\infty)$. The integration with respect to the Lebesgue measure will be replaced by the integration with respect to a real valued square integrable continuous martingale $\{X_t\}_{t\in I}$ whose quadratic variation $\{\langle X\rangle_t\}_{t\in I}$ is deterministic. The real numbers a and b will be replaced by two real valued deterministic (measurable) functions g and $\tilde{g}$. The Fourier transform will be replaced by a unitary operator $\mathcal{G}$ that will be defined in the next section.

2. The $\mathcal{G}$-transform

Let $(\Omega,\mathcal{F},P)$ be a probability space and $\{\mathcal{F}_t\}_{t\in I}$ a filtration over $\mathcal{F}$. Let $\{X_t\}_{t\in I}$, where $I=[0,T]$ or $I=[0,\infty)$, be a real valued square integrable continuous martingale defined on Ω, adapted to $\{\mathcal{F}_t\}_{t\in I}$, whose quadratic variation process $\{\langle X\rangle_t\}_{t\in I}$ is deterministic. That means, for all $t\in I$, $\langle X\rangle_t$ is constant a.s.(almost surely). Thus, as a process, X_t is equal in distribution to a time-changed Brownian motion $B_{h(t)}$ with a deterministic time change $h(t)$. We assume that $X_0=0$ a.s.. Let $f:\mathbb{R}^2\to\mathbb{R}$ be a twice differentiable function with continuous second order partial derivatives. Applying Itô's formula to the semimartingales $X_t^1=X_t$ and $X_t^2=\langle X\rangle_t$ we obtain:

$$\begin{aligned} f(X_t,\langle X\rangle_t)-f(X_0,\langle X\rangle_0) &= \int_0^t \frac{\partial f}{\partial x}(X_s,\langle X\rangle_s)dX_s \\ &+ \int_0^t \frac{\partial f}{\partial y}(X_s,\langle X\rangle_s)d\langle X\rangle_s \\ &+ \frac{1}{2}\int_0^t \frac{\partial^2 f}{\partial x^2}(X_s,\langle X\rangle_s)d\langle X\rangle_s. \end{aligned} \tag{2}$$

From relation (2), we see that, if f satisfies the differential equation:

$$\frac{1}{2}\frac{\partial^2 f}{\partial x^2}+\frac{\partial f}{\partial y}=0, \tag{3}$$

then $f(X_t,\langle X\rangle_t)$ is a local martingale (see Durrett[1], page 70).

The function $f_c(x,y) = e^{cx-\frac{1}{2}c^2y}$ satisfies equation (3), for all $c \in \mathbb{C}$. Thus $f_c(X_t, \langle X\rangle_t)$ is a local martingale, for all $c \in \mathbb{C}$. Since $X_t \overset{\mathcal{D}}{=} B_{h(t)}$ we have $E(e^{cX_t}) < \infty$, for all $c \in \mathbb{R}$ and $t \in I$. Thus the process $\mathcal{E}_{c,t} := e^{cX_t - \frac{1}{2}c^2\langle X\rangle_t}$ is a martingale (c is fixed).

Lemma 2.1. *For all $t \geq 0$ and all c and d complex numbers, we have:*

$$\mathcal{E}_{c,t}\mathcal{E}_{d,t} = e^{cd\langle X\rangle_t}\mathcal{E}_{c+d,t}. \tag{4}$$

Proof.

$$\begin{aligned}\mathcal{E}_{c,t}\mathcal{E}_{d,t} &= e^{cX_t-\frac{1}{2}c^2\langle X\rangle_t}e^{dX_t-\frac{1}{2}d^2\langle X\rangle_t}\\ &= e^{cd\langle X\rangle_t}e^{(c+d)X_t-\frac{1}{2}(c+d)^2\langle X\rangle_t}\\ &= e^{cd\langle X\rangle_t}\mathcal{E}_{c+d,t}.\end{aligned}$$

□

For any $t \in I$, let V_t be the vector space spanned by all exponential functions $\{\mathcal{E}_{c,t}\}_{c\in\mathbb{C}}$. Lemma 2.1 proves that V_t is closed under multiplication. The constant process 1 belongs to V_t since $1 = \mathcal{E}_{0,t}$. Since X_t is a real valued function, we have $\overline{\mathcal{E}_{c,t}} = \mathcal{E}_{\bar{c},t} \in V_t$, for all $c \in \mathbb{C}$. Thus V_t is an algebra containing the constant functions and closed under conjugation.

Lemma 2.2. *For all s, $t \geq 0$ and all c and d complex numbers we have:*

$$E\left[\mathcal{E}_{c,s}\mathcal{E}_{d,t}\right] = e^{cd\langle X\rangle_{s\wedge t}}, \tag{5}$$

where $s \wedge t$ denotes the minimum between s and t.

Proof. Let us assume that $s \leq t$. We have:

$$\begin{aligned}E\left[\mathcal{E}_{c,s}\mathcal{E}_{d,t}\right] &= E\left[E\left[\mathcal{E}_{c,s}\mathcal{E}_{d,t}|\mathcal{F}_s\right]\right]\\ &= E\left[\mathcal{E}_{c,s}E\left[\mathcal{E}_{d,t}|\mathcal{F}_s\right]\right]\\ &= E\left[\mathcal{E}_{c,s}\mathcal{E}_{d,s}\right]\\ &= E\left[e^{cd\langle X\rangle_s}\mathcal{E}_{c+d,s}\right]\\ &= e^{cd\langle X\rangle_s}E\left[\mathcal{E}_{c+d,s}\right]\\ &= e^{cd\langle X\rangle_s}E\left[\left[\mathcal{E}_{c+d,s}|\mathcal{F}_0\right]\right]\\ &= e^{cd\langle X\rangle_s}E\left[\mathcal{E}_{c+d,0}\right]\\ &= e^{cd\langle X\rangle_s}.\end{aligned}$$

□

We define the function: $\mathcal{G} : \cup_{t\in I} V_t \to \cup_{t\in I} V_t$, $\mathcal{G}\mathcal{E}_{c,t} := \mathcal{E}_{-ic,t}$. We can see that $\mathcal{G}$ preserves the inner product. Indeed, we have:

$$\begin{aligned}
E\left[\mathcal{G}\mathcal{E}_{c,s}\overline{\mathcal{G}\mathcal{E}_{d,t}}\right] &= \left[\mathcal{E}_{-ic,s}\overline{\mathcal{E}_{-id,t}}\right] \\
&= \left[\mathcal{E}_{-ic,s}\mathcal{E}_{i\bar{d},t}\right] \\
&= e^{(-ic)(i\bar{d})X_{s\wedge t}} \\
&= e^{c\bar{d}X_{s\wedge t}} \\
&= E\left[\mathcal{E}_{c,s}\mathcal{E}_{\bar{d},t}\right] \\
&= E\left[\mathcal{E}_{c,s}\overline{\mathcal{E}_{d,t}}\right].
\end{aligned}$$

Thus $\mathcal{G}$ can be uniquely extended to a unitary operator, that we denote also by $\mathcal{G}$, from the Hilbert space $\mathcal{H}$ into itself, where $\mathcal{H}$ is the closure of the vector space spanned by $\{\mathcal{E}_{c,t} \mid c \in \mathbb{C}, t \in I\}$ in $L^2(\Omega, \mathcal{F}, P)$.

3. The Operator of Multiplication by X_t

Let $\{X_t\}_{t\in I}$ be a square integrable continuous martingale process adapted to the filtration $\mathcal{F}_t$ and having a deterministic quadratic variation process $\{\langle X\rangle_t\}_{t\in I}$. We know that for each $t \in I$, $X_t \stackrel{\mathcal{D}}{=} B_{h(t)}$.

We note the following. If t is fixed, then $h(t)$ is fixed, and X_t is normally distributed with mean 0 and variance $h(t)$. Thus,

$$e^{c|X_t|} \in L^2(\Omega, \mathcal{F}, P). \tag{6}$$

Since $|X_t| < e^{|X_t|}$, we have

$$\begin{aligned}
|X_t\mathcal{E}_{c,t}| &= |X_t| \cdot |e^{cX_t - \frac{c^2}{2}\langle X\rangle_t}| \\
&\le e^{|X_t|}e^{|c|\cdot|X_t|}e^{-\frac{\mathrm{Re}(c^2)}{2}\langle X\rangle_t} \\
&= e^{-\frac{\mathrm{Re}(c^2)}{2}\langle X\rangle_t}e^{(|c|+1)|X_t|}.
\end{aligned}$$

Since, according to (6), $e^{(|c|+1)|X_t|} \in L^2(\Omega, \mathcal{F}, P)$, we conclude that $X_t\mathcal{E}_{c,t} \in L^2(\Omega, \mathcal{F}, P)$, for all $c \in \mathbb{C}$ and $t \in I$.

Lemma 3.1. *For any $t \in I$ and $c \in \mathbb{C}$, we have:*

$$X_t\mathcal{E}_{c,t} = \lim_{r\to 0} \frac{\mathcal{E}_{r,t} - 1}{r}\mathcal{E}_{c,t}, \quad \text{in } L^2 - \text{sense}. \tag{7}$$

Proof. We have:

$$\lim_{r\to 0}\frac{\mathcal{E}_{r,t}-1}{r} = \lim_{r\to 0}\frac{e^{rX_t-\frac{r^2}{2}\langle X\rangle_t}-1}{r}$$
$$= \frac{d}{dr}\left(e^{rX_t-\frac{r^2}{2}\langle X\rangle_t}\right)\Big|_{r=0}$$
$$= X_t.$$

Thus $X_t\mathcal{E}_{c,t} = \lim_{r\to 0}\frac{\mathcal{E}_{r,t}-1}{r}\mathcal{E}_{c,t}$ pointwise. By Taylor's formula with Lagrange's remainder, for any fixed $\omega \in \Omega$, there exists a real number $u(\omega)$ between 0 and $rX_t(\omega) - \frac{r^2}{2}\langle X\rangle_t$ such that

$$e^{rX_t(\omega)-\frac{r^2}{2}\langle X\rangle_t} = 1 + rX_t(\omega) - \frac{r^2}{2}\langle X\rangle_t + \frac{1}{2}e^{u(\omega)}\left[rX_t(\omega) - \frac{r^2}{2}\langle X\rangle_t\right]^2.$$

This relation can be rewritten as:

$$\frac{e^{rX_t(\omega)-\frac{r^2}{2}\langle X\rangle_t}-1}{r} - X_t(\omega) = -\frac{r}{2}\langle X\rangle_t + \frac{r}{2}e^{u(\omega)}\left[X_t(\omega) - \frac{r}{2}\langle X\rangle_t\right]^2.$$

From here, using the inequalities $e^{u(\omega)} \le e^{|r|\cdot|X_t(\omega)|}$ and $|X_t(\omega)| < e^{|X_t(\omega)|}$, and (6), it follows from Lebesgue's dominated convergence theorem, that $\frac{e^{rX_t(\omega)-\frac{r^2}{2}\langle X\rangle_t}-1}{r} - X_t(\omega) \to 0$ in $L^2(\Omega, \mathcal{F}, P)$, as $r \to 0$. □

4. The Differential Operator

Definition 4.1. For any $t \in I$ and $c \in \mathbb{C}$, we define $D_t\mathcal{E}_{c,t} := c\langle X\rangle_t\mathcal{E}_{c,t}$.

We extend D_t by linearity to the vector space V_t spanned by the exponential functions $\{\mathcal{E}_{c,t}\}_{c\in\mathbb{C}}$. We must check that D_t is well-defined.

Lemma 4.1. *$D_t : V_t \to V_t$ is a well-defined operator.*

Proof. To check that D_t is a well-defined operator from V_t into itself we must check that if a function f can be expressed in two different ways as a linear combination of exponential functions, then calculating D_tf using these two linear combinations we obtain the same result. This reduces to checking that if $\lambda_1, \lambda_2, \ldots, \lambda_N \in \mathbb{C}$ and $c_1, c_2, \ldots, c_N \in \mathbb{C}$, $c_i \neq c_j$, for $i \neq j$, such that:

$$\sum_{k=1}^{N}\lambda_k\mathcal{E}_{c_k,t} = 0,$$

then

$$\sum_{k=1}^{N} \lambda_k c_k \langle X \rangle_t \mathcal{E}_{c_k,t} = 0.$$

If $\sum_{k=1}^{N} \lambda_k \mathcal{E}_{c_k,t} = 0$, then for any $n \in \{0, 1, \ldots, N-1\}$ we have: $E\left[\left(\sum_{k=1}^{N} \lambda_k \mathcal{E}_{c_k,t}\right) \mathcal{E}_{n,t}\right] = 0$. This means, according to formula (5), that:

$$\sum_{k=1}^{N} \lambda_k e^{nc_k \langle X \rangle_t} = 0, \quad \forall n \in \{0, 1, \ldots, N-1\}.$$

The above relations represent a homogenous linear system of N equations and N unknowns $\lambda_1, \lambda_2, \ldots, \lambda_N$, having a Vandermonde determinant. Since $c_i \neq c_j$, for $i \neq j$, the Vandermonde determinant of the above system is different from zero if $\langle X \rangle_t \neq 0$. Thus if $\langle X \rangle_t \neq 0$, then $\lambda_1 = \lambda_2 = \cdots = \lambda_N = 0$. Hence, either $\langle X \rangle_t = 0$ or $\lambda_1 = \lambda_2 = \cdots = \lambda_N = 0$. Therefore, $\sum_{k=1}^{N} \lambda_k c_k \langle X \rangle_t \mathcal{E}_{c_k,t} = 0$. □

We call D_t *the differential operator.* Its initial domain is V_t which is a dense subspace of the Hilbert space $\mathcal{H}_t$, where $\mathcal{H}_t$ is the closure of V_t in $L^2(\Omega, \mathcal{F}, P)$. Observe that $\mathcal{H}_t \subset L^2(\Omega, \sigma(X_t), P) \subset L^2(\Omega, \mathcal{F}_t, P)$, where $\sigma(X_t)$ is the smallest σ-field on Ω with respect to which X_t is measurable. Since $X_t \stackrel{\mathcal{D}}{=} B_{h(t)}$, we have $\mathcal{H}_t = L^2(\Omega, \sigma(X_t), P)$ (see Janson[4], page 19).

We denote the adjoint of D_t by D_t^*.

Lemma 4.2. *For any $t \in I$, V_t is contained in the domain of D_t^* and for any $\varphi \in V_t$ we have:*

$$X_t \varphi = D_t \varphi + D_t^* \varphi. \tag{8}$$

Proof. Since V_t is spanned by the exponential functions we have to check that for all $d \in \mathbb{C}$, $\mathcal{E}_{d,t}$ belongs to the domain of D_t^* and for any $c \in \mathbb{C}$ we have:

$$\langle D_t \mathcal{E}_{c,t}, \mathcal{E}_{d,t} \rangle = \langle \mathcal{E}_{c,t}, (X_t - D_t) \mathcal{E}_{d,t} \rangle, \tag{9}$$

where $\langle , \rangle$ denotes the L^2-inner product. Let us observe that, for any complex number c, the function $g_c : \mathbb{R}^2 \to \mathbb{R}$, $g_c(x, y) = (x - cy)e^{cx - \frac{c^2}{2} y}$ satisfies the differential equation:

$$\frac{1}{2} \frac{\partial^2 g_c}{\partial x^2} + \frac{\partial g_c}{\partial y} = 0.$$

It follows from Itô's formula that the process $\{g_c(X_t, \langle X\rangle_t)\}_{t\in I}$ is a martingale. We denote this martingale process by $\{\mathcal{X}_t\mathcal{E}_{c,t}\}_{t\in I}$. So, $\mathcal{X}_t\mathcal{E}_{c,t} = (X_t^{\bullet} - c\langle X\rangle_t)e^{cX_t - \frac{c^2}{2}\langle X\rangle_t}$. From this definition we obtain the formula:

$$X_t\mathcal{E}_{c,t} = \mathcal{X}_t\mathcal{E}_{c,t} + c\langle X\rangle_t\mathcal{E}_{c,t}. \tag{10}$$

Thus we have:

$$\begin{aligned}\langle \mathcal{E}_{c,t}, (X_t - D_t)\mathcal{E}_{d,t}\rangle &= E[X_t\mathcal{E}_{c,t}\mathcal{E}_{\bar{d},t}] - E[\mathcal{E}_{c,t}D_t\mathcal{E}_{\bar{d},t}] \\ &= E[X_t\mathcal{E}_{c+\bar{d},t}e^{c\bar{d}\langle X\rangle_t}] - E[\mathcal{E}_{c,t}\bar{d}\langle X\rangle_t\mathcal{E}_{\bar{d},t}] \\ &= e^{c\bar{d}\langle X\rangle_t}E[X_t\mathcal{E}_{c+\bar{d},t}] - \bar{d}\langle X\rangle_t E[\mathcal{E}_{c,t}\mathcal{E}_{\bar{d},t}].\end{aligned}$$

Using relations (10) and (5) we obtain:

$$\begin{aligned}&\langle \mathcal{E}_{c,t}, (X_t - D_t)\mathcal{E}_{d,t}\rangle \\ &= e^{c\bar{d}\langle X\rangle_t}E[\mathcal{X}_t\mathcal{E}_{c+\bar{d},t} + (c+\bar{d})\langle X\rangle_t\mathcal{E}_{c+\bar{d},t}] - \bar{d}\langle X\rangle_t E[\mathcal{E}_{c,t}\mathcal{E}_{\bar{d},t}] \\ &= e^{c\bar{d}\langle X\rangle_t}E[\mathcal{X}_t\mathcal{E}_{c+\bar{d},t}] + (c+\bar{d})\langle X\rangle_t e^{c\bar{d}\langle X\rangle_t}E[\mathcal{E}_{c+\bar{d},t}] - \bar{d}\langle X\rangle_t e^{c\bar{d}\langle X\rangle_t}.\end{aligned}$$

Because $\{\mathcal{X}_t\mathcal{E}_{c+\bar{d},t}\}_{t\in I}$ and $\{\mathcal{E}_{c+\bar{d},t}\}_{t\in I}$ are martingale processes we have:

$$\begin{aligned}E[\mathcal{X}_t\mathcal{E}_{c+\bar{d},t}] &= E[\mathcal{X}_0\mathcal{E}_{c+\bar{d},0}] \\ &= 0\end{aligned}$$

and

$$\begin{aligned}E[\mathcal{E}_{c+\bar{d},t}] &= E[\mathcal{E}_{c+\bar{d},0}] \\ &= 1.\end{aligned}$$

Thus, we obtain:

$$\begin{aligned}&\langle \mathcal{E}_{c,t}, (X_t - D_t)\mathcal{E}_{d,t}\rangle \\ &= e^{c\bar{d}\langle X\rangle_t}\cdot 0 + (c+\bar{d})\langle X\rangle_t e^{c\bar{d}\langle X\rangle_t}\cdot 1 - \bar{d}\langle X\rangle_t e^{c\bar{d}\langle X\rangle_t} \\ &= c\langle X\rangle_t e^{c\bar{d}\langle X\rangle_t} \\ &= c\langle X\rangle_t E[\mathcal{E}_{c,t}\mathcal{E}_{\bar{d},t}] \\ &= \langle c\langle X\rangle_t\mathcal{E}_{c,t}, \mathcal{E}_{d,t}\rangle \\ &= \langle D_t\mathcal{E}_{c,t}, \mathcal{E}_{d,t}\rangle.\end{aligned}$$

□

Corollary 4.1. *For all $t \in I$, the differential operator D_t admits a closed extension.*

Proof. Let W_t be the subspace of all functions φ in $\mathcal{H}_t$ for which there exists a sequence $\{\varphi_n\}_{n\geq 1} \subset V_t$ such that $\varphi_n \to \varphi$ in $L^2(\Omega, \mathcal{F}, P)$ and the sequence $\{D_t\varphi_n\}_{n\geq 1}$ is Cauchy in $L^2(\Omega, \mathcal{F}, P)$. We define the operator $\widetilde{D}_t$ on W_t in the following way: if $\varphi \in W_t$ and $\{\varphi_n\}_{n\geq 1} \subset V_t$ such that $\varphi_n \to \varphi$ in $L^2(\Omega, \mathcal{F}, P)$ and $\{D_t\varphi_n\}_{n\geq 1}$ is Cauchy in $L^2(\Omega, \mathcal{F}, P)$, then $\widetilde{D}_t\varphi = u$, where $u = \lim_{n\to\infty} D_t\varphi_n$ in $L^2(\Omega, \mathcal{F}, P)$. We notice that $u \in \mathcal{H}_t$.

We need to check that $\widetilde{D}_t\varphi$ is well defined. Let $\{\varphi_n\}_{n\geq 1} \subset V_t$ such that $\varphi_n \to \varphi$ in $L^2(\Omega, \mathcal{F}, P)$ and $\{D_t\varphi_n\}_{n\geq 1}$ is Cauchy in $L^2(\Omega, \mathcal{F}, P)$, and $\{\psi_n\}_{n\geq 1} \subset V_t$ such that $\psi_n \to \varphi$ in $L^2(\Omega, \mathcal{F}, P)$ and $\{D_t\psi_n\}_{n\geq 1}$ is Cauchy in $L^2(\Omega, \mathcal{F}, P)$. Let $u := \lim_{n\to\infty} D_t\varphi_n$ and $v := \lim_{n\to\infty} D_t\psi_n$. Both limits are in the L^2-sense. For any $\phi \in V_t$ we have:

$$\begin{aligned}\langle u, \phi\rangle &= \lim_{n\to\infty} \langle D_t\varphi_n, \phi\rangle \\ &= \lim_{n\to\infty} \langle \varphi_n, D_t^*\phi\rangle \\ &= \langle \varphi, D_t^*\phi\rangle.\end{aligned}$$

In the same way, we can see that

$$\langle v, \phi\rangle = \langle \varphi, D_t^*\phi\rangle.$$

Thus $\langle u, \phi\rangle = \langle v, \phi\rangle$, for all $\phi \in V_t$. Since V_t is dense in $\mathcal{H}_t$ and u, $v \in \mathcal{H}_t$ it follows that $u = v$. □

From now on we will denote the closure of D_t by D_t, too.

5. Commutation Relationships

Lemma 5.1. *For any $t \in I$, we have:*

$$[D_t, X_t] = \langle X\rangle_t I. \tag{11}$$

Proof. We will check that for any exponential function $\mathcal{E}_{c,t}$ we have:

$$D_tX_t\mathcal{E}_{c,t} - X_tD_t\mathcal{E}_{c,t} = \langle X\rangle_t\mathcal{E}_{c,t}. \tag{12}$$

Indeed we have:

$$D_tX_t\mathcal{E}_{c,t} = D_t \lim_{s\to 0} \frac{\mathcal{E}_{s,t} - 1}{s}\mathcal{E}_{c,t}.$$

We have seen that, because of (6), the limit from the right-hand side of the last equality is both pointwise and in the L^2-sense. The last equality can be written now as:

$$\begin{aligned}D_tX_t\mathcal{E}_{c,t} &= D_t \lim_{s\to 0} \frac{\mathcal{E}_{s,t}\mathcal{E}_{c,t} - \mathcal{E}_{c,t}}{s} \\ &= D_t \lim_{s\to 0} \frac{\mathcal{E}_{s+c,t}e^{sc\langle X\rangle_t} - \mathcal{E}_{c,t}}{s}.\end{aligned}$$

Because we are working with the closure of the operator D_t and the limit from the last relation is in the L^2-sense, we can commute D_t with the limit if $D_t \frac{\mathcal{E}_{s+c,t} e^{sc\langle X \rangle_t} - \mathcal{E}_{c,t}}{s}$ converges in $L^2(\Omega, \mathcal{F}, P)$, as $s \to 0$. Commuting D_t with the limit we get:

$$\begin{aligned}
D_t X_t \mathcal{E}_{c,t} &= \lim_{s \to 0} D_t \frac{\mathcal{E}_{s+c,t} e^{sc\langle X \rangle_t} - \mathcal{E}_{c,t}}{s} \\
&= \lim_{s \to 0} \frac{e^{sc\langle X \rangle_t} D_t \mathcal{E}_{s+c,t} - D_t \mathcal{E}_{c,t}}{s} \\
&= \lim_{s \to 0} \frac{e^{sc\langle X \rangle_t} (s+c) \langle X \rangle_t \mathcal{E}_{s+c,t} - c \langle X \rangle_t \mathcal{E}_{c,t}}{s} \\
&= \langle X \rangle_t \lim_{s \to 0} \frac{(s+c) e^{sc\langle X \rangle_t} \mathcal{E}_{s+c,t} - c \mathcal{E}_{c,t}}{s} \\
&= \langle X \rangle_t \lim_{s \to 0} \frac{(s+c) \mathcal{E}_{c,t} \mathcal{E}_{s,t} - c \mathcal{E}_{c,t}}{s} \\
&= \langle X \rangle_t \left[\lim_{s \to 0} \mathcal{E}_{s,t} + c \lim_{s \to 0} \frac{\mathcal{E}_{s,t} - 1}{s} \right] \mathcal{E}_{c,t} \\
&= \langle X \rangle_t [\mathcal{E}_{0,t} + c X_t] \mathcal{E}_{c,t} \\
&= \langle X \rangle_t (1 + c X_t) \mathcal{E}_{c,t} \\
&= \langle X \rangle_t \mathcal{E}_{c,t} + X_t (c \langle X \rangle_t \mathcal{E}_{c,t}) \\
&= \langle X \rangle_t \mathcal{E}_{c,t} + X_t D_t \mathcal{E}_{c,t}.
\end{aligned}$$

Because of (6) all the above limits are not only pointwise, but also in the L^2-sense. Thus

$$D_t X_t \mathcal{E}_{c,t} - X_t D_t \mathcal{E}_{c,t} = \langle X \rangle_t \mathcal{E}_{c,t}. \qquad \square$$

Corollary 5.1. *For any $t \in I$, we have:*

$$[D_t, D_t^*] = \langle X \rangle_t I. \tag{13}$$

Proof. According to formula (8) we have $X_t = D_t + D_t^*$. Thus:

$$\begin{aligned}
[D_t, D_t^*] &= [D_t, X_t - D_t] \\
&= [D_t, X_t] - [D_t, D_t] \\
&= \langle X \rangle_t I - 0 \\
&= \langle X \rangle_t I.
\end{aligned} \qquad \square$$

Lemma 5.2. *For any $t \in I$, we have:*

$$D_t \mathcal{G} = -i \mathcal{G} D_t. \tag{14}$$

Proof. For any exponential function $\mathcal{E}_{c,t}$, we have:

$$\begin{aligned} D_t\mathcal{G}\mathcal{E}_{c,t} &= D_t\mathcal{E}_{-ic,t} \\ &= -ic\langle X\rangle_t\mathcal{E}_{-ic,t} \\ &= -ic\langle X\rangle_t\mathcal{G}\mathcal{E}_{c,t} \\ &= -i\mathcal{G}\left(c\langle X\rangle_t\mathcal{E}_{c,t}\right) \\ &= -i\mathcal{G}D_t\mathcal{E}_{c,t}. \end{aligned}$$

□

Lemma 5.3. *For any $t \in I$, we have:*

$$D_t^*\mathcal{G} = i\mathcal{G}D_t^*. \tag{15}$$

Proof. For any two exponential functions $\mathcal{E}_{c,t}$ and $\mathcal{E}_{d,t}$, we have:

$$\begin{aligned} \langle D_t^*\mathcal{G}\mathcal{E}_{c,t}, \mathcal{E}_{d,t}\rangle &= \langle \mathcal{G}\mathcal{E}_{c,t}, D_t\mathcal{E}_{d,t}\rangle \\ &= \langle \mathcal{E}_{-ic,t}, d\langle X\rangle_t\mathcal{E}_{d,t}\rangle \\ &= \bar{d}\langle X\rangle_t\langle \mathcal{E}_{-ic,t}, \mathcal{E}_{d,t}\rangle \\ &= \bar{d}\langle X\rangle_t e^{-ic\bar{d}\langle X\rangle_t} \\ &= i(-i\bar{d})\langle X\rangle_t e^{c\overline{(id)}\langle X\rangle_t} \\ &= i(-i\bar{d})\langle X\rangle_t\langle \mathcal{E}_{c,t}, \mathcal{E}_{id,t}\rangle \\ &= i\langle \mathcal{E}_{c,t}, id\langle X\rangle_t\mathcal{E}_{id,t}\rangle \\ &= i\langle \mathcal{E}_{c,t}, D_t\mathcal{E}_{id,t}\rangle \\ &= i\langle D_t^*\mathcal{E}_{c,t}, \mathcal{E}_{id,t}\rangle. \end{aligned}$$

Since $\mathcal{G}$ preserves the inner product, being a unitary operator, we obtain:

$$\begin{aligned} \langle D_t^*\mathcal{G}\mathcal{E}_{c,t}, \mathcal{E}_{d,t}\rangle &= i\langle D_t^*\mathcal{E}_{c,t}, \mathcal{E}_{id,t}\rangle \\ &= i\langle \mathcal{G}D_t^*\mathcal{E}_{c,t}, \mathcal{G}\mathcal{E}_{id,t}\rangle \\ &= i\langle \mathcal{G}D_t^*\mathcal{E}_{c,t}, \mathcal{E}_{-i(id),t}\rangle \\ &= \langle i\mathcal{G}D_t^*\mathcal{E}_{c,t}, \mathcal{E}_{d,t}\rangle. \end{aligned}$$

Thus $D_t^*\mathcal{G} = i\mathcal{G}D_t^*$. □

6. Heisenberg Inequality

Lemma 6.1. *Let $t \in I$ and $Y_t \in L^2(\Omega, \sigma(X_t), P)$. Let c_t, $\tilde{c}_t \in \mathbb{R}$. If Y_t belongs to the domains of the operators D_t, D_t^*, $D_tD_t^*$, and $D_t^*D_t$, then*

$$E[(X_t - c_t)^2|Y_t|^2]^{\frac{1}{2}}E[(X_t - \tilde{c}_t)^2|\mathcal{G}Y_t|^2]^{\frac{1}{2}} \geq \langle X\rangle_t E[|Y_t|^2]. \tag{16}$$

Proof. Since $X_t = D_t + D_t^*$, we have:

$$
\begin{aligned}
&E[(X_t - c_t)^2|Y_t|^2]^{\frac{1}{2}} E[(X_t - \tilde{c}_t)^2|\mathcal{G}Y_t|^2]^{\frac{1}{2}} \\
&= E[|(D_t + D_t^* - c_t)Y_t|^2]^{\frac{1}{2}} E[|(D_t + D_t^* - \tilde{c}_t)\mathcal{G}Y_t|^2]^{\frac{1}{2}} \\
&= E[|(D_t + D_t^* - c_t)Y_t|^2]^{\frac{1}{2}} E[|(D_t\mathcal{G} + D_t^*\mathcal{G} - \tilde{c}_t\mathcal{G})Y_t|^2]^{\frac{1}{2}}.
\end{aligned}
$$

Using the commutation relationships (14) and (15) we obtain:

$$
\begin{aligned}
&E[(X_t - c_t)^2|Y_t|^2]^{\frac{1}{2}} E[(X_t - \tilde{c}_t)^2|\mathcal{G}Y_t|^2]^{\frac{1}{2}} \\
&= E[|(D_t + D_t^* - c_t)Y_t|^2]^{\frac{1}{2}} E[|(-i\mathcal{G}D_t + i\mathcal{G}D_t^* - \tilde{c}_t\mathcal{G})Y_t|^2]^{\frac{1}{2}} \\
&= E[|(D_t + D_t^* - c_t)Y_t|^2]^{\frac{1}{2}} E[|i\mathcal{G}(-D_t + D_t^* + i\tilde{c}_t)Y_t|^2]^{\frac{1}{2}}.
\end{aligned}
$$

Since $\mathcal{G}$ is a unitary operator and i has modulus 1, we have $E[|i\mathcal{G}(-D_t + D_t^* + i\tilde{c}_t)Y_t|^2] = E[|(-D_t + D_t^* + i\tilde{c}_t)Y_t|^2]$. Thus we obtain:

$$
\begin{aligned}
&E[(X_t - c_t)^2|Y_t|^2]^{\frac{1}{2}} E[(X_t - \tilde{c}_t)^2|\mathcal{G}Y_t|^2]^{\frac{1}{2}} \\
&= E[|(D_t + D_t^* - c_t)Y_t|^2]^{\frac{1}{2}} E[|(-D_t + D_t^* + i\tilde{c}_t)Y_t|^2]^{\frac{1}{2}} \\
&= \frac{1}{2}\left\{E[|(D_t + D_t^* - c_t)Y_t|^2]^{\frac{1}{2}} E[|(-D_t + D_t^* + i\tilde{c}_t)Y_t|^2]^{\frac{1}{2}}\right. \\
&+ \left. E[|(-D_t + D_t^* + i\tilde{c}_t)Y_t|^2]^{\frac{1}{2}} E[|(D_t + D_t^* - c_t)Y_t|^2]^{\frac{1}{2}}\right\}.
\end{aligned}
$$

Applying Schwarz inequality we obtain:

$$
\begin{aligned}
&E[(X_t - c_t)^2|Y_t|^2]^{\frac{1}{2}} E[(X_t - \tilde{c}_t)^2|\mathcal{G}Y_t|^2]^{\frac{1}{2}} \\
&= \frac{1}{2}\left\{E[|(D_t + D_t^* - c_t)Y_t|^2]^{\frac{1}{2}} E[|(-D_t + D_t^* + i\tilde{c}_t)Y_t|^2]^{\frac{1}{2}}\right. \\
&+ \left. E[|(-D_t + D_t^* + i\tilde{c}_t)Y_t|^2]^{\frac{1}{2}} E[|(D_t + D_t^* - c_t)Y_t|^2]^{\frac{1}{2}}\right\}. \\
&\geq \frac{1}{2}\left\{|\langle(D_t + D_t^* - c_t)Y_t, (-D_t + D_t^* + i\tilde{c}_t)Y_t\rangle|\right. \\
&+ \left.|\langle(-D_t + D_t^* + i\tilde{c}_t)Y_t, (D_t + D_t^* - c_t)Y_t\rangle|\right\}.
\end{aligned}
$$

Applying the triangle inequality we get:

$$
\begin{aligned}
&E[(X_t - c_t)^2|Y_t|^2]^{\frac{1}{2}} E[(X_t - \tilde{c}_t)^2|\mathcal{G}Y_t|^2]^{\frac{1}{2}} \\
&\geq \frac{1}{2}\,|\langle(D_t + D_t^* - c_t)Y_t, (-D_t + D_t^* + i\tilde{c}_t)Y_t\rangle \\
&+ \langle(-D_t + D_t^* + i\tilde{c}_t)Y_t, (D_t + D_t^* - c_t)Y_t\rangle|\,.
\end{aligned}
$$

Using the linearity in the first argument and the conjugate linearity in the second argument of the inner product, and the definition of the adjoint operator, because c_t and $\tilde{c}_t$ are real numbers, the last expression can be simplified and we obtain finally:

$$\begin{aligned}
&E[(X_t - c_t)^2 |Y_t|^2]^{\frac{1}{2}} E[(X_t - \tilde{c}_t)^2 |\mathcal{G} Y_t|^2]^{\frac{1}{2}} \\
&\geq \frac{1}{2} \left| -2\langle D_t Y_t, D_t Y_t\rangle + 2\langle D_t^* Y_t, D_t^* Y_t\rangle \right| \\
&= \left| -\langle D_t Y_t, D_t Y_t\rangle + \langle D_t^* Y_t, D_t^* Y_t\rangle \right| \\
&= \left| -\langle D_t^* D_t Y_t, Y_t\rangle + \langle D_t D_t^* Y_t, Y_t\rangle \right| \\
&= \left| \langle (D_t D_t^* - D_t^* D_t) Y_t, Y_t\rangle \right| \\
&= \left| \langle [D_t, D_t^*] Y_t, Y_t\rangle \right| .
\end{aligned}$$

Using the commutation relationship (13) we obtain:

$$\begin{aligned}
&E[(X_t - c_t)^2 |Y_t|^2]^{\frac{1}{2}} E[(X_t - \tilde{c}_t)^2 |\mathcal{G} Y_t|^2]^{\frac{1}{2}} \\
&\geq \left| \langle \langle X\rangle_t Y_t, Y_t\rangle \right| \\
&= \langle X\rangle_t \langle Y_t, Y_t\rangle \\
&= \langle X\rangle_t E[|Y_t|^2].
\end{aligned}$$

□

Theorem 6.1. *Let $\{Y_t\}_{t\in[0,T]}$ be a stochastic process such that for each $t \in [0, T]$, $Y_t \in L^2(\Omega, \sigma(X_t), P)$ and Y_t belongs to the domain of D_t, D_t^*, $D_t D_t^*$, and $D_t^* D_t$. Then for any Borel measurable functions g, $\tilde{g} : [0,T] \to \mathbb{R}$, we have:*

$$\begin{aligned}
&E\left[\left|\int_0^T (X_t - g(t)) Y_t dX_t\right|^2\right]^{\frac{1}{2}} E\left[\left|\int_0^T (X_t - \tilde{g}(t)) \mathcal{G} Y_t dX_t\right|^2\right]^{\frac{1}{2}} \\
&\geq E\left[\int_0^T |Y_t|^2 \langle X\rangle_t d\langle X\rangle_t\right] .
\end{aligned} \tag{17}$$

We assume that all the measurability and square integrability conditions, necessary for the existence of the above stochastic integrals, hold.

Proof. Using the isomorphism through which the stochastic integral is extended from the simple processes to the square integrable processes we obtain:

$$\begin{aligned}
&E\left[\left|\int_0^T (X_t - g(t)) Y_t dX_t\right|^2\right] E\left[\left|\int_0^T (X_t - \tilde{g}(t)) \mathcal{G} Y_t dX_t\right|^2\right] \\
&= \int_0^T E\left[|(X_t - g(t)) Y_t|^2\right] d\langle X\rangle_t \int_0^T E\left[|(X_t - \tilde{g}(t)) \mathcal{G} Y_t|^2\right] d\langle X\rangle_t .
\end{aligned}$$

Applying Schwarz inequality we obtain:

$$
\begin{aligned}
& E\left[\left|\int_0^T (X_t - g(t))Y_t dX_t\right|^2\right] E\left[\left|\int_0^T (X_t - \tilde{g}(t))\mathcal{G}Y_t dX_t\right|^2\right] \\
&= \int_0^T E\left[|(X_t - g(t))Y_t|^2\right] d\langle X\rangle_t \int_0^T E\left[|(X_t - \tilde{g}(t))\mathcal{G}Y_t|^2\right] d\langle X\rangle_t \\
&\geq \left\{\int_0^T E\left[|(X_t - g(t))Y_t|^2\right]^{\frac{1}{2}} E\left[|(X_t - \tilde{g}(t))\mathcal{G}Y_t|^2\right]^{\frac{1}{2}} d\langle X\rangle_t\right\}^2 .
\end{aligned}
$$

Applying inequality (16), for $c_t := g(t)$ and $\tilde{c}_t := \tilde{g}(t)$, we obtain:

$$
\begin{aligned}
& E\left[\left|\int_0^T (X_t - g(t))Y_t dX_t\right|^2\right]^{\frac{1}{2}} E\left[\left|\int_0^T (X_t - \tilde{g}(t))\mathcal{G}Y_t dX_t\right|^2\right]^{\frac{1}{2}} \\
&\geq \int_0^T E\left[|(X_t - g(t))Y_t|^2\right]^{\frac{1}{2}} E\left[|(X_t - \tilde{g}(t))\mathcal{G}Y_t|^2\right]^{\frac{1}{2}} d\langle X\rangle_t \\
&\geq \int_0^T \langle X\rangle_t E[|Y_t|^2] d\langle X\rangle_t \\
&= E\left[\int_0^T |Y_t|^2 \langle X\rangle_t d\langle X\rangle_t\right] .
\end{aligned}
$$

□

If we take, $Y_t := \varphi(X_t)$, where $\varphi : \mathbb{R} \to \mathbb{C}$ is a Borel measurable function, then we obtain the following:

Corollary 6.1. *Let $\varphi : \mathbb{R} \to \mathbb{C}$ be a Borel measurable function such that for each $t \in [0, T]$, $\varphi(X_t)$ belongs to the domain of D_t, D_t^*, $D_t D_t^*$, and $D_t^* D_t$. Then for any Borel measurable functions g, $\tilde{g} : [0,T] \to \mathbb{R}$, we have:*

$$
\begin{aligned}
& E\left[\left|\int_0^T (X_t - g(t))\varphi(X_t) dX_t\right|^2\right]^{\frac{1}{2}} E\left[\left|\int_0^T (X_t - \tilde{g}(t))\mathcal{G}\varphi(X_t) dX_t\right|^2\right]^{\frac{1}{2}} \\
&\geq E\left[\int_0^T |\varphi(X_t)|^2 \langle X\rangle_t d\langle X\rangle_t\right] . \qquad (18)
\end{aligned}
$$

We assume that all the measurability and square integrability conditions, necessary for the existence of the above stochastic integrals, hold.

Suppose $X_t = B_t$ is a Brownian motion process. Then for each $t \in [0, T]$, $\langle X\rangle_t = t$ is deterministic. The inequality (17) becomes:

Theorem 6.2. *Let $\{Y_t\}_{t\in[0,T]}$ be a stochastic process such that for each $t \in [0, T]$, Y_t is measurable with respect to the sigma-field $\sigma(B_t)$ and Y_t belongs to the domain of D_t, D_t^*, $D_tD_t^*$, and $D_t^*D_t$. Then for any Borel measurable functions g, $\tilde{g} : [0,T] \to \mathbb{R}$, we have:*

$$E\left[\left|\int_0^T (B_t - g(t))Y_t dB_t\right|^2\right]^{\frac{1}{2}} E\left[\left|\int_0^T (B_t - \tilde{g}(t))\mathcal{G}Y_t dB_t\right|^2\right]^{\frac{1}{2}}$$
$$\geq E\left[\int_0^T t|Y_t|^2 dt\right]. \tag{19}$$

We assume that all the measurability and square integrability conditions, necessary for the existence of the above stochastic integrals, hold.

In all the inequalities from this section we may replace $\int_0^T$ by $\int_0^\infty$ and obtain the same results.

References

1. Durrett, R.: *Brownian Motion and Martingales in Analysis.* The Wadsworth Mathematics Series, 1984.
2. Dym, H. and McKean, H. P.: *Fourier Series and Integrals.* Academic Press, 1972.
3. Folland, G. B. and Sitaram, A.: The Uncertainty Principle: A Mathematical Survey, *The Journal of Fourier Analysis and Applications*, Vol. 3, Nr. 3, 1997, 207-238.
4. Janson, S.: *Gaussian Hilbert Spaces.* Cambridge Tracts in Mathematics 129, Cambridge University Press, 1997.
5. Strichartz, R.: *A Guide to Distribution Theory and Fourier Transforms.* CRC Press, 1994.
6. Von Neumann, J.: *Mathematical Foundations of Quantum Mechanics.* Princeton University Press, 1955.

AN INFINITE DIMENSIONAL STOCHASTIC PROCESS AND THE LÉVY LAPLACIAN ACTING ON WND-VALUED FUNCTIONS

KENJIRO NISHI, KIMIAKI SAITÔ

Department of Information Sciences
Meijo University
Tenpaku, Nagoya 468-8502, Japan
E-mails: knishi@ccmfs.meijo-u.ac.jp, ksaito@ccmfs.meijo-u.ac.jp

In this paper we give an infinite dimensional stochastic process and prove that the infinitesimal generator of the process is given by the Lévy Laplacian acting on some domain consisting of WND(White Noise Distribution)-valued functions.

1. Introduction

An infinite dimensional Laplacian was introduced by P. Lévy in his book [22]. This Laplacian has been studied by many authors. (See Refs. 1-8,12,17-21,23,25,26,28,29,31-37 and others) The stochastic process associated with the Lévy Laplacian has been discussed in Refs. 1,3,6,21,24,32-34,37.

In this paper, extending the methods in Refs. 21,24,29,33,34, we present a new approach to construct a domain of the Lévy Laplacian acting on WND-valued functions and give an infinite dimensional stochastic process generated by the Laplacian. This is a generalization of the result in Ref. 29. The construction of stochastic process is breifly as follows:

Let $E \equiv \mathcal{S}(\mathbf{R})$ be the Schwartz space consisting of rapidly decreasing functions defined on $\mathbf{R}$, where $\mathbf{R}$ means the field of all real numbers. Fix a finite interval $T \subset \mathbf{R}$ and take $\eta_T \in E$ satisfying $\eta_T = \frac{1}{\sqrt{|T|}}$ on T. Let $\{X_t\}_{t\geq 0}$ be a stochastic process such that each characteristic function of X_t is given in the form $E[e^{izX_t}] = e^{-th(z)}$ for all $z \in \mathbf{R}$. Set $Y_t^\lambda = X_{\lambda t}$ for $\lambda \geq 0$. Then an infinite dimensional stochastic process $\{(\xi_\lambda + Y_t^\lambda \eta_T)_{\lambda>0}; t \geq 0\}, \xi_\lambda \in E_{\mathbf{C}}, \lambda > 0$, has the Lévy Laplacian as the infinitesimal generator on some domain consisting of S-transforms of WND-valued functions.

The paper is organized as follows. In Section 2 we summarize some basic definitions and results in white noise theory. In Section 3 we introduce the definition of the Lévy Laplacian acting on WND-valued functions and give an equi-continuous semigroup of class (C_0) generated by the Laplacian. In Section 4, based on infinitely many one-dimensional stochastic processes, we give an infinite dimensional stochastic process generated by the Lévy Laplacian.

2. Standard setup of white noise calculus

In this section we assemble some basic notations of white noise analysis following Refs. 7,10,16,19,26,30.

We take the space $E^* \equiv \mathcal{S}'(\mathbf{R})$ of tempered distributions with the standard Gaussian measure μ such that

$$\int_{E^*} \exp\{i\langle x, \xi\rangle\}\, d\mu(x) = \exp\left(-\frac{1}{2}|\xi|_0^2\right), \qquad \xi \in E \equiv \mathcal{S}(\mathbf{R}),$$

where $\langle \cdot, \cdot \rangle$ is the canonical bilinear form on $E^* \times E$ and $|\cdot|_0$ is the $L^2(\mathbf{R})$-norm.

Let $A = -(d/du)^2 + u^2 + 1$. Then this is a densely defined self-adjoint operator on $L^2(\mathbf{R})$ and there exists an orthonormal basis $\{e_\nu; \nu \geq 0\} \subset E$ for $L^2(\mathbf{R})$ such that $Ae_\nu = 2(\nu+1)e_\nu$. Define the norm $|\cdot|_p$ by $|f|_p = |A^p f|_0$ for $f \in E$ and $p \in \mathbf{R}$, and let E_p be the completion of E with respect to the norm $|\cdot|_p$. Then E_p is a real separable Hilbert space with the norm $|\cdot|_p$ and the dual space E_p' of E_p is the same as E_{-p} (see Ref.14). The space E is the projective limit space of $\{E_p; p \geq 0\}$ and the space E^* is the dual space of E. We denote the complexifications of $L^2(\mathbf{R})$, E and E_p by $L^2_{\mathbf{C}}(\mathbf{R})$, $E_{\mathbf{C}}$ and $E_{\mathbf{C},p}$, respectively.

The space $(L^2) = L^2(E^*, \mu)$ of complex-valued square-integrable functionals defined on E^* admits the well-known Wiener-Itô decomposition:

$$(L^2) = \bigoplus_{n=0}^{\infty} H_n,$$

where H_n is the space of multiple Wiener integrals of order $n \in \mathbf{N}$ and $H_0 = \mathbf{C}$. Let $L^2_{\mathbf{C}}(\mathbf{R})^{\hat{\otimes} n}$ denote the n-fold symmetric tensor product of $L^2_{\mathbf{C}}(\mathbf{R})$. If $\varphi \in (L^2)$ is represented by $\varphi = \sum_{n=0}^{\infty} \mathbf{I}_n(f_n)$, $f_n \in L^2_{\mathbf{C}}(\mathbf{R})^{\hat{\otimes} n}$, then the (L^2)-norm $\|\varphi\|_0$ is given by

$$\|\varphi\|_0 = \left(\sum_{n=0}^{\infty} n! |f_n|_0^2\right)^{1/2},$$

where $|\cdot|_0$ means also the norm of $L^2_{\mathbf{C}}(\mathbf{R})^{\hat{\otimes}n}$.

For $p \in \mathbf{R}$, let $\|\varphi\|_p = \|\Gamma(A)^p\varphi\|_0$, where $\Gamma(A)$ is the second quantization operator of A. If $p \geq 0$, let $(E)_p$ be the domain of $\Gamma(A)^p$. If $p < 0$, let $(E)_p$ be the completion of (L^2) with respect to the norm $\|\cdot\|_p$. Then $(E)_p$, $p \in \mathbf{R}$, is a Hilbert space with the norm $\|\cdot\|_p$. It is easy to see that for $p > 0$, the dual space $(E)_p^*$ of $(E)_p$ is given by $(E)_{-p}$. Moreover, for any $p \in \mathbf{R}$, we have the decomposition

$$(E)_p = \bigoplus_{n=0}^{\infty} H_n^{(p)},$$

where $H_n^{(p)}$ is the completion of $\{\mathbf{I}_n(f); f \in E_{\mathbf{C}}^{\hat{\otimes}n}\}$ with respect to $\|\cdot\|_p$. Here $E_{\mathbf{C}}^{\hat{\otimes}n}$ is the n-fold symmetric tensor product of $E_{\mathbf{C}}$. We also have $H_n^{(p)} = \{\mathbf{I}_n(f); f \in E_{\mathbf{C},p}^{\hat{\otimes}n}\}$ for any $p \in \mathbf{R}$, where $E_{\mathbf{C},p}^{\hat{\otimes}n}$ is also the n-fold symmetric tensor product of $E_{\mathbf{C},p}$. The norm $\|\varphi\|_p$ of $\varphi = \sum_{n=0}^{\infty} \mathbf{I}_n(f_n) \in (E)_p$ is given by

$$\|\varphi\|_p = \left(\sum_{n=0}^{\infty} n!|f_n|_p^2\right)^{1/2}, \qquad f_n \in E_{\mathbf{C},p}^{\hat{\otimes}n},$$

where the norm of $E_{\mathbf{C},p}^{\hat{\otimes}n}$ is denoted also by $|\cdot|_p$.

The projective limit space (E) of spaces $(E)_p$, $p \in \mathbf{R}$ is a nuclear space. The inductive limit space $(E)^*$ of spaces $(E)_p, p \in \mathbf{R}$ is nothing but the dual space of (E). The space $(E)^*$ is called the space of *generalized white noise functionals* or *white noise distributions.* We denote by $\ll \cdot,\cdot \gg$ the canonical bilinear form on $(E)^* \times (E)$. Then we have

$$\ll \Phi, \varphi \gg = \sum_{n=0}^{\infty} n!\langle F_n, f_n\rangle$$

for any $\Phi = \sum_{n=0}^{\infty} \mathbf{I}_n(F_n) \in (E)^*$ and $\varphi = \sum_{n=0}^{\infty} \mathbf{I}_n(f_n) \in (E)$, where the canonical bilinear form on $(E_{\mathbf{C}}^{\hat{\otimes}n})^* \times (E_{\mathbf{C}}^{\hat{\otimes}n})$ is denoted also by $\langle\cdot,\cdot\rangle$.

Since $\varphi_\xi \equiv \exp\{\langle\cdot,\xi\rangle - \frac{1}{2}\langle\xi,\xi\rangle\} \in (E)$, the *S-transform* is defined on $(E)^*$ by

$$S[\Phi](\xi) = \ll \Phi, \varphi_\xi \gg, \qquad \xi \in E_{\mathbf{C}}.$$

Theorem 2.1. *(see e.g. Refs.10,19,26,31) A complex-valued function F on $E_{\mathbf{C}}$ is the S-transform of an element in $(E)^*$ if and only if F satisfies the conditions:*

1) *For every $\xi, \eta \in E_{\mathbf{C}}$, the function $z \to F(\xi + z\eta)$, $z \in \mathbf{C}$, is an entire function of z.*

2) *There exist non-negative constants K, a and p such that*

$$|F(\xi)| \leq K \exp\{a|\xi|_p^2\}, \quad \xi \in E_{\mathbf{C}}.$$

3. The Lévy Laplacian acting on WND-valued functions

Let $F \in S[(E)^*]$. Then, by Theorem 2.1, we see that for any $\xi, \eta \in E_{\mathbf{C}}$ the function $F(\xi + z\eta)$ is an entire function of $z \in \mathbf{C}$. Hence we have the series expansion:

$$F(\xi + z\eta) = \sum_{n=0}^{\infty} \frac{z^n}{n!} F^{(n)}(\xi)(\eta, \dots, \eta),$$

where $F^{(n)}(\xi) : E_{\mathbf{C}} \times \cdots \times E_{\mathbf{C}} \to \mathbf{C}$ is a continuous n-linear functional.

Fix a finite interval $T \subset \mathbf{R}$. Take an orthonormal basis $\{\zeta_n\}_{n=0}^{\infty} \subset E$ for $L^2(T)$ satisfying the equally dense and uniform boundedness property (see e.g. Refs. 19,20). Let $\mathcal{D}_L$ denote the set of all $\Phi \in (E)^*$ such that the limit

$$\widetilde{\Delta}_L S[\Phi](\xi) = \lim_{N \to \infty} \frac{1}{N} \sum_{n=0}^{N-1} S[\Phi]''(\xi)(\zeta_n, \zeta_n)$$

exists for any $\xi \in E_{\mathbf{C}}$ and is in $S[(E)^*]$. The Lévy Laplacian Δ_L is defined by

$$\Delta_L \Phi = S^{-1} \widetilde{\Delta}_L S \Phi$$

for $\Phi \in D_L$. We denote the set of all functionals $\Phi \in \mathcal{D}_L$ such that $S[\Phi](\eta) = 0$ for all $\eta \in E$ with $\mathrm{supp}(\eta) \subset T^c$ by $\mathcal{D}_L^T$.

Fix a continuous function $h : \mathbf{R} \to [0, \infty)$ satisfying the condition:

> There exists a stochastic process $\{X_t\}_{t \geq 0}$ such that each characteristic function of X_t is given by $E[e^{izX_t}] = e^{-th(z)}$ for all $z \in \mathbf{R}$.

Using the Wick ordering $: \cdot :$ we write a generalized white noise functional with the S-transform $S[\Phi]$ given by

$$S[\Phi](\xi) = \int_{T^n} f(\mathbf{u}) e^{ia_1\xi(u_1)} \dots e^{ia_n\xi(u_n)} d\mathbf{u}, \quad f \in E_{\mathbf{C}}^{\hat{\otimes} n}, \quad a_1, \dots, a_n \in \mathbf{R},$$

by

$$\Phi = \int_{T^n} f(u_1, \dots, u_n) : e^{ia_1x(u_1)} \cdots e^{ia_nx(u_n)} : d\mathbf{u}.$$

For any $\lambda > 0$ and $n \in \mathbf{N}$, we put

$$\mathbf{D}_{n,\lambda} = LS\left\{ \int_{T^n} f(\mathbf{u}) : \prod_{k=1}^{n} e^{ia_k x(u_k)} : d\mathbf{u}; f \in E_{\mathbf{C}}^{\hat{\otimes} n},\ \mathrm{supp} f \subset T^n \right.$$

$$\left. 0 < a_1 \leq a_2 \leq \cdots \leq a_n,\ \sum_{k=1}^{n} a_k = \sqrt{|T|} n\lambda,\ \sum_{k=1}^{n} a_k^2 = |T|\lambda h(n\lambda) \right\}.$$

Put $\mathbf{D}_{0,\lambda} = \mathbf{C}$. Then $\mathbf{D}_{n,\lambda}$ is a linear subspace of $(E)_{-p}$ for any $p > \frac{5}{12}$. (See Ref. 21) We define a space $\overline{\mathbf{D}_{n,\lambda}}$ by the completion of $\mathbf{D}_{n,\lambda}$ in $(E)_{-p}$ with respect to $\|\cdot\|_{-p}$. Then for each $n \in \mathbf{N} \cup \{0\}$ and $\lambda > 0$, $\overline{\mathbf{D}_{n,\lambda}}$ becomes a Hilbert space with the inner product of $(E)_{-p}$. Using the simillar method in Ref. 36, we get the following:

Theorem 3.1. *(cf.Ref. 36, see also Refs. 21,34) For each $n \in \mathbf{N} \cup \{0\}$ and $\lambda > 0$, the operator Δ_L becomes a continuous linear operator from $\overline{\mathbf{D}_{n,\lambda}}$ into itself such that*

$$\Delta_L \Phi = -\lambda h(n\lambda)\Phi \text{ for any } \Phi \in \overline{\mathbf{D}_{n,\lambda}}.$$

Proposition 3.1. *(cf. Ref.36) Let $\Phi = \sum_{n=0}^{\infty} \Phi_n$ and $\Psi = \sum_{n=0}^{\infty} \Psi_n$ be generalized white noise functionals such that Φ_n and Ψ_n are in $\overline{\mathbf{D}_{n,\lambda}}$ for each $\lambda > 0$ and $n \in \mathbf{N} \cup \{0\}$. If $\Phi = \Psi$ in $(E)^*$, then $\Phi_n = \Psi_n$ for all $n \in \mathbf{N} \cup \{0\}$.*

The operator Δ_L is a self-adjoint operator on $\overline{\mathbf{D}_{n,\lambda}}$ for each $\lambda > 0$ and $n \in \mathbf{N} \cup \{0\}$.

Put $\alpha_N^\lambda(n) = \sum_{\ell=0}^{N} (\lambda h(n\lambda))^{2\ell}$ and define a space $\mathbf{E}_{-p,N}^\lambda$ by

$$\mathbf{E}_{-p,N}^\lambda = \left\{ \sum_{n=1}^{\infty} \Phi_n \in (E)^*; \sum_{n=1}^{\infty} \alpha_N^\lambda(n) \|\Phi_n\|_{-p}^2 < \infty, \Phi_n \in \overline{\mathbf{D}_{n,\lambda}}, n = 1, 2, \ldots \right\}$$

with the norm $|||\cdot|||_{-p,N,\lambda}$ given by

$$|||\Phi|||_{-p,N,\lambda} = \left(\sum_{n=1}^{\infty} \alpha_N^\lambda(n) \|\Phi_n\|_{-p}^2 \right)^{1/2}, \quad \Phi = \sum_{n=1}^{\infty} \Phi_n \in \mathbf{E}_{-p,N}^\lambda$$

for each $N \in \mathbf{N}, p > \frac{5}{12}$ and $\lambda > 0$. Then for any $N \in \mathbf{N}, p > \frac{5}{12}$ and $\lambda > 0$, $\mathbf{E}_{-p,N}^\lambda$ is in $(E)_{-p}$ and is a Hilbert space with respect to the norm $|||\cdot|||_{-p,N,\lambda}$.

Put $\mathbf{E}^{\lambda}_{-p,\infty} = \bigcap_{N=1}^{\infty} \mathbf{E}^{\lambda}_{-p,N}$ with projective limit topology. Then we have the following inclusion relations:

$$\mathbf{E}^{\lambda}_{-p,\infty} \subset \cdots \subset \mathbf{E}^{\lambda}_{-p,N+1} \subset \mathbf{E}^{\lambda}_{-p,N} \subset \cdots \subset \mathbf{E}^{\lambda}_{-p,1} \subset (E)_{-p}.$$

The space $\mathbf{E}^{\lambda}_{-p,\infty}$ includes $\cup_{n=1}^{\infty}\overline{\mathbf{D}_{n,\lambda}}$ for any $\lambda > 0$. The operator Δ_L becomes a continuous linear operator defined on $\mathbf{E}^{\lambda}_{-p,2}$ into $\mathbf{E}^{\lambda}_{-p,1}$ satisfying $|||\Delta_L\Phi|||_{-p,N,\lambda} \leq |||\Phi|||_{-p,N+1,\lambda}$, $N = 1, 2, 3, \dots,\quad \Phi \in \mathbf{E}^{\lambda}_{-p,\infty}$. With these properties, we have the following:

Theorem 3.2. *(Refs. 21,36) The operator Δ_L is a self-adjoint operator densely defined on $\mathbf{E}^{\lambda}_{-p,N}$ for each $N \in \mathbf{N}$, $p > \frac{5}{12}$ and $\lambda > 0$.*

For each $t \geq 0$ and $\lambda > 0$ we consider an operator G_t^{λ} on $\mathbf{E}^{\lambda}_{-p,\infty}$ defined by

$$G_t^{\lambda}\Phi = \sum_{n=1}^{\infty} e^{-t\lambda h(n\lambda)}\Phi_n$$

for $\Phi = \sum_{n=1}^{\infty}\Phi_n \in \mathbf{E}^{\lambda}_{-p,\infty}$. Then we have the following:

Theorem 3.3. *(cf. Refs. 21,35) For each $\lambda > 0$ the family $\{G_t^{\lambda}; t \geq 0\}$ is an equi-continuous semigroup of class (C_0) generated by Δ_L as a continuous linear operator defined on $\mathbf{E}^{\lambda}_{-p,\infty}$.*

Take a function h such that there exists a finite Borel measure ν on $(0,\infty)$ satisfying

$$\sum_{n=1}^{\infty}\int_{(0,\infty)} \frac{d\nu(\lambda)}{\alpha_N^{\lambda}(n)} < \infty, \quad N = 1, 2, 3, \dots .$$

For example, $h(z) = |z|$ is such a function. In fact, if we take a finite Borel measure ν on $(0,\infty)$ satisfying

$$\int_{(0,\infty)} \frac{d\nu(\lambda)}{\lambda^4} < \infty,$$

then ν satisfies the above condition.

Proposition 3.2. *(cf. Refs. 29) Let $N \in \mathbf{N}$. If ν is absolutely continuous with respect to the Lebesgue measure on $(0,\infty)$, then $\mathbf{E}^{\lambda_1}_{-p,N} \cap \mathbf{E}^{\lambda_2}_{-p,N} = \{0\}$ holds for $\nu \times \nu$-almost all (λ_1, λ_2) in $(0,\infty) \times (0,\infty)$.*

Proof. Take an element Φ in $\mathbf{E}^{\lambda_1}_{-p,N} \cap \mathbf{E}^{\lambda_2}_{-p,N}$. Then the functional Φ has two expressions:

$$\Phi = \sum_{n=1}^{\infty} \Phi_n = \sum_{n=1}^{\infty} \Psi_n, \quad \Phi_n \in \overline{\mathbf{D}_{n,\lambda_1}}, \Psi_n \in \overline{\mathbf{D}_{n,\lambda_2}}.$$

For each $n \in \mathbf{N}$, Φ_n and Ψ_n are expressed in the forms

$$\Phi_n = \lim_{N\to\infty} \int_{T^n} f^{[N]}_{\lambda_1,n}(u_1, \ldots, u_n) : \prod_{\nu=1}^{n} e^{ia^{[N]}_\nu x(u_\nu)} : d\mathbf{u},$$

$$\sum_{\nu=1}^{n} a^{[N]}_\nu = \sqrt{|T|}n\lambda_1, \sum_{\nu=1}^{n} (a^{[N]}_\nu)^2 = |T|\lambda_1 h(n\lambda_1),$$

and

$$\Psi_n = \lim_{N\to\infty} \int_{T^n} g^{[N]}_{\lambda_2,n}(u_1, \ldots, u_n) : \prod_{\nu=1}^{n} e^{ib^{[N]}_\nu x(u_\nu)} : d\mathbf{u},$$

$$\sum_{\nu=1}^{n} b^{[N]}_\nu = \sqrt{|T|}n\lambda_2, \sum_{\nu=1}^{n} (b^{[N]}_\nu)^2 = |T|\lambda_2 h(n\lambda_2),$$

respectively, where $f^{[N]}_{\cdot,n} \in E_{\mathbf{C}}^{\hat{\otimes} n+1}$ and $g^{[N]}_{\cdot,n} \in E_{\mathbf{C}}^{\hat{\otimes} n+1}$. Take $\xi_T \in E_{\mathbf{C}}$ with $\xi_T = 1$ on T. Put $\xi = \alpha\xi_T + \eta$ for arbitrarily fixed $\alpha \in \mathbf{C}$. Then, by taking the S-transforms, we have

$$\begin{aligned}&\sum_{n=1}^{\infty} e^{in\lambda_1\alpha} \lim_{N\to\infty} \int_{T^n} f^{[N]}_{\lambda_1,n}(u_1, \ldots, u_n) \prod_{\nu=1}^{n} e^{ia^{[N]}_\nu \eta(u_\nu)} d\mathbf{u}\\ &= \sum_{n=1}^{\infty} e^{in\lambda_2\alpha} \lim_{N\to\infty} \int_{T^n} g^{[N]}_{\lambda_2,n}(u_1, \ldots, u_n) \prod_{\nu=1}^{n} e^{ib^{[N]}_\nu \eta(u_\nu)} d\mathbf{u}.\end{aligned}$$

Since $\cup_{r\in\mathbf{Q}}\{(\lambda, r\lambda); \lambda > 0\}$, $\mathbf{Q}$: the field of all rational numbers, is a null set in $\mathbf{R}^2$ with respect to the Lebesgue measure, this implies $\Phi_n = \Psi_n = 0$ for almost all (λ_1, λ_2) in $\mathbf{R}^2$ and for each $n \in \mathbf{N}$. Consequently we obtain $\Phi = 0$ for almost all (λ_1, λ_2) in $\mathbf{R}^2$. □

By Proposition 3.2, for $N \in \mathbf{N}$ we can define a space $\mathcal{E}_{-p,N}$ by

$$\mathcal{E}_{-p,N} = \left\{ (\Phi^\lambda)_{\lambda>0}; \Phi^\lambda = \sum_{n=1}^{\infty} \Phi^\lambda_n \in \mathbf{E}^\lambda_{-p,N}, \forall\lambda > 0, \right.$$

$$\sum_{n=1}^{\infty} \int_{(0,\infty)} |||\Phi_n^\lambda|||_{-p,N,\lambda}^2 d\nu(\lambda) < \infty \Bigg\}$$

with norm given by

$$|||\Phi|||_{\mathcal{E}_{-p,N}} = \left(\sum_{n=1}^{\infty} \int_{(0,\infty)} |||\Phi_n^\lambda|||_{-p,N,\lambda}^2 d\nu(\lambda) \right)^{1/2}$$

for $\Phi \in \mathcal{E}_{-p,N}$. Then for each $N \in \mathbf{N}$, we have $\mathcal{E}_{-p,N+1} \subset \mathcal{E}_{-p,N}$. Put $\mathcal{E}_{-p,\infty} = \bigcap_{N=1}^{\infty} \mathcal{E}_{-p,N}$ with projective limit topology.

The Lévy Laplacian Δ_L is defined on the space $\mathcal{E}_{-p,\infty}$ by

$$\Delta_L \Phi = (\Delta_L \Phi^\lambda)_{\lambda>0} \text{ for each } \Phi = (\Phi^\lambda)_{\lambda>0} \in \mathcal{E}_{-p,\infty}.$$

The operator Δ_L becomes a continuous linear operator from $\mathcal{E}_{-p,\infty}$ into itself. For any $t \geq 0$, we introduce also an operator G_t on $\mathcal{E}_{-p,\infty}$ by

$$G_t \Phi = (G_t^\lambda \Phi^\lambda)_{\lambda>0} \text{ for each } \Phi = (\Phi^\lambda)_{\lambda>0} \in \mathcal{E}_{-p,\infty}.$$

Then we have the following:

Theorem 3.4. *The family $\{G_t; t \geq 0\}$ is an equi-continuous semigroup of class (C_0) generated by Δ_L as a continuous linear operator defined on $\mathcal{E}_{-p,\infty}$.*

Proof. Let $p > \frac{5}{12}$ be a number arbitrarily fixed. For any $t \geq 0$ and $N \in \mathbf{N}$, the norm $|||G_t \Phi|||_{\mathcal{E}_{-p,N}}$ for $\Phi = (\Phi^\lambda)_{\lambda>0} \in \mathcal{E}_{-p,N}$ can be estimated as follows:

$$\begin{aligned}
|||G_t\Phi|||_{\mathcal{E}_{-p,N}}^2 &= \sum_{n=1}^{\infty} \int_{(0,\infty)} |||G_t^\lambda \Phi_n^\lambda|||_{-p,N,\lambda}^2 d\nu(\lambda) \\
&= \sum_{n=1}^{\infty} \int_{(0,\infty)} \alpha_N^\lambda(n) ||e^{-t\lambda h(n\lambda)} \Phi_n^\lambda||_{-p}^2 d\nu(\lambda) \\
&\leq \sum_{n=1}^{\infty} \int_{(0,\infty)} \alpha_N^\lambda(n) ||\Phi_n^\lambda||_{-p}^2 d\nu(\lambda) \\
&= \sum_{n=1}^{\infty} \int_{(0,\infty)} |||\Phi_n^\lambda|||_{-p,N,\lambda}^2 d\nu(\lambda) \\
&= |||\Phi|||_{\mathcal{E}_{-p,N}}^2.
\end{aligned}$$

Hence the family $\{G_t; t \geq 0\}$ is equi- continuous in t. By Theorem 3.3, for any $t, s \geq 0$, and $\Phi = (\Phi^\lambda)_{\lambda>0} \in \mathcal{E}_{-p,N}$, we get

$$\begin{aligned} G_t G_s \Phi &= (G_t^\lambda G_s^\lambda \Phi^\lambda)_{\lambda>0} \\ &= (G_{t+s}^\lambda \Phi^\lambda)_{\lambda>0} \\ &= G_{t+s}\Phi \end{aligned}$$

and $G_0\Phi = (G_0^\lambda \Phi^\lambda)_{\lambda>0} = (\Phi^\lambda)_{\lambda>0} = \Phi$. Since $|||(G_t^\lambda - G_s^\lambda)\Phi^\lambda|||_{-p,N,\lambda}^2 \leq 4|||\Phi^\lambda|||_{-p,N,\lambda}^2, \lambda > 0$, $\sum_{n=1}^\infty \int_{(0,\infty)} |||\Phi_n^\lambda|||_{-p,N,\lambda}^2 d\nu(\lambda) < \infty$, and by Theorem 3.4, we have

$$|||G_t\Phi - G_{t_0}\Phi|||_{\mathcal{E}_{-p,N}}^2 = \sum_{n=1}^\infty \int_{(0,\infty)} |||(G_t^\lambda - G_s^\lambda)\Phi^\lambda|||_{-p,N,\lambda}^2 d\nu(\lambda) \to 0,$$

as $t \to t_0$ for each $t, t_0 \geq 0, N \in \mathbf{N}$ and $\Phi = (\Phi^\lambda)_{\lambda>0} \in \mathcal{E}_{-p,\infty}$. Thus the family $\{G_t; t \geq 0\}$ is an equi-continuous semigroup of class (C_0). We next prove that the infinitesimal generator of the semigroup is given by Δ_L. For $N \in \mathbf{N}$ we see that

$$\begin{aligned} &\left|\left|\left| \frac{G_t\Phi - \Phi}{t} - \Delta_L \Phi \right|\right|\right|_{\mathcal{E}_{-p,N}}^2 \\ &= \sum_{n=1}^\infty \int_{(0,\infty)} \alpha_N^\lambda(n) \left|\left| \left(\frac{e^{-t\lambda h(n\lambda)} - 1}{t} + \lambda h(n\lambda) \right) \Phi_n^\lambda \right|\right|_{-p}^2 d\nu(\lambda). \end{aligned} \tag{3.1}$$

Since $\Phi = (\Phi^\lambda)_{\lambda>0} \in \mathcal{E}_{-p,\infty}$, we have

$$\sum_{n=1}^\infty \int_{(0,\infty)} \alpha_{N+1}^\lambda(n) ||\Phi_n^\lambda||_{-p}^2 d\nu(\lambda) < \infty. \tag{3.2}$$

By the mean value theorem, for any $t > 0$ there exists a constant $\theta \in (0,1)$ such that

$$\left| \frac{e^{-t\lambda h(n\lambda)} - 1}{t} \right| = \lambda h(n\lambda) e^{-t\theta\lambda h(n\lambda)} < \lambda h(n\lambda).$$

Therefore we can estimate each term in (3.1) as follows.

$$\begin{aligned} &\alpha_N^\lambda(n) \left|\left| \left(\frac{e^{-t\lambda h(n\lambda)} - 1}{t} + \lambda h(n\lambda) \right) \Phi_n^\lambda \right|\right|_{-p}^2 \\ &= \alpha_N^\lambda(n) \left| \frac{e^{-t\lambda h(n\lambda)} - 1}{t} + \lambda h(n\lambda) \right|^2 ||\Phi_n^\lambda||_{-p}^2 \leq 4\alpha_{N+1}^\lambda(n) ||\Phi_n^\lambda||_{-p}^2. \end{aligned}$$

By (3.2),

$$\lim_{t\to 0}\left|\frac{e^{-t\lambda h(n\lambda)}-1}{t}+\lambda h(n\lambda)\right|=0$$

and the Lebesgue convergence theorem, we obtain

$$\lim_{t\to 0}\left|\left|\left|\frac{G_t\Phi-\Phi}{t}-\Delta_L\Phi\right|\right|\right|^2_{\mathcal{E}_{-p,N}}=0.$$

Thus the proof is completed. □

4. An infinite dimensional stochastic process generated by the Lévy Laplacian

For any $p\in\mathbf{R}$ we define a space $E^{(0,\infty)}_{\mathbf{C},p}$ by

$$E^{(0,\infty)}_{\mathbf{C},p}=\{(\xi_\lambda)_{\lambda>0};\xi_\lambda\in E_{\mathbf{C},p},\xi_\lambda \text{ is strongly measurable in } \lambda.\}.$$

This space is a linear space and a complete metric space with metric d_p given by

$$d_p(\boldsymbol{\xi},\boldsymbol{\eta})=\int_{(0,\infty)}\frac{|\xi_\lambda-\eta_\lambda|_p}{1+|\xi_\lambda-\eta_\lambda|_p}d\nu(\lambda)$$

for $\boldsymbol{\xi}=(\xi_\lambda)_{\lambda>0},\boldsymbol{\eta}=(\eta_\lambda)_{\lambda>0}\in E^{(0,\infty)}_{\mathbf{C},p}$.

We also define a space $\mathbf{C}^{(0,\infty)}$ by

$$\mathbf{C}^{(0,\infty)}=\{(z_\lambda)_{\lambda>0};z_\lambda\in\mathbf{C},z_\lambda \text{ is measurable in } \lambda.\}$$

with metric

$$\rho(\mathbf{z},\mathbf{u})=\int_{(0,\infty)}\frac{|z_\lambda-u_\lambda|}{1+|z_\lambda-u_\lambda|}d\nu(\lambda)$$

for $\mathbf{z}=(z_\lambda)_{\lambda>0}$, $\mathbf{u}=(u_\lambda)_{\lambda>0}\in\mathbf{C}^{(0,\infty)}$. This space is also a linear space and a complete metric space with metric ρ.

Since $d_p\le d_q$ if $p\ge q$, we introduce the projective limit topology to $E^{(0,\infty)}_{\mathbf{C}}\equiv\bigcap_p E^{(0,\infty)}_{\mathbf{C},p}$. The S-transform can be extended to a continuous linear operator on $\mathcal{E}_{-p,\infty}$ by

$$S\Phi(\boldsymbol{\xi})=(S\Phi^\lambda(\xi_\lambda))_{\lambda>0},\quad \boldsymbol{\xi}=(\xi_\lambda)_{\lambda>0}\in E^{(0,\infty)}_{\mathbf{C}},$$

for any $\Phi=(\Phi^\lambda)_{\lambda>0}\in\mathcal{E}_{-p,\infty}$. The space $S[\mathcal{E}_{-p,\infty}]$ is endowed with the topology induced from $\mathcal{E}_{-p,\infty}$ by the S-transform. Then the S-transform becomes a homeomorphism from $\mathcal{E}_{-p,\infty}$ onto $S[\mathcal{E}_{-p,\infty}]$. The transform $S\Phi$ of $\Phi \in \mathcal{E}_{-p,\infty}$ is a continuous operator from $E_{\mathbf{C}}^{(0,\infty)}$ into $\mathbf{C}^{(0,\infty)}$. We denote the operator by the same notation $S\Phi$.

Let $\widetilde{G_t}$ be an operator defined on $S[\mathcal{E}_{-p,\infty}]$ by

$$\widetilde{G_t} = SG_tS^{-1}$$

for $t \geq 0$. Then by Theorem 3.4, $\{\widetilde{G_t};\, t \geq 0\}$ is an equi-continuous semi-group of class (C_0) generated by the operator $\widetilde{\Delta_L}$.

Let $\{X_t\}_{t\geq 0}$ be a stochastic process such that for any $t \geq 0$ the characteristic function of X_t is given by $E[e^{izX_t}] = e^{-th(z)}$, $z \in \mathbf{R}$. Set $Y_t^\lambda = X_{\lambda t}$. Take a smooth function $\eta_T \in E$ with $\eta_T = 1/\sqrt{|T|}$ on T. Define an $E_{\mathbf{C}}^{(0,\infty)}$-valued stochastic process $\{\mathbf{Y}_t; t \geq 0\}$ by

$$\mathbf{Y}_t = (Y_t^\lambda \eta_T)_{\lambda>0}, \quad t \geq 0.$$

Then we have the following:

Theorem 4.1. *Let F be the S-transform in $S[\mathcal{E}_{-p,\infty}]$. Then the equality*

$$\widetilde{G_t}F(\boldsymbol{\xi}) = E[F(\boldsymbol{\xi} + \mathbf{Y}_t)]$$

holds for $t \geq 0$.

Proof. We first consider the case when $F \in S[\mathcal{E}_{-p,\infty}]$ is given by

$$F(\boldsymbol{\xi}) = (F^\lambda(\xi_\lambda))_{\lambda>0}, \quad F^\lambda(\xi_\lambda) = \int_{T^n} f(\mathbf{u}) e^{ia_1\xi_\lambda(u_1)} \cdots e^{ia_n\xi_\lambda(u_n)} d\mathbf{u}$$

with $f \in E_{\mathbf{C}}^{\hat{\otimes} n}$. Then we have

$$\begin{aligned}
E[F(\boldsymbol{\xi} + \mathbf{Y}_t)] &= (E[F^\lambda(\xi_\lambda + Y_t^\lambda \eta_T)])_{\lambda>0} \\
&= \left(\int_{T^n} f(\mathbf{u}) e^{ia_1\xi_\lambda(u_1)} \cdots e^{ia_n\xi_\lambda(u_n)} E[e^{in\lambda Y_t^\lambda}] d\mathbf{u} \right)_{\lambda>0} \\
&= \left(e^{-t\lambda h(n\lambda)} F^\lambda(\xi_\lambda) \right)_{\lambda>0} \\
&= (\widetilde{G_t^\lambda} F^\lambda(\xi_\lambda))_{\lambda>0} \\
&= \widetilde{G_t}F(\boldsymbol{\xi}).
\end{aligned}$$

Next let $F = \left(\sum_{n=1}^\infty F_n^\lambda\right)_{\lambda>0} \in S[\mathcal{E}_{-p,\infty}]$. Then for almost all $\lambda > 0$ and for any $n \in \mathbf{N}$, F_n^λ is expressed in the following form:

$$F_n^\lambda(\xi_\lambda) = \lim_{N\to\infty} \sum_{\mathbf{a}} \int_{T^n} f_{\lambda,\mathbf{a}}^{[N]}(\mathbf{u}) e^{ia_1\xi_\lambda(u_1)} \cdots e^{ia_n\xi_\lambda(u_n)} d\mathbf{u},$$

where $\sum_{\mathbf{a}}$ means a sum of finitely many terms on $\mathbf{a}$ moving on $\sum_{j=1}^{n} a_j = \sqrt{|T|}n\lambda, \sum_{j=1}^{n} a_j^2 = |T|\lambda h(n\lambda)$ and $(f_{\cdot,\mathbf{a}}^{[N]})_N$ is a sequence of functions in $E_{\mathbf{C}}^{\hat{\otimes} n+1}$. Hence, for each $\lambda > 0$, we have

$$\begin{aligned}
&\sum_{n=1}^{\infty} E[|F_n^{\lambda}(\xi_\lambda + Y_t^{\lambda}\eta_T)|] \\
&= \sum_{n=1}^{\infty} E\left[\lim_{N\to\infty}\left|\sum_{\mathbf{a}}\int_{T^n} f_{\lambda,\mathbf{a}}^{[N]}(\mathbf{u})\prod_{j=1}^{n} e^{ia_j\xi_\lambda(u_\nu)}e^{ia_jY_t^{\lambda}\eta_T(u_\nu)}d\mathbf{u}\right|\right] \\
&= \sum_{n=1}^{\infty}\lim_{N\to\infty}\left|\sum_{\mathbf{a}}\int_{T^n} f_{\lambda,\mathbf{a}}^{[N]}(\mathbf{u})\prod_{j=1}^{n} e^{ia_j\xi_\lambda(u_\nu)}d\mathbf{u}\right| \\
&= \sum_{n=1}^{\infty}|F_n^{\lambda}(\xi_\lambda)|.
\end{aligned}$$

Since $F_n^{\lambda} \in S[\mathbf{E}_{-p,\infty}]$, there exists some $\Phi_n^{\lambda} \in \mathbf{E}_{-p,\infty}$ such that $F_n^{\lambda} = S[\Phi_n^{\lambda}]$ for almost all λ and each n. By the Schwarz inequality, we see that

$$\begin{aligned}
\sum_{n=1}^{\infty}|F_n^{\lambda}(\xi_\lambda)| &\leq \sum_{n=1}^{\infty}||\Phi_n^{\lambda}||_{-p}||\varphi_{\xi_\lambda}||_p \\
&\leq \left\{\sum_{n=1}^{\infty}\alpha_N^{\lambda}(n)^{-1}\right\}^{1/2}\left\{\sum_{n=1}^{\infty}\alpha_N^{\lambda}(n)||\Phi_n^{\lambda}||_{-p}^2\right\}^{1/2}||\varphi_{\xi_\lambda}||_p < \infty,
\end{aligned}$$

for almost all $\lambda > 0$, for each $\xi_\lambda \in E_{\mathbf{C}}$ and $N \in \mathbf{N}$. Therefore by the continuity of $\widetilde{G_t^{\lambda}}, \lambda > 0$, we get that

$$\begin{aligned}
E[F(\boldsymbol{\xi} + \mathbf{Y}_t)] &= (E[F^{\lambda}(\xi_\lambda + Y_t^{\lambda}\eta_T)])_{\lambda>0} \\
&= \left(\sum_{n=1}^{\infty} E[F_n^{\lambda}(\xi_\lambda + Y_t^{\lambda}\eta_T)]\right)_{\lambda>0} \\
&= \left(\sum_{n=1}^{\infty}\widetilde{G_t^{\lambda}}F_n^{\lambda}(\xi_\lambda)\right)_{\lambda>0} \\
&= \left(\widetilde{G_t^{\lambda}}\sum_{n=1}^{\infty}F_n^{\lambda}(\xi_\lambda)\right)_{\lambda>0} \\
&= \widetilde{G_t}F(\xi).
\end{aligned}$$

Thus we obtain the assertion. □

The translation $x \mapsto x+\eta$, $x \in E^*$, can be lifted to the space $(E)^*$ of generalized functions whenever $\eta \in E$, i.e., $\tau_\eta \Phi(x) = \Phi(x+\eta)$ is defined for $\Phi \in (E)^*$. More precisely, a continuous linear operator τ_η from $(E)^*$ into itself is uniquely specified by $S[\tau_\eta \Phi](\xi) = S[\Phi](\xi+\eta)$, $\xi \in E_{\mathbf{C}}$.

For $\boldsymbol{\eta} \in E_{\mathbf{C}}^{(0,\infty)}$ we introduce a continuous linear operator $\tau_{\boldsymbol{\eta}}$ from $\mathcal{E}_{-p,\infty}$ into itself by $S[\tau_{\boldsymbol{\eta}}\Phi](\boldsymbol{\xi}) = S[\Phi](\boldsymbol{\xi}+\boldsymbol{\eta})$, $\boldsymbol{\xi} \in E_{\mathbf{C}}^{(0,\infty)}$. Then we have the following:

Theorem 4.2. *For any $\Phi \in \mathcal{E}_{-p,\infty}$ we have*

$$G_t\Phi = E[\tau_{\mathbf{Y}_t}\Phi], \quad t \geq 0.$$

Acknowledgments

This work was supported by the Research Project "White Noise Theory and Quantum Information" promoted by the Ministry of Education in Japan and Meijo University and was also partially supported by JSPS grant 15540141. The authors are grateful for the support.

References

1. Accardi, L. and Bogachëv, V.: The Ornstein–Uhlenbeck process associated with the Lévy Laplacian and its Dirichlet form, *Prob. Math. Stat.* **17** (1997), 95–114.
2. Accardi, L., Gibilisco, P. and Volovich, I.V.: Yang-Mills gauge fields as harmonic functions for the Lévy Laplacian, *Russian J. Math. Phys.* **2** (1994), 235–250.
3. Accardi, L., Smolyanov, O. G.: Trace formulae for Levy-Gaussian measures and their application, in *Mathematical Approach to Fluctuations Vol. II*, pp. 31–47, World Scientific, 1995.
4. Chung, D. M., Ji, U. C. and Saitô, K.: Cauchy problems associated with the Lévy Laplacian in white noise analysis, *Infin. Dimen. Anal. Quantum Probab. Rel. Top.* **2** (1999), 131–153.
5. Feller, M. N.: Infinite-dimensional elliptic equations and operators of Lévy type, *Russian Math. Surveys* **41** (1986), 119–170.
6. Hasegawa K.: Lévy's Functional Analysis in terms of an infinite dimensional Brownian motion I, *Osaka J. Math*, **19** (1982), 405–428. .
7. Hida, T.: *Analysis of Brownian Functionals*, Carleton Math. Lect. Notes, No. 13, Carleton University, Ottawa, 1975.

8. Hida, T.: A role of the Lévy Laplacian in the causal calculus of generalized white noise functionals, *Stochastic Processes*, pp. 131–139, Springer–Verlag, 1993.
9. Hida, T., Kuo, H.-H. and Obata, N.: Transformations for white noise functionals, *J. Funct. Anal.* **111** (1993), 259–277.
10. Hida, T., Kuo, H.-H., Potthoff, J. and Streit, L.: *White Noise: An Infinite Dimensional Calculus*, Kluwer Academic, 1993.
11. Hida, T., Obata, N. and Saitô, K.: Infinite dimensional rotations and Laplacian in terms of white noise calculus, *Nagoya Math. J.* **128** (1992), 65–93.
12. Hida, T. and Saitô, K.: White noise analysis and the Lévy Laplacian, in *Stochastic Processes in Physics and Engineering*, S. Albeverio et al. Eds., pp. 177–184, 1988.
13. Hille, E and Phillips R. S.: *Functional Analysis and Semi-Groups*, AMS Colloq. Publ. Vol. 31, Amer. Math. Soc, 1957.
14. Itô, K.: Stochastic analysis in infinite dimensions, in *Proc. International Conference on Stochastic Analysis*, pp. 187–197, Academic Press, 1978.
15. Kubo, I.: A direct setting of white noise calculus, in *Stochastic Analysis on Infinite Dimensional Spaces*, pp. 152–166, Pitman Res. Notes in Math. Vol. 310, 1994.
16. Kubo, I. and Takenaka, S.: Calculus on Gaussian white noise I–IV, *Proc. Japan Acad.* **56A** (1980) 376–380; **56A** (1980) 411–416; **57A** (1981) 433–436; **58A** (1982) 186–189.
17. Kuo, H.-H.: On Laplacian operators of generalized Brownian functionals, in *Lecture Notes in Math. Vol. 1203*, pp. 119–128, Springer–Verlag, 1986.
18. Kuo, H.-H.: Lectures on white noise calculus, *Soochow J. Math.***18** (1992), 229–300.
19. Kuo, H.-H.: *White Noise Distribution Theory*, CRC Press, 1996.
20. Kuo, H.-H., Obata, N. and Saitô, K.: Lévy Laplacian of generalized functions on a nuclear space, *J. Funct. Anal.* **94** (1990), 74–92.
21. Kuo, H.-H., Obata, N. and Saitô, K.: Diagonalization of the Lévy Laplacian and related stable processes, to appear in *Infin. Dimen. Anal. Quantum Probab. Rel. Top.* **5** (2002).
22. Lévy, P.: *Leçons d'Analyse Fonctionnelle*, Gauthier–Villars, Paris, 1922.
23. R.Léandre and I. A. Volovich: The stochastic Lévy Laplacian and Yang-Mills equation on manifolds, *Infin. Dimen. Anal. Quantum Probab. Rel. Top.* **4** (2001) 161–172.
24. Nishi, K., Saitô, K. and Tsoi, A. H.: A stochastic expression of a semi-group generated by the Lévy Laplacian, in *Quantum Information III*, T. Hida and K. Saitô, Eds., pp. 105–117, World Scientific, 2000.
25. Obata, N.: A characterization of the Lévy Laplacian in terms of infinite dimensional rotation groups, *Nagoya Math. J.* **118** (1990), 111–132.
26. Obata, N.: *White Noise Calculus and Fock Space*, Lect. Notes in Math. Vol. 1577, Springer–Verlag, 1994.
27. Obata, N.: Integral kernel operators on Fock space —Generalizations and applications to quantum dynamics, *Acta Appl. Math.* **47** (1997), 49–77.
28. Obata, N.: Quadratic quantum white noises and Lévy Laplacian, *Nonlinear*

Analysis **47** (2001), 2437–2448.

29. Obata,N. and Saitô, K.: Cauchy processes and the Lévy Laplacian, *QP-PQ: Quantum Probability and White Noise Analysis* **14** *–Non-Commutativity, Infinite-Dimensionality, and Probability at the Crossroads* (2002), 360–373.
30. Polishchuk, E. M.: *Continual Means and Boundary Value Problems in Function Spaces*, Birkhäuser, Basel/Boston/Berlin, 1988.
31. Potthoff, J. and Streit, L.: A characterization of Hida distributions, *J. Funct. Anal.* **101** (1991), 212–229.
32. Saitô, K.: Itô's formula and Lévy's Laplacian I and II, *Nagoya Math. J.* **108** (1987), 67–76; *ibid.* **123** (1991), 153–169.
33. Saitô, K.: A (C_0)-group generated by the Lévy Laplacian II, *Infin. Dimen. Anal. Quantum Probab. Rel. Top.* **1** (1998) 425–437.
34. Saitô, K.: A stochastic process generated by the Lévy Laplacian, *Acta Appl. Math.* **63** (2000), 363–373.
35. Saitô, K.: The Lévy Laplacian and stable processes, *Chaos, Solitons and Fractals* **12** (2001), 2865–2872.
36. Saitô, K., Tsoi, A. H.: The Lévy Laplacian as a self-adjoint operator, in *Quantum Information*, T. Hida and K. Saitô, Eds., pp. 159–171, World Scientific, 1999.
37. Saitô, K., Tsoi, A. H.: The Lévy Laplacian acting on Poisson noise functionals, *Infin. Dimen. Anal. Quantum Probab. Rel. Top.* **2** (1999), 503–510.
38. Saitô, K., Tsoi, A. H.: Stochastic processes generated by functions of the Lévy Laplacian, in *Quantum Information II*, T. Hida and K. Saitô, Eds., pp. 183–194, World Scientific, 2000.
39. Yosida, K.: *Functional Analysis (3rd Edition)*, Springer–Verlag, 1971.

FOUNDATION OF CHAOS THROUGH OBSERVATION

MASANORI OHYA

We propose a new approach to define chaos in dynamical systems from the point of view of Information Dynamics. Observation of chaos in reality depends on how to observe it, that is, it is adaptive with means to observe it. For instance, the apperance of chaos depends on how to take the scale in space and time. Therefore such an observation sholud not be related with some mathematical limiting procedures. We will explain our new treatment of chaos based on information dynamics in this paper.

1. Introduction

There exist several trials to describe chaos appeared in classical or quantum dynamical systems. One of the present authors introduced Information Dynamics(ID for short) [18] as a frame to discuss complexity and chaos appeared in various fields, in which he tried to find a common basis by synthesizing the state change (dynamics) and the complexity associated with dynamical systems. Since then, ID has been applied to several different topics [9,19], among which a chaos degree, a quantity measuring the degree of chaos associated with a dynamics, was introduced by means of the complexities in ID and its entropic version (called Entropic Chaos Degree (EDC for short)) has been computed numerically for rather famous chaotic dynamics such as logistic map, baker's transformation, Tinkerbel map. It is surprised that the result of the ECD exactly macthes to that of Lyapunov exponent in the case that the later can be computed. Moreover the algorithm computing the ECD is much easier than that of Lyapunov exponent, so that the ECD can be almost always computable even when the Lyapunov exponent can not be so. However there are some unclear points in both conceptually and mathematically why the ECD could be so successful for computational experiments. In this paper we study these points and propose a new description of chaos. The key philosophy is that the outcome of observing chaotic system in reality depends upon how to observe it, for instance, how to take the scale in space and time. This motivation leads us naturally to abandon taking several mathematical limiting procedures. From such a point of view, the chaos degree is redefined in this paper.

In Section 2, we briefly review Information Dynamics and Chaos Degree, and in Section 3 the entropic chaos degree and its algorithm are recalled with a computational result. In section 4 and 5, quantum chaos degree is briefly reviewd. In Section 6, a new way judging chaos from a given dynamics is discussed based on the ECD, that is, we propose a new view to define chaos of dynamical systems.

2. Information Dyanamics and Chaos Degree

We briefly review what ID is. Let $(\mathcal{A}, \mathcal{S}, \alpha(G))$ be an input (or initial) system and $(\overline{\mathcal{A}}, \overline{\mathcal{S}}, \overline{\alpha}(\overline{G}))$ be an output (or final) system. Here $\mathcal{A}$ is a set of some objects to be observed and $\mathcal{S}$ is a set of some means to get the observed value, $\alpha(G)$ describes a certain evolution of system with a parameter g in a certain set G. Often we have $\mathcal{A} = \overline{\mathcal{A}}$, $\mathcal{S} = \overline{\mathcal{S}}$, $\alpha = \overline{\alpha}$, $G = \overline{G}$. Therefore it can be said

[Giving a mathematical structure to input and output triples
$\equiv$ Having a theory]

The dynamics of state change is described by a channel, that is, a map Λ^*: $\mathcal{S} \to \overline{\mathcal{S}}$ (sometimes $\mathcal{S} \to \mathcal{S}$). The fundamental point of ID is that ID contains two complexities in itself. Let $(\mathcal{A}_t, \mathcal{S}_t, \alpha^t(G^t))$ be the total system of $(\mathcal{A}, \mathcal{S}, \alpha)$ and $(\overline{\mathcal{A}}, \overline{\mathcal{S}}, \overline{\alpha})$, and S be a subset of $\mathcal{S}$ in which we are measuring observables (e.g., S is the set of all KMS or stationary states in C*-system). Two complexities are denoted by C and T. C is the complexity of a state φ measured from a reference system S, in which we actually observe the objects in $\mathcal{A}$ and T is the transmitted complexity associated with a state change $\varphi \to \Lambda^*\varphi$, both of which should satisfy the following properties :
⟨**Aximos of complexities**⟩

(i) For any $\varphi \in S \subset \mathcal{S}$,

$$C^S(\varphi) \geq 0, \; T^S(\varphi; \Lambda^*) \geq 0$$

(ii) For any orthogonal bijection $j : ex\mathcal{S} \to ex\mathcal{S}$, the set of all extremal points of $\mathcal{S}$,

$$C^{j(S)}(j(\varphi)) = C^S(\varphi)$$

$$T^{j(S)}(j(\varphi); \Lambda^*) = T^S(\varphi; \Lambda^*)$$

(iii) For $\Phi \equiv \varphi \otimes \psi \in S_t \subset \mathcal{S}_t$,

$$C^{S_t}(\Phi) = C^S(\varphi) + C^{\overline{S}}(\psi)$$

(iv) $0 \le T^{\mathcal{S}}(\varphi;\Lambda^*) \le C^{\mathcal{S}}(\varphi)$
(v) $T^{\mathcal{S}}(\varphi; id) = C^{\mathcal{S}}(\varphi)$, where "$id$" is an identity map from $\mathcal{S}$ to $\mathcal{S}$.

Instead of (iii), when "(iii') $\Phi \in \mathcal{S}_t \subset \mathcal{S}_t$, put $\varphi \equiv \Phi|\mathcal{A}$ (i.e., the restriction of Φ to $\mathcal{A}$), $\psi \equiv \Phi|\overline{\mathcal{A}}$, $C^{\mathcal{S}_t}(\Phi) \le C^{\mathcal{S}}(\varphi) + C^{\overline{\mathcal{S}}}(\psi)$ " is satisfied, C and T is called a pair of strong complexity. Therefore ID is defined as follows:

Definition 2.1. : Information Dynamics is described by

$$\left(\mathcal{A}, \mathcal{S}, \alpha(G); \overline{\mathcal{A}}, \overline{\mathcal{S}}, \overline{\alpha}(\overline{G}); \Lambda^*; C^{\mathcal{S}}(\varphi), T^{\mathcal{S}}(\varphi;\Lambda^*)\right)$$

and some relations R among them.

In the framework of ID, we have to

(i) mathematically determine

$$\mathcal{A}, \mathcal{S}, \alpha(G); \overline{\mathcal{A}}, \overline{\mathcal{S}}, \overline{\alpha}(\overline{G}),$$

(ii) choose Λ^* and R, and
(iii) define $C^{\mathcal{S}}(\varphi)$, $T^{\mathcal{S}}(\varphi;\Lambda^*)$.

In ID, several different topics can be treated on a common standing point so that we can find a new clue bridging several fields.

We assume $\overline{\mathcal{A}} = \mathcal{A}$ for simlicity in the sequel. For a certain subset $\mathcal{S}$ (called the reference space) of $\mathcal{S}$ and a state $\varphi \in \mathcal{S}$, there exists a decomposition of the state φ into a mixture of extreme (pure) states such that

$$\varphi = \int_{\mathcal{S}} \omega d\mu$$

This extremal decomposition of φ describes the degree of mixture of φ in the reference space $\mathcal{S}$. The measure μ is not always unique, so that the set of all such measures is denoted by $M_{\varphi}(\mathcal{S})$.

For instance, when $(\mathcal{A}, \mathcal{S})$ and is a C*-system containing both classical and quantum systems; that is, $\mathcal{A}$ and is a C* algebra and $\mathcal{S}$ is the set of all states on $\mathcal{A}$, the reference space $\mathcal{S}$ is a weak* compact convex subset of $\mathcal{S}$ and the measure μ is not uniquely determined unless $\mathcal{S}$ is the Schoque simplex. In this paper we will not go to the details of such general mathematical discussion.

A measure of chaos produced by dynamics Λ^* is defined in [21,22]:

Definition 2.2. (1)ψ is more chaotic than φ if $C(\psi) \ge C(\varphi)$.

(2)When $\varphi \in \mathcal{S}$ changes to $\Lambda^*\varphi$, the *chaos* degree associated to this state change (dynamics) Λ^* is given by

$$D^{\mathcal{S}}(\varphi;\Lambda^*) = \inf\left\{\int_{\mathcal{S}} C^{\mathcal{S}}(\Lambda^*\omega)\, d\mu; \mu \in M_\varphi(\mathcal{S})\right\}.$$

Definition 2.3. A dynamics Λ^* produces chaos iff $D^{\mathcal{S}}(\varphi;\Lambda^*) > 0$.

It is important to note here that the dynamics Λ^* in the definition is not necessarily same as original dynamics (channel) but is one reduced from the original one such that it causes an evolution for a certain observed value like orbit. However for simplicity we often use the same notation in this paper. In some cases, the above chaos degree $D^{\mathcal{S}}(\varphi;\Lambda^*)$ can be expressed as

$$D^{\mathcal{S}}(\varphi;\Lambda^*) = C^{\mathcal{S}}(\Lambda^*\varphi) - T^{\mathcal{S}}(\varphi;\Lambda^*).$$

3. Entropic Chaos Degree and its Algorithm

Although there exist several complexities [20], one of the most useful examples of C and T are Shannon's entropy and mutual entropy in classical systems (von Neumann entropy and quantum mutual entropy in quantum systems [23]), respectively.

The concept of entropy was introduced and developed to study the topics such as irreversible behavior, symmetry breaking, amount of information transmission, so that it originally describes a certain chaotic property of state.

Let us recall the simplest case of C and T, that is, Shannon's entropy and mutual entropy. In classical communication systems, an input state φ is a probability distribution $p = (p_k) = \sum_k p_k \delta_k$ and a channel Λ^* is a transition probability $(t_{i,j})$, so that the compound state of φ and its output $\overline{\varphi}\ (\equiv \overline{p} = (\overline{p}_i) = \Lambda^* p)$ is the joint distribution $r = (r_{i,j})$ with $r_{i,j} \equiv t_{i,j} p_j$. Then the complexities C and T are given as

$$\begin{aligned} C(p) &= S(p) = -\textstyle\sum_k p_k \log p_k, \\ T(p;\Lambda^*) &= I(p;\Lambda^*) = \textstyle\sum_{i,j} r_{i,j} \log \frac{r_{i,j}}{p_j \overline{p}_i}. \end{aligned}$$

Thus the entropic chaos degree of the channel Λ^* becomes

Definition 3.1.

$$D(p;\Lambda^*) = S(\Lambda^* p) - I(p;\Lambda^*).$$

Quantum version of the above entropic chaos degree was discussed in [10,22], on which we will briefly review here in the case of usual Hilbert space expression. Let ρ be a quantum state, namely, a density operator on a Hilbert space $\mathcal{H}$, and Λ^* be a channel sending the set $\mathcal{S}$ of all states on $\mathcal{H}$ into itself. Then the entropic chaos degree is defined by

$$D(\rho;\Lambda^*) = \inf\left\{\sum_k \lambda_k S(\Lambda^* E_k); \{E_k\} \in \mathcal{E}\right\},$$

where $\mathcal{E}$ is the set of all Schatten decompositions (i.e., one dimensional spectral decompositions) of the state $\rho := \sum_k \lambda_k E_k$, and S is the von Neumann entropy.

3.1. *Algorithm Computing Chaos Degree*

In order to observe a chaos produced by a dynamics, one often looks at the behaivor of orbits made by that dynamics, more generally, looks at the behavior of a certain observed value. Therefore in our scheme we directly compute the chaos degree once a dynamics is explicitly given as a state change of a system. However even when the direct calculation does not show a chaos, a chaos will appear if one forcuses to some aspect of the state change, e.g., a certain observed value which may be called orbit as usual. The algorithm computing the chaos degree for a dynamic is the following two cases [21,22,12,10]:

(1) *Dynamics is given by* $\frac{dx}{dt} = \Gamma_t(x)$ *with* $x \in I \equiv [a,b]^{\mathbf{N}} \subset \mathbf{R}^{\mathbf{N}}$: First find a difference equation $x_{n+1} = \Gamma(x_n)$ with a map Γ on $I \equiv [a,b]^{\mathbf{N}} \subset \mathbf{R}^{\mathbf{N}}$ into itself, secondly let $I \equiv \bigcup_k A_k$ be a finite partation with $A_i \cap A_j = \emptyset\ (i \neq j)$. Then the state $\varphi^{(n)}$ of the orbit determined by the difference equation is defined by the probabilty distribution $\left(p_i^{(n)}\right)$, that is, $\varphi^{(n)} = \sum_i p_i^{(n)} \delta_i$, where for a given initial value $x \in I$ and the characteristic function 1_A

$$p_i^{(n)} \equiv \frac{1}{n+1} \sum_{k=m}^{m+n} 1_{A_i}\left(\Gamma^k x\right).$$

Now when the initial value x is distributed due to a measure ν on I, the above $p_i^{(n)}$ is given as

$$p_i^{(n)} \equiv \frac{1}{n+1} \int_I \sum_{k=m}^{m+n} 1_{A_i}\left(\Gamma^k x\right) d\nu.$$

The joint distribution $\left(p_{ij}^{(n,n+1)}\right)$ between the time n and $n+1$ is defined by

$$p_{ij}^{(n,n+1)} \equiv \frac{1}{n+1} \sum_{k=m}^{m+n} 1_{A_i}\left(\Gamma^k x\right) 1_{A_j}\left(\Gamma^{k+1} x\right)$$

or

$$p_{ij}^{(n,n+1)} \equiv \frac{1}{n+1} \int_I \sum_{k=m}^{m+n} 1_{A_i}\left(\Gamma^k x\right) 1_{A_j}\left(\Gamma^{k+1} x\right) d\nu.$$

Then the channel Λ_n^* at n is determined by

$$\Lambda_n^* \equiv \left(\frac{p_{ij}^{(n,n+1)}}{p_i^{(n)}}\right) \Longrightarrow \varphi^{(n+1)} = \Lambda_n^* \varphi^{(n)},$$

and the entropic chaos degree is given by the definition 3.1;

$$D_A(x;F) = D_A\left(p^{(n)};\Lambda_n^*\right) = \sum_i p_i^{(n)} S(\Lambda_n^* \delta_i) = \sum_{i,j} p_{ij}^{(n,n+1)} \log \frac{p_i^{(n)}}{p_{ij}^{(n,n+1)}}. \tag{1}$$

We can judge whether the dynamics causes a chaos or not by the value of D as the definition 2.2

$$D > 0 \Longleftrightarrow \textit{chaotic}$$
$$D = 0 \Longleftrightarrow \textit{stable.}$$

This chaos degree was applied to several dynamical maps such logistic map, Baker's transformation and Tinkerbel map, and it could explain their chaotic characters. This chaos degree has several merits compared with usual measures such as Lyapunov exponent as explained below.

(2) *Dynamics is given by* $\varphi_t = \Gamma_t^* \varphi_0$ *on a Hilbert space*: Similarly as making a difference equation for state, the channel Λ_n^* at n is first deduced from Γ_t^*, which should satisfy $\varphi^{(n+1)} = \Lambda_n^* \varphi^{(n)}$. By means of this constructed channel ($\boldsymbol{\alpha}$) we compute the chaos degree D directly according

to the definition 3.2 or (β) we take a proper observable X and put $x_n \equiv \varphi^{(n)}(X)$, then go back to the algorithm (1).

Note that the chaos degree D does depend on a partition A taken, which is somehow different from usual degree of chaos (cf., dynamical entropy [1,4,3,14]). This is a key point of our understanding of chaos, which will be discussed in the next section.

3.2. *Logistic Map*

Let us explain how the entropy chaos degree (ECD) well describes to the chaotic behaivor of logistic map.

The logistic map is defined by

$$x_{n+1} = a x_n \left(1 \quad x_n\right), x_n \in [0,1], 0 \leq a \leq 4$$

The solution of this equation bifurcates as shown in Fig.1.

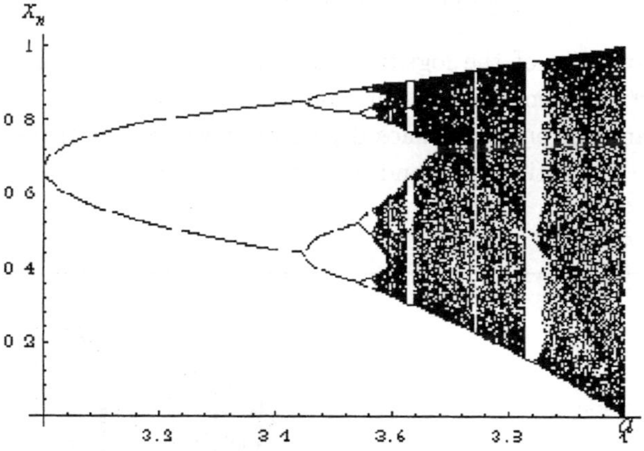

Figure 1. The bifurcation diagram for logistic map

<Lyapunov exponent>

(1) Let f be a map on $\mathbf{R}$, and let $x_0 \in \mathbf{R}$. Then the Lyapunov exponent $\lambda_{\mathcal{O}}(f)$ for the orbit $\mathcal{O} \equiv \{f^n(x_0); n = 0, 1, 2, \cdots\}$is defined by

$$\lambda_{\mathcal{O}}(f) = \lim_{n\to\infty} \lambda_{\mathcal{O}}^{(n)}(f), \quad \lambda_{\mathcal{O}}^{(n)}(f) = \frac{1}{n} \log \left| \frac{df^n}{dx}(x_0) \right|$$

(2) Let $f = (f_1, f_2, \cdots, f_m)$ be a map on $\mathbf{R}^m$, and let $x_0 \in \mathbf{R}^m$. The Jacobi matrix $J_n = Df^n(r_0)$ at r_0 is defined by

$$J_n = Df^n(r_0) = \begin{pmatrix} \frac{\partial f_1^n}{\partial x_1}(r_0) & \cdots & \frac{\partial f_1^n}{\partial x_m}(r_0) \\ \vdots & & \vdots \\ \frac{\partial f_m^n}{\partial x_1}(r_0) & \cdots & \frac{\partial f_m^n}{\partial x_m}(r_0) \end{pmatrix}.$$

Then, the Lyapunov exponent $\lambda_{\mathcal{O}}(f)$ of f for the orbit $\mathcal{O} \equiv \{f^n(x_0); n = 0, 1, 2, \cdots\}$ is defined by

$$\lambda_{\mathcal{O}}(f) = \log \tilde{\mu}_1, \quad \tilde{\mu}_k = \lim_{n \to \infty} (\mu_k^n)^{\frac{1}{n}} \; (k = 1, \cdots, m).$$

Here, μ_k^n is the kth largest square root of the m eigenvalues of the matrix $J_n J_n^T$.

$$\lambda_{\mathcal{O}}(f) > 0 \Rightarrow \text{Orbit } \mathcal{O} \text{ is chaotic.}$$
$$\lambda_{\mathcal{O}}(f) \leq 0 \Rightarrow \text{Orbit } \mathcal{O} \text{ is stable.}$$

The properties of the logistic map depend on the parameter a. If we take a particular constant a, for example, $a = 3.71$, then the Lyapunov exponent and the entropic chaos degree are positive, the trajectory is very sensitive to the initial value and one has the chaotic behavior.

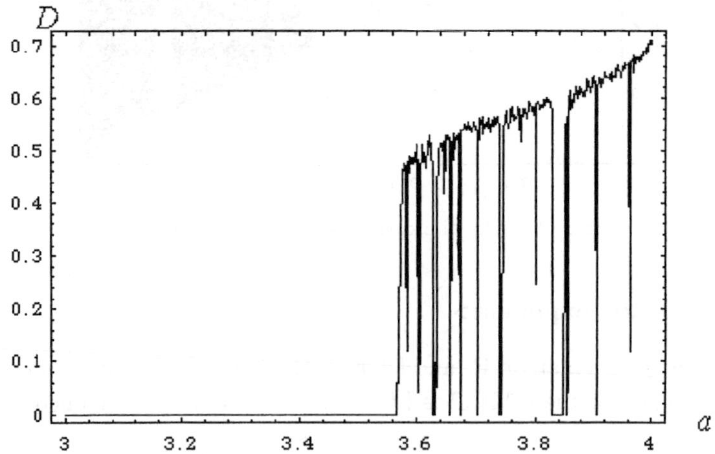

Figure 2. Chaos degree for logistic map

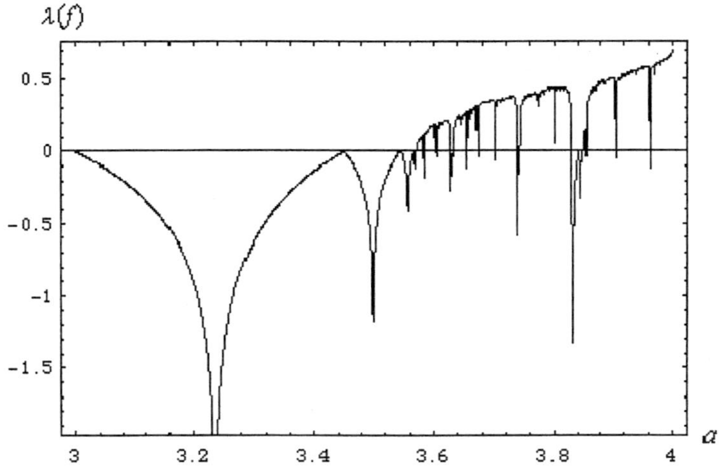

Figure 3. Lyapunov exponent for logistic map

From the above example and some other maps (see the paper [11]), Lyapunov exponent and the entropic chaos degree have clear correspondence, but the ECD can resolve some inconvenient properties of the Lyapunov exponent as follows:

(1) Lyapunov exponent takes negative value and sometimes ∞ , but the ECD is always positive for any $a \geq 0$.
(2) It is difficult to compute the Lyapunov exponent for some maps like Tinkerbell map f because it is difficult to compute f^n for large n. On the other hand, the ECD of f is easily computed.
(3) Generally, the algorithm for the ECD is much easier than that for the Lyapunov exponent.

4. Entropic chaos degree for quantum dynamics

Let us consider von Neumann-Liouville equation

$$i\frac{d\rho_t}{dt} = [H(t), \rho_t] \tag{2}$$

with the initial condition

$$\rho_s = \rho \tag{3}$$

The solution of (2) is given in the form

$$\rho_{t,s} = \; ^{*}_{t,s}\rho = U_{t,s}\rho U^{*}_{t,s} \tag{4}$$

where

$$U_{t,s} = T\exp\left(-i\int_s^t H\left(t'\right)dt'\right) \tag{5}$$

and it follows from (5) and (4) that the relations

$$U_{t,s}U_{s,u} = U_{t,u}, \qquad U_{t,t} = I, \qquad t \geq s \geq u \tag{6}$$

and

$$\Gamma^*_{t,s}\Gamma^*_{s,u} = \Gamma^*_{t,u}, \qquad \Gamma^*_{t,t} = id, \qquad t \geq s \geq u \tag{7}$$

hold. From (6) and (7) one finds that

$$U_{t+\Delta,s} = U_{t+\Delta,t}U_{t,s} \tag{8}$$

and

$$\Gamma^*_{t+\Delta,s} = \Gamma^*_{t+\Delta,t}\Gamma^*_{t,s}. \tag{9}$$

So the relation between $U_{t+\Delta,s}\left(\Gamma^*_{t+\Delta,s}\right)$ and $U_{t,s}\left(\Gamma^*_{t,s}\right)$ is linear. Let us put

$$s = 0, \quad t = n\tau,$$
$$U_n = U_{n\tau,0}, \quad \Gamma^*_n = \Gamma^*_{n\tau,0}, \tag{10}$$
$$V_n = T\exp\left(-i\int_{(n-1)\tau}^{n\tau} H\left(t'\right)dt'\right). \tag{11}$$

Then one finds

$$U_n = V_nV_{n-1}\cdot\dots\cdot V_1 \tag{12}$$
$$\Gamma^*_n\rho = V_nV_{n-1}\cdot\dots\cdot V_1\rho\left(V_nV_{n-1}\cdot\dots\cdot V_1\right)^* = U_n\rho U_n^* \tag{13}$$

Let us consider as an example the following case. Let $F_1,\dots,F_n$ be selfadjoint operators such that $I,F_1,\dots,F_n$ are linearly independent. Suppose that

$$H\left(t\right) = \sum_{k=1}^{N} c_k\left(t\right)F_k + H_0 \qquad for \quad t \geq 0,$$

where $c_1\left(t\right),\dots,c_N\left(t\right)$ are solutions of the equations

$$\frac{dc_k\left(t\right)}{dt} = f_k\left(c_1\left(t\right),\dots,c_N\left(t\right)\right) \qquad k = 1,\dots,N \tag{14}$$

with some function f and initial conditions

$$c_k\left(0\right) = c_k^0$$

and it is assumed that

$$H(t) = H_0 \qquad for \quad t < 0.$$

The equations (14) may lead to chaos. However its discrete version is quite complicated. Another example can be given in the form

$$H(t) = H_0 + H_n \qquad for \quad (n-1)\tau \leq t < n\tau, \quad n = 1, 2, \ldots \tag{15}$$

$$H(t) = H_0 \qquad for \quad t < 0$$

and suppose that each H_n is determined as follows

$$H_{n+1} = G(H_n) \tag{16}$$

or more explicitly

$$H_n = \sum_{k=1}^{N} c_n^{(k)}(t) F_k$$

where

$$c_{n+1}^{(k)}(t) = g_k\left(c_n^{(1)}, \ldots, c_n^{(N)}\right), \quad n = 1, 2, \ldots$$

and F_k are as above. The above functions G and g are properly defined and may cause a chaos. Our treatment discussed up to here in this section can be easily extended for the case of non-Hamiltonian dynamics described in terms of dynamical semi-groups. The channel describing a discrete dynamics of a quantum system as in the above examples is written as

$$\Gamma_{n+1}^* = \Lambda_n^* \Gamma_n^* \tag{17}$$

hence

$$\rho_{n+1} = \Gamma_{n+1}^* \rho = \Lambda_n^* \rho_n.$$

It follows from the above examples that operators V_n and consequently maps Λ_n^* may inherit some chaotic properties of the equations (14) or (16). However it seems that one way to investigate the properties of Λ_n^* is to calculate the expectation values of some observables. The simple choice is to consider one observable. Let $X \in B(\mathcal{H})$ be such an observable and $\rho \in \mathcal{S}(\mathcal{H})$ be a state of the system. Consider the sequence (orbit)

$$x_n = \mathrm{tr} X \Lambda_n^* \rho, \tag{18}$$

where

$$\Lambda_n^* \rho = V_n \rho V_n^* \tag{19}$$

with

$$V_n = T\exp\left(-i\int_{(n-1)\tau}^{n\tau} H\left(t'\right)dt'\right).$$

In the special case (15) one has

$$V_n = \exp\left\{-i\tau\left(H_0 + H_n\right)\right\}, \quad n = 1, 2, \ldots$$

The sequence $\{x_n\}$ characterize the change of Λ_n^*. Let X and ρ be fixed and take a proper natural number N. Let r, R be

$$r = \inf_n \mathrm{tr} X\Lambda_n^*\rho = \inf_n x_n, \qquad R = \sup_n \mathrm{tr} X\Lambda_n^*\rho = \sup_n x_n, \qquad n = 1, 2, \cdots, N. \tag{20}$$

Remark here that in the same vein as above, the quantum entropic chaos degree was applied to analyze quantum baker's map by taking special observable of the system.

5. Properties of entropic chaos degree for quantum dynamics

In this section we examine that the entropic chaos degree for quantum dynamics really gives zero value to some non-chaotic dynamics.

Theorem 5.1. *For any $\rho \in \mathcal{S}(\mathcal{H})$, $\Gamma_n^* : \mathcal{S}(\mathcal{H}) \to \mathcal{S}(\mathcal{H})$, $n = 1, 2, \cdots$, we have:*

Let U be a unitary operator. If $\Gamma_n^*\rho = U^n\rho U^{*n}$, then $D(\rho; \Gamma_n^*) = 0$.

If Γ^* is a channel such that $\Gamma_n^* = (\Gamma^*)^n$, then $D(\rho; \Gamma_n^*) = 0$.

Let $\{\Lambda_n^*\}$ be a family of channels with common invariant state ρ_0, that is, $\Xi_n^*\rho_0 = \rho_0$ for any n. If $\Gamma_n^* = \Xi_n^* \cdots \Xi_1^*$, then $D(\rho_0; \Gamma_n^*) = 0$.

Let ρ_0 be a fixed state on $\mathcal{H}$. If $\Lambda_n^*\rho \xrightarrow{w} \rho_0$, then there exists an equi-partition B_e such that $\lim_{n\to\infty} D\left(p_{B^e}^{(n)}; \Gamma_{n,B^e}^*\right) = 0$.

Let $\{P_n\}$ be one dimensional orthogonal projections such that $\sum_k P_k = I$. If $\Gamma_n^*\rho = \sum_{k_1 \ldots k_n} P_{k_n}P_{k_{n-1}} \cdots P_{k_1}\rho P_{k_1} \cdots P_{k_{n-1}}P_{k_n}$, then $D(\rho; \Gamma_n^*) = 0$.

Proof: Let X be an observable and r, R be real numbers given by (20). Let $B = \{B_i\}$ be any partition of the interval $\mathbf{I} = [r, R]$, and for suitable large natural number M , let $B^e = \{B_i^e\}$ be an equi-partition given by

$$B_i^e = \left[\frac{i}{M}(R - r) + r, \frac{i+1}{M}(R - r) + r\right)$$

for $i = 0, \cdots, M - 2$,

$$B^{e}_{M-1} = \left[\frac{M-1}{M}(R-r)+r, R\right].$$

(1) It follows from the condition $\Gamma^*_n\rho = U^n\rho U^{*n}$, that is, $\Lambda^*_k(\cdot) = U\cdot U^*$

$$x_k = \mathrm{tr}X\Lambda^*_k\rho = \mathrm{tr}XU\rho U^*, \qquad k=1,2,\cdots.$$

For a sufficiently large n_0 and a given partition B, there exists B_{n_0} of B including the point x_k. As x_k is independent of k, we have

$$x_k \in B_{n_0}$$

for any $k \in \mathbf{N}$. One finds that

$$p^{(n)}_{n_0,B} = \frac{1}{m+1}\sum_{k=n}^{m+n} 1_{B_{n_0}}(x_k) = 1, \tag{21}$$

$$p^{(n,n+1)}_{n_0 n_0,B} = \frac{1}{m+1}\sum_{k=n}^{m+n} 1_{B_{n_0}}(x_k)\, 1_{B_{n_0}}(x_{k+1}) = 1. \tag{22}$$

The (21) and (22) imply

$$D\left(p^{(n)}_B; \Gamma^*_{n,B}\right) = 0$$

for any partition B, hence

$$D(\rho; \Gamma^*_n) = 0.$$

(2), (3) In these cases, it follows the definition of Γ^*_n that the corresponding Λ^*_k are Λ^* and Ξ^*_k, respectively. Thus the corresponding sequences x_k are

$$x_k = trX\Lambda^*\rho \text{ and } x_k = trX\Xi^*_k\rho_0 = trX\rho_0, \text{respectively,}$$

which are independent of k. So the same argument of (1) is applied to conclude the statements. (4)Suppose that $\Lambda^*_n\rho$ weakly converges to a fixed state ρ_0 in the limit as $n\to\infty$. Then it can be shown that $\|\Lambda^*_n\rho - \rho_0\|_1 \to 0$ $(n\to\infty)$ (see [9]). Hence, for any $\varepsilon>0$, there exists an n_0, such that for all $n \geq n_0$, we have $\|\Lambda^*_n\rho - \rho_0\|_1 < \varepsilon$.

Let x_* be $\mathrm{tr}(X\rho_0)$. Setting $\varepsilon_1 = \varepsilon\|X\|$ for any observable X, for all $n \geq n_0$, we have

$$
\begin{aligned}
|x_n - x_*| &= |\mathrm{tr}\,(X\Lambda_n^*\rho) - \mathrm{tr}\,(X\rho_0)| \\
&= |\mathrm{tr}X\,(\Lambda_n^*\rho - \rho_0)| \\
&\le \|\Lambda_n^*\rho - \rho_0\|_1 \,\|X\| \\
&< \varepsilon \,\|X\| \\
&= \varepsilon_1.
\end{aligned}
$$

Let $I(\varepsilon_1, x_*)$ be the interval $(x_* - \varepsilon_1, x_* + \varepsilon_1)$. For a sufficiently large n_0 and a given equi-partition B^e, there exists a component B_j^e of B^e including $I(\varepsilon_1, x_*)$. It is clear that for all $n \ge n_0$, we have $x_n \in B_j^e$. Thus, we obtain

$$
x_{n+k} \in B_j^e, \quad (x_{n+k}, x_{n+k+1}) \in B_j^e \times B_j^e, \quad k = 1, 2, \dots. \tag{23}
$$

for large n.

¿From eq.(23), one finds that

$$
\lim_{n\to\infty} p_{j,B^e}^{(n)} = 1, \quad \lim_{n\to\infty} p_{jj,B^e}^{(n,n+1)} = 1. \tag{24}
$$

Eq.(24) implies that

$$
\lim_{n\to\infty} D\left(p_{B^e}^{(n)}; \Gamma_{n,B^e}^*\right) = 0
$$

for the equi-partition B^e.
(5) Applying n times measurement of PVM $\{P_k\}$, we have

$$
x_k = \mathrm{tr}X\Lambda_k^*\rho = \mathrm{tr}X\left(\sum_j P_j\rho P_j\right), \qquad k = 1, 2, \cdots.
$$

which is again independent of k, so that

$$
D\,(\rho; \Gamma_n^*) = 0.Q.E.D.
$$

6. New Description of Chaos

First of all we examine carefully when we say that a certain dynamics produces a chaos. Let us take the logistic map as an example. The original differential equation of the logistic map is

$$\frac{dx}{dt} = ax(1-x), 0 \le a \le 4 \tag{25}$$

with initial value x_0 in $[0, 1]$. This equation can be easily solved analytically, whose solution (orbit) does not have any chaotic behavior. However once we make the equation above discrete such as

$$x_{n+1} = ax_n(1-x_n), 0 \le a \le 4. \tag{26}$$

This difference equation produces a chaos. Taking the discrete time is necessary not only to make a chaos but also to observe the orbits drawn by the dynamics. Similarly as quantum mechanics, it is not possible for human being to understand any object without observing it, for which it will not be possible to trace a orbit continuously in time.

Now let us think about finite partition $A=\{A_k; k=1,\cdots,N\}$of a proper set $I \equiv [a,b]^{\mathbf{N}} \subset \mathbf{R}^{\mathbf{N}}$ and equi-partition $B^e = \{B_k^e; k=1,\cdots,N\}$ of I. Here "equi" means that all elements B_k^e are identical. We denote the set of all partitions by $\mathcal{P}$ and the set of all equi-partitions by $\mathcal{P}^e$. In the section 3, we specify a special partition, in particular, an equi-partition for computer experiment calculating the ECD. Such a partition enables to observe the orbit of a given dynamics, and moreover it provides a criterion for observing chaos. There exist several reports saying that one can observe chaos in nature, which are very much related to how one observes the phenomena, for instance, scale, direction, aspect. It has been difficult to find a satisfactory theory (mathematics) to explain such chaotic phenomena. In the difference equation () we take some time interval τ between n and $n+1$, if we take $\tau \to 0$, then we have a complete different dynamics. If we take coarse graining to the orbit of x_t in () for time during τ; $x_n \equiv \frac{1}{\tau}\int_{(n-1)\tau}^{n\tau} x_t dt$, we again have a very different dynamics. Moreover it is important for mathematical consistency to take the limits $n \to \infty$ or N (the number of equi-partitions)$\to \infty$, i.e., making the partition finer and finer, and consider the limits of some quantities as describing chaos, so that mathematical terminologies such as "lim", "sup", "inf" are very often used to define such quantities. In this paper we take the opposite position, that is, any observation will be unrelated or even contradicted to such limits. Observation of chaos is a result due to taking suitable scales of, for example, time, distance or domain, and it will not be possible in the limiting cases.

We claim in this paper that most of chaos are scale-dependent phenomena, so the definition of a degree measuring chaos should dependes on certain scales taken.

Taking into cosideration of this view we modify the definitions of the chaos degree given in the previous sections as below.

Going back to a triple $(\mathcal{A}, \mathcal{S}, \alpha(G))$ considered in Section 2 and we use this triple both for an input and an output systems. Let a dynamics be described by a mapping Γ_t with a parameter $t \in G$ from $\mathcal{S}$ to $\mathcal{S}$ and let an observation be described by a mapping $\mathcal{O}$ from $(\mathcal{A}, \mathcal{S}, \alpha(G))$ to a triple $(\mathcal{B}, \mathcal{T}, \beta(G))$. The triple $(\mathcal{B}, \mathcal{T}, \beta(G))$ might be same as the original one or its subsystem and the obserevation map $\mathcal{O}$ may contains several different types of observations, that is, it can be decomposed as $\mathcal{O} = \mathcal{O}_m \cdots \mathcal{O}_1$.Let us list some examples of observations.

For a given dynamics $\frac{d\varphi}{dt} = F(\varphi_t)$, equivalently, $\varphi_t = \Gamma_t^* \varphi$, one can take several observations.

Example 6.1. Time Scaling (Discretizing): $\mathcal{O}_\tau$: t$\rightarrow n$, $\frac{d\varphi}{dt}(t) \rightarrow \varphi_{n+1}$, so that $\frac{d\varphi}{dt} = F(\varphi_t) \Rightarrow \varphi_{n+1} = F(\varphi_t)$ and $\varphi_t = \Gamma_t^* \varphi \Rightarrow \varphi_n = \Gamma_n^* \varphi$. Here τ is a unit time needed for the observation.

Example 6.2. Size Scaling (Conditional Expectation, Partition): Let $(\mathcal{B}, \mathcal{T}, \beta(G))$ be a subsystem of $(\mathcal{A}, \mathcal{S}, \alpha(G))$, both of which have a certain algebraic structure such as C*-algebra or von Neumann algebra. As an example, the subsystem $(\mathcal{B}, \mathcal{T}, \beta(G))$ has abelian structure describing a macroscopic world which is a subsystem of a non-abelian (non-commutative) system $(\mathcal{A}, \mathcal{S}, \alpha(G))$ describing *a* micro-world. A mapping $\mathcal{O}_C$ preserving norm (when it is properly defined) from $\mathcal{A}$ to $\mathcal{B}$ is, in some cases, called a conditional expectation. A typical example of this conditional expectation is according to a projection valued measure $\{P_k;\ P_k P_j = P_k \delta_{kj} = P_k^* \delta_{kj} \geq 0,\ \sum_k P_k = I\ \}$ associated with quantum measurement (von Neumann measurement) such that $\mathcal{O}_C(\rho) = \sum_k P_k \rho P_k$ for any quantum state (density operator) ρ. When $\mathcal{B}$ is a von Neumann algebra generated by $\{P_k\}$, it is an abelian algebra isometrically isomorphic to $L^\infty(\Omega)$ with a certain Hausdorff space Ω, so that in this case $\mathcal{O}_C$ sends a general state φ to a probability measure (or distribution) p. Similar example of $\mathcal{O}_C$ is one coming from a certain representation (selection) of a state such as a Schatten decomposition of ρ ; $\rho = \mathcal{O}_R \rho = \sum_k \lambda_k E_k$ by one-dimensional orthgonal projections $\{E_k\}$ associated to the eigenvalues of ρ with $\sum_k E_k = I$. Another important example of the size scaling is due to a finite partition of an underlining space Ω, e.g., space of orbit, defined as $\mathcal{O}_P(\Omega)=\{P_k; P_k \cap P_j = P_k\delta_{kj}(k,j=1,\cdots N),\ \cup_{k=1}^N P_k = \Omega\}$.

We go back to the discussion of th entropic chaos degree. Starting from a given dynamics $\varphi_t = \Gamma_t^*\varphi$, it becomes $\varphi_n = \Gamma_n^*\varphi$ after handling the operation $\mathcal{O}_\tau$. Then by taking proper combinations $\mathcal{O}$ of the size scaling operations like $\mathcal{O}_C$, $\mathcal{O}_R$ and $\mathcal{O}_P$, the equation $\varphi_n = \Gamma_n^*\varphi$ changes to $\mathcal{O}(\varphi_n) = \mathcal{O}(\Gamma_n^*\varphi)$, which will be written by $\mathcal{O}\varphi_n = \mathcal{O}\Gamma_n^*\mathcal{O}^{-1}\mathcal{O}\varphi$ or $\varphi_n^{\mathcal{O}} = \Gamma_n^{*\mathcal{O}}\varphi^{\mathcal{O}}$. Then our entropic chaos degree is redifined as follows:

Definition 6.1. The entropic chaos degree of Γ^* with an initial state φ and observation $\mathcal{O}$ is defined by $D^{\mathcal{O}}(\varphi;\Gamma^*) = \int_{\mathcal{O}(\mathcal{S})} S\left(\Gamma^{*\mathcal{O}}\omega^{\mathcal{O}}\right) d\mu^{\mathcal{O}}$, where $\mu^{\mathcal{O}}$ is the measure operated by $\mathcal{O}$ to a extremal decomposition measure of φ selected by of the observation $\mathcal{O}$ (its part $\mathcal{O}_R$).

Definition 6.2. The entropic chaos degree of Γ^* with an initial state φ is defined by $D(\varphi;\Gamma^*) = \inf\left\{D^{\mathcal{O}}(\varphi;\Gamma^*);\mathcal{O}\in\mathcal{SO}\right\}$, *where* $\mathcal{SO}$ is a proper set of observations natually determined by a given dynamics.

Then one judges whether a given dynamics causes a chaos or not by the following way.

Definition 6.3. (1) A dynamics Γ^* is chaotic for an initial state φ in an observation $\mathcal{O}$ iff $D^{\mathcal{O}}(\varphi;\Gamma^*) > 0$. (2)A dynamics Γ^* is totally chaotic for an initial state φ iff $D(\varphi;\Gamma^*) > 0$.

In Definition , $\mathcal{SO}$ is determined by a given dynamics and some conditions attached to the dynamics, for instance, if we start from a difference equation with a special representation of an initial state, then $\mathcal{SO}$ excludes $\mathcal{O}_\tau$ and $\mathcal{O}_R$.

The idea introducing in this paper to understand chaos can be applied not only to the entropic chaos degree but also to some other degrees such as dynamical entropy, whose applications and the comparison of several degrees will be discussed in the forthcoming paper.

In the caase of logistic map, $x_{n+1} = ax_n(1-x_n) \equiv F(x_n)$, we obtain this difference equation by taking the observation $\mathcal{O}_\tau$ and take an observation $\mathcal{O}_P$ by equi-partition of the orbit space $\Omega = \{x_n\}$ so as to define a state (probability distribution). Thus we can compute the entropic chaos degree as is discussed in Section 3.

It is important to notice here that the chaos degree does depend on the choice of observations. As an example, we consider a circle map

$$\theta_{n+1} = f_\nu(\theta_n) = \theta_n + \omega \pmod{2\pi}, \tag{27}$$

where $\omega = 2\pi v (0 < v < 1)$. If v is a rational number N/M, then the orbit $\{\theta_n\}$ is periodic with the period M. If v is irrational, then the orbit $\{\theta_n\}$ densely fills the unit circle for any initial value θ_0; namely, it is a quasiperiodic motion.

We proved in [10] the following theorem.

Let $\mathbf{I} = [0, 2\pi]$ be partioned into L disjoint components with equal length; $\mathbf{I} = B_1 \cap B_2 \cap \ldots \cap B_L$.

Theorem 6.1.

- (1) *If v is rational number N/M , then the finite equi-partition $P = \{B_k; k = 1, \cdots, M\}$ implies$D^{\mathcal{O}}(\theta_0; f_\nu) = 0$.*
  (2) *If v is irrational, then $D^{\mathcal{O}}(\theta_0; f_\nu) > 0$ for any finite partition P=$\{B_k\}$.*

Note that our entropic chaos degree shows a chaos to quasiperiodic circle dynamics by the observation due to a partition of the orbit, which is different from usual understanding of chaos. However usual belief that quasiperiodic circle dynamics will not cause a chaos is not at all obvious, but is realized in a special limiting case as shown in the following proposition.

Proposition 6.1. *For the above circle map, if v is irrational, then $D(\theta_0; f_\nu) = 0$.*

Proof. Let take an equipartition $P = \{B_k\}$ as □

$$B_k \equiv \left\{ x; 2\pi \frac{k-1}{l} \leq x < 2\pi \frac{k}{l} \pi \right\}, \quad k = 1, 2, \ldots, l,$$

where l is a certain integer and $B_{k+l} = B_k$. When ν is irrational, put $\nu_0 \equiv [l\nu]$ with Gaussian $[\cdot]$. Then $f_\nu(B_k)$ intersects only two intervals $B_{k+\nu_0}$ and $B_{k+\nu_0+1}$, so that denote the ration of the Lebesgue measure of $f_\nu(B_k) \cap B_{k+\nu_0}$ and that of $f_\nu(B_k) \cap B_{k+\nu_0+1}$ by $1-s : s$. This s is equal to $l\nu - [l\nu]$ and the entropic chaos degree becomes

$$D^P = -s \log s - (1-s) \log (1-s) .$$

Take the continued fraction expasion of ν and denote its j-th approximate by $\frac{b_j}{c_j}$. Then it holds

$$\left|\nu - \frac{b_j}{c_j}\right| \leq \frac{1}{c_j^2}.$$

For the above equi-partition $B = \{B_k\}$ with $l = c_j$, we find

$$|l\nu - b_j| \leq \frac{1}{k}$$

and

$$[l\nu] = \begin{cases} b_j \left(\text{when } \nu - \frac{b_j}{c_j} > 0\right) \\ b_{j-1} \left(\text{when } \nu - \frac{b_j}{c_j} < 0\right) \end{cases}.$$

It implies

$$D^P \to \frac{\log c_j}{c_j},$$

which goes to 0 as $j \to \infty$. Hence $D = \inf\left\{D^P; P\right\} = 0$. *Q.E.D.*

Such a limiting case will not take place in real observation of natural objects, so that we claim that chaos is a phenomenon depending on observations, which results the definition of chaos above.

In the forthcoming paper [24], we will discuss in more detailed our idea "adaptivity" how to reach to chaos dynamics by starting from general differential dynamics. That is, it is demonstrated how we can get to chaos dynamics by considering observations introduced in this paper, and we calculate the entropic chaos degrees in each dynamics.

Acknowledgments

The authors thank IIAS and SCAT for finatial supports in this work.

References

1. L. Accardi, M. Ohya and N. Watanabe, Dynamical entropy through quantum Markov chain, Open System and Information Dynamics, 4, 71-87, (1997).
2. K.T. Alligood, T.D. Sauer and J.A. Yorke, *Chaos-An Introduction to Dynamical Systems-*, Textbooks in Mathematical Sciences, Springer (1996).
3. R. Alicki and M. Fannes, *Defining quantum dynamical entropy,* Lett. Math. Physics, 32, 75-82, (1994).

4. F. Benatti, *Deterministic Chaos in Infinite Quantum Systems*, Springer (1993).
5. R.L. Devaney, *An Introduction to Chaotic dynamical Systems*, Benjamin (1986).
6. G. Casati and B. Chirikov,*Quantum Chaos: Between Order and Disorder*, Cambridge University Press (1995)
7. G.G.Emch, H.Narnhofer, W.Thirring and G.L.Sewell, Anosov actions on noncommutative algebras, J.Math.Phys., **35**, No.11, 5582-5599 (1994).
8. H. Hasegawa, Dynamical formulation of quantum level statistics, Open Systems and Information dynamics, **4**, 359-377 (1997).
9. R.S. Ingarden, A. Kossakowski and M. Ohya, *Information Dynamics and Open Systems*, Kluwer Academic Publishers (1997).
10. K.Inoue, M.Ohya and A.Kossakowski, A Description of Quantum Chaos, preprint (2002)
11. K. Inoue, M. Ohya and K. Sato, Application of chaos degree to some dynamical systems, Chaos, Solitons & Fractals, **11**, 1377-1385 (2000).
12. K.Inoue, M.Ohya and .V.Volovich, Semiclassical properties and chaos degree for the quantum baker's map, Journal of Mathematical Physics, **43**, No.1, 734-755, 2002.
13. K.Inoue, M.Ohya and I.V.Volovich, On quantum-classical correspondence for baker's map, quant-ph/0108107.
14. A. Kossakowski, M. Ohya and N. Watanabe, Quantum dynamical entropy for completely positive maps, Infinite Dimensional Analysis, Quantum Probability and Related Topics, **2**, No.2, 267-282 (1999).
15. W.A. Majewski, Does quantum chaos exitst? a quantum Lyapunov exponents approach,quant-ph/9805068.
16. N. Muraki and M. Ohya, Entropy functionals of Kolmogorov Sinai type and their limit theorems, Lett. in Mathematical Physics, 36, 327-335, (1996).
17. M.Ohya, Some aspects of quantum information theory and their applications to irreversible processes, Reports on Mathematical Physics, **27**, No.2, 19-47 (1989).
18. M. Ohya, Information dynamics and its applications to optical communication processes, Lecture Note in Physics, **378**, 81-92 (1991).
19. M. Ohya, State change, complexity and fractal in quantum systems, Quantum Communications and Measurement, Plenum Press, New York, 309-320 (1995).
20. M. Ohya, Complexity and fractal dimensions for quantum states, Open Systems and Information Dynamics, **4**, 141-157 (1997).
21. M. Ohya, Complexities and their applications to characterization of chaos, International Journal of Theoretical Physics, **37**, No.1, 495-505 (1998).
22. M.Ohya, Complexity in Quantum System and its Application to Brain Function, **Quantum Information II**, ed. T.Hida and K.Saito, World Scientific (2000)
23. M. Ohya and D. Petz, *Quantum Entropy and Its Use*, Springer-Verlag, TMP (1993).
24. A. Kossakowski, M. Ohya and Y. Togawa, in preparation

NOTE ON POISSON NOISE

SI SI

Faculty of Information Science and Technology
Aichi Prefectural University,
Aichi-ken 480-1198, Japan
e-mail: sisi.ist.aichi-pu.ac.jp

In this paper, we have established a useful fuctional equation of the characteristic functional of Poisson noise, that is

$$C_P(\xi) = \int_0^\infty \lambda \exp[-\lambda t + i\xi(t)] C_P(S_t\xi) dt,$$

where $C_P(\xi)$ is the characteristic functional of Poisson noise. And then the conditional characteristic functional, for which n points are present, is obtained as

$$C_P^n(\xi) = E[e^{i\langle \dot{P}, \xi \rangle} | P(1) = n] = \left(\int_0^1 e^{i\xi(t)} dt \right)^n, \quad \xi \in C^\infty[0,1].$$

AMS Mathematics Subject Classification 2000 : 60H40

1. Background

A Brownian motion may be said to be one of the elemental additive process as with stationary increments and, indeed, most important process. . Another elemental additive process is a Poisson process. Let it be denoted by $P(t), t \geq 0$. Remind that the probability that $P(t)$ takes the value k (non-negative integer) is $\frac{\lambda t^k}{k!} e^{-\lambda t}$, where λ is the *intensity*.

The time derivative $\dot{P}(t)$ of a Poisson process is a *Poisson noise.* A poisson noise is a generalized stochastic process, with the characteristic functional$C_P(\xi)$ such that

$$C_P(\xi) = \exp[\lambda \int_R (e^{i\xi(t)} - 1) dt]. \tag{1.1}$$

It is often convenient to let it be centered; $\dot{P}(t)-\lambda$, which is a generalized stochastic process with the characteristic functional of the form

$$C_P(\xi) = \exp[\lambda \int_R (e^{i\xi(t)} - 1 - i\xi(t))dt)]. \tag{1.2}$$

It can be easily obtained that the characteristic functional $C_P(\xi)$ of a Poisson noise with R^d-parameter is

$$C_P(\xi) = \exp[\lambda \int_{R^d} (e^{i\xi(t)} - 1)dt^d].$$

To discriminate from R^1-parameter case, the notation $\dot{P}^d(t)$ is often used.

We can introduce a class of functionals of Poisson noise and find that the analysis has much similarity to the case of Gaussian white noise.

The probability distribution μ_P is introduced on the space $E^* \supset L^2(R^d)$, so that we are given a probability space (E^*, μ_P). The complex Hilbert space $(L^2_P) = L^2(E^*, \mu_P)$ is the basic space, on which our analysis starts.

To fix the idea, we consider the case $d = 1$. By the form of the characteristic functional (1.1) it is easily proved that

$$\int_{E^*} \langle x, \xi \rangle^2 d\mu_P(x) = \lambda \int_R \xi(u)^2 du + \lambda^2 (\int_R \xi(u)du)^2,$$

which is finite. We can also prove that

$$\int_{E^*} |\langle x, \xi \rangle|^n d\mu_P(x) < \infty.$$

Thus, all the polynomials in $\langle x, \xi \rangle$ belong to the Hilbert space $(L^2)_P = L^2(E*, \mu_P)$.

The U-transform is defined by

$$U\varphi(\xi) = C_P(\xi)^{-1} \int_{E^*} e^{i\langle x, \xi \rangle} \varphi(x) d\mu_P(x), \ \xi \in E,$$

for $\varphi \in (L^2)_P$.

Set

$$K(\xi, \eta) = \exp[\langle e^{i\xi} - 1, e^{i\eta} - 1 \rangle], \ \xi, \eta \in E.$$

Then, the kernel $K(\xi, \eta)$ is proved to be positive definite, so that we can form a minimal reproducing kernel Hilbert space $\mathbf{F} = \mathbf{H}(K)$ with kernel K.

Let the inner products in $(L^2)_P$ and $\mathbf{F}$ be denoted by $(\cdot,\cdot)$ and $(\cdot,\cdot)_{\mathbf{F}}$, respectively. Then, we have

$$(U\varphi, U\psi)_{\mathbf{F}} = (\varphi, \psi), \quad \varphi, \psi \in (L^2)_P.$$

As in the case of Gaussian white noise, we have the Fock space, that is a direct sum decomposition of $(L^2)_P$:

$$(L^2)_P = \bigoplus_{n=0}^{\infty} H_{pn},$$

where H_{pn} is the space of the discrete chaos of degree n.

Under the U-transform the subspace H_{pn} is transformed to a subspace $\mathbf{F}_n$ of $\mathbf{F}$:

$$\mathbf{F}_n = \{U\varphi(\xi) = \int_{R^n} F(u_1, u_2, ..., u_n) \prod_1^n P(\xi(u_j)) du^n, \quad \varphi \in H_{pn}\},$$

where $P(\xi(u)) = e^{i\xi(u)} - 1$. The kernel F is a symmetric $L^2(R^n)$-function (that is $F \in \hat{L}^2(R^n)$) and

$$\|\varphi\|^2 = n! \|F\|_{\mathbf{F}}^2.$$

Actually, the U-transform gives a bijection between H_n and $\mathbf{F_n}$.

$$\mathbf{F}_n \cong \widehat{L^2(R^d)}. \tag{1.3}$$

In addition,

$$\mathbf{F}_n \perp \mathbf{F}_m, \; n \neq m \tag{1.4}$$

holds.

Proposition 1.1. *The subspace* $\mathbf{F}_n$ *is spanned by*

$$\frac{\partial}{\partial Q(\eta_1(t_1))} \frac{\partial}{\partial Q(\eta_1 2(t_2))} \cdots \frac{\partial}{\partial Q(\eta_n(t_n))} C_P(\xi),$$

where $Q(x) = e^{ix} - 1, \eta_i \in E$, *and* $t_i; i = 1, 2, ..., n$, *are different.*

In order to define *generalized Poisson noise functionals* we follow two steps. First the subspace H_{pn} is extended. Namely, kernel functions F can be taken to be a generalized function in the symmetric Sobolev space of order $-(n+1)/2$. Thus we have a larger space H_n^{-n}. The next step is to have a weighted sum of the H_n^{-n}: Take a decreasing sequence c_n of positive numbers to define

$$(L^2)_P^- = \bigoplus c_n H_n^{-n}. \tag{1.5}$$

This is the space of *generalized Poisson noise functionals.*

Accordingly in the RKHS $\mathbf{F}$, we have

$$\mathbf{F}^- = \bigoplus_n c_n \mathbf{F}_n^{(-n)}, \tag{1.6}$$

$$\mathbf{F}_n^{(-n)} \cong H_n^{-n}, \text{ under } U \text{ transform.}$$

Example Remind that a stochastic bilinear form $\langle x, I(t)\rangle$ ia a version of a Poisson process $P(t)$. The U-transform of a centered Poisson process is given by a kernel $\chi_{I(t)}(u) - \lambda t$. Hence the centered Poisson noise $\dot{P}(t) - \lambda$ corresponds to $\delta_t(u)$. Its p-th power has the kernel $\delta_t(u)^{p\otimes}$ and defines a generalized Poisson noise functionals; nit is a polynomial in $\dot{P}(t)$.

The generalized functional for the case $p = 2$ helps to have a rigorous definitionof the path integral using the Poisson bridge.

Proposition 1.2.

i) $\frac{\partial}{\partial P(\eta)}$ *is a derivation acting on* $\mathbf{F}^-$. *It is an annihilation operator and the domain involves* $\mathbf{F}_n$.

ii) The adjoint operator $(\frac{\partial}{\partial P(\eta)})^*$ *is defined and it is a creation operator.*

Compound Poisson process and compound Poisson noise (the time derivative of a compound Poisson process; it is often called a Lévy noise) shall be discussed in our comming paper. The case where the parameter space is taken to be $R^d, d \geq 1$, will also be discussed elsewhere.

We also see good examples in quantum optics (see e.g. [4]), where the intensity λ is randomized.

The projection operator $E(t)$ in (L_p^2) can be defined by

$$E(t)\varphi = E[\varphi|\mathbf{B}_t(P)],$$

where $\mathbf{B}_t(P)$ is the σ-field, with respect to which all the $\langle x, \chi_{I(s)} \rangle, s < t$, are measurable in the measure space $(E^*, \mathbf{B}, \mu_p)$.

Proposition 1.3. *$E(t)$ commutes with the projection $P_n : (L^2_p) \to \mathbf{F}_n$.*

Remark *The same property holds for the time operator $\int tdE(t)$.*

2. A functional equation for $C_P(\xi)$

We assume that $d = 1$.

Theorem 2.1. *A characterization of Poisson noise is given by the functional equation for the characteristic functional C_P :*

$$C_P(\xi) = \int_0^\infty \lambda \exp[-\lambda t + i\xi(t)] C_P(S_t \xi) dt, \tag{2.1}$$

where S_t is the shift operator:

$$(S_t\xi)(u) = \xi(u + t).$$

The converse is true under the assumption that C_P is continuous in $L^2(R)$ and associated with independent values at each point t_i.

Proof. Let the jumping times be $t_0, t_1, \cdots, t_n$ such that $0 = t_0 < t_1 < \cdots < t_n = 1$. Set $\tau_i = t_i - t_{i-1}$, then $\{\tau_i\}$ are independent identically distributed exponential distribution with mean $\lambda^{-1}, \lambda > 0$.

Thus we can write

$$\dot{P}(t) = \sum \delta_{t_j},$$

and so we have

$$\begin{aligned}C_p(\xi) &= E\left[\exp(i\langle\dot{P},\xi\rangle)\right] = E\left[\exp(i\sum\langle\delta_{t_j},\xi\rangle)\right]\\&= E\left[\exp(i\xi(t_1))\exp(i\sum_{j=2}^{\infty}\xi(t_1+\sum_{k=2}^{j}\tau_k))\right]\\&= E\left[E\left(\exp(i\xi(t)\exp(i\sum_{j=2}^{\infty}\xi(t+\sum_{k=2}^{j}\tau_k))|t_1=t\right)\right]\\&= E\left[\int\exp(i\xi(t))\exp(i\sum_{2}^{\infty}\xi(t+\sum_{1}^{j-1}\tau_k)\lambda e^{-\lambda t}dt\right]\\&= \int\exp(i\xi(t))\lambda e^{-\lambda t}E\left[\exp(i\sum\xi(t+\sum_{1}^{j-1}\tau_k)dt\right]\\&= \int\exp(i\xi(t))\lambda e^{-\lambda t}C_p(S_t\xi)dt\end{aligned}$$

For the proof of the converse.

We know from the assumption of the continuity of $C_p(\xi)$ that there is a measure, in the Sobolev space E_1, of order -1, which defines the $C_p(\xi)$

The additivity of Z follows from the fact that $\dot{Z}(t)$ has independent values at every t.

$$\begin{aligned}\langle\dot{z}(\cdot),S_t\xi(\cdot)\rangle &= \int\dot{z}(s)\xi(s+t)ds\\&= \int\dot{z}(s-t)\xi(s)ds\\&= \langle\dot{z}(\cdot-t),\xi\rangle\end{aligned}$$

Since

$$E\left[e^{i\langle\xi,\delta_t\rangle}\right] = e^{i\xi(t)},$$

the product $e^{i\xi(t)}C_p(S_t\xi)$ can be considered as the characteristic functional of $\dot{z}(\cdot-t)+\delta_t$.

We can say that $\dot{z}(\cdot-t)$ and δ_t are independent since δ_t is non-random function. The δ_t corresponds to the epoch of the end of exponential holding time. The shift S_t means the change to the next exponential distribution. It means that the right hand side of (2.1) can be written as

$$C_P(\xi) = E\left[e^{i\left(\sum_j\langle\delta_{t_j},\xi_j\rangle\right)}\right],$$

which is the characteristic functional of Poisson noise. Thus the assertion has proved.

The Poisson noise with randomized intensity plays an important role in quantum optics. It is also characterized by the functional equation for the charcateristic functional. Assume that the intensity is subject to the exponential distribution with parameter α and let $\widehat{C}_P$ denote the Laplace transform of $C_P = C_P^\lambda, \lambda \geq 0$. Then, we have

$$\widehat{C}_P(\alpha, \xi) = -\frac{d}{d\alpha} \int e^{i\xi(t)} \widehat{C_P}(\alpha + t, S_t \xi) dt.$$

This functional is no more than that of a generalized stochastic process with independent values at every moment, however it is still subject to a lattice distribution as is seen from the form of the characteristic functional. Further properties of a Poisson noise will be discussed in Section 4.

3. Multi-dimensional parameter Poisson sheet

The characteristic functional $C_P(\xi)$ of a Poisson noise $\dot{P}^d(t), t \in R^d (d > 1)$, has the same form except that the domain of the integral in its expression extends over an unbounded region as was mentioned before. The Poisson sheet is useful to observe the properties of a Poisson noise in this case, too. The R^d parameter Poisson sheet $P^d(t), t = (t_1, t_2, \cdots, t_d)$, is defined by a linear functional of R^d parameter Poisson noise $\dot{P}^d(t)$ in such a way that

$$P^d(t) = \langle \dot{P}^d, \chi_{I(t)} \rangle, \quad I(t) = \prod_{j=1}^{d} [0, t_j].$$

It has independent variations and the probability distribution is of Lattice type.

4. Invariance of Poisson noise

It is obvious that the time shift leaves the probability distribution μ_p of a Poisson noise invariant, since

$$C_p(S_t^k \xi) = C_p(\xi), \quad (S_t \xi)(u) = \xi(u - te_k), \ u \in R^d, \tag{4.1}$$

where e_k is the unit vector in the k-th direction of R^d.

Obviously, orthogonal group $O(d)$ bacting on the probability space R^d presents invariance of Poisson measure.

Proposition 4.1. *The probability distribution μ_P of an R^d-parameter Poisson noise is invarian t under*

1) the orthogonal group $O(d)$, and

2) the shifts of R^d.

Interesting property can be seen in the dilation of R^d-vectors:

$$\tau_t : u \to ue^{at}, \ u \in R^d, \ a > 0.$$

Then characteristic functional $C_P(\xi)$ changes into

$$e^{\lambda \int (e^{\xi(ue^{at})}-1)du^d} = e^{-\lambda atd \int (e^{\xi v}-1)dv^d}.$$

Namely, the intensity changes from λ to λe^{-dat}, but remains to be at the Poisson distribution.

Some invariance of the probability distribution of the conditional Poisson noises have already been discused in Section

Before functionals of Poisson noise and their analysis is discussed, it is worth mentioning some properties related to the transformations of the distribution. Given n delta functions or equivalently, given an event A_n, the conditional probability concerning the positions (supports) of n delta functions, is invariant under the permutation of those delta functions. More rigorously, the positions of the delta functions are represented by the random variables (random vectors in the R^d-parameter case) $X_1, X_2, \cdots, X_n$ which are distributed independently and identically. This means the probability distribution of the random vectors is invariant under the action of the permutation group $S(n)$. This property implies the invaiance under the transformations given below. Let $V(t), t \in D$, be a Poisson noise. Its probability distribution μ_P^D is given on the space $E(D)^*$, which is the dual space of a nuclear spaace $E(D)$ dense in $L^2(D)$. The event A_n corresponds to a subset $B(n)$ of $E(D)^*$ by the measurable mapping:

$$\omega \longrightarrow x \in E(D)^*, \quad \omega \in A_n,$$

where almost all x (with respect to the measure μ_P) is viewed as a sample $V(t,\omega)$ of the Poisson noise. Such a transformation can be done for every n and defines a mapping from Ω into $E(D)^*$. Under this mapping the action of the group $S(n)$ naturally defines a linear transformation g acting on $E(D)^*$ which leaves the measure μ_P invariant.

If only the characteristic functional is observed, it is hard to imagine a reasonable transformation group even in the case $d = 1$, under which the

Poisson measure is kept invariant. However, some conditional probability distribution of μ_P enjoys interesting invariance given by a certain group.

Here is an example. Let the unit time interval be fixed. Consider the event A_n where n delta measures of Poisson noise exist. In other words, A_n is the event, where a Poisson process $P(t)$ jumps at t_j : $P(t_j - 0) = j$ and $P(t_j + 0) = j + 1,\ j = 1, 2, \cdots, n$. Then we can prove

Proposition 4.2. *The probability density of* $(t_0, t_1, \cdots, t_n)$ *is uniform on the symplex* $\Delta_n : \sum_{j=0}^{n} t_j = 1, t_j \geq 0, j = 0, 1, \cdots n,$

Corollary 4.1. *The probability distribution of* $(t_0, t_1, \cdots, t_n)$ *is invariant under the permutation of coordinates, and the distribution has maximum entropy among the distributions are the symplex* Δ_n.

If only the characteristic functional is observed, it is hard to imagine a reasonable transformation group even in the case $d = 1$, under which the Poisson measure is kept invariant. However, some conditional probability distribution of μ_P enjoys interesting invariance by a certain group.

Here is an example. Let the unit time interval be fixed. Consider the event A_n where n delta measures of Poisson noise exist. In other words, A_n is the event, where a Poisson process $P(t)$ jumps : $P(t_j - 0) = j$ and $P(t_j + 0) = j + 1$. Then we can prove

Proposition 4.3. *The probability distribution of* $(t_0, t_1, \cdots, t_n)$ *is uniform on the symplex* $\sum_{j=0}^{n} t_j = 1, t_j \geq 0, j = 0, 1, \cdots n.$

Let $P(t)$ have jump points $\tau_1, \tau_2, \cdots, \tau_n$. Consider the characteristic functional under the condition $P(1) = n$;

$$C_P^n(\xi) = E[e^{i\langle \dot{P}, \xi \rangle} | P(1) = n], \quad \xi \in C^\infty[0, 1].$$

Obviously for $n = 0$,

$$C_P^0(\xi) = 1,$$

and for $n = 1$

$$\begin{aligned} C_P^1(\xi) &= E[e^{i\langle \dot{P}, \xi \rangle} | P(1) = 1] \\ &= E[e^{i\xi(\tau_1)} | P(1) = 1] \\ &= \int_0^1 e^{i\xi(t)} dt, \end{aligned}$$

since τ_1 has uniform distribution under the assumption $P(1) = 1$.

For the case $n = 2$, consider the conditional characteristic functional C_P^2. There are two jump points τ_1 and τ_2. If $\tau_1 = x_1$, $\tau_2 - \tau_1 = x_2$ and $1 - \tau_2 = x_3$, then $x_i \geq 0$ and $x_1 + x_2 + x_3 = 1$. We have the uniform distribution on the symplex $x_1 + x_2 + x_3 = 1$.

Assume $\tau_1 = x_1$. Then the conditional distribution of $\tau_2 - \tau_1$ is uniform on $[0, 1 - x_1]$ and the density is $\frac{1}{1-x_1}$.

Hence, $E(e^{i<\dot{P},\xi>}|\tau_1 = x_1, P(1) = 2)$ is the same conditional expectation of $e^{i<\dot{P}.\xi>}$ under the assumption $P(1) = 1$ on the time interval $[1, 1 - x_1]$. The density is $2(1 - x_1)$.

Hence we have

$$E(e^{i<\dot{P},\xi>}|\tau_1 = x_1, P(1) = 2) = \int_0^{1-x_1} e^{i\xi(x_2)} \frac{1}{1 - x_1} dx_2.$$

Now we prove

$$\begin{aligned}
&E(e^{i<\dot{P},\xi>}|P(2) = 2) \\
&= \int_0^1 E_{\tau_2}(e^{i(\xi(\tau_1)+\xi(\tau_2)}|\tau_1 = x_1, P(1) = 2) \cdot 2(1 - x_1)dx_1 \\
&= \int_0^1 2(1 - x_1)e^{i\xi(x_1)} E_{\tau_2}(e^{i\xi(\tau_2)}|\tau_1 = x_1, P(1) = 2)dx_1
\end{aligned}$$

Now $\tau_2 = x_1 + (\tau_2 - x_1), \tau_2 - x_1 = x_2$.

Hence $E_{\tau_2}(e^{i\xi(\tau_2)}| \ \tau_1 = x_1, P(1) = 2)$ has the same value as $E_\tau(e^{i\xi(x_1+\tau)}|P(1) = 1)$, where $\tau = \tau_2 - x_1$.

Thus the above expectation is equal to

$$\begin{aligned}
&\int_0^1 \frac{dx_1}{1 - x_1} \int_0^{1-x_1} e^{i(\xi(x_1)+\xi(x_1+x_2)} 2(1 - x_1)dx_1 dx_2 \\
&= 2 \int_0^1 \int_x^1 e^{i\xi(x)} e^{i\xi(y)} dx_1 dy,
\end{aligned}$$

where $x = x_1$ and $y = x_1 + x_2$. By the symmetry of the distribution of (τ_1, τ_2) we have

$$2 \int_0^1 \int_x^1 e^{i\xi(x)+i\xi(y)} dydx = 2 \int_0^1 \int_y^1 e^{i(\xi(x)+\xi(y))} dxdy.$$

The average of these integrals is in agreement with

$$(\int_0^1 e^{i\xi(x)} dx)^2,$$

which is $C_P^2(\xi)^2$.

For general n, we can prove the following proposition.

Proposition 4.4.

$$C_P^n(\xi) = E[e^{i\langle \dot{P},\xi\rangle}|P(1) = n] = \left(\int_0^1 e^{i\xi(t)}dt\right)^n, \quad \xi \in C^\infty[0,1] \qquad (4.2)$$

Remark If $P(1)$ is free then $C_P(\xi) = \exp[\lambda \int (e^{i\xi(t)-1)dt}]$, as usual.

We now consider the conditional characteristic functional $C_P^n(\xi)$ for the case $d = 2$. It is obvious that

$$C_P^0(\xi) = 1,$$

and

$$C_P^1(\xi) = \int_0^1 e^{i\xi(t)}dt,$$

as is above.

For $n = 2$, we can prove that

$$C_P^1(\xi) = \int_0^1 \left(e^{i\xi(t)}dt\right)^2,$$

under the assumption that the two points τ_1 and τ_2 are symmetric in $[0,1] \times [0,1]$, by using in the same way as above .

For a general n, we have

$$C_P^1(\xi) = \int_0^1 \left(e^{i\xi(t)}dt\right)^n,$$

under the assumption that the two points τ_1, τ_2 and τ_n are symmetric in $[0,1] \times [0,1]$,

5. Compound Poisson noise

A compound Poisson noise appears in general innovation, for which the Lévy decomposition is applicable. Hence, many interesting properties are obtained by reducing the given compound Poisson noise to individual elemental Poisson noise, which is the component of the innovation of the random complex system in question. The Lévy decomposition of the innovation in the standard case is explained as follows. To make the matters, somewhat simpler, we still take $\alpha = 1$. The innovation is a system of idealized elemental random variable, so that it is reasonable to assume that it is a time derivative of an additive process which is continuous in probability, although this is some restriction, we may consider the increment of the additive process is stationary. Hence, we have a Lévy process $L(t)$.

Under these assumptions, the innovation, which is now denoted by $\dot{L}(t)$, has the characteristic functional

$$C_L(\xi) = \exp\left[-\frac{\sigma^2}{2}\|\xi\|^2 + \int\int (e^{i\xi(t)}u - 1 - i\xi(t)u)dn(u)\right], \tag{5.1}$$

where $dn(u)$ is the Lévy measure satisfying the condition

$$\int \frac{u^2}{1+u^2} dn(u) < \infty. \tag{5.2}$$

If there is no Gaussian part $-\frac{\sigma^2}{2}$, we are given a compound Poisson noise.

From the expression of $C_L(\xi)$, we understand $L(t)$ is a sum of Brownian motion (up to constant) and a superposition of independent Poisson process with different jumps.

The reduction to an elemental Poisson noise is done in the standard manner, although computability (or possibility of measurement) problem is involved. While, one must meet a difficulty in constructing the superposition of Poisson noises, which are mutually independent. In reality, the quasi-convergence of the sum (the integral) of them is the basic method to have a compound Poisson noise. (See e.g. P. Lévy [4].)

The time derivative $\dot{L}(t)$ is also a candidate of innovation, although it is not elemental.

6. The space (P)

We introduce a space $(\mathbf{P})$ or more precisely denoted by $(\mathbf{P}(X))$, which contains $\mathbf{B}(X)-$ measurable, complex valued, random variables. The $X(C)$ is now viewed as a continuous curve in $(\mathbf{P})$ with parameter C.

It is possible to define a complex Hilbert space $L^2(\Omega, \mathbf{B}(X), P) = (L^2)$. Then a continuous injection

$$(L^2) \subset (\mathbf{P})$$

is established.

The analysis on (L^2) is a powerful tool to investigate the probablistic structure of a white noise $x(\cdot)$, but we do not expect the existence of the variance of $X(C)$, so we can not expect the analysis on (L^2).

Having motivated by many problems in applications and reminded our original idea of innovation theory, we are going to introduce a new class of functionals of white noise either Gaussian or Poisson type. We also try to find available tools for the new analysis on the space $(\mathbf{P})$.

Take a probability space $(\Omega, \mathbf{B}, P)$ on which a vector space (P) of complex-valued random variables is given. The toplogy which is to be introduced to this space is defined by the convergence in probability. It is often useful to take a metric $d(X, Y)$ for $X, Y \in (P)$, defined by

$$d(X,Y) = E(\frac{|X-Y|}{1+|X-Y|}).$$

This metric, as is well known, defines the same topology as that defined by convergence in probability.

The following assertion is easily proved, since the mean square convergence defines stronger topology than that defined by convergence in probability.

Proposition 6.1. *The inclusions*

$$(L^2) \subset (P), \quad (L^2_P) \subset (P)$$

hold and they define continuous injections.

The analysis of (compound) Poisson functionals, in particular, various operations acting on them can be dealt with in this framework without much restrictions.

We take a random field $X(C) = X(C, \omega), C \in \mathbf{C},\ \omega \in \Omega(P)$, and introduce a space $(\mathbf{P})$ of random variables which are the functions of $X(C)$'s.

Let $\mathbf{B}(X)$ be the smallest sigma-field of subsets of Ω, with respect to which all the $X(C)$'s are measurable. Let $\mathbf{C}$ be topologized by the Euclidean metric ρ, namely the distance $\rho(C, C')$ is defined by

$$\rho(C, C') = \max_{u \in C, u' \in C'} \rho(u, u')$$

with respect to this metric, we assume that $X(C)$ is continuous in probability. Given a random field $X(C)$ which is continuous in probability in the variable C.

Limit theorems for $X(C)$ can be discussed within the frame work of (**P**).

Our most important direction is the variational calculus. This can be established in the following manner. To avoid non-essential complex description, we assume that $d = 2$. By assumption C can be represented by a smooth vector valued function, say $f(t) = (f_1(t), f_2(t)), t \in [0,1]$.

We can therefore define a characteristic functional $C_X(\xi), \xi = (\xi_1, \xi_2) \in E \times E$. Hence, the reproducing kernel Hilbert space $\mathbf{H}(C)$ can be defined so that Hilbert space technique can be used.

On the other hand, but still in line with the characteristic functional method, we are given the $T-$ transform of $X(C) = X(f)$, to have a functional $U(\xi, f), \xi \in E \times E$, depending on f. Hence, the theory of (non-random) functional analysis is ready to be applied.

A stochastic process $X(t)$ and random field $X(C)$ that will be discussed in this section are assumed to live in (P) and continuous. The most powerful tool to analyze them would be the charcateristic functional.

i) For a stochastic process, we have the characteristic functional

$$C_X(\xi) = E(e^{i\langle X, \xi \rangle}), \xi \in E.$$

For the case of R^d parameter random fields, we have the same expression to give charcateristic functionals.

ii) For a random field $X(C)$, C being an ovaloid in the plane. The parameter C is represented by a pair of functions, say $(a_C(t), b_C(t))$, and the discussion is reduced to the case i) with parameter C.

References

1. L. Accardi et al eds., Selected papers of Takeyuki Hida, World Sci. Pub. Co. Ltd.2001.
2. L. Accardi, T. Hida and Si Si, Innovation approach to some stochastic processes and random fields, Univ. Roma II, Volterra Center.
3. T. Hida, Brownian motion, Springer-Verlag. 1980 (Original Japanese edition, 1975).
4. P. Lévy, Processus stochastiques et mouvement browniwn. Gauthier-Villars, 1948. 2ème ed. 1965.
5. John R. Klauder and E.C.G. Dudarshan, Fundamentals of Quantum Optics, Math. Phys. Monograph series W. A Benjamin, Inc. 1968.

6. Si Si, A.Tsoi and Win Win Htay, Jump finding of a stable process, Quantum Information V, T. Hida and K. Saito World Scientific (in press).

FOUNDATION OF QUANTUM CAPACITY

NOBORU WATANABE
Department of Information Sciences
Tokyo University of Science
Yamazaki 2641, Noda City, Chiba 278-8510, Japan
E-mail: watanabe@is.noda.tus.ac.jp

In communication process, a channel has an activity to communicate information of input system to the output system. The mutual entropy denotes an amount of information correctly transmitted from the input system to the output system through a channel. The (semi-classical) mutual entropies for classical input and quantum output were defined by several researchers [5,4,6]. The fully quantum mutual entropy for quantum input and output by means of the relative entropy of Umegaki [25] was defined by Ohya [11] in 1983, and he extended it [13] to general quantum systems by using the relative entropy of Araki [2] and Uhlmann [26]. The purely quantum capacity is formulated by taking the supremum of the fully quantum mutual entropy with respect to a certain subset of the initial state space. Namely, the (semi-classical) capacity are defined by taking the supremum of the (semi-classical) mutual entropy according to the subsets of the input state space. The capacity shows the ability of the information transmission of the channel, which is used as a measure for construction of channels.

In this paper, we briefly review quantum channels for several models and we briefly explain the quantum mutual entropy and the quantum capacity for quantum channels. We concretely calculate the quantum capacity for the quantum channels and discuss the purely quantum capacity of fully quantum mutual entropy which generalizes the Holevo bound by using [19].

1. Quantum Channels

In development of quantum information theory, the concept of channel has been played an important role. In particular, an attenuation channel introduced in [11] has been paid much attention in optical communication. A quantum channel is a map describing the state change from an initial system to a final system, mathematically. Let us consider the construction of the quantum channels.

Let $\mathcal{H}_1, \mathcal{H}_2$ be the complex separable Hilbert spaces of an input and an output systems, respectively, and let $\mathbf{B}(\mathcal{H}_k)$ be the set of all bounded linear operators on $\mathcal{H}_k$. $\mathfrak{S}(\mathcal{H}_k)$ is the set of all density operators on $\mathcal{H}_k$ $(k = 1, 2) : \mathfrak{S}(\mathcal{H}_k) \equiv \{\rho \in \mathbf{B}(\mathcal{H}_k) ; \rho \geq 0,\ \rho = \rho^*,\ tr\rho = 1\}$

(1) A map Λ^* from the input system to the output system is called a **(purely) quantum channel.**

(2) The quantum channel Λ^* satisfying the affine property (i.e., $\sum_k \lambda_k = 1\ (\forall \lambda_k \geq 0)\ \Rightarrow\ \Lambda^*\left(\sum_k \lambda_k \rho_k\right) = \sum_k \lambda_k\ \Lambda^*(\rho_k),\ \forall \rho_k \in \mathfrak{S}(\mathcal{H}_1)$) is called a **linear channel.**

A map Λ from $\mathbf{B}(\mathcal{H}_2)$ to $\mathbf{B}(\mathcal{H}_1)$ is called the dual map of $\Lambda^* : \mathfrak{S}(\mathcal{H}_1) \to \mathfrak{S}(\mathcal{H}_2)$ if Λ satisfies

$$tr\rho\Lambda(A) = tr\Lambda^*(\rho) A \tag{1.1}$$

for any $\rho \in \mathfrak{S}(\mathcal{H}_1)$ and any $A \in \mathbf{B}(\mathcal{H}_2)$.

(3) Λ^* from $\mathfrak{S}(\mathcal{H}_1)$ to $\mathfrak{S}(\mathcal{H}_2)$ is called a **completely positive (CP) channel** if its dual map Λ satisfies

$$\sum_{j,k=1}^{n} B_j^* \Lambda\left(A_j^* A_k\right) B_k \geq 0 \tag{1.2}$$

for any $n \in \mathbf{N}$, any $B_j \in \mathbf{B}(\mathcal{H}_1)$ and any $A_k \in \mathbf{B}(\mathcal{H}_2)$.

(4) Λ^* is **Schwarz type** if $\Lambda(\bar{A}^*) = \Lambda(\bar{A})^*$ and $\Lambda(\bar{A})^* \Lambda(\bar{A}) \leq \Lambda(\bar{A}^* \bar{A})$.

One can consider the quantum channel for more general systems. The input system denoted by $(\mathcal{A}, \mathfrak{S}(\mathcal{A}))$ and an output system by $(\overline{\mathcal{A}}, \overline{\mathfrak{S}}(\overline{\mathcal{A}}))$, where $\mathcal{A}$ (resp. $\overline{\mathcal{A}}$) is a C^*-algebra, $\mathfrak{S}(\mathcal{A})$ (resp. $\overline{\mathfrak{S}}(\overline{\mathcal{A}})$) is the set of all states on $\mathcal{A}$ (resp. $\overline{\mathcal{A}}$).

When some outside effects should be considered in a certain physical process such as noise and an effect of reservoir, it is convenient to extend the system $\mathcal{A}$ to $\mathcal{A} \otimes \mathcal{B}$, where $\mathcal{B}$ describes the outside system. In such cases, the concept of lifting introduced in [1] is useful.

(5) A **lifting** from $\mathcal{A}$ to $\mathcal{A} \otimes \mathcal{B}$ is a continuous map

$$\mathcal{E}^* : \mathfrak{S}(\mathcal{A}) \to \mathfrak{S}(\mathcal{A} \otimes \mathcal{B}). \tag{1.3}$$

(6) A lifting $\mathcal{E}^*$ is **linear** if it is affine.

(7) A lifting $\mathcal{E}^*$ is **nondemolition for a state** $\varphi \in \mathfrak{S}(\mathcal{A})$ if $\mathcal{E}^*\varphi(A \otimes I) = \varphi(A)$ for any $A \in \mathcal{A}$.

Given a channel $\Lambda^* : \mathfrak{S}(\mathcal{A}) \to \overline{\mathfrak{S}}(\overline{\mathcal{A}})$ and $\varphi \in \mathcal{S} \subset \mathfrak{S}(\mathcal{A})$, take a certain decomposition such that

$$\varphi = \int_{\mathcal{S}} \omega \, d\mu. \tag{1.4}$$

Then

$$\mathcal{E}^*\varphi = \int_{\mathcal{S}} \omega \otimes \Lambda^*\omega \, d\mu \qquad \text{(compound state [11,12])} \tag{1.5}$$

is a **nondemolition lifting.** This compound state (lifting) can be extensively used in the sequel sections.

The concept of lifting came from the above cmpound state (nonlinear lifting) [11] and the dual of a transition expectation (linear lifting) [1], hence it is a natural generalization of these concepts.

Let $(\Omega, \mathfrak{F}_\Omega, P(\Omega)), \left(\tilde{\Omega}, \mathfrak{F}_{\tilde{\Omega}}, P\left(\tilde{\Omega}\right)\right)$ be input and output probability spaces, where $\mathfrak{F}_\Omega$ (resp. $\mathfrak{F}_{\tilde{\Omega}}$) is a σ-algebra of Ω (resp. $\tilde{\Omega}$) and $P(\Omega)$, $P\left(\tilde{\Omega}\right)$ are the sets of regular probability measures on Ω and $\tilde{\Omega}$.

A channel Ξ^* transmitted from a probability measure to a quantum state is called a classical-quantum (CQ) channel, and a channel $\tilde{\Xi}^*$ from a quantum state to a probability measure is called a quantum-classical (QC) channel. The capacity of both CQ and QC channels have been discussed in several papers [5], [14], [19]. A channel from the classical input system to the classical output system through CQ, Q and QC channels is now denoted by

$$P(\Omega) \overset{\Xi^*}{\to} \mathfrak{S}(\mathcal{H}) \overset{\Lambda^*}{\to} \mathfrak{S}\left(\tilde{\mathcal{H}}\right) \overset{\tilde{\Xi}^*}{\to} P\left(\tilde{\Omega}\right)$$

One of the examples of the CQ channel Ξ^* is given in [22] as follows: Let $C(\Omega)$ be the set of all continuous functions on Ω. For each $\omega \in \Omega$, we assume that $\rho_\omega \in \mathfrak{S}(\mathcal{H})$ is $\mathfrak{F}_\Omega$ - measurable. Then the CQ channel Ξ^* is denoted by

$$tr\Xi^*(\mu)(\cdot) = \int_\Omega (tr\rho_\omega(\cdot))\, \mu(d\omega) \tag{1.6}$$

for any $\mu \in P(\Omega)$. Namely, one of the QC channel is given in [22]. Let $\{\Pi_n\}$ be the set of all non negative Hermite operators with $\sum_n \Pi_n = I$, which is called a positive operator valued measure (POV). The QC channel $\tilde{\Xi}^*$ given by the measurement process using $\{\Pi_n\}$ is obtained by

$$\tilde{\Xi}^*(\sigma) = \sum_n \Pi_n^{\frac{1}{2}} \sigma \Pi_n^{\frac{1}{2}} \tag{1.7}$$

for any $\sigma \in \mathfrak{S}\left(\tilde{\mathcal{H}}\right)$.

1.1. *Noisy optical channel*

To discuss the communication system using the laser signal mathematically, it is necessary to formulate the quantum communication theory being able to treat the quantum effects of signals and channels. In order to discuss influences of noise and loss in communication processes, one needs the following two systems [11]. Let $\mathcal{K}_1$, $\mathcal{K}_2$ be the separable Hilbert spaces for the noise and the loss systems, respectively.

Quantum communication process is described by the following scheme [11].

$$\begin{array}{ccc}
 & \xi \in \mathfrak{S}(\mathcal{K}_1) & \\
 & \downarrow & \\
\mathfrak{S}(\mathcal{H}_1) \ni \rho & \longrightarrow & \overline{\rho} = \Lambda^* \rho \in \mathfrak{S}(\mathcal{H}_2) \\
 & & \downarrow \\
 & & \text{Loss}
\end{array}$$

$$\begin{array}{lcl}
\mathfrak{S}(\mathcal{H}_1) & \rightarrow \Lambda^* \rightarrow & \mathfrak{S}(\mathcal{H}_2) \\
\downarrow & & \uparrow \\
\gamma^* & & a^* \\
\downarrow & & \uparrow \\
\mathfrak{S}(\mathcal{H}_1 \otimes \mathcal{K}_1) & \rightarrow \pi^* \rightarrow & \mathfrak{S}(\mathcal{H}_2 \otimes \mathcal{K}_2)
\end{array}$$

The quantum channel Λ^* is given by the composition of three mappings a^*, π^*, γ^* such as

$$\Lambda^* = a^* \circ \pi^* \circ \gamma^*. \tag{1.8}$$

a^* is a CP channel from $\mathfrak{S}(\mathcal{H}_2 \otimes K_2)$ to $\mathfrak{S}(\mathcal{H}_2)$ defined by

$$a^*(\sigma) = tr_{\mathcal{K}_2}\sigma \tag{1.9}$$

for any $\sigma \in \mathfrak{S}(\mathcal{H}_2 \otimes \mathcal{K}_2)$, where $tr_{\mathcal{K}_2}$ is a partial trace with respect to $\mathcal{K}_2$. π^* is the CP channel from $\mathfrak{S}(\mathcal{H}_1 \otimes \mathcal{K}_1)$ to $\mathfrak{S}(\mathcal{H}_2 \otimes \mathcal{K}_2)$ depending on the physical property of the device. γ^* is the CP channel from $\mathfrak{S}(\mathcal{H}_2)$ to $\mathfrak{S}(\mathcal{H}_1 \otimes \mathcal{K}_1)$ with a certain noise state $\xi \in \mathfrak{S}(\mathcal{K}_2)$ defined by

$$\gamma^*(\rho) = \rho \otimes \xi \tag{1.10}$$

for any $\rho \in \mathfrak{S}(\mathcal{H}_1)$. The quantum channel Λ^* with the noise ξ is written by

$$\Lambda^*(\rho) = tr_{\mathcal{K}_2}\pi^*(\rho \otimes \xi) \tag{1.11}$$

for any $\rho \in \mathfrak{S}(\mathcal{H}_1)$.

Here we briefly review noisy optical channel [20]. A channel Λ^* is called a noisy optical channel if π^* and ξ above are given by

$$\xi \equiv |m\rangle\langle m| \text{ and } \pi^*(\cdot) \equiv V(\cdot)V^*, \tag{1.12}$$

where $|m\rangle\langle m|$ is m photon number state in $\mathcal{H}_1$ and V is a linear mapping from $\mathcal{H}_1 \otimes \mathcal{K}_1$ to $\mathcal{H}_2 \otimes \mathcal{K}_2$ given by

$$V(|n\rangle \otimes |m\rangle) \equiv \sum_{j=0}^{n+m} C_j^{n,m} |j\rangle \otimes |n+m-j\rangle, \tag{1.13}$$

$$\begin{aligned} C_j^{n,m} \equiv & \sum_{r=L}^{K} (-1)^{n-r} \frac{\sqrt{n!m!}}{r!(n-r)!} \alpha^{m-j+2r} \\ & \times \frac{\sqrt{j!(n+m-j)!}}{(j-r)!(m-j+r)!} \left(-\bar{\beta}\right)^{n+j-2r} \end{aligned} \tag{1.14}$$

for any $|n\rangle$ in $\mathcal{H}_1$ and $K \equiv \min\{j,n\}$, $L \equiv \max\{j-m,0\}$, where α and β are complex numbers satisfying $|\alpha|^2 + |\beta|^2 = 1$, and $\eta = |\alpha|^2$ is the transmission rate of the channel. In particular, $\rho \otimes \xi$ is given by the tensor products of two coherent states $|\theta\rangle\langle\theta| \otimes |\kappa\rangle\langle\kappa|$, then $\pi^*(\rho \otimes \xi)$ is obtained by

$$\begin{aligned} \pi^*(\rho \otimes \xi) = & |\alpha\theta + \beta\kappa\rangle\langle\alpha\theta + \beta\kappa| \\ & \otimes \left|-\bar{\beta}\theta + \bar{\alpha}\kappa\right\rangle\left\langle-\bar{\beta}\theta + \bar{\alpha}\kappa\right|. \end{aligned}$$

π^* is called a generalized beam splittings. It means that one beam comes and two beams appear after passing through π^*. Here we remark that the attenuation channel Λ_0^* [11] is derived from the noisy optical channel with $m = 0$. That is, the attenuation channel Λ_0^* was formulated in[11] on 1983 such as

$$\Lambda_0^*(\rho) = tr_{\mathcal{K}_2}\pi_0^*(\rho \otimes \xi_0) \tag{1.15}$$

$$= tr_{\mathcal{K}_2} V_0 (\rho \otimes |0\rangle \langle 0|) V_0^*, \tag{1.16}$$

where $|0\rangle \langle 0|$ is the vacuum state in $\mathfrak{S}(\mathcal{K}_1)$, V_0 is the mapping from $\mathcal{H}_1 \otimes \mathcal{K}_1$ to $\mathcal{H}_2 \otimes \mathcal{K}_2$ given by

$$V_0(|n_1\rangle \otimes |0\rangle) = \sum_j^{n_1} C_j^{n_1} |j\rangle \otimes |n_1 - j\rangle, \tag{1.17}$$

$$C_j^{n_1} = \sqrt{\frac{n_1!}{j!(n_1-j)!}\eta^j (1-\eta)^{n_1-j}}. \tag{1.18}$$

This attenuation channel is most important channel for discussing the optical communication processes. After that, Accardi and Ohya [1] reformulated it by using liftings, which is the dual map of the transition expectation by mean of Accardi. It contains the concept of beam splittings, which is extended by Fichtner, Freudenberg and Libsher [3] concerning the mappings on generalized Fock spaces.

Assume that $\mathcal{H}_1 = \mathcal{H}_2 = \mathcal{H}$ and $\mathcal{K}_1 = \mathcal{K}_2 = \mathcal{K}$. Let $W_0 : \mathcal{H} \to \mathcal{H} \otimes \mathcal{K}$ be a mapping given by

$$W_0 |\theta\rangle \equiv |\alpha\theta\rangle \otimes \left|-\bar{\beta}\theta\right\rangle, \qquad \left(|\alpha|^2 + |\beta|^2 = 1\right). \tag{1.19}$$

A lifting $\mathcal{E}_0^*$ from $\mathfrak{S}(\mathcal{H})$ to $\mathfrak{S}(\mathcal{H} \otimes \mathcal{K})$ is defined by

$$\mathcal{E}_0^*(\rho) \equiv W_0 \rho W_0^* \tag{1.20}$$

for any $\rho \in \mathfrak{S}(\mathcal{H})$. Then the attenuation channel Λ_0^* is described by

$$\Lambda_0^*(\rho) = tr_{\mathcal{K}}\mathcal{E}_0^*(\rho), \qquad (\forall \rho \in \mathfrak{S}(\mathcal{H})). \tag{1.21}$$

2. Quantum Mutual Entropy

The quantum entropy was introduced by von Neumann around 1932 [9], which is defined by

$$S(\rho) \equiv -tr\rho \log \rho$$

for any density operators ρ in $\mathfrak{S}(\mathcal{H}_1)$. It denotes the amount of information contained in the quantum state given by the density operator.

Now, the spectral set of ρ is discrete, so that we write the spectral decomposition of ρ as

$$\rho = \sum_n \lambda_n P_n,$$

where λ_n is an eigenvalue of ρ and P_n is the projection from $\mathcal{H}$ onto the eigenspace associated with λ_n. Therefore, if every eigenvalue λ_n is non-degenerate, then the dimension of the range of P_n is one (we denote this by $dimP_n = 1$). If a certain eigenvalue, say λ_n, is degenerate, then P_n can be further decomposed into one dimensional projections:

$$P_n = \sum_{j=1}^{dimP_n} E_j^{(n)}.$$

where $E_j^{(n)}$ is a one-dimensional projection expressed by $E_j^{(n)} = |x_j^{(n)} >< x_j^{(n)}|$ with the eigenvector $x_j^{(n)}$ $(j = 1, 2, \cdots, \dim P_n)$ for λ_n. By relabelling the indices j, n of $\{E_j^{(n)}\}$, we write

$$\rho = \sum_n \lambda_n E_n \tag{2.1}$$

with

$$\lambda_1 \geq \lambda_2 \geq \cdots \lambda_n \geq \cdots,$$

$$E_n \perp E_m (n \perp m).$$

We call this decomposition the Schatten decomposition [24].

In order to discuss some physical phenomena, for instance, phase transitions, we had better start without Hilbert space, so that we need to formulate the entropy of a state in a C* system [13,15,?].

Let $(\mathcal{A}, \mathfrak{S}(\mathcal{A}), \alpha(R))$ be a C*-dynamical system and $\mathcal{S}$ be a weak* compact and convex subset of $\mathfrak{S}(\mathcal{A})$. For instance, $\mathcal{S} = \mathfrak{S}(\mathcal{A})$; $\mathcal{S} = I(\alpha)$, the set of all invatiant states for α; $\mathcal{S} = K(\alpha)$, the set of all KMS states.

Every state $\varphi \in \mathcal{S}$ has a maximal measure μ pseudosupported on $ex\mathcal{S}$ (the set of all extreme points of $\mathcal{S}$) such that

$$\varphi = \int_{\mathcal{S}} \omega d\mu. \tag{2.2}$$

The measure μ giving the above decomposition is not unique unless $\mathcal{S}$ is a Choquet simplex, so that we denote the set of all such measures by $M_\varphi(\mathcal{S})$.

Put

$$D_{\varphi}(S)=\{M_{\varphi}(S);\exists\mu_k\subset\mathbb{R}^+\text{ and }\{\varphi_k\}\subset exS$$
$$\text{s.t. }\sum_k\mu_k=1,\mu=\sum_k\mu_k\delta(\varphi_k)\Big\},\tag{2.3}$$

where $\delta(\varphi)$ is the Dirac measure concentrated on an state φ, and define

$$H(\mu)=-\sum_k\mu_k\log\mu_k\tag{2.4}$$

for a measure $\mu\in D_{\varphi}(\mathcal{S})$. Then the **entropy of a state** $\varphi\in\mathcal{S}$ w.r.t. $\mathcal{S}$ (**$\mathcal{S}$-mxinig entropy**) is defined by

$$S^S(\varphi)=\begin{cases}\inf\{H(\mu);\ \mu\in D_{\varphi}(S)\}\\ +\infty & \text{if}\quad D_{\varphi}(S)=\emptyset.\end{cases}$$

This entropy is an extension of von Neumann's entropy [9], and it depends on the set $\mathcal{S}$ chosen. Hence it represents the uncertainty of the state measured from the reference system $\mathcal{S}$. When $\mathcal{S}=\mathfrak{S}(\mathcal{A})$, we simply denote $\mathrm{S}^{\mathfrak{S}(\mathcal{A})}(\varphi)$ by $\mathrm{S}(\varphi)$ in the sequel.

The classical mutual entropy is determied by an input state and a channel, so that we denote the quantum mutual entropy with respect to the input state ρ and the quantum channel Λ^* by $I(\rho;\Lambda^*)$. This quantum mutual entropy $I(\rho;\Lambda^*)$ should satisfy the following three conditions:

(1) If the channel Λ^* is identity map, then the quantum mutual entropy equals to the von Neumann entropy of the input state, that is, $I(\rho;id)=S(\rho)$.
(2) If the system is classical, then the quantum mutual entropy equals to the classical mutual entropy.
(3) The following fundamental inequalities are satisfied:

$$0\le I(\rho;\Lambda^*)\le S(\rho).$$

In order to define such a quantum mutual entropy, we need the quantum relative entropy and the joint state, which is called a compound state, describing the correlation between an input state ρ and the output state $\Lambda^*\rho$ through a channel Λ^*. The compound state σ_E (corresponding to joint state in CS) of ρ and $\Lambda^*\rho$ was introduced by Ohya in [11,12], which is given by

$$\sigma_E=\sum_n\lambda_nE_n\otimes\Lambda^*E_n,\tag{2.5}$$

where E derives from a Schatten decomposition $\{\rho = \sum_n \lambda_n E_n\}$ of ρ. The quantum mutual entropy with respect to the input state ρ and the quantum channel Λ^* was defined by Ohya [11] such as

$$I(\rho; \Lambda^*) \equiv \sup_E S(\sigma_E, \sigma_0), \tag{2.6}$$

where $\sigma_0 = \rho \otimes \Lambda^* \rho$ and $S(\sigma_E, \sigma_0)$ is the quantum relative entropy defined by Umegaki [25] and Lindblad [7] as follows:

$$S(\rho, \sigma) \equiv \begin{cases} tr\rho(\log\rho - \log\sigma) & (\text{when } \overline{ran\rho} \subset \overline{ran\sigma}) \\ \infty & (\text{otherwise}) \end{cases}$$

There were several trials to extend the relative entropy to more general quantum systems and apply it to some other fields [2,26,15]. This mutual entropy $I(\rho; \Lambda^*)$ satisfies all conditions (1)~(3) mentioned above. It represents the amount of information correctly transmitted from the input quantum state ρ to the output quantum system through the quantum channel Λ^*.

The mutual entropy was first discussed by Shannon to study the information transmission in classical systems and its fully general quantum version was formulated in [13].

Let $(\mathcal{A}, \mathfrak{S}(\mathcal{A}), \alpha(G))$ be a unital C^*-system and $\mathcal{S}$ be a weak* compact convex subset of $\mathfrak{S}(\mathcal{A})$. For an initial state $\varphi \in \mathcal{S}$ and a channel Λ^* : $\mathfrak{S}(\mathcal{A}) \to \mathfrak{S}(\mathcal{B})$, two fundamental compound states are

$$\Phi_\mu^{\mathcal{S}} = \int_S \omega \otimes \Lambda^* \omega d\mu \quad (= \mathcal{E}^* \varphi \text{ in } ((1.3))) \tag{2.7}$$

$$\Phi_0 = \varphi \otimes \Lambda^* \varphi. \tag{2.8}$$

The compound state $\Phi_\mu^{\mathcal{S}}$ expresses the correlation between the input state φ and the output state $\Lambda^*\varphi$. The mutual entropy w.r.t. $\mathcal{S}$ and μ is

$$\mathrm{I}_\mu^{\mathcal{S}}(\varphi\,;\,\Lambda^*) = \mathrm{S}(\Phi_\mu^{\mathcal{S}},\, \Phi_0) \tag{2.9}$$

and the mutual entropy w.r.t. $\mathcal{S}$ is defined as

$$\mathrm{I}^{\mathcal{S}}(\varphi\,;\Lambda^*) = \sup\left\{\mathrm{I}_\mu^{\mathcal{S}}(\varphi\,;\Lambda^*)\ ;\ \mu \in M_\varphi(\mathcal{S})\right\}. \tag{2.10}$$

When a state φ is expressed as $\varphi = \sum_k \mu_k \varphi_k$ (fixed), the mutual entropy is given by

$$\mathrm{I}(\varphi\,;\Lambda^*) = \sum_k \mu_k \mathrm{S}(\Lambda^*\omega, \Lambda^*\varphi). \tag{2.11}$$

If $\Lambda : \mathcal{B} \to \mathcal{A}$ is a unitial completely positive mapping between the algebras $\mathcal{A}$ and $\mathcal{B}$, that is, the dual Λ^* is a channeling transformation from the state space of $\mathcal{A}$ into that of $\mathcal{B}$, then

$$S(\Lambda^*\varphi_1, \Lambda^*\varphi_2) \leq S(\varphi_1, \varphi_2). \tag{2.12}$$

Let $\Lambda : \mathcal{B} \to \mathcal{A}$ be completely positive unitial mapping and φ be a state of $\mathcal{B}$. So φ is an initial state of the channel Λ^*. The quantum mutual entropy is defined after [11] as

$$I(\varphi; \Lambda^*) = \sup \left\{ \sum_j \lambda_j S(\Lambda^*\varphi_j, \Lambda^*\varphi); \right. \\ \left. \sum_j \lambda_j \varphi_j = \varphi \right\}, \tag{2.13}$$

where the least upper bound is over all orthogonal extremal decompositions. Note that the definition (2.6) of the mutual entropy is written as

$$I\left(\rho; \Lambda^*\right) = \sup \left\{ \sum_k \lambda_k S\left(\Lambda^*\rho_k, \Lambda^*\rho\right); \right. \\ \left. \rho = \sum_k \lambda_k \rho_k \in F_o\left(\rho\right) \right\}, \tag{2.14}$$

where $F_o\left(\rho\right)$ is the set of all orthogonal finite decompositions of ρ. The proof of the above equality is given in [14] by means of fundamental properties of the quantum relative entropy. For the case of probability distribution $\rho = \sum_k \lambda_k \delta_k$ and classical -quantum channel γ^*, the mutual entropy can be denoted by

$$I\left(\rho; \gamma^*\right) = \sum_k \lambda_k S\left(\gamma^*\delta_k, \gamma^*\rho\right), \tag{2.15}$$

where δ_k is the delta measure. When the following minus term is well-defined, it equals to

$$I\left(\rho; \gamma^*\right) = S\left(\gamma^*\rho\right) - \sum_k \lambda_k S\left(\gamma^*\delta_k\right), \tag{2.16}$$

which has been taken as the definition of the semi-classical mutual entropy for a classical-quantum channel [5,4,6].

Holevo proved the following inequality in 1973 [5].

Theorem 2.1. *When* $\mathcal{A} = \mathbb{C}^k$ *and* $\mathcal{B} = \mathbb{C}^m$ *of the above notation*

$$I_{cl} = \sum_{i,j} p_{ji} \log \frac{p_{ji}}{\lambda_i q_j} \leq S(\gamma^* \varphi) - \sum_k \lambda_i S(\gamma^* \varphi_k)$$

holds.

Holevo's upper bound can now be expressed by

$$S(\gamma^* \varphi) - \sum_i \lambda_i S(\gamma^* \varphi_i) = \sum_i \lambda_i S(\gamma^* \varphi_i, \gamma^* \varphi) . \qquad (2.17)$$

Yuen and Ozawa [27] propose to call the above Theorem the fundamental theorem of quantum communication. The theorem bounds the performance of the detecting scheme. For general quantum case, we have the following inequality according to the lemma of [11].

Theorem 2.2. [16] *When the Schatten decomposition (i.e., one dimensional spectral decomposition)* $\varphi = \sum_i \lambda_i \varphi_i$ *is unique,*

$$I_{cl} \leq I = \sum_i \lambda_i S(\Lambda^* \varphi_i, \Lambda^* \varphi).$$

for any quantum channel Λ^*.

We see that in most cases the bound can not be achieved. Namely, the bound may be achieved in the only case when the output states $\Lambda^* \varphi_i$ *have commuting densities.*

Proposition 2.1. [16] *If the states* $\Lambda^* \varphi_i$, $1 \leq i \leq m$, *do not commute, then*

$$I_{cl} < S(\Lambda^* \varphi) - \sum_i \lambda_i S(\Lambda^* \varphi_i)$$

is a strict inequality.

Now we show two theorems proved in [19] according to the Holevo bound. Let us define an extension of the functional of the relative entropy. If A and B are two positive Hermitian operators (not necassarily the states, i.e. not necessarily with unit traces) then we set

$$S(A, B) = trA\,(\log A - \log B) . \qquad (2.18)$$

There is the following *Bogoliubov inequality.*

Theorem 2.3. [19] *One has*

$$S(A, B) \geq trA\,(\log trA - \log trB) \qquad (2.19)$$

The following theorem gives us the bound of the mutual entropy $I(\rho;\ \Lambda^* \circ \gamma^*)$, where γ^* is a classical -quantum channel.

Theorem 2.4. [19] *For a probability distribution $\rho = \{\lambda_k\}$ and a quantum coded state $\sigma = \gamma^* \rho \equiv \sum_k \lambda_k \sigma_k$, $\lambda_k \geq 0$, $\sum_k \lambda_k = 1$, one has the following inequality for any quantum channel decomposed as $\Lambda^* = \Lambda_1^* \circ \Lambda_2^*$ such that $\Lambda_1^* \sigma \equiv \sum_i E_i \sigma E_i$ by a projection valued measure $\{E_i\}$:*

$$\begin{aligned}
&\sum_k \lambda_k S(\sigma_k, \sigma) \\
&\geqq I(\rho;\ \Lambda^* \circ \gamma^*) \qquad (2.20)\\
&= \sum_k \lambda_k S(\Lambda^* \sigma_k, \Lambda^* \sigma) \\
&\geqq \sum_i \Big[-tr(\Lambda_2^* \sigma E_i) \log tr(\Lambda_2^* \sigma E_i) + \\
&\quad + \sum_k \lambda_k tr(\Lambda_2^* \sigma_k E_i) \log tr(\Lambda_2^* \sigma_k E_i) \Big] \qquad (2.21)
\end{aligned}$$

In the case that the channel Λ_2^* is trivial; $\Lambda_2^* \sigma = \sigma$, the above inequality reduces to the bound obtained by Holevo [5]:

$$\begin{aligned}
&\sum_k \lambda_k S(\sigma_k, \sigma) \\
&= -tr\sigma \log \sigma + \sum_k \lambda_k tr\sigma_k \log \sigma_k \\
&\geq I(\rho;\ \Lambda^* \circ \gamma^*) = I(\rho;\ \Lambda_1^* \circ \gamma^*) \\
&\geqq \sum_i \Big[-tr(\sigma E_i) \log tr(\sigma E_i) \\
&\quad + \sum_k \lambda_k tr(\sigma_k E_i) \log tr(\sigma_k E_i) \Big] \qquad (2.22)
\end{aligned}$$

3. Quantum Capacity

The capacity of purely quantum channel was studied in [16,14,?,17,?,?].

Let $\mathcal{S}$ be the set of all input states satisfying some physical conditions. Let us consider the ability of information transmition for the quantum channel Λ^*. The answer of this question is the capacity of quantum channel Λ^* for a certain set $\mathcal{S} \subset \mathfrak{S}(\mathcal{H}_1)$ defined by

$$C_q^{\mathcal{S}}(\Lambda^*) \equiv \sup \{ I(\rho; \Lambda^*) ; \rho \in \mathcal{S} \}. \qquad (3.1)$$

When $\mathcal{S} = \mathfrak{S}(\mathcal{H}_1)$, the capacity of quantum channel Λ^* is denoted by $C_q(\Lambda^*)$. Then the following theorem for the attenuation channel was proved in [16].

Theorem 3.1. [16] *For a subset* $\mathcal{S}_n \equiv \{\rho \in \mathfrak{S}(\mathcal{H}_1); \dim s(\rho) = n\}$, *the capacity of the noisy optical channel* Λ^* *satisfies*

$$C_q^{\mathcal{S}_n}(\Lambda^*) = \log n,$$

where $s(\rho)$ *is the support projection of* ρ.

When the mean energy of the input state vectors $\{|\tau\theta_k\rangle\}$ can be taken infinite, i.e.,

$$\lim_{\tau\to\infty} |\tau\theta_k|^2 = \left|x^{(\infty)}\right|^2 = \infty$$

the above theorem tells that the quantum capacity for the noisy optical channel Λ^* with respect to $\mathcal{S}_n$ becomes $\log n$. It is a natural result, however it is impossible to take the mean energy of input state vector infinite. Therefore we have to compute the quantum capacity

$$C_q^{\mathcal{S}_e}(\Lambda^*) = \sup\{I(\rho;\Lambda^*); \rho \in \mathcal{S}_e\} \tag{3.2}$$

under some constraint $\mathcal{S}_e \equiv \{\rho \in \mathcal{S}; E(\rho) < e\}$ on the mean energy $E(\rho)$ of the input state ρ. In [13,16,8], we also considered the pseudo-quantum capacity $C_{pq}^{\mathcal{S}_e}(\Lambda^*)$ defined by

$$C_{pq}^{\mathcal{S}_e}(\Lambda^*) = \sup\{I_p(\rho;\Lambda^*); \rho \in \mathcal{S}_e\} \tag{3.3}$$

with the pseudo-mutual entropy $I_{pq}(\rho;\Lambda^*)$

$$I_{pq}(\rho;\Lambda^*) = \sup\left\{\sum_k \lambda_k S(\Lambda^*\rho_k, \Lambda^*\rho); \rho = \sum_k \lambda_k \rho_k, \right.$$
$$\left. \text{finite decomposition}\right\}, \tag{3.4}$$

where the supremum is taken over all finite decompositions instead of all orthogonal pure decompositions for purely quantum mutual entropy. A *pseudo-quantum code* is a probability distribution on $\mathfrak{S}(\mathcal{H})$ with finite support in the set of product states. So $\{(\lambda_k), (\rho_k)\}$ is a pseudo-quantum code if (λ_k) is a probability vector and ρ_k are product states of $B(\mathcal{H})$. The quantum states ρ_k are sent over the quantum mechanical media, for example, optical fiber, and yield the output quantum states $\Lambda^*\rho_k$. The performance of coding and transmission is measured by the *pseudo-mutual entropy (information)*

$$I_{pq}((\lambda_k),(\rho_k);\Lambda^*) = I_{pq}(\rho;\Lambda^*) \tag{3.5}$$

with $\rho = \sum_k \lambda_k \rho_k$. Taking the supremum over certain classes of pseudo-quantum codes, we obtain various capacities of the channel. The supremum is over product states because we have mainly product (that is, memoryless) channels in our mind. Here we consider a subclass of pseudo-quantum codes. A *quantum code* is defined by the additional requirement that $\{\rho_k\}$ is a set of pairwise orthogonal pure states [11]. However the pseudo-mutual entropy is not well-matched to the conditions explained in Sec.3, and it is difficult to compute numerically [17]. From the monotonicity of the mutual entropy [15], we have

$$0 \leq C_q^{\mathcal{S}_0}(\Lambda^*) \leq C_{pq}^{\mathcal{S}_0}(\Lambda^*) \leq \sup\{S(\rho); \rho \in \mathcal{S}_0\}.$$

Let us consider the quantum capacity for a general system [16] given by N-fold tensor products. Let $\mathcal{H}_j^{(N)}$, $B(\mathcal{H}_j^{(N)})$ and $\mathfrak{S}(\mathcal{H}_j^{(N)})$ be the N-fold tensor products of $\mathcal{H}_j$, $B(\mathcal{H}_j)$ and $\mathfrak{S}(\mathcal{H}_j)$ given by

$$\mathcal{H}_j^{(N)} = \bigotimes_{i=1}^{N} \mathcal{H}_j, \; B(\mathcal{H}_j^{(N)}) = \bigotimes_{i=1}^{N} B(\mathcal{H}_j), \; \mathfrak{S}(\mathcal{H}_j^{(N)}) = \bigotimes_{i=1}^{N} \mathfrak{S}(\mathcal{H}_j) \quad (j = 1, 2).$$

The quantum channel Λ_N^* from $\mathfrak{S}(\mathcal{H}_1^{(N)})$ to $\mathfrak{S}(\mathcal{H}_2^{(N)})$ defined by

$$\Lambda_N^* = \Lambda^* \otimes \cdots \otimes \Lambda^* \tag{3.6}$$

is called a memoryless channel, where Λ^* is the quantum channel from $\mathfrak{S}(\mathcal{H}_1)$ to $\mathfrak{S}(\mathcal{H}_2)$. The quantum mutual entropy with respect to the input state $\rho^{(N)} \in \mathfrak{S}(\mathcal{H}_1^{(N)})$ and the channel Λ_N^* is decribed by

$$I(\rho^{(N)}; \; \Lambda_N^*) = \sup\left\{\sum_j p_j^{(N)} S(\Lambda_N^* \rho_j^{(N)}, \Lambda_N^* \rho^{(N)}); \rho^{(N)} = \sum_j p_j^{(N)} \rho_j^{(N)}\right\}, \tag{3.7}$$

where the supremum is taken over all von Neumann - Schatten decompositions $\sum_j p_j^{(N)} \rho_j^{(N)}$ of $\rho^{(N)}$. The pair $((p_j^{(N)}),\ (\rho_j^{(N)}))$ of $(p_j^{(N)})$ and $(\rho_j^{(N)})$ is called a quantum code of order N if $(p_j^{(N)})$ is a priori probability vector and $(\rho_j^{(N)})$ is the orthogonal pure states in $\mathfrak{S}(\mathcal{H}_1^{(N)})$. Moreover $((p_j^{(N)}),\ (\rho_j^{(N)}))$ is called a pseudo-quantum code of order N if $(p_j^{(N)})$ is a priori probability vector and $(\rho_j^{(N)})$ is pure states in $\mathfrak{S}(\mathcal{H}_1^{(N)})$. We denote the set of all quantum codes (resp. pseudo-quantum codes) of order N by $\mathcal{C}_q^{(N)}$ (resp. $\mathcal{C}_{pq}^{(N)}$). The quantum mutual entropy with respect to the quantum codes (resp. pseudo-quantum codes) and the quantum channel Λ_N^* is defined by

$$I(((p_j^{(N)}), (\rho_j^{(N)})), \Lambda_N^*) = I(\rho^{(N)}; \; \Lambda_N^*) \tag{3.8}$$

where $\rho^{(N)} = \sum_j p_j^{(N)} \rho_j^{(N)}$. Based on the quantum mutual entropy, the quantum capacity and the pseudo quantum capacity for the quantum channel Λ_N^* are formulated as

$$C_q(\Lambda_N^*) = \sup\left\{I(((p_j^{(N)}),(\rho_j^{(N)})),\Lambda_N^*);\ ((p_j^{(N)}),(\rho_j^{(N)})) \in \mathcal{C}_q^{(N)}\right\},\tag{3.9}$$

$$C_{pq}(\Lambda_N^*) = \sup\left\{I(((p_j^{(N)}),(\rho_j^{(N)})),\Lambda_N^*);\ ((p_j^{(N)}),(\rho_j^{(N)})) \in \mathcal{C}_{pq}^{(N)}\right\}\tag{3.10}$$

From the above definitions, the inequality

$$C_q(\Lambda_N^*) \leq C_{pq}(\Lambda_N^*)\tag{3.11}$$

is held. Moreover the mean quantum capacity and the mean pseudo-quantum capacity per single use are given by

$$C_q^{(\infty)}(\Lambda^*) = \lim_{N\to\infty} \frac{1}{N} C_q(\Lambda_N^*)\tag{3.12}$$

$$C_{pq}^{(\infty)}(\Lambda^*) = \lim_{N\to\infty} \frac{1}{N} C_{pq}(\Lambda_N^*)\tag{3.13}$$

In order to estimate the quantum mutual entropy , we introduce the concept of *divergence center.* Let $\{\omega_i : i \in I\}$ be a family of states and $R > 0$. We say that the state ω is a divergence center for $\{\omega_i : i \in I\}$ with radius $\leq R$ if

$$S(\omega_i, \omega) \leq R \qquad \text{for every } i \in I.$$

In the following discussion about the geometry of relative entropy (or divergence as it is called in information theory) the ideas of [?] can be recognized very well.

Lemma 3.1. [16] *Let $((\lambda_k),(\rho_k))$ be a quantum code for the channel Λ^* and ω a divergence center with radius $\leq R$ for $\{\Lambda^*\rho_k\}$. Then*

$$I_{pq}((\lambda_k),(\rho_k);\Lambda^*) \leq R.$$

Lemma 3.2. [16] *Let ψ_0, ψ_1 and ω be states of $B(\mathcal{K})$ such that the Hilbert space $\mathcal{K}$ is finite dimensional and set $\psi_\lambda = (1-\lambda)\psi_0 + \lambda\psi_1$ $(0 \leq \lambda \leq 1)$. If $S(\psi_0,\omega)$, $S(\psi_1,\omega)$ are finite and*

$$S(\psi_\lambda,\omega) \geq S(\psi_1,\omega) \quad (0 \leq \lambda \leq 1)$$

then

$$S(\psi_1,\omega) + S(\psi_0,\psi_1) \leq S(\psi_0,\omega).$$

Lemma 3.3. [16] *Let $\{\omega_i : i \in I\}$ be a finite set of states of $B(\mathcal{K})$ such that the Hilbert space $\mathcal{K}$ is finite dimensional. Then the exact divergence center is unique and it is in the convex hull on the states ω_i.*

Theorem 3.2. [16] *Let $\Lambda^* : \mathfrak{S}(\mathcal{H}) \to \mathfrak{S}(\mathcal{K})$ be a channel with finite dimensional $\mathcal{K}$. Then the capacity $C_p(\Lambda^*)$ is the divergence radius of the range of Λ^*.*

3.1. *Numerical Computation for Capacity of Noisy Optical Channel*

Now, we compute the capacity of the noisy optical channel for input coherent states with a coherent noise state and we will discuss the inequalities of quantum capacity based on [18].

Proposition 3.1. [21] *For any states ρ given by $\rho = \lambda|x\rangle\langle x| + (1-\lambda)|y\rangle\langle y|$ with any nonorthogonal pair $x, y \in \mathcal{H}$ and any $\lambda \in [0,1]$, the Schatten decomposition of ρ is uniquely determined by*

$$\rho = \lambda_0 E_0 + \lambda_1 E_1, \tag{3.14}$$

where two eigenvalues λ_0 and λ_1 of ρ are

$$\begin{aligned}\lambda_j &= \frac{1}{2}\left\{1 + (-1)^j \sqrt{1 - 4\lambda(1-\lambda)\left(1 - |\langle x, y\rangle|^2\right)}\right\} \\ &= (1-j)\,\|\rho\| + j\,(1 - \|\rho\|), \quad (j = 0, 1).\end{aligned}$$

Moreover two projections E_0, E_1 are constructed by the eigenvectors $|e_j\rangle$ with respect to λ_j $(j = 0, 1)$

$$E_j = |e_j\rangle\langle e_j| = (a_j|x\rangle + b_j|y\rangle)\left(\bar{a}_j\langle x| + \bar{b}_j\langle y|\right),$$

where the constants a_j, b_j $(j = 0, 1)$ are given as follows:

$$\begin{aligned}|b_j|^2 &= \frac{1}{\tau_j^2 + 2|\langle x, y\rangle|\tau_j + 1} \quad (j = 0, 1), \\ |a_j|^2 &= \tau_j^2|b_j|^2, a_j\bar{b}_j = \bar{a}_j b_j = \tau_j|b_j|^2, \\ \tau_j &= \frac{-(1-2\lambda)}{2(1-\lambda)|\langle x, y\rangle|} \\ &\quad + (-1)^j \frac{\sqrt{1 - 4\lambda(1-\lambda)(1 - |\langle x, y\rangle|^2)}}{2(1-\lambda)|\langle x, y\rangle|},\end{aligned}$$

Let ρ be an input state given by

$$\rho = \lambda|0\rangle\langle 0| + (1-\lambda)|\theta\rangle\langle\theta| \tag{3.15}$$

where $|0\rangle$ is a vacuum state vector in $\mathcal{H}$ and $|\theta\rangle$ is a coherent state vector in $\mathcal{H}$. From the above proposition, the Schatten decomposition of ρ is obtained by

$$\rho = \lambda_0 E_0^{0,\theta} + \lambda_1 E_1^{0,\theta}, \tag{3.16}$$

where the eigenvalues λ_j $(j = 0, 1)$ of ρ are obatined by

$$\lambda_j = \frac{1}{2}\left\{1 + (-1)^j \sqrt{1 - 4\lambda(1-\lambda)\left(1 - \exp(-|\theta|^2)\right)}\right\}$$

and the two projections $E_0^{0,\theta}$, $E_1^{0,\theta}$ and the eigenvector $|e_j^{0,\theta}\rangle$ of λ_j are

$$\begin{aligned} E_j^{0,\theta} &= |e_j^{0,\theta}\rangle\left\langle e_j^{0,\theta}\right| \quad (j = 0, 1), \\ |e_j^{0,\theta}\rangle &= a_{j,0,\theta}|0\rangle + b_{j,0,\theta}|\theta\rangle \quad (j = 0, 1), \end{aligned}$$

where

$$\begin{aligned} |b_{j,0,\theta}|^2 &= \frac{1}{\tau_{j,0,\theta}^2 + 2\exp(-\frac{1}{2}|\theta|^2)\tau_{j,0,\theta} + 1} \quad (j = 0, 1), \\ |a_{j,0,\theta}|^2 &= \tau_{j,0,\theta}^2 |b_{j,0,\theta}|^2, \\ a_{j,0,\theta}\bar{b}_{j,0,\theta} &= \bar{a}_{j,0,\theta} b_{j,0,\theta} = \tau_{j,0,\theta}|b_{j,0,\theta}|^2 \\ \tau_{j,0,\theta} &= \frac{-(1-2\lambda)}{2(1-\lambda)\exp(-\frac{1}{2}|\theta|^2)} \\ &\quad + (-1)^j \frac{\sqrt{1 - 4\lambda(1-\lambda)(1-\exp(-|\theta|^2))}}{2(1-\lambda)\exp(-\frac{1}{2}|\theta|^2)}. \end{aligned}$$

In order to compute the quantum capacity, we use the following two subsets of $\mathfrak{S}(\mathcal{H}_1)$ according to the energy constraint:

$$\begin{aligned} \mathcal{S}_e &\equiv \{\rho = \lambda|0\rangle\langle 0| + (1-\lambda)|\theta\rangle\langle\theta| \in \mathfrak{S}(\mathcal{H}_1); \\ &\qquad \lambda \in [0,1], \theta \in \mathbf{C}, \mathbf{E}(\rho) = |\theta|^2 \le e\Big\}, \end{aligned} \tag{3.17}$$

$$\begin{aligned} \mathcal{S}_e' &\equiv \{\rho = \lambda|0\rangle\langle 0| + (1-\lambda)|\theta\rangle\langle\theta| \in \mathfrak{S}(\mathcal{H}_1); \\ &\qquad \lambda \in [0,1], \theta \in \mathbf{C}, \mathbf{E}'(\rho) = (1-\lambda)|\theta|^2 \le e\Big\}. \end{aligned} \tag{3.18}$$

3.2. *Noisy optical channel:*

The exact value of the capacity for a physical channel is a very important to be found. In this section, we compute the exact capacity C_q for a noisy optical channel Λ^* (1.11) under a certain energy constraint. Note that the pseudo-quantum capacity C_{pq} is difficult to compute.

Let $\mathcal{S}$ be the subset of $\mathfrak{S}(\mathcal{H}_1)$ given by

$$\mathcal{S} = \{\rho = \lambda \left|0\right\rangle \left\langle 0\right| + (1-\lambda) \left|\theta\right\rangle \left\langle \theta\right| ;\ \lambda \in [0,1],\ \theta \in \mathbb{C}\}, \tag{3.19}$$

where $\left|0\right\rangle \left\langle 0\right|$ is the vacuum state and $\left|\theta\right\rangle \left\langle \theta\right|$ is a coherent state on $\mathcal{H}_1$.

From proposition 3.1, the von Neumann - Schatten decomposition of ρ is uniquely written as

$$\rho = \|\rho\| E_0^{0,\theta} + (1 - \|\rho\|) E_1^{0,\theta}, \tag{3.20}$$

where $E_j^{0,\theta} = \left|z_j^{0,\theta}\right\rangle \left\langle z_j^{0,\theta}\right|$ $(j = 0,1)$. We assume that a noise state ξ of Λ^* is expressed as a chorerent state $\xi = \left|\kappa\right\rangle \left\langle \kappa\right|$. For the above input state $\rho \in \mathcal{S}$, the output state $\Lambda^*\rho$ is

$$\begin{aligned}\Lambda^*(\rho) = \lambda &\left|\sqrt{1-\eta}\gamma\right\rangle \left\langle \sqrt{1-\eta}\gamma\right| \\ &+(1-\lambda) \left|\sqrt{\eta}\theta + \sqrt{1-\eta}\gamma\right\rangle \left\langle \sqrt{\eta}\theta + \sqrt{1-\eta}\gamma\right|.\end{aligned}$$

Let $\Lambda^* E_j^{0,\theta}$ be the decomposition denoted by

$$\Lambda^* E_j^{0,\theta} = \tilde{\lambda}_j \tilde{E}_{j0} + (1 - \tilde{\lambda}_j) \tilde{E}_{j1}, \tag{3.21}$$

where $\tilde{\lambda}_j$ $(j = 0,1)$ are the constants such that

$$\begin{aligned}\tilde{\lambda}_0 &= \frac{1}{2}(1 + \exp(-\frac{1}{2}(1-\eta)|\theta|^2)) \\ &\quad \times \frac{\tau_{0,\theta}^2 + 2\exp(-\frac{1}{2}|\theta_\eta|^2)\tau_{0,\theta} + 1}{\tau_{0,\theta}^2 + 2\exp(-\frac{1}{2}|\theta|^2)\tau_{0,\theta} + 1} \\ \tilde{\lambda}_1 &= \frac{1}{2}(1 + \exp(-\frac{1}{2}(1-\eta)|\theta|^2)) \\ &\quad \times \frac{t_{0,\theta}^2 + 2\exp(-\frac{1}{2}|\theta_\eta|^2)t_{0,\theta} + 1}{t_{0,\theta}^2 + 2\exp(-\frac{1}{2}|\theta|^2)t_{0,\theta} + 1}.\end{aligned}$$

When one dimemsional projection $\tilde{E}_{jk}$ is written as $\tilde{E}_{jk} = |\tilde{x}_{jk}\rangle\langle\tilde{x}_{jk}|$, the vectors $\tilde{x}_{jk}$ $(j, k = 0, 1)$ satisfy the following conditions:

$$\langle \tilde{x}_{00}, \tilde{x}_{01} \rangle = \frac{\tau^2 - 1}{\sqrt{(\tau^2+1)^2 - 4\exp(-|\theta_\eta|^2)\tau^2}} \neq 0,$$

$$\langle \tilde{x}_{10}, \tilde{x}_{11} \rangle = \frac{t^2 - 1}{\sqrt{(t^2+1)^2 - 4\exp(-|\theta_\eta|^2)t^2}} \neq 0.$$

From Proposition 3.1, the eigenvalues of $\Lambda^* E_j^{0,\theta}$ $(j = 0, 1)$ are

$$\tilde{\lambda}_{j0}^{\tilde{x}_{j0},\tilde{x}_{j1}} = \frac{1}{2}\left\{1 + \sqrt{1 - \tilde{\lambda}_j(1-\tilde{\lambda}_j)(1 - |\langle \tilde{x}_{j0}, \tilde{x}_{j1}\rangle|^2)}\right\},$$

$$\tilde{\lambda}_{j1}^{\tilde{x}_{j0},\tilde{x}_{j1}} = \frac{1}{2}\left\{1 + \sqrt{1 - \tilde{\lambda}_j(1-\tilde{\lambda}_j)(1 - |\langle \tilde{x}_{j0}, \tilde{x}_{j1}\rangle|^2)}\right\}.$$

The quantum mutual entropy (2.6) with respect to an input state $\rho \in \mathcal{S}$ and the noisy optical channel Λ^* defined by (1.11) with the coherent noise ξ is determined by

$$\mathrm{I}(\rho; \Lambda^*) = \mathrm{S}(\Lambda^*\rho) - \|\rho\|\mathrm{S}(\Lambda^* E_0^{0,\theta}) - (1 - \|\rho\|)\mathrm{S}(\Lambda^* E_1^{0,\theta}) \tag{3.22}$$

where

$$\mathrm{S}(\Lambda^*\rho) = -\sum_{j=0}^{1} \tilde{\lambda}_j^{\sqrt{1-\eta}\gamma, \sqrt{\eta}\theta + \sqrt{1-\eta}\gamma} \log \tilde{\lambda}_j^{\sqrt{1-\eta}\gamma, \sqrt{\eta}\theta + \sqrt{1-\eta}\gamma}$$

$$\mathrm{S}(\Lambda^* E_j^{0,\theta}) = -\sum_{i=0}^{1} \tilde{\lambda}_{ji}^{\tilde{x}_{j0},\tilde{x}_{j1}} \log \tilde{\lambda}_{ji}^{\tilde{x}_{j0},\tilde{x}_{j1}}.$$

From the above result of the quantum mutual entropy $I(\rho; \Lambda^*)$, we expricitly compute the quantum capacity for the noisy optical channel Λ^* with the coherent noise state $|\kappa\rangle\langle\kappa|$ such as

$$C_q^{(1)}(\Lambda^*)_e = C_q^{\mathcal{S}_e}(\Lambda^*) = \sup_{\lambda,\theta}\left\{\mathrm{I}(\rho;\Lambda^*); |\theta|^2 \le e\right\}, \tag{3.23}$$

$$C_q^{(2)}(\Lambda^*)_e = C_q^{\mathcal{S}'_e}(\Lambda^*) = \sup_{\lambda,\theta}\left\{\mathrm{I}(\rho;\Lambda^*); (1-\lambda)|\theta|^2 \le e\right\} \tag{3.24}$$

as shown below.

Theorem 3.3. [18]

$$C_q^{\mathcal{S}_e}(\Lambda^*) < C_q^{\mathcal{S}'_e}(\Lambda^*).$$

(1) For each $\eta \in \{0.2, 0.5, 0.8\}$, $\lambda \in [0,1]$ and $|\theta|^2 \leq 3$, we have the quantum capacity of the noisy quantum channel Λ^* with the coherent noise ξ such as Fig.1:

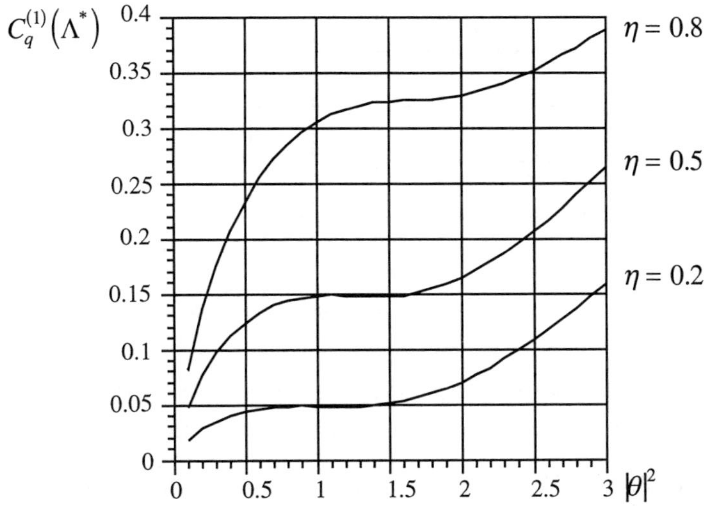

Figure 1. Quantum Capacity $C_q^{(1)}(\Lambda^*)$.

(2) For each $\eta \in \{0.2, 0.5, 0.8\}$ and fix $E = (1-\lambda)|\theta|^2 \leq 3$, we obtain the quantum capacity of the quantum mechanical channel Λ^* with the coherent noise ξ such as Fig.2:

Next we discuss what is the most suitable modulation in OOK, PPM, PWM for the noisy optical channel. The subsets with respect to the optical modulations OOK, PPM, PWM are given by

$$\begin{aligned} \mathcal{S}_e^{OOK} &\equiv \{\rho = \lambda|0\rangle\langle 0| + (1-\lambda)|\theta\rangle\langle\theta| \in \mathfrak{S}(\mathcal{H}_1)\,; \\ &\quad \lambda \in [0,1]\,, \theta \in \mathbb{C}, |\theta|^2 \leq e\Big\} \end{aligned} \tag{3.25}$$

$$\begin{aligned} \mathcal{S}_e^{PPM} &\equiv \{\rho = \lambda|0\rangle\langle 0| \otimes |\theta\rangle\langle\theta| + (1-\lambda)|\theta\rangle\langle\theta| \otimes |0\rangle\langle 0| \\ &\quad \in \mathfrak{S}(\mathcal{H}_1) \otimes \mathfrak{S}(\mathcal{H}_1)\,; \lambda \in [0,1]\,, \theta \in \mathbb{C}, |\theta|^2 \leq e\Big\} \end{aligned} \tag{3.26}$$

$$\begin{aligned} \mathcal{S}_e^{PWM} &\equiv \{\rho = \lambda|0\rangle\langle 0| \otimes |\theta\rangle\langle\theta| + (1-\lambda)|\theta\rangle\langle\theta| \otimes |\theta\rangle\langle\theta| \\ &\quad \in \mathfrak{S}(\mathcal{H}_1) \otimes \mathfrak{S}(\mathcal{H}_1)\,; \lambda \in [0,1]\,, \theta \in \mathbb{C}, |\theta|^2 \leq e\Big\} \end{aligned} \tag{3.27}$$

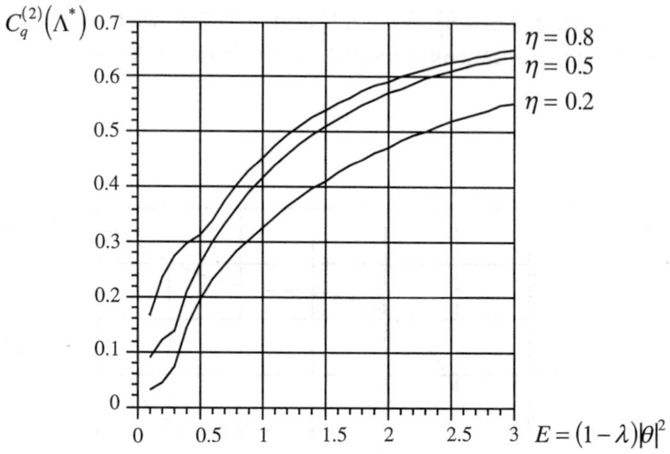

Figure 2. Quantum Capacity $C_q^{(2)}(\Lambda^*)$.

Calculating the capacity of the noisy optical channel for the above subsets consisted by the optical modulations, we have the following theorem.

Theorem 3.4. [18] *The capacities of the noisy optical channel for the subsets* $\mathcal{S}_e^{OOK}$, $\mathcal{S}_e^{PPM}$ *and* $\mathcal{S}_e^{PWM}$ *satisfy the following inequalities*

$$C_q^{\mathcal{S}_e^{OOK}}(\Lambda^*) = C_q^{\mathcal{S}_e^{PWM}}(\Lambda^*) \leq C_q^{\mathcal{S}_e^{PPM}}(\Lambda^*).$$

Since the above quantum capacity is a quantity to measure the ability for information transmission of a quantum channel, it is important to investigate a logical gate for quantum computer. Fredkin - Toffoli - Milburn (FTM) gate is one of logical gates for quantum computer. This gate can be rigorously formulated as a quantum channel and the conservation of information through the FTM gate is shown by means of the quantum mutual entropy. These facts are presented in [23].

References

1. L. Accardi and M. Ohya, *Compond channnels, transition expectation and liftings*, Appl. Math, Optim. **39**, 33–59, 1999.
2. H. Araki, *Relative entropy for states of von Neumann algebras*, Publ. RIMS Kyoto Univ. **11**, 809–833, 1976.

3. K.H. Fichtner, W. Freudenberg and V. Liebscher, *Beam splittings and time evolutions of Boson systems*, Fakultät für Mathematik 0xdf und Informatik, **Math/ Inf/96/ 39**, Jena, 105, 1996.
4. R.S.Ingarden, *Quantum information theory*, Rep. Math. Phys., **10**, pp.43-73, 1976.
5. A.S.Holevo, *Some estimates for the amount of information transmittable by a quantum communication channel (in Russian)*, Problemy Peredachi Informacii, **9**, pp.3-11, 1973.
6. L.B.Levitin, *Physical information theory for 30 years: basic concepts and results*, Springer Lect. Note in Phys., **378**, pp.101-110, 1991.
7. G.Lindblad, *Entropy, information and quantum measurements*, Commun. Math. Phys., **33**, pp.111-119, 1973.
8. N.Muraki, M.Ohya and D.Petz, Note on entropy of general quantum systems, Open Systems and Information Dynamics, **1**, No.1, pp.43–56, 1992.
9. von Neumann, *Die Mathematischen Grundlagen der Quantenmechanik*, Springer-Berlin, 1932.
10. M.Ohya, *Quantum ergodic channels in operator algebras*, J. Math. Anal. Appl. **84**, pp.318-327, 1981.
11. M. Ohya, *On compound state and mutual information in quantum information theory*, IEEE Trans. Information Theory, **29**, 770–777, 1983.
12. M.Ohya, *Note on quantum probability*, L. Nuovo Cimento, **38**, pp.402–406, 1983.
13. M. Ohya, *Some aspects of quantum information theory and their applications to irreversible processes*, Rep. Math. Phys. **27**, 19–47, 1989.
14. M. Ohya, *Fundamentals of quantum mutual entropy and capacity*, Open Systems and Information Dynamics, 6. No.1, 69-78, 1999.
15. M. Ohya and D. Petz, *Quantum Entropy and Its Use*, Springer, 1993.
16. M. Ohya, D. Petz and N. Watanabe, *On capacity of quantum channels*, Probability and Mathematical Statistics, **17**, 179-196, 1997.
17. M. Ohya, D. Petz and N. Watanabe, *Numerical computation of quantum capacity*, International J. Theor. Phys., **38** , No.1, pp.507-510, 1998.
18. M. Ohya, D. Petz and N. Watanabe, *Quantum capacity of a noisy optical channel*, preprint.
19. M. Ohya, I. Volovich, *On quantum capacity and its bound,* preprint.
20. M. Ohya and N. Watanabe, *Construction and analysis of a mathematical model in quantum communication processes*, Electronics and Communications in Japan, Part 1, **68**, No.2, 29-34, 1985.
21. M. Ohya and N. Watanabe, *Quantum capacity of noisy quantum channel*, Quantum Communication and Measurement, **3**, 213-220, 1997.
22. M. Ohya and N. Watanabe, *Foundatin of Quantum Communication Theory (in Japanese)*, Makino Pub. Co., 1998.
23. M. Ohya and N. Watanabe, *On the mathematical treatment of the Fredkin - Toffoli - Milburn gate*, Physica D, **120**, 206-213, 1998.
24. R. Schatten; *Norm Ideals of Completely Continuous Operators*, Springer-Verlag, 1970.
25. H. Umegaki, *Conditional expectations in an operator algebra IV (entropy and*

information), Kodai Math. Sem. Rep. **14**, pp. 59–85, 1962.

26. A.Uhlmann, *Relative entropy and the Wigner-Yanase-Dyson-Lieb concavity in interpolation theory*, Commun. Math. Phys., **54**, pp.21–32, 1977.
27. H.P.Yuen and M.Ozawa, *Ultimate information carrying limit of quantum systems*, Phys. Rev. Lett., **70**, pp.363–366, 1993.

NOTE ON LINEAR PROCESS

WIN WIN HTAY

Department of Computational mathematics
University of Computer Studies
Yangon, Myanmar

White noise analysis that discusses the analysis of functionals of (Gaussian) white noise, has extensively developed in these years. It is now the time to study the analysis of functionals of other noises, in particular those of Poisson noise as well as combinations of Gaussian and Poisson noises. In this paper we shall first prepare some background on the analysis of Poisson noise functionals. Then, linear functionals of Poisson noise will be discussed. There one can see both similarity and dissimilarity to Gaussian case. It is noted that dissimilarity is more interesting and significant. The path-wise integral based on Poisson noise and more generally functionals of sample path of Poisson noise (or Pisson process) have significant probabilistic characters different from Gaussian case. To study these properties we introduce a class (**P**) of random functions; in fact, it is different from L^2-space. A new methods of analysis should be introduced. Having been suggested by P. Lévy [9], we shall define a linear process, which can be expressed as a sum of linear functions of Gaussian white noise and Poisson noise. Linear processes are discussed in the space (**P**), where the innovation approach plays an important role. We shall also establish a characterization theorem of a linear process, where the canonical representation theory for Gaussian processes and its generalization are basic tools. For the purpose of applications to information science and technology, we shall metion an efficient use of the ergodic theorem.

Subject classfication 2000 : 60H40

1. Introduction

First we review of the theory on some functionals of a Poisson noise which is a generalized stationary stochastic process with independent value at every point. For convenience we may assume that t runs through the whole real line. In fact, it is easy to get such a noise. The characteristic functional of the centered Poisson white noise is of the form

$$C_P(\xi) = \exp[\int_{-\infty}^{\infty} (e^{i\xi(t)} - 1 - i\xi(t))dt], \tag{1.1}$$

where $\xi \in E$.

There is the associated measure space (E^*, μ_P), and the Hilbert space $L^2(E^*, \mu_P) = (L^2)_P$ is defined.

Many results of the analysis on $(L^2)_P$ have been obtained, however most of them are studied by analogy with the Gaussian case or its modifications, as far as the construction of the space of generalized functionals. Here we only note that the $(L^2)_P$ admits the direct sum decomposition of the form

$$(L^2)_P = \bigoplus_n H_{P,n}. \tag{1.2}$$

This decomposition can be obtained as follows. Let $P(t)$ denote a Poisson process as before. The subspace $H_{P,n}$ is spanned by the Poisson-Charlier polynomials defined as follows. Let $p(x, \lambda)$ be the Poisson distribution with intensity $\lambda > 0$:

$$p(x, \lambda) = \frac{\lambda^x}{x!} e^{-\lambda}, \ x = 0, 1, 2, \ldots.$$

Set

$$p_n(x; \lambda) = \frac{\lambda^{\frac{n}{2}}}{n} (-1)^n \frac{\Delta^n p(x; \lambda)}{p(x; \lambda)},$$

where Δ is the difference operator:

$$\Delta f(x) = f(x+1) - f(x)$$

The polynomials $p_n(x; \lambda), n = 0, 1, 2, \ldots$, form an orthogonal system in the following sense:

$$\sum_{x=0}^{\infty} p_n(x; \lambda) p_m(x; \lambda) p(x; \lambda) = \delta_{n,m}. \tag{1.3}$$

In the case of Poisson noise functionals the Poisson-Charlier polynomials play the same role of Hermite polynomials in Gaussian case, and we come to the decomposition of $(L^2)_P$ in question. We now come to a linear functional of the $\dot{P}(u)$.

2. Linear functionals of Poisson noise

Let a stochastic process $X(t)$ be given by an integral

$$X(t) = \int_0^t F(t,u)\dot{P}(u)du. \tag{2.1}$$

It may be viewed simply as a linear functional of $\dot{P}(t)$, however there are two ways of understanding the meaning of the integral in such a way that

i) the integral is defined in the Hilbert space by taking $\dot{P}(t)dt$ to be a *random measure*, and the stochastic integral is defined. Like the homogeneous Chaos, multiple integral can also be defined.
ii) on the other hand the integral is understood as a continuous bilinear form of a test function $F(t,\cdot)$ and a sample function of $\dot{P}(t)$ (the *path-wise* integral). This can be done if the kernel is a smooth function of u over the interval $[0,t]$ since a sample function of $\dot{P}(\cdot)$ is a generalized function. Namely, the integral is defined as a bilinear form, and $X(t)$ is called a linear process.

Assume that $F(t,t)$ never vanishes and that it is not a canonical kernel, that is, it is not a kernel function of an invertible integral operator. Then, we can claim that for the integral in the first sense $X(t)$ has less information compared to $P(t)$. Because there is a linear function of $P(s), s \leq t$ which is orthogonal to $X(s), s \leq t$. On the other hand, if $X(t)$ is defined in the second sense, we can prove the following proposition.

Note that we always assume that the variation $\delta X(t)$ exists.

Proposition 1 Under the assumptions stated above, if the $X(t)$ above is defined sample function-wise, we have the following equality for sigma-fields:

$$\mathbf{B}_t(X) = \mathbf{B}_t(P), \quad t \geq 0. \tag{2.2}$$

Proof By assumption it is easy to see that $X(t)$ and $P(t)$ share the same jump points, which means that the information is fully transfered from $P(t)$ to $X(t)$.This proves the equality

The above argument tells us that we are led to introduce a space $(\mathbf{P})$ of random variables that come from separable stochastic processes for which existence of variance may not be expected. This sounds to be a vague statement, however we can rigorously defined. There the topology is defined by either the almost sure convergent or the convergence in probability, and there is no need to think of mean square topology. On the space (**P**) filtering and prediction for strictly stationary process can naturally be discussed.

For the linear process given by (2.1), we can define $N-$ple Markov property in the same manner as in the Gaussian case.

Proposition 2 If $X(t)$ given by (2.1), we can define $N-$ple Markov then the kernel $F(t,u)$ is a Goursat kernel of degree N.

Also we prove that the conditional expectation is given by the formula

$$E[X(t)|\mathbf{B}_s(X)] = \int_0^s F(t,u)\dot{P}(u)du, \; s \le t. \tag{2.3}$$

The rest of the proof is exactly the same as in the Gaussian case.

Remark In the above proposition the kernel F is not necessarily a canonical (an invertible) kernel, unlike the Gaussian case.

3. Linear processes

We now come to a more general case. Following P. Lévy [9], we define a system of linearly correlated and a linear process.

Definition Given a system of random variables $\{X, \{X_\alpha\}\}$. The X is said to be *linearly correlated* with $\{X_\alpha\}$, if X is expressed as a sum

$$X = U + V,$$

where V is a linear function of the X_α's and where U and V are independent.

Definition A stochastic process $\{X(t)\}$ is called a *linear process* if for any t and $h > 0$, the variable $X(t+h)$ is linealy correlated with $\{X(s), s \le t\}$.

Important class of linear processes

Let $F(t,u)$ and $G(t,u)$ be in C^2 as functions of u for any fixed t. Define $X(t)$ by

$$X(t) = \int_0^t F(t,u)\dot{B}(u)du + \int_0^t G(t,u)\dot{Y}(u)du \tag{3.1}$$

where $B(t)$ is Brownian motion and $Y(u)$ is a compound Poisson with stationary increments.

The sample functions $\dot{B}(u,\omega)$ and $\dot{Y}(u,\omega)$ for fixed ω are generalized functions in the sense of Gel'fand. However, as we observed in Section 2, the integrals in (3.1) are considered as the path-wise integrals.

Now, assume that $G(t,t) \neq 0$ for every t. Observing the sample functions of $X(t,\omega)$ we can obtain all the jumps of $Y(t,\omega)$ by noting the equation:

$$\begin{aligned}\delta X(t) &= F(t,t)\dot{B}(t)dt + dt\int_0^t \frac{\partial}{\partial t}F(t,u)\dot{B}(u)du \\ &\quad + G(t,t)\dot{Y}(t)dt + dt\int_0^t \frac{\partial}{\partial t}G(t,u)\dot{Y}(u)du. \end{aligned} \tag{3.2}$$

Hence, we have, for $t > s$,,

$$E[\dot{Y}(s)X(t)] = G(t,s). \tag{3.3}$$

Thus $G(t,u)$ is determined and so

$$X(t) - \int_0^t G(t,u)\dot{Y}(u)du$$

is obtained. That is, we know

$$\int_0^t F(t,u)\dot{B}(u)du.$$

1) If F is canonical then $\dot{B}(t)$ is obtained.

2) If F is non-canonical, then it is known how to obtain the innovation $B_1(t)$ for $\int_0^t F(t,u)\dot{B}(u)du$. (See Accardi-Hida-Si Si [1].)

Thus, we obtain the innovation $\dot{B}(t) + \dot{Y}(t)$ of $X(t)$. Summing up

Proposition 3 For a linear process given by (3.1) we have

1) The innovation of the process $X(t)$ is given by $\dot{B}(t)+\dot{Y}(t)$, or equivalently by the system $\{\dot{B}(t), \dot{Y}(t)\}$.

2) $\int_0^t F(t,u)\dot{B}(u)du$ and $\int_0^t G(t,u)\dot{Y}(u)du$ can be descriminated.

Multiple Markov property can be defined by using the conditional expectation the canonical kernels F and G are Goursat kernel.

4. Characterization theorem

Let $X(t)$ be a linear process as in the last section. Assume that Gaussian process has a canonical distribution. Namely, $X(t)$ can be written as

$$X(t) = \int_0^t F(t,u)\dot{B}(u)du + \int_0^t G(t,u)\dot{Y}(u)du,$$

by using the same notation.

We pause to note some different probabilistic characters between the two terms of $X(t)$.

1) Gaussian term. Gaussian system is linearly correlated in both direction, but Poisson is in only one direction, namely for the future. Details can be found in [8] and in our forthcoming paper.
2) The characteristic functional of $X(t)$ is expressed in the form

$$\begin{aligned} C_X(\xi) &= E[e^{i\langle X,\xi\rangle}] \\ &= \exp\left[-\frac{1}{2}\int (F*\xi)(s))^2 ds + \int\int \left(e^{i(G*\xi)(s)u} - 1 - iu(G*\xi)(s)\right) dsdn(u)\right], \end{aligned}$$

where ξ is a test function, dn is the Lévy measure and

$$(F*\xi)(s) = \int F(t,s)\xi(t)dt.$$

Consider the case where $Y(t)$ consist only of single Poisson process, and where both F and G are canonical kernels and are smooth with $G(t,t) \neq 0$. Note that the canonical property is originaly defined for the case of Gaussian process, but this concept is still used in the compound Poisson case, so far as L^2-theory is concerned.

Obviously, we have

$$C_X(\xi) = C_G(\xi).C_P(\xi),$$

where $C_G(\xi)$ and $C_P(\xi)$ are characteristic functionals of a Gaussian white noise and a Poisson noise, respectively.

Theorem Given the characteristic functional C_X of $X(t)$. Then, canonical kernels F and G are obtained by observing a sample function.

Proof. The theorem can be proved by actual computation for the characteristic functional. In fact, once we know that $X(t)$ is a linear process expressed by a Gaussian white noise and a Poisson noise, we can obtain the second functional derivative in ξ and know the second functional derivative has no singular part. There explicit forms are found for $(F * \xi)(s)$ and $(G * \xi)(s)$. Then, by using the canonical property of the kernels. we find F and G.

Acknowledgments

The author would like to express her deep thanks to Professor T. Hida for his valuable suggestions and the invitation to the Winter school held at Meijo University, Nagoya in January 2003.

References

1. L. Accardi, T. Hida and Si Si, Innovations for some stochastic processes. Volterra Center Notes, 2002.
2. L. Accardi, T. Hida and Win Win Htay, Boson Fock representation of stochastic process (in Russian) Math. Notes 67, 2000, pp 3-14.
3. L.Accardi, H-H. Kuo, N. Obata, K. Saito, Si Si, L. Streit (ed), Recent Developments in Infinite Dimensional Analysis and Quantum Probability (Special Issue of Acta Applicandae Mathematicae) Kluwer Academic Pub. Co. 2000D
4. T. Hida, Canonical representation of Gaussian processes and their applications, Memoirs Coll. Sc., Kyoto A33 (1960) 109-155
5. T. Hida, and Si Si, Innovations for random fields, Infinite Dimensional Analysis, Quantum Probability and Related Topics Vol 1 (1998), World Scientific, 499-509.
6. T. Hida and Si Si, Elemental random variables in White Noise Theory: beyond Reductionism, Quantum Information III, ed. T. Hida et. World Scientific 2001, pp59-66
7. T. Hida and Si Si, Innovation approach to some random fields, Application of white noise theory, World Scientific (to appear)
8. T. Hida, Si Si and Win Win Htay, Variational calculus for random fields parametrized by a curve or surface, to appear in g Infinite Dimensional Analysis, Quantum Probability and Related Topics,World Scientific
9. P. Lévy, Fonctions aléatoires à corrélation linéaire, Illinois J. of Math. 1 (1957), 217-258.
10. Si Si, Topics on random fields, Quantum Information I, eds. T. Hida and K. Saito, World Scientific,1999, pp. 179—194.
11. Si Si, Gaussian processes and Gaussian random fields, Quantum Information III, ed. T. Hida and K. Saito. World Scientific 2001, pp. 195-204.

12. Si Si and Win Win Htay, Entropy in subordination and Filtering, Acta Applicandae Mathemathecae Vol. 63, Kluwer Academic Publishers, pp. 433-439.
13. Si Si, A. Tsoi and Win Win Htay, Jump finding of a stable process, Quantum Information V, World Scientific, (in press).
14. Win Win Htay, Optimalities for Random functions: Lee Wienerfs network and non-canonical representations of stationary Gaussian Processes, Nagoya Math. J. Vol. 149, 1998, pp. 9-17.